Environmental Standards

Springer

Berlin
Heidelberg
New York
Hong Kong
London
Milan
Paris
Tokyo

C. Streffer • J. Bücker • A. Cansier • D. Cansier •
C. F. Gethmann • R. Guderian • G. Hanekamp •
D. Henschler • G. Pöch • E. Rehbinder • O. Renn •
M. Slesina • K. Wuttke

Environmental Standards

Combined Exposures and Their Effects on Human Beings and Their Environment

Mit herzlichen Grüßen
Dein
Vetter ×ten Grades
Christian.

Springer

FOR THE AUTHORS:

Professor Dr. Dr. h.c. Christian Streffer
Auf dem Sutan 12
45239 Essen
Germany

EDITING:

Friederike Wütscher
Europäische Akademie GmbH
Wilhelmstraße 56
53474 Bad Neuenahr-Ahrweiler
Germany

ISBN 3-540-44097-6   Springer-Verlag Berlin Heidelberg New York

Library of Congress Cataloging-in-Publication Data Applied For

A catalog record for this book is available from the Library of Congress.
Bibliographic information published by Die Deutsche Bibliothek
Die Deutsche Bibliothek lists this publication in die Deutsche Nationalbibliographie; detailed
bibliographic data is available in the Internet at <http://dnb.ddb.de>.

Springer-Verlag Berlin Heidelberg New York
a member of BertelsmannSpringer Science+Business Media GmbH

http://www.springer.de

© Springer-Verlag Berlin Heidelberg 2003
Printed in Germany

Cover Design: Erich Kirchner, Heidelberg
Typesetting: Köllen Druck+Verlag GmbH, Bonn + Berlin

Printed on acid free paper   30/3141 – 5 4 3 2 1 0

Preface to the Translation

The *Europäische Akademie zur Erforschung von Folgen wissenschaftlich-technischer Entwicklungen Bad Neuenahr-Ahrweiler GmbH* is concerned with the scientific study of the consequences of scientific and technological advance both for the individual and social life and for the natural environment. The Europäische Akademie intends to contribute to a rational way of society of dealing with the consequences of scientific and technological developments. This aim is mainly realised in the development of recommendations for options to act from the point of view of a long-term societal acceptance. The work of the Europäische Akademie mostly is carried out in temporary interdisciplinary project groups whose members are notable scientists from various European universities. Overarching issues, e.g. from the fields of Technology Assessment or Ethics of Science, are dealt with by the staff of the Europäische Akademie.

The results of the work of the Europäische Akademie is published in the series "Wissenschaftsethik und Technikfolgenbeurteilung" (Ethics of Science and Technology Assessment), Springer Verlag. The academy's study report "Umweltstandards. Kombinierte Expositionen und ihre Wirkungen auf den Menschen und seine Umwelt" was published in January 2000. For the first time it addressed the issue of combined exposures in an interdisciplinary manner reaching concrete recommendations for policy making. Since there is still no such treatment available in English, the academy decided to provide for an English translation that is published in the present volume.

Special thanks is due to Friederike Wütscher for the editorial work in preparing the text for print.

Bad Neuenahr-Ahrweiler, May 2003 — Carl Friedrich Gethmann

Preface

The rapid growth of the world population – nearly six-fold over the last hundred years – combined with the rising number of technical installations especially in the industrialized countries has lead to ever tighter and more strained living spaces on our planet. Because of the inevitable processes of life, man was at first an exploiter rather than a careful preserver of the environment. Environmental awareness with the intention to conserve the environment has grown only in the last few decades. Environmental standards have been defined and limit values have been set largely guided, however, by scientific and medical data on single exposures, while public opinion, on the other hand, now increasingly calls for a stronger consideration of the more complex situations following combined exposures. Furthermore, it turned out that environmental standards, while necessarily based on scientific data, must also take into account ethical, legal, economic, and sociological aspects. A task of such complexity can only be dealt with appropriately in the framework of an inter-disciplinary group.

Hence in fall 1996, Professor Cansier (Tübingen), Professor Gethmann (Essen/Bad Neuenahr), Professor Guderian (Essen), Professor Henschler (Würzburg), Professor Pöch (Graz), Professor Rehbinder (Frankfurt a.M.), Professor Renn (Stuttgart), and Professor Streffer (Essen) met to form a project group of the Europäische Akademie GmbH that should concern itself, in an interdisciplinary manner, with environmental standards and limits on combined exposures. They agreed to carry out an exemplary study on combinations of chemical substances with each other as well as with ionizing radiation in terms of their effects on human health. The scope of the study would particularly include ecotoxicological effects of the combination of significant air pollutants. Finally, the consequences of the individual results for the setting of environmental standards were to be developed in an interdisciplinary manner.

From the point of view of their respective disciplines, the members of the project group drew up so-called seed texts, in some cases co-authored by scientific staff and collaborators. For the Europäische Akademie zur Erforschung von Folgen wissenschaftlich-technischer Entwicklungen Bad Neuenahr-Ahrweiler GmbH, first Dr. Saupe and then Dr. Hanekamp attended to the project group. Dr. Hanekamp contributed strongly as a co-author, too, and as a collaborator in the final shaping of the project. The seed texts served as drafts, which were discussed and then ratified by the plenum of the project group, including the co-authors, as the final text that we now present. The project group met for nineteen sessions in all, and it invited other scientists for discussions, including Professor Dr. Müller-Herold (Zürich), for whose suggestions we are particularly grateful.

In spring 1998, the work of the project group had progressed to a point where the results could be submitted for critical discussion to a group of experienced specialists. These discussions, held with Professor Dr. Bolt (Dortmund), Professor Dr. Burkhart (München), Dr. Gallas (Bonn), Professor Dr. Jäger (Gießen), Professor Dr. Janich (Marburg), Professor Dr. Jöckel (Essen), Professor Dr. Jungermann (Berlin), Professor Dr. Michaelis (Bernburg) and Professor Dr. Winter (Bremen), were extremely fruitful and have enhanced the text in its further revision. For this, our thanks are due to all participants. We also would like to thank Ms Uhl of the Europäische Akademie, who carried out the final arrangement of the complete text.

Working with our colleagues and with the academy was a good and enjoyable experience. With this volume, the project group intends to demonstrate ways of dealing with the complex issues of combined exposures, for the protection of humans and their environment.

Essen, September 1999   Christian Streffer

Contents

Summary

Introduction

Increasing technicalisation and population density have led to the situation where mankind and its environment are exposed to a multitude of factors of which very often only little is known about their mid- and longterm health consequences. The development, operation and disposal of technical equipment and products release harmful substances and physical agents like ionising radiation and noise that could have an undesired effect on mankind and its environment. Such side effects of technical development and the intensified public discussion about such topics have contributed to the fact that a mainly unreflective, optimistic view about progress, like the popular belief in the 1950s and 1960s, can hardly be upheld nowadays. Instead, far more often the pendulum swings in the opposite direction, most apparent in the demands for a "zero exposure". One of the means to pave a future and acceptable way amongst such extreme views is the creation and use of environmental standards.

The knowledge about harmful effects of diverse individual exposures on the environment and human health has resulted in the definition of environmental standards for individual chemical and physical agents. This individual substance orientated procedure is in accordance with the fundamental medical research analysis of contextual responses, which concentrates mainly on individual exposures. The examination of combined lowest dose exposures, which correspond much more realistically to exposure conditions in the actual environment than the analysis of single substances, entails major methodological difficulties in the experimentation and evaluation procedure. However, the potential risk of combined exposures and public discussion thereof make the analysis of this topic area a desideratum of the utmost priority.

From a scientific, medical, sociological, economic, legal and philosophical perspective, this study will examine the necessity, realisability and consequences of *environment standards for combined exposures*. The main focus will concentrate on human exposure, especially to carcinogens and genotoxicological effects, and on the exposure of a certain selection of plants, which will serve to answer the most pragmatically urgent questions.

Taking account of the effect mechanisms involved a categorisation of combined exposures will be made, on the basis of which criteria will be established, which, regardless of the complexity of the context of exposition, will enable us to set limit values in order to maintain or achieve concrete environment quality standards. The study shall help to recognise and reveal knowledge gaps and possible solutions. The results of this project shall also form the basis for improving the evaluation of combined exposures, in order to enable the creation of an efficient framework for the relevant legal regulations and economic procedures.

Methodological Foundations

Environmental standards or limit values are understood as the concrete, quantitative expression of environmental quality targets. They are the means to their realisation. Environmental quality targets in their turn represent normative concepts with regard to the quality of man's environment or that of other life forms. In some respects, environmental quality targets are, in condensed terms, the guiding concepts of environmental policy, expressing general environment(s) goals in such areas as health, sustainability and system functionality etc.

Environmental standards do not represent a natural but a cultural phenomenon, i.e. conventional settings. In the first instance they are prescriptive and not descriptive objects. However, natural phenomena play a "decisive" role in the determination of environmental standards. The inter-relationship between the observation of nature and the regulatory stipulation of procedures can be illustrated by differentiating between "threshold value" and "limit value". Threshold values are indeed natural phenomena that can be shown on a graph as curve peculiarities ("thresholds"). A threshold value represents a certain relationship between a cause – i.e. the exposure – and its effect. It is a measured value that indicates, for example, a dose or dose range upon which a certain effect occurs or does not occur. In contrast to this, the *limit* value indicates at which point in the cause-effect relationship a course of action (also including the use of a device, apparatus or large-scale industrial plant) shall have its "limit". Finally an *environmental quality target* stipulates what shape or form a certain sector of the environment *shall* or *shall not* have. In order to achieve such a target, the limit will be determined by a limit value, the observance of which being the (assumed) minimum prerequisite that has to be fulfilled in order to achieve this target.

Threshold values, i.e. the corresponding threshold range, are determined in scientific experiments and can be displayed on a graph as curve properties. However, it is by no means possible to give threshold values for every kind of effect. In contrast, limit values are set according to certain aims. They may possibly correspond to or be orientated to threshold values. But even the use of safety margins, for example, already leads to a difference between threshold value and limit value. By the very fact that stochastic values are characterised by the lack of a threshold, they represent a two-fold problem. Firstly because of the effect even at the lowest dose and secondly due to the lack of a threshold for setting a limit value within a dose-response diagram. Therefore, different characteristic points have to be found. When measuring the risk acceptability of stochastic effects, it is possible to orient to the mean bandwidth of the background exposure, assuming there is pragmatic "normative" indifference to this.

In analysing combined effects, the terminology used is of particular importance. There is no standard terminology to be found in the relevant literature on this subject. Indeed, one may speak of a "confusion of terms". Therefore, one of the aims of this study is by way of formulation to create a standard terminology to facilitate and enable communication within and between the individual disciplines concerned.

A *phenomenological terminology* of combined effects in a case when several substances also individually applied are combined could look as follows:

Example of two substances A and B (x = A or x = B)

$E_{A+B} > E_X$ synergism relating to x   S (A,B) or S(B,A) resp.

$E_{A+B} < E_X$ antagonism relating to x   A(A,B) or A(B,A) resp.

Synergism and antagonism have to be qualified, i.e. a reference substance x has to be indicated. The choice of a reference substance however may not be interpreted in a mechanistic fashion, e.g. that an antagonism in relation to x would mean that the other substance or substances would have an effect on x. Effect in a chemical context has to be determined as a symmetrical relationship. In addition, it must be taken into account that a synergism and antagonism of two substances do not have to be effective over the entire range of concentration. If a reference substance is not explicitly indicated, the substance that singularly shows the strongest effect will be taken as a reference point.

To extend the phenomenological terminology, additivity models may be used as comparative tools. However, the description of the deviations from such models have to be clearly differentiated from a terminological point of view from the phenomenological terminology.

The starting point for the formulation of the dose-additivity model are equal increments in the dose-response relationship of individual substances. It is assumed that the doses of single substances, possibly provided with a constant factor, can be added up in order to determine the effect of a combined dose on the basis of a common dose-response relationship. The effect of one substance will thus be equated with that of the other component of the mixture, or as a dilution of same. In practice, the sum of the individual doses is calculated and the effect is read from the common dose-effect curve on the graph. In the effect additivity model, single effects are simply added together to determine the combined effect. The effect additivity model can therefore be applied to any combinations of the same effect.

The model involving independent or relative effect additivity is based on the assumption that the combined effect of single doses is the result of a fictitiously assumed consecutive exposure and that the individual effects are related to the overall effect, in such a way that the effects of the individual doses relative to one another each correspond to a part of the overall effect (actual effect subtracted from overall effect).

The possible consequences of choosing a model in the order of zero approximation can be illustrated using the example of doses comprising combined substances, which when administered alone show no effect. Assuming effect additivity involving a combination of any given number of substances the concentrations of which individually show no effect also shows no effect. However, if one assumes dose additivity, the sum of individually ineffective doses may well result in a relative dose which is effective. At present environmental law often applies the "more cautious" dose additivity model subject to more precise scientific results.

Limit values, as boundaries for political decision making, require legitimation, therefore the rational criteria for defining limit values has to be discussed. Especially important is the question as to how rational it is to set limit values which are not, or not directly, based upon threshold values. To justify limit values, environmental quality targets play an important role. In the interest of a democratically acceptable environmental policy it has to be demanded that the determination of environmental quality targets and the setting up of corresponding limit values is accomplished in a rational manner.

The rationality of an action is measured against its desired ends. Environmental standards have to fulfil the role of protecting man and nature from harm, i.e. when envisaging potentially harmful action to stipulate risk limits. Hence the key term risk is latently at the core of the discussion on environment standards. When it is a matter of rationality in determining environmental standards and the establishment of limit values, it is important to address the question as to whether and, as the case may be, in what way it is possible to weigh the risks rationally.

When setting limit values not only shall certain risks be excluded but also certain risks are often tolerated or accepted. Therefore all risk comparisons, which are the core element for setting limit values, contain a normative element, in so far as a limit value implies that anyone, who is subject to this limit value, may have to tolerate certain harmful consequences and may be permitted to subject others to the same consequences, i.e. such that correspond to a lower degree of risk than that stipulated in the limit value concerned. However, current political discussion makes it abundantly clear that such tolerant behaviour is in no way always readily taken for granted. The normative problem can therefore be boiled down to the question as to which risks can be accepted at all and by what right. The cause of controversial discussions in this context lies in the differing evaluation of commensurateness and risk acceptability.

The question of what others can be expected to accept has always been a fundamental problem in society. However, at the moment, in the context of establishing limit values, this is an even more controversial topic. Firstly, it must be remembered that the question of de facto acceptance fundamentally does not provide any reliable criteria for establishing a legitimate set of limit values. De facto acceptance is as a rule in content vague and over a longer period of time unstable. The particular interests of individuals and groups often lead to diverging expectations. However limit values should principally be acceptable to anyone and be of long-term validity. Therefore, they have to be the instrumental result of mutual objectives. To resolve disagreement with regard to set purposes it is necessary to reach understanding regarding the pursued ends, which in many cases may be realised through various, different purposes. However, discursive conflict solving may be easier, if common objectives are already explicitly or implicitly shared. Conclusions can therefore be drawn from discovering which purposes and objectives actors in society, by virtue of their own actions, implicitly already accept ("revealed preferences"). On the basis of such a set of purposes an actor can be required to accept that a certain degree of risk may be involved in a given action if he otherwise accepts risk on basis of the same subjective benefit/harm assessment (principle of pragmatic consistence). Such a rule would constitute a criterion for risk acceptability.

The principle of pragmatic consistency therefore assumes that from the actions of the agents it is possible to infer their willingness to accept risks. This process can be reconstructed as a fictitious discourse. In this a statement can be incorporated, which, though never actually expressed, still represents an "eloquent testimony".

Not acceptance but acceptability is therefore the criterion for justifying valid demands expressed in the form of environmental standards. The question of acceptability therefore raises the problem of justifying norms, i.e. universal requirements (requirements applying to everybody). Justification problems of this kind are the topic of the philosophical discipline Ethics, while the factually propagated or executed imperatives make up elements of morals currently in force.

In order to be of purpose, environmental standards have to relate to proven effects or at least scientifically plausible effect estimates. In principle one has to accept a broad mechanism, consisting of a chain of steps along the path to the final effect. In addition, effects cannot always be defined as isolated effects, e.g. phenomenal health damage. Even the accumulation in the human body of a certain chemical which in itself is not harmful, may produce a damaging effect. Thus certain substances having accumulated in the body's fatty tissue may find their way into breast milk and have a harmful effect on the infant. In this context it also becomes clear that to ascertain effects and causes implies considerable requirements with regard to the measurability of dependencies. Only when exposure to a harmful substance is accessible to experimental analysis and only when the effect can be observed, can one speak of cause and effect in a *meaningful* way.

In the environmental legislation in Germany and other European countries there is no valid, uniform conception of how to set up environmental standards. On the contrary, current environmental legislation is guided by a variety of very different principles. Thus, the principle of protection, the precautionary principle with its large array of various derived principles, the principle of minimising total cost, the principle of cost-effectiveness or the principle of sustainability are applied.

This kind of plurality of principles is unsatisfactory for methodological reasons. Simple examples will help to explain that the different principles will lead to different results. This applies for example to the alternative use of the precautionary principle and cost effectiveness.

For every principle, however, areas can be cited in which application is plausible. The principle of precaution for example is meaningful if one assumes harmfulness on the basis of scientific criteria, and when little is known about the extent of cause and effect.

To the same extent as knowledge grows about the quantitative relationships the more the application of the principle of precaution loses its plausibility, since the growing level of knowledge achieved enables more precisely targeted procedures.

Precise knowledge about the quantitative relationships may under certain circumstances allow preventive measures to be entirely abandoned in favour of aftercare procedures (e.g. repair of material damage). Finally one should not lose sight of the fact that it is nonsensical from a pragmatic point of view and legally not permissible in view of the principle of proportionality to undertake extensive precautionary measures against extremely small risks. If sufficient knowledge is available, targeted measures can be undertaken under consideration of the relevant risks.

Considerations with the objective of choosing an environmental policy principle within fundamental, environmental policy framework conditions, such as the quality of existing knowledge or the type of risks involved, provide a selection criterion, which will permit a selection commensurate to each context. As a meta-principle we shall call this the *weighing principle*. The application of individual environmental policy principles will be decided upon in accordance with this weighing principle. Depending on the scope of the relevant knowledge, the weighing principle will also help to derive the most urgent research imperatives.

The principles of trans-subjectivity and consistency are subsidiary to the weighing principle, since the result of weighing should be dependent on consistent conditions and not on the weighing person or context by whom or in which the weighing is conducted.

Scientific Foundations

Dose-Response Curves and Models

The biological effects of chemical substances and physical factors can be represented by dose-response curves, which are very valuable for the analysis and evaluation of toxic effects of agents, singly or combined and in the context of environmental standards especially in the lower dose range.

For certain effect mechanisms it is possible to develop mathematical dose-response relationship models for reversible and also for irreversible effects. However, the extrapolation of the medium and high dose range into the environmentally relevant lower or lowest dose range is unsatisfactory, since there is either no or very little available experimental data in this area. The application of a model to sparse data points within the lower dose range is also hampered by the fact that the differences, for example, between sigmoid and exponential curves are very small.

An analysis of numerous experimental and epidemiological data in this lower dose range revealed an almost identical shape for sigmoid and polynome function curves with a linear or linear-quadratic equation, which is an interesting fact for environmental standards and makes it desirable to further analyse the applicability of sigmoid curves in this context. The analyses and comparisons showed that mutagenic effects or tumour frequencies could also be well represented by sigmoid curves.

Given the combined effect of two or more agents, it is possible that stronger or even weaker effects can occur than in the case of effects of a single agent, whereby possible effects within the lower dose range are especially interesting with regard to environmental standards, which, for example, occur frequently if the agents reveal unidirectional toxic effects. However, numerically those combined effects prevail in which different agents show different biological effects, in as far as these have a biological effect at all. In such cases one should not expect any combined effects or interactions. On the contrary one should expect a parallel or consecutive sequence of different effects.

Combined effects can on the one hand be compared either with the corresponding effect of the single agents or with the pharmaco-/toxicodynamic models of combined effects. Due to the differing terminology ("additive", "synergism"), test approaches (e.g. fixed dose or mixture experiments), and the various types of graphical presentation (e.g. doseresponse curve or isobolograms), it is a difficult task for the non-expert to find his way in this field. With regard to environmental standards for the combined effect of agents, the mixed approach and the analysis of doseresponse curves appear to be the most expressive. In this procedure the curves of the components will be created either in a single or mixed application (with a fixed dose ratio) and the "mixture curves" can be compared with theoretical combined effects.

The dose-additivity and independent model, which can be based on certain mechanisms, are of special interest for the present study. In the doseadditivity model each agent in the combination reacts as a dilution of one particular substance. In the case of independence the agents act according to different mechanisms but lead to the same (toxic) effect independently. The mathematical effect-additivity model calculates combined effects that equal the sum of all single effects.

Differences or similarities in the combined effects of these models explain themselves by the type of doseeffect relationship and are dependent on the analysed dose range.

Whilst there can well be differences regarding model effects in the medium to higher dose range, these will mainly disappear in the lower dose range. Furthermore, examples of empirical combined effects show a match of observed and calculated effects in the lower dose range. In the medium and higher dose range not seldomly stronger (or also weaker) effects than were expected on the basis of the models are observed.

The medium and higher dose range may show combined effects that cannot be observed in the lower dose range, i.e. deviations from the model effects dose-additivity and independence. Of these effects those combined effects are interesting which exceed corresponding model effects e.g. super-additive effects or larger effects than would be expected for independent combined effect. The latter is due to special toxico-kinetic or toxico-dynamic processes, which explain the rarer cases of spectacular increases in combination. In these few cases the combinations react partly with much stronger damaging effect than the individual single components, which has to be considered for corresponding regulatory procedures.

Wherever combined effects established on the basis of models differ remarkably from each other e.g. in the case of multiple exposures, one will, given toxic effects, observe stronger effects in the dose-additivity model compared to the independent model. If the type of combined effect is unknown, the dose-additivity model can be considered as a "worst case" scenario when evaluating the effect of serious or dangerous substances.

Due to the theoretical considerations and the analysis of empirical combined effects of *unidirectional agents* e.g. medicine, alcohol and malformation causing substances the following situation evolves:

Frequently effects of combined exposures in the medium and higher dose range are larger than the effect of the component with the stronger effect. Generally we do not find any amplification effects in the lower dose range. For anti-tumour substances and also teratogenes with a different molecular target the threshold doses of a substance A in the presence of B is not reduced. Dose-additive combined effects are to be expected (e.g. in the case of hormones and partly with regard to carcinogenic agents).

In addition to the described toxico-dynamic combined effects, toxicokinetic interactions are of significance, e.g. one substance halting the reduction of another substance. The models described above are not suitable for the description of kinetic combined effects, but these interactions deserve major interest, since they can lead in rare cases to spectacular amplification effects.

Combined Effect of Radiation and Substances

In the case of exposure to ionising radiation, it is very often the case that an assessment of the dose in the target tissues and cells can be better conducted than in the case of chemical substances. Since, in addition, there is a relatively high amount of experimental data regarding combined exposures with the inclusion of ionising radiation, these investigations have been given exemplary treatment in this study. The dose limits for ionising radiation at the workplace and in the environment are set out in such a way that non-stochastic, deterministic effects of radiation can be avoided. Therefore, only stochastic effects of radiation (genetic mutation and cancer) have to be taken into account in the area of dose limits and lower doses. This criterion also has to be fulfilled in the case of combined effects. Consequently, those agents are of interest which can have an interaction on the various stages of the development of stochastic effects of radiation following corresponding exposure. The numerous experimental investigations and epidemiological studies have led to the following results with regard to combined effects between ionising radiation and multiple chemical substances:

- Genotoxic substances, which like ionising radiation cause the initial DNA-damage in the case of development of cancer and mutations, generally lead to additive effects following combined exposures. This is particularly true in the low dose range.
- Substances, which impair the repair of DNA damage following exposure to ionising radiation, can cause super-additive effects. This is true, for example, of heavy metals, caffeine and various chemo-therapeutics. However, relatively high concentrations of substances have to be reached for this kind of suppression of enzyme systems of DNA repair to occur.
- Substances which reduce the primary radical reactions following radioactive impact can reduce the effect of the exposure. This is particularly true in the case of radicalscavengers, e.g. substances containing sulphhydryl groups.
- Substances which alter the regulation of cell proliferation subsequent to radiation exposure can cause super-additive effects. In these cases substances can shorten radiation-dependent delay processes in the proliferation cycle of the cells, which normally allow repair processes and thus these substances reduce DNA repair. In addition, substances which above all have a hormonal effect, can stimulate cell proliferation and on the basis of mechanisms of this kind, amplify the development of cancer following radioactive exposure.

For interactions between the development of the effects of radiation and substances, the sequence of exposure in time is of considerable significance. For the influencing of primary processes of damage development the exposures to ionising radiation and to the appropriate chemical substances have to be close together. This is also true for those substances which reduce DNA repair. They have to have an effect on the exposed cells within only a few hours of radiation exposure. In contrast, substances which lead to amplification of the development of cancer by stimulation of cell proliferation can still cause a super-additive effect following substantially later exposure. The varying experimental investigations, as part of which dose-response relationships have been worked out, have shown that super-additive effects in general, only occur in the medium to high dose range. In the low dose range mainly additive effects result. It must also be taken into account here that all exposures to chemical substances are accompanied by low dose ionising radiation from natural sources.

Investigations of the combined effects following exposure to densely ionising radiation (neutrons, alpha-radiation and others) with toxic substances have led to the result that substances which inhibit DNA-repair lead to a lesser extent to super-additive effects given these combinations. In contrast, substances which stimulate cell proliferation cause superadditive effects even with densely ionising radiation. Therefore, in general super-additive effects between chemical pollutants and ionising radiation only have an effect if specific interactions occur between the pollutant and the various stages of development of the effect of radiation. In general, a relative increase in the radiation risk by a factor of 2–3 is reached. In very few cases does this effect extend beyond a factor of 5.

The analyses of epidemiological investigations have uncovered superadditive effects in the case of persons exposed to radiation who are also heavy cigarette smokers. Here too, high levels of exposure are necessary for a super-additive effect to occur in the proportion of lung tumours.

Overall the investigations of combined exposures with ionising radiation indicate a whole series of possible exposures which can lead to superadditive effects. A number of ideas with regard to the effect mechanisms, which permit extrapolation for further possible combinations, could be worked out. However, these super-additive effects have only been observed in the medium and high dose ranges. Therefore the extrapolation of this data in low dose ranges would have to be validated further.

Combined Effect of Substances

The number of possible combinations of chemical substances is legion, the portion of exposures which actually occur in the everyday life of man and in the environment is still enormously big. This clearly places emphasis on the purely chemical combinations compared to more simple systems such as radiation and medication, as well as exposure at work and requires differing strategies for determination, evaluation and regulation.

The fact that most of the limits for risk minimisation or avoidance have been fixed with regard to chemical substances, is also an expression of the variety which is different from other systems. However, the criteria used in the determination of these limits vary to an extraordinary degree. Individual categories, perhaps those having existed longer, are orientated along the lines of "effect thresholds", others follow minimisation principles, whereby the scope of "precaution" to be shown moves within broad limits with the following result: The "safety margins" vary from zero to close to a million. The reasons for this are numerous: Differences in the traditions of academic disciplines dealing with the matter, uncertainties in the evaluation of scientific risk data, in socio-political demands, in the acceptance of (residual) risks, and so on. Given broad margins of increased factors of safety in various types of exposure limits, the enticing attempts presented again and again on the part of administration to use simple calculation models to regulate combined effects, which additively aggregate limiting value fragments according to their share in mixtures, are out of the question.

Literature available on toxic effects of chemical substance mixtures reveals that the subject has been processed in a non-systematic manner not covering all areas. Very heterogeneous formulation of questions and above all the use of very high doses irrelevant to questions regarding limit values, render the comparison of data for the purposes of working out mutual regularities more difficult. For this reason, we attempt in this article to sort the description of known facts relating to the combined effects of chemical substances according to mechanistic models – whilst not claiming completeness. Substances in mixtures may attack the same or different target organs, function according to the same or differing mechanisms and become effective in an unchanged state or following enzymatic transformation; and this independent from each other or under mutual influence (interaction).

Interactions, by far the more important category for combined effects, can take place at all stages of the movement of substances within the organism: in resorption and distribution, metabolisation, infiltration into cells, at the (biochemical) target structure in the cell, in the repair of damage incurred. Depending on the direction of the influence of action, completely independent, additive, sub-additive or super-additive effects are to be expected in combination. Super-additive effects are of particular significance with regard to the question of setting limit values for combinations. It is possible to derive different kinds of scenarios from the mechanistic basic factors as the basis for possible determinations of limit values, which permit predictions with regard to interactions and their consequences for combined effects:

- Substances in combination attack as such, i.e. in unchanged form various target structures independently of each other and generate various damaging effects. With regard to the relative specificity of the effects of harmful substances this could well be the most frequently occurring case. In this regard, effect-related limit values for individual substances do also have validity in combination.
- In unchanged form and independent of each other, substances generate the same effects at the same target structures. Addition of the effects is assumed here. Limit values for individual substances potentially do not provide sufficient protection if they are effect-threshold orientated. A reduction thus might be necessary.
- Substances have effect only after enzymatic biotransformation. If one component restricts the development of the other's form of effect a weakening of effect is to be expected. Existing limit values for individual substances afford sufficient protection.
- If one substance restricts the enzymatic inactivation of another, it is possible that a strong increase in effect can come about – provided the degree of detoxification is high. This combination, although rarely occurring, is the most important for practical regulations since it is possible that super-additive increases may occur.

It is theoretically possible that one active substance can, through induction of the bio-activation enzymes of the other, increase its effect. Since limit values for these substances, such as tumour promoters, are generally set below the threshold values for induction, this case does not play a fundamental role in the practical field.

The description of the existing literature, according to this guideline, separates conventional toxic substances ("reversible" effect) from genotoxic ("irreversible" effect causing mutations and cancer), since the latter follow other laws with regard to dose and effect: No threshold doses can be estimated. For example, the results of the examination of complex, undefined mixtures are presented and compared with those purposefully combined. These are checked with regard to the correctness of predefined hypotheses. The forecasts theoretically derived according to the above schema are essentially confirmed by existing literature; apparent contradictions can be explained. Valid predictions, however, can only be made if the toxico-kinetic and toxico-dynamic characteristics of the components of a mixture have been compiled to a sufficient degree beforehand. Depending on the place and type of attack the rules regarding independence or interaction are confirmed as sound. In the case of interactions the essential general rule emerges: As the dose decreases, additive/superadditive effects become weaker, in the area of limit values they are mostly no longer detectable. Exceptions are based solely on insufficient knowledge about the effect characteristics of the action for individual components within mixtures.

Combined Effects on Plants

Ozone, the sulphur oxide compounds and nitrogen emissions currently pose the greatest dangers for vegetation in Central Europe and in North America. Extensive experimental and epidemiological findings also exist with regard to the effects of these contaminants as individual components. At the same time, there are guideline values for the aforementioned pollutants as individual components which have been derived by various expert committees with an aim to protecting vegetation. For these reasons these three components have been selected as examples for the evaluation of combined effects in ecotoxicology. As well as sub-additive reactions, combinations of these pollutants cause additive and super-additive reactions in plants. In the case of weak damaging effects it is possible that additive effects can give an underestimation. Almost all combinations result in combined effects, which (to a large extent) correspond to independent effects, thus indicating a lack of any particular interactions. One exception is the combination of SO_2 and NO_2, where often a synergism in the context of a super-additive effect has been found. The cause of this intensifying interaction is being discussed.

The differing effect characteristics of the selected components (SO_2 and O_3 phytotoxic, SO_2 with lack of sulphur and the N-compounds, trophic up to a certain load level) have the consequence that even given low shifts in concentration, qualitative changes occur in the reaction of the plants.

The derived guideline values for the protection of vegetation against mixed emissions are based on the results of three methodological approaches:

- Comparative experimental investigations on the effects of two or three components either individually or combined;
- Comparative epidemiological investigations in areas with high exposure values with regard to the effect of filtered and unfiltered air;
- Models to derive patterns of concentration over time which are still tolerable.

The results of these experiments do not suffice to determine guideline values for combined exposures with the components selected, which are below the existing guideline values for the individual components:

- When determining the guideline values for individual components the responsible expert committees more or less explicitly took results from experiments with combined exposures into account.
- The lowest concentrations of pollutant gases applied in the combination experiments which still led to sub-additive, additive or super-additive effects were each time over and above the guideline values derived for the individual pollutants to protect vegetation.

- Plants, when subjected to exposure in chamber systems are mostly exposed to a higher speed of deposition, i.e. a higher effective dose given the same concentration of pollutant as outdoors. Because the derivation of guideline values for individual components is essentially based on experimental and epidemiological experiments in chamber systems, a safety margin should naturally be contained in guideline values obtained in this manner.

Risk perception

From a psychological and sociological point of view, the structures and processes of individual perception of risks and the ways these are dealt with socially, should be examined more closely. Attention is directed here at the level of perceptions and associations with the help of which people understand their environment and on the basis of which they carry out their actions; one speaks of 'constructed reality'. The following patterns of conception characterise the scope of significance of risk as a part of intuitive perception:

- *Risk as a threat:* The notion that an event can affect large portions of the population concerned, at any point in time, creates the feeling of being threatened and powerlessness. The extent of the perceived risk in this case is a function of three factors: of the randomness of the event, of the maximum extent of the damage expected and of the time span available to avoid the damage.
- *Risk as a stroke of fate:* Natural disasters are mostly viewed as unavoidable events, which admittedly have devastating consequences but are seen to be "whims of nature" or "the will of God" (in many cases also as mythological punishment from God for collective sinful behaviour) and are therefore beyond human intervention.
- *Risk as a challenge of one's own powers:* In this understanding of risk, people take risks in order to challenge their own powers and provide them with a taste of triumph after successfully winning a battle against the powers of nature or other risk factors. The fundamental incentive to participate is to ignore nature or co-competitors and to master self-created risk situations through one's own behaviour.
- *Risk as a game of chance:* If the principle of chance is recognised as a component of risk, then the perception of stochastic distribution of payouts is the closest to the technical-scientific concept of risk. Now this concept is almost never used in the perception and evaluation of technical risks. It refers to financial risks.
- *Risk as an early indicator of dangers:* According to this understanding of risk, scientific studies help to discover insidious dangers early and uncover hidden relationships between activities and/or occurrences and their latent effects. Examples of this usage of the notion of risk can be found in the cognitive coming to terms with low doses of radioactivity, food additives, chemical pesticides or the genetic manipulation of plants and animals.

Many people perceive risks as a complex, multi-dimensional phenomenon in which subjective loss expectations (not to mention those calculated statistically) only play a subordinate role, whilst the context of the risky situation, which has an effect in the different meanings of risk perception, substantially influences the degree of the risk perceived. This dependency of risk evaluation on context is the decisive factor. This dependency on accompanying circumstances is not random but follows certain regularities. These can be investigated by using specific psychological investigations. Investigations with regard to the context conditions of risk perception have thus been able to identify the following factors, amongst others, as being relevant:

- Familiarity with the risk source;
- Willingness to take risk;
- Personal ability to control the degree of risk;
- Ability of the risk source to cause disaster;
- Impression of an even distribution of benefit and risk;
- Impression of reversibility of risk consequences;
- Trustworthiness of the information sources.

The significance of these qualitative characteristics for the evaluation of risks offers an obvious explanation for the fact that precisely those risk sources which were deemed to be particularly low risk in the technical risk analysis, met with the greatest resistance amongst the population. Risk sources which are viewed as controversial, such as nuclear energy, are particularly often associated with negatively laden attributes, whilst in contrast leisure time risks carry more positive attributes.

Until now it has not been investigated empirically how the indicated reaction patterns change in the case of combined risks. In order to fill this research gap a representative survey was carried out within the framework of the project in 1998 amongst the population of the federal German state of Baden-Württemberg. The survey involved a representative crosssection of 1,500 adults. The main emphasis of the survey was the perception and evaluation of technology within the population. Within the scope of the survey, the significance of environmental risks in general and of combined environmental risks in particular were researched.

In the evaluation of combined environmental risks a broad gulf was revealed between the opinion of most experts and the lay individuals surveyed. Over two-thirds of those surveyed were convinced that super-additive effects are generally to be expected given the interaction of several pollutants. Just under a quarter (23,8%) were indifferent in their answer and only 9,6% rejected this assumption. The reason for this clear answer can on the one hand be derived from everyday experiences with medium or high doses of medication or stimulants. On the other the typical risk aversion for the risk type "indicator of damage" plays a fundamental role, particularly the intolerance of any further uncertainty which exists. Because of this attitude of expectation, it is indeed understandable that most people believe reports about damage to health through combined environmental pollutants and that the scientific experts and regulatory authorities will find it difficult to convince the sceptical public of the opposite.

Possible effects on health are not only triggered by the pollutants but arise as the result of psychological or psychosomatic processes. Of special significance here are those psychosomatic reactions which have come to be known in literature as part of the category "multiple chemical sensitivity syndrome". Public perception of this syndrome is contributing towards the popularisation of an effect connection. A

vicious circle thus begins: The perception of exposures, above all to combined pollutants, causes fear and uneasiness amongst the observers. In turn, this uneasiness
leads to psychosomatic reactions in some individual cases. These *reactions* are
observed and are deemed as evidence of the supposed causal connection between
emissions and damage to health.

In addition to this, combined effects are suitable as topics of political mobilisation.
For one thing, they are the subject of press coverage since health and the environment
are favourite media topics. Due to the perception of real syndromes, the decline in
expert credibility and the discrepancy between risk researchers and lay persons, public pressure is mounting for politics to regulate more vigorously. This pressure leads
of course to counter pressure from those groups which would be negatively affected
by more stringent regulation. The row has the effect of polarising society.

Economic Aspects

Cost-efficient standards: Environmental policy should on the one hand guarantee
that certain load limits be adhered to with regard to man and the environment. On
the other hand it should also pay heed to the aspects of cost-effectiveness. Environmental economics emphasises the significance of the principle of cost-effectiveness. It has worked out in the case of single pollutants that economic instruments
(environmental taxes and tradeable permits) enable a more cost-efficient form of
environmental protection than legal regulation and should therefore be applied
there. In the case of multiple pollutants, however, the objective of economic efficiency has additional significance. Under these conditions, at least two substances
are responsible for one specific damage. If a load limit has been determined, it then
has to be clarified as to which of the pollutants is to be regulated against and to
which degree. From an economical viewpoint this decision should be made according to the criterion of cost-effectiveness. The limit values should therefore be determined in such a manner that the total economic costs are as low as possible.

Damage to the environment may cause damage to health, material damages
(including losses in production), as well as ecological damage. Load limits are
derived from objective notions with regard to the extent of permissible damage in
these areas. Guideline targets of this kind provide the measure for limit values in
environmental policy. Environmental policy's central guideline target is securing
the health of the nation, whilst health itself is subject to delimitation in a certain
way and can be measured in the light of the relative frequency or severity of an illness. When considering cost effectiveness, the guideline targets of environmental
policy are taken for granted.

It is assumed that one guideline target is compatible with varying combinations
of agents. The possibility of substituting the agents (A and B) should therefore
exist. A certain additional quantity of A would, according to the relative specific
harmfulness of the substances, require a certain reduction in the quantity of B. The
type of combined effect (additivity, super- and sub-additivity) has bearing on the
relationship of the specific harmfulness of the substances. The opportunity presents
itself to make a selection according to aspects of cost from amongst the bundle of
ecologically equivalent combinations. So as to keep the costs of environmental protection as low as possible, agents where the reduction thereof is associated with rel-

atively low costs and/or relatively strong relief effect should be reduced to a greater extent than those substances with relatively high abatement costs and/or relatively low relief effect. Because costs typically increase progressively with increasing levels of pollutant abatement it would mostly be appropriate to impose limitations on all substances concerned. One should not concentrate on one particular substance even though it may be the main cause of damage. In the exceptional case of a substance which is very cheap to reduce and provides high levels of relief on reduction, it can, however, also be efficient to concentrate on this substance.

Whether the effects of the agents in combination become stronger or weaker or whether they are the same as when occurring separately, has bearing on the level of cost of environmental policy. Environmental protection costs increase in the case of super-additivity of effects in comparison to additivity. The strengthening of individual effects is the same as an increase in costs. In order to meet the same guideline target as for additivity, it is necessary to set stricter environmental standards for the agents, which in turn causes higher costs. There is then a tendency in the political process for it to become more difficult to meet any set objective. If, on the other hand, two substances combined actually weaken damaging effects, this is tantamount to cost-free abatement. Given an additive relationship, the limiting values would have to be set lower to achieve the same protective effect as for sub-additivity. Meeting a guideline target is linked with less strict requirements and consequently with lower environmental costs. High costs of environmental protection are a fundamental obstacle to enforcing strict environmental standards. Because of these varying implications with regard to cost, it is important to establish clarity with regard to the type of effects.

Cost comparisons are only possible in the case of quantitatively formulated guideline targets (maximum permissible degree of health risk, upper limits for global warming, avoidance of summer smog, amongst others) and standards. This underlines the general demand on the part of the economy that legislative organs should set targets with a greater emphasis than to date on quantitative rather than qualitative criteria ("avoidance of damaging environmental effects" or "state of the art of environmental precautions").

Measures: With the aid of an appropriate set of instruments (laws, charges, tradeable permits, voluntary agreements and subsidies) it must be ensured that the limit values set are adhered to. The special requirement of the use of these instruments in the multiple pollutant case is to bring about the efficient abatement structure. Not all instruments are always suitable for achieving this in the same way. Tradeable permits have advantages in the case of combined effects with linear additivity. If environmental charges or legal standards are used, the efficient limiting values are to be directly set as the target values. The task becomes easier in the case of tradeable permits and linear, additive exposure. Knowledge of the structure of abatement costs is not necessary here. All that needs to be known is a combination of emission caused damages which is compatible with the guideline target and the specific harmfulness of the substances. By issuing tradeable permits to the amount of emissions recognised as permissible which can be exchanged amongst themselves in the ratio of the specific levels of harmfulness it can be achieved that an efficient structure will result through adjustments within the economy itself.

In the case of non-linear combined effects this advantage with regard to tradeable permits solutions no longer exists. In these cases no special instruments are immediately evident. In which way certain protective and precautionary targets can best be achieved then depends, as is the case with single exposures, on the respective problem.

Decisions under uncertainty: In those areas where relatively little is still known about combined effects, any decisions with regard to limit values are to be made under uncertain conditions. Economic-decision-making models help reveal the relevant determining factors for economically rational decisions (relative harmfulness and relative abatement costs, extent and measurement of uncertainty as well as risk evaluation). This is shown for differing degrees of uncertainty (stochastic model, fuzzymodel and uncertainty model). The analysis clearly shows that even given a limited level of knowledge, it is possible to make rational decisions. Therefore, in the process of determining limit values decision making theory should be applied.

Political conclusions: Because in practice to date the single substance consideration has predominated and cost-efficiency considerations hardly play a role at all in multi-substance phenomena, opportunities present themselves for an improvement in policy. In the course of a reform it can prove particularly advantageous to extend a policy which until now has only been intended for one pollutant to include others (for example climate protection policy) and to alter the relationship of existing (single substance) limit values for all agents contributing towards any specific damage. Even if very little information is available about the effects of ecologically equivalent combinations of agents, the economic approach is indeed meaningful because if, for example, only three ecologically equivalent combinations of several constituent pollutants are known, it is nevertheless possible to make a selection according to their economic costs.

If several pollutants occur both in combination as well as separately, a policy of differentiation should be pursued, which in the case of combined 30 exposures works with specific limit values and which otherwise uses individual values (see the part on legal questions as well).

Generally it will be most efficient to include all significant co-causative agents in the policy. In special cases, however, concentration on a single pollutant may be the most cost-advantageons. In this case, there is a possible conflict with the objective of just distribution of burden, according to which parties causing environmental damage are to be called to account. The generality-demand is part of the normative reason for the polluter-pays-principle. One-sided allocations of burden are particularly problematic if companies competing directly with each other are affected differently thus distorting the competition. One possible solution of the conflict between efficiency and just distribution is that financial compensation be introduced among burdened and exempted parties. All exempted parties would have to be committed for side-payments.

Legal Questions regarding Combined Effects

Until now, legal questions regarding combined effects of substances and radiation do not play a central role in German environmental law. In particular, there is a lack of any discussion in jurisprudence of this question. Because of its constitutional obligations to protect, the state is bound to provide comprehensive protection of the individual against dangers and risks caused by environment pollution, which itself also includes combined effects. However, there is a broad margin of political discretion particularly in the area of purely precautionary measures.

The individual media-related environmental laws contain no express regulation pertaining to combined effects. However, with regard to the general formulations of legal purpose and powers to regulate they do not preclude consideration of such effects. In laws governing the use of chemical substances the law governing the manufacture and marketing of drugs expressly demands consideration of the combined effect of multiple substances. The law on chemicals relates to the individual substance with regard to determining dangerous properties but in contrast, with regard to regulation, combined effects may be taken into consideration (bans and restrictions, classification, labelling and packaging). The same is true of the food and consumer goods law.

Sub-legal regulations in the form of limit or guideline values, trigger values and classification rules often take combined effects into account. In doing so they include such effects in the safety margin, view a substance as characterising, or assume concentration or dose-additive effects; often there are correction possibilities should there be a basis for superadditive or antagonostic effects, however, without therebeing any obligation to do so. Ambient quality values and emission and/or trigger values for carcinogenic and certain other particularly dangerous substances, maximum workplace concentration values in industrial safety and regulations regarding the classification of preparations as well as mixtures posing a threat to water, are of particular significance.

Questions with regard to allocating the regulatory burdens of tackling combined effects are the subject of discussion particularly in environmental liability law. Private liability law has developed two allocation models, namely proportional liability and joint liability with internal contribution rules, both of which are transferable to other cases of "retrospective" liability; this can be seen in the regulation of polluter responsibility under the federal soil conservation act. In the case of "prospective" regulation, one mainly follows the principle of priority. The concept of a "community of emitters" with joint responsibility is fundamentally rejected.

In the United States, Switzerland and the Netherlands combined effects are given more extensive consideration in parts of environmental law than in Germany.

It is evident that knowledge with regard to the type and extent of combined effects is as yet insufficient. The legal regulations regarding the determination of combined effects should therefore be improved. This requires, above all, an amendment of chemical legislation at European Union level. The manufacturer should be obligated to investigate any indication of combined effects in the base set of testing; these findings would then have to be expanded upon if necessary on the basis of supplementary test requirements. The testing of old substances should also be amended accordingly.

It must be taken into account from the legal point of view that when regulating combined effects, constitutional requirements do not compel the state to seek a solution which is "optimal" in substance. The application of the principle of protection (defending against danger) and of the precautionary principle (taking precautions against risk) permit appropriate solutions to be found if their constitutional limitations are heeded. Fundamentally, a distinction has to be made between proven and assumed combined effects. Given mere suspicion of combined effects it is possible to proceed along the lines of the precautionary principle. However, this presupposes a minimum degree of plausibility with regard to the suspicion. Purely speculative assumptions about combined effects do not suffice and, in any case, there have to be correction possibilities in place in case no combined but instead completely different kinds of effects arise from joint exposure to several substances. Consequently, global risk minimisation strategies are problematic without any basis for assuming combined effects. In so far as there is reason to believe that combined effects exist at all, it is essentially permissible to assume the dose-additive effect model or else independent combined effects. Sufficient correction possibilities are, however, required in case other types of effects, in particular those of the super-additive kind, are established.

Particularly worthy of consideration as regulative strategies are individual case evaluations, safety margins, limit values for combined effects and limitations of dispersion. A broad degree of executive discretion exists with regard to the selection and tailoring of these strategies. Out of all possible strategies for the regulation of combined effects it is individual case assessment which represents the fundamentally superior solution. However, for pragmatic reasons it should not hinder the search for global solutions.

Special (additional) safety margins for combined effects added to single substance-related limit values are in principle an acceptable strategy if we are dealing with substances with a threshold value. However, this requires that the safety factor affords an appropriate level of protection should the substance occur on its own and that the substances very frequently appear in a certain combination. In principle, as a typifying solution, they meet the requirements of the principle of proportionality. However, in the low dose range, they can lead to an excess of precaution.

If one component of a mixture is so dominant that it characterises the whole mixture, no actual regulation for combined effects of the mixture is required; the limit value for the individual substance shall prevail.

Generally speaking, it is limit values (and values approximate to limit values, such as trigger values and classification regulations) which should be preferred for combined effects. This is particularly true in the area of protection of occupational health as well as in the case of substances without a threshold value. In the case of known combined effects it is entirely possible, given the present level of scientific knowledge, to perform limited quantification. As for the rest, the regulator can assume the prescriptive model of dose-additive effects or else independent combined effects, so far as appropriate correction possibilities are made a proviso.

Global limitations on emissions (e.g. minimisation) are acceptable if they are used on their own or in addition to other strategies, even for single substance risks.

With regard to the allocation of the reduction burden (addressing measures to particular actors), other measures are to be considered in addition to the option solution tailored to combined effects caused by a single source. These would include in particular, quota systems, compensation solutions (burden of reduction for one source with obligation to pay compensation on the part of the other), as well as the priority principle. Since the Constitution does not call for a "balanced" environmental policy and the principles of equality and proportionality only set extreme limits, there is a broad margin of political discretion with regard to allocation.

Wherever possible an option solution should be chosen which allows the source to decide with regard to allocation. This of course presupposes that several combinations of the substances in question are acceptable from point of view of both health and the environment. As for the rest, the postulate of equitable distribution demands that more emphasis be placed on the polluter-pays principle – perhaps through the use of quotas with guaranteed "burden margin". As an alternative, which at the same time takes aspects of economic efficiency into consideration, obligations to reduce emission levels combined with compensation on the part of other sources involved could also be tried. Regulatory models related to the point in time at which dangerous activities were started (e.g. priority principle) should only be considered if other solutions fail.

With regard to the relationship between regulation and allocation, it has to be borne in mind that the "logic of regulation" (environmental policy decision-making) limits the "logic of attribution" (decision-making regarding as to whom the measures are addressed). Not all of the possible regulatory models are compatible with the attribution models considered. It is individual evaluations and special limit values which are most favourable in this regard. These can be linked with a multitude of attribution models.

Introduction

The need to set environmental standards is a result of the technological domination of nature by mankind. Increasing technicalization has led to mankind and its environment being exposed to a multitude of factors, of which is often only little known with regard to their medium- and long-term effects. In the course of the development, operation and disposal of technical installations and products, harmful substances and physical agents such as ionizing radiation and noise are released that could have undesired effects on human health and the environment. Such side effects of technological developments and the intensified public discussion of such topics have contributed to the almost complete demise of the, to a large extent, blind belief in progress that was the popular, optimistic view in the 1950s and 1960s. Today, we often observe instead that the pendulum swings in the opposite direction, most apparent in the demands for "zero exposure". One way to open a sustainable and acceptable path between such extreme views is the establishment and application of environmental standards understood as quantitative specifications of target values concerning the environment, referred to as environmental quality targets.

Findings about the harmful effects of diverse individual exposures on the environment and human health have led to the definition of environmental standards.[1] This single-substance oriented approach is in accordance with the fundamental and medical research analysis of the effect context, which focuses mainly on individual exposures. The examination of combined lowest-dose exposures, which correspond much more realistically to exposure conditions in the actual environment, entails major methodological difficulties making this area of research appear hardly attractive. However, the potential risks of combined exposures as well as the intensity of their public discussion make the analysis of this topic area a desideratum of the utmost priority.

The spectacular case of a supposedly highly synergistic, endocrine effect of two substances acting only slightly endocrine in single exposure may serve as an example for how explosive this issue is. The results referred to were published in June 1996.[2] Endocrine effects of exogenous substances have been discussed for some time under the catchphrase "feminization of nature".[3] The possibility, however, that these effects could be enhanced many times over in combination with other substances gave rise to heated discussions about the practicality of regulatory measures

---

[1] See e.g. *Akademie der Wissenschaften zu Berlin* (1992) and *Der Rat von Sachverständigen für Umweltfragen* (1996).
[2] See Science, 272: 1489 (1996).
[3] See e.g. Colburn et al. (1996).

to limit the concentration of substances in all kinds of environmental media to the end of limiting harmful effects, because the limit values stipulated by these measures are usually derived from investigations of individual substances [in single exposure]. The spectacular results could not be reproduced by other scientists and ultimately not even by the authors themselves[4] who thereupon withdrew the publication[5]. This example may illustrate – quite independently of the outcome of further investigations – the urgency of solving the problems surrounding the establishment of limit values for combined exposures.

Hence, this study will examine the necessity, realizability and consequences of *environmental standards for combined exposures* from a scientific, medical, sociological, economic, legal and philosophical perspective.

By presenting selected examples of highest practical relevance, criteria will be developed which should allow classifying and assessing the plethora of conceivable combinations with regard to their health effects in a meaningful and understandable manner, based on effect mechanisms of general applicability. The task of working out methods for defining environmental standards for the case of combined exposures is a particular challenge of interdisciplinary collaboration, especially since complex problems regarding the comparability and acceptability of risks as well as of their just distribution have to be solved. Apart from the interaction of ionizing radiation with harmful substances, the area best researched in the empirical fashion, combinations of chemical exposures will be treated too. From pharmacology, where combined effects in the medium and higher dose range have been examined for a relatively long time already, models and findings on mechanisms can be included, even if they have to be transposed to the lower dose range relevant for environmental effects.

The focus of this study will be on human exposure especially to carcinogenic and genotoxic agents, and on the exposure of certain, selected plants, which will serve to answer the most pragmatically urgent questions.

The complexity of the toxicokinetics and toxicodynamics of combined exposures suggests a recursive approach starting from single exposures. Consequently, knowledge of the effect mechanisms for the individual damage patterns and noxae is extremely relevant. Even if the diagnosis has to be that there is a lack of sound data, a number of experimental approaches have been developed, and some epidemiological investigations on combined exposures and their effects have been performed in model studies. This is equally true for a small number of combined exposures of particular practical importance.

On the basis of a categorization of combined exposures founded on a mechanistic understanding, we will develop criteria which make feasible, regardless of the complexity of the individual effect concepts, the establishment of limit values in order to maintain or achieve specific environmental quality targets. Even more than those due to single exposures, risks posed by combined exposures are characterized by certain attributes which are viewed as threatening both by affected and spectator groups. Furthermore, they are apt to be used as instruments for the mobilization for or against certain interests, since the suspicion of negative effects of

---

[4]   See e.g. Gaido et al. (1997).
[5]   See Science, 277: 462.

combined exposures can hardly ever be falsified. At the same time, claims of negative effects oblige the cautionary principle of intuitive perception ("better safe than sorry"). The relevance of psychosomatic aspects will be discussed too. Another supposition prevalent in this context – that the pessimistic expert embodies the highest degree of credibility – is still reinforced by the experience that, occasionally in the past, environmental exposures have been depicted as less harmful than they later turned out to be. The perception of combination risks may not have been widely investigated yet; however, the assumption appears to be justified that a number of factors, which have proven to be relevant for the perception of risks posed by single exposures, are even stronger here. Therefore, public information must not be limited to the scientific explanation of the effect chain; it must consciously include in the communication process the psychological and social mechanisms of risk amplification.

Such "social amplification" of risk perceptions increases the pressure on policy to take countermeasures and enforce statutory regulations, independent of scientifically justifiable standards. Politicians feel compelled to take action. However, such action can have further consequences: It can lead to considerable follow-up costs, distract from the real issues and undermine the trust in the competence of politics in the long term. These unintentional consequences are a crucial motive for dealing with the risks from combined exposures, and their perception, in the study presented here.

To answer the question which substance should be regulated to which extent, when a combination of two or more agents causes harm, also requires economic knowledge. For, if regulation aims at keeping the national costs of environmental protection as low as possible, one has to know which criteria the environmental standards for the individual agents must meet in order to achieve this objective. Since there still are large gaps in our knowledge on combined exposures, the political definition of combined standards involves decisions under conditions of uncertainty – decisions that call for special analysis. In environmental economy up to now, phenomena involving several harmful agents have not been paid much attention yet.

On the part of jurisprudence, one has to clarify at first if and to what extent the legal framework (the constitutional duty of protection and the legislation, relating to media and substances, of health and environmental protection) demands consideration of combined effects in the assessment of substance risks, and if it calls for regulation on those effects. In the tradition of such legislation, but also due to the requirements of the economy of administration, the tendency seems to be to treat substances separately, even if there are some statutory limits and similar values taking into account combined effects. The claim of protection in constitutional law rather points to the comprehensive approach. If one follows the latter view, according to current law or *de lege ferenda*, then the question arises what the appropriate strategies of regulation could be. Here, against the background of considerable gaps in the information basis, the foremost task is to explore the potential and limitations of the precautionary principle for managing the risks caused by combined effects. In addition, it needs to be examined, by way of principles of legal imputation, which of the involved parties have to bear the corresponding reduction burden and the costs incurred thereby.

From the perspective of philosophy, apart from the aspects covered by the philosophy of science, concerning terminology and methods one has to deal mainly with ethical questions, in the first place with the weighing of options under risk. Risks have to be compared and actions in risk contexts have to be examined for consistency. Fundamental for every kind of environmental policy is the distinction between risk acceptance and risk acceptability, which is closely linked to the differentiation between risk and risk perception. Environmental policy, as a transsubjective endeavour, has to focus on the risk itself and its acceptability; otherwise regulation will not achieve the desired long-term validity beyond the ups and downs of de facto "acceptance fluctuations".

The ethical questions demonstrate that the scientific foundations of setting environmental standards are of fundamental importance also from the perspective of the moral responsibility of science. Both by undertaking a research project and by not doing so, facts can be generated that may initiate, amplify or attenuate social processes. In certain cases, the sciences produce social effects, which demand reflection with regard to their justifiability, just by the choice of a subject to be studied. At present, this is the case for all questions related to harm to or protection of the environment.

This study shall help to close identifiable knowledge gaps.[6] The results of this project shall also form the basis for improving the evaluation of combined exposures, in order to enable the creation of an efficient framework for the relevant legal regulations and economic procedures.

---

[6] See e.g. Gemeinsames Aktionsprogramm "Umwelt und Gesundheit" (*Joint Action Programme "Environment and Health"*) of the German Federal Ministry for the Environment and the German Federal Health Ministry, Section 2.2, "Entscheidungen auf solider Basis" (*"Decisions on a Sound Basis"*).

1 Methodological Foundations of Defining Environmental Standards

1.1 Environmental Standards and Human Attitudes to Nature

In the following, *"environmental standards"* will mean quantitative definitions of environmental quality targets. They are instruments for achieving these targets. Environmental quality targets are statements expressing normative concepts about the quality of the environment of human or other living beings. In a sense, environmental quality targets are a condensate of paradigms of environmental policy with respect to the demands on the environment(s) towards general goals in such areas as health, sustainability, system functionality etc.[1] The issue of environmental standards can be approached with various interests in mind, which may be broadly classified as descriptive and prescriptive interests.

Descriptive interests are pursued where one considers environmental standards as an expression of the communal shaping of the environment and of risks faced by society. According to these goals, one investigates, for instance, which phenomena are de facto classed as damage, what damage is regarded as unacceptable, how to appraise probabilities, how to perform calculations with those probabilities and which sensitivities of which groups or individuals are assumed.

The importance of these questions, which essentially belong to the realm of sociological research, is beyond any doubt. However, in the following we will insist that there is, apart from the descriptive interests, a prescriptive scientific interest in environmental standards, where the central question is: How *shall* we define environmental standards in the context of both certain risk acceptances and certain expectations of risk acceptances, and how can we distinguish defining environmental standards as a *rational* activity? Questions of this kind are of extraordinary importance, if only because the answers will have significant consequences for life in societies and for the shape of individual lives. For this reason, the political debates about environmental standards are led with such engagement. Especially for issues surrounded by considerable political conflict, it is desirable to have some rational criteria for guidance through the discourse.[2]

---

[1] See *Der Rat von Sachverständigen für Umweltfragen* (1996) and *Akademie der Wissenschaften zu Berlin* (1992, p. 33).

[2] The study by the *Akademie der Wissenschaften* in Berlin, *Umweltstandards*, has illustrated this problem using the paradigm of radiation protection as an example. A comprehensive account of the problem was published by the *Rat von Sachverständigen für Umweltfragen* (Expert Council for the Environment): *Der Rat von Sachverständigen für Umweltfragen* (1996); also see Gethmann (1992a).

The subject of the study presented in this volume is the definition of environmental standards for combined exposures. The methodological foundations are the same as for single exposures, with some exceptions, which will be pointed out clearly. However, this does not dispense us from the necessity to discuss the methodological foundations in detail, since they will set the principal course for our investigation.

1.1.1
Environmental Standards as Cultural Achievements[3]

Many laypeople, but also some scientists seem to believe that environmental standards are natural phenomena that need to and can be discovered. In their opinion, it would only require some smartly equipped (laboratory) expedition to find them somewhere, some day. Such "naturalism", under which term we will sum up this position[4], includes the believe that environmental standards can be taken directly from certain graphical features of some curves, from zero values, inflection points, saddle points etc. In reality, environmental standards, as entities determining actions, do not simply follow from the observation of nature, but from communal objectives, for instance the goal of protecting the environment (in a certain, qualified way).

Hence, *environmental standards do not represent natural phenomena but cultural ones i.e. conventional settings. In the first instance they are prescriptive and not descriptive objects.* Still, natural phenomena play a "decisive" role in the establishment of environmental standards. The inter-relationship between the observation of nature and the regulatory stipulation of procedures can be illustrated by the distinction between "threshold values" and "limit value".[5,6] Threshold values are indeed natural phenomena that can be graphically determined from curve features ("thresholds"). A *threshold value* represents a certain relationship between a cause – i.e. an exposure – and its effect. It is a measured quantity that indicates, for example, a dose or a dose range upon which a certain effect occurs. In contrast to this, the

---

[3]   For this section, also see *Akademie der Wissenschaften zu Berlin* (1992, chapter 1) and Gethmann (1992a).

[4]   In principle, one differentiates between ontological (metaphysical) and epistemic naturalism. An ontological naturalism is characterized by the view that every event is a natural event. It always is an epistemic naturalism as well that can be characterized by the claim on the methodological exclusivity of the scientific explanation. An epistemic naturalism, on the other hand, is compatible with ontologies other than the naturalistic one, too (see Hartmann and Janich 1996, Mittelstraß 1995). Both of the two forms of naturalism exclude a genuinely cultural area in the scientific approach. Hence, they are incompatible with the thesis advocated here, which is that one can argue rationally for decisions not based on natural-scientific knowledge. These two forms, which are primarily of epistemological relevance, are joined by another form: ethical naturalism, which is characterized by a reduction of normative to descriptive expressions.

[5]   In the following, the terms "environmental standard" and "limit value" will be used synonymously; one could also say *limit value in a wider sense*. In the juristic discussion, a *limit value in the narrower sense* is distinguished from guideline values and private norms as special types of environmental standards.

[6]   The fact that the result of a statistical evaluation of experiments often is an interval rather than a definite threshold value does not affect the relevance of our argument regarding the difference between a "threshold" and a "limit".

limit value indicates at which point in the cause-effect relationship a course of action (including the use of an appliance, apparatus or large-scale industrial plant) shall reach its "limit". Finally an *environmental quality target* stipulates what shape or form a certain sector of the environment *shall* or *shall not* have. In order to achieve such a target, the limit will be determined by means of a limit value, the observance of which being the (assumed) minimum prerequisite to be fulfilled in order to achieve the target. Hence, one must always distinguish precisely between the concepts of a threshold value, a limit value or an environmental standard and of an environmental quality target. A threshold value does not *per se* provide a limit value, because a normative basis is required for its establishment. Such a basis can be an environmental quality target, which again does not allow deriving a limit value in the absence of empirical information. Nevertheless, limit values are a proven instrument for realizing environmental quality targets.

Now one could *demand* that a limit value must always be based on a threshold value. This demand cannot be maintained, however, if only for the reason that threshold values cannot be found in many cases. This is true e.g. for ionizing radiation, which might have carcinogenic and mutagenic effects even at the most miniscule doses. The same applies for a number of chemical noxae. In such cases, one refers to stochastic effects (since the causes behind the effects can only be described in a stochastic fashion). Accordingly, non-stochastic effects are such effects for which threshold values exist. However, for stochastic effects, too, limit values can be established, as has been shown by the well-tried practice of radiation protection.[7]

Furthermore, the case can occur that there may be a threshold value, which, for good reasons, should still not be used as a basis for a limit value. In many cases (e.g. when there are uncertainties in the knowledge) a safety margin around a threshold value will be observed, meaning that a stricter [lower] limit value will be set. On the other hand it is conceivable to set a limit value higher than the corresponding threshold value. In ecotoxicology, this approach is actually the rule. For instance, the introduction of a limit value could incur unjustifiable costs in the light of a marginal risk reduction achieved by it. The breaking of threshold values is tolerated in large-scale emergency situations, in particular, e.g. in technological or natural disasters, at least for short periods. For security and emergency services, limit values may be applied that range above certain threshold values. For plants, exposures beyond a threshold value are accepted for longer periods too. However, this toleration of exceeding threshold values is always fraught with a considerable burden of justification. A threshold value is regarded as such a powerful phenomenon that it must only be broken for very good reasons. At the same time, such examples demonstrate the necessity to distinguish between limit value and threshold value in the sense explained here.

In summary, we can state: A threshold value *can* be set as a limit value, but limit values do not necessarily refer to threshold values. They represent values that shall not be exceeded for certain reasons, and they limit the allowable exposure. The criteria for establishing limit values generally stem from the objective to prevent or minimize certain injuries. Hence, limit values may be set below, at or indeed above

---

[7]   For the distinctions stochastic/non-stochastic and probabilistic/deterministic, respectively, see section 1.2 and chapter 2.

a threshold value. Moreover, there are certain cases where limit values have to be set for noxae for which a threshold value cannot be established, as is the case for cancerogenic effects.

Thus, limit values, in contrast to threshold values, are by no means natural phenomena[8], but (in the case of restrictions) communal limits on a course of action. This may be illustrated by the following comparison: In Germany, the maximum permissible concentration of alcohol in the blood of a driver of a road vehicle is the 0.5-per mil limit. This is a limit value laid down by convention; it is not a threshold value.[9] Nobody would believe that the observable effect of alcohol on the driver would considerably change at 0.6 per mil compared to 0.4 per mil. Nevertheless, it would be mistaken to conclude from the conventionality of the setting that it is arbitrary, for there are (more or less good) medical and statistical reasons for the limit value. Apart from that, however, we are dealing with considerations on an opportune course of action regarding the objective to allow only drivers with a certain, minimum fitness to drive on the roads. This objective is constitutive for setting the limit value. If another goal were pursued, for instance a healthy lifestyle or the establishment of toxicity limits, one would lay down quite different limit values. The important point in these considerations is that conventionality does not mean arbitrariness, but practicality with regard to an objective. Whether a means fulfils its end, can be a hard criterion. At least in most cases it will *not* be a question of leaving the answer to an arbitrary decision.

The preceding line of argument is not limited to single exposures; it equally applies to combined exposures. The particular difficulties constituted by combined exposures for setting limit values are treated in section 1.2, in terms of laying a methodological foundation, and in chapter 2 from the perspective of medicine and science.

The conceptual assignation of the terms "environmental quality target", "environmental standard" or "limit value", respectively, and "threshold value" obviously assumes a certain interpretation of the relationship between *humans and nature*, which has been discussed controversially from the age of Greek philosophy till today. This controversy also dominates the present debates on environmental policies. The ideas about nature facing each other in these debates[10] can be classed – at the price of considerable simplification – under two distinct positions going back to Plato, on the one hand, and to Aristotle, on the other. The platonic variant is characterized by the effort to describe nature as a whole, an all-embracing entity of which human beings are just one part (the *holistic* view of nature). In the context of a

---

[8]   If one describes threshold phenomena as "natural" phenomena, one must not forget that even the establishment of a "natural" threshold is inconceivable without human input. It may require expensive and strenuous laboratory practice supported by complex measuring equipment. What science calls "nature" is in fact the result of scientific practice i.e. a cultural achievement. Thus, limit values and threshold values may be cultural phenomena, but in a different sense: In measuring a threshold, the scientist establishes a value, which he can, however, not dispose of. Limit values, on the other hand, can always be disposed of by society (even if not arbitrarily, but – at least supposedly – towards a certain objective).

[9]   The conventional nature of the alcohol limit already becomes clear from the recent lowering of the limit from 0.8 to 0.5 per mil – and from the debate preceeding this decision.

[10]   See Schäfer and Ströker (1993ff.).

methodical shaping of scientific languages, the introduction of such a concept of nature becomes difficult, because it either does not hold any power of differentiation or implies very strong premises, following which a course of action shall be judged on its being or not being "in accordance with nature". The concept of nature introduced by formulating an ideal of nature – hence by laying down how things should be according to nature – acquires a normative power and, at the same time, requires justification with regard to the normative premises.[11]

Aristotle, on the other hand, proceeds from the individual things and formulates criteria for their naturalness – an approach that shall be called *methodical*. Natural things are distinguished from artefacts, *physis* from *techne*. The natural holds the principle of movement in itself; artefacts are determined in their movement by human beings.[12] The concept of nature acquires normative power only in the context of a teleology that assigns to each natural being its natural movement towards its appropriate natural aim. It has to be noted that the distinction between the natural and the artificial starts from human action, and that this introduction of the concept does not imply the obligation of human action keeping to a certain order of the whole of nature.[13]

The holistic understanding of nature proves to be secondary to the methodical one, for two methodological reasons:

1. Even the qualification as a whole must be done by means of predicators, the author of which is the human being making distinctions. For it is the human being, in the end, who assigns a place in nature for himself and experiences himself as the acting entity: "Man is part of nature", says man. The role of the preserver is equally rooted in an (implicit) self-definition of the acting human and in his role of the exploiter. This shows that the role of the actor, from the perspective of his self-experience, can be denied neither operatively nor cognitively. The actor is always at the centre of the field of action, wherever he may move. It is one of the incontestable elements of his self-experience that the "I"-author of a (speech) action cannot be reduced, without semantic loss, to the actor of a corresponding description of the action ("pragmacentrism").[14] With the (by no means unproblematic) premise that just the individuals of the species homo sapiens sapiens fulfil this structure of centrality, this results in a structural anthropocentrism, which must not be confused with a (practical) human egoism. The term "anthropocentrism" stands for a structural statement that is strictly distinct from the self-empowerment of humans in an exploitative relationship to nature.
2. The normative content of the understanding of nature is due to a teleological interpretation ultimately based on the self-experience of the actor as a planning

---

[11] A holistic view of this kind is taken by (though with considerable differences in some details) H. Jonas, K.-M. Meyer-Abich, G. Altner and F. Mathews. – The normative premises can remain implicit and call upon normative designs that, as it were, are disguised in this way. In the view of nature proposed in Plato's Timaios, for instance, L. Schäfer diagnoses an ideological foundation going back to Plato's Politeia (Schäfer 1993, p. 81).

[12] Aristoteles understood the concept of movement in a far wider sense than today's common understanding, which is guided by mechanics. According to Aristoteles, the qualitative transformation of a thing is movement too.

[13] See Janich (1996).

[14] See Gethmann (1997).

actor i.e. as an actor who defines objectives and strives for their achievement. Talk of "nature" as an actor, in the sense of nature holding a claim to the rank of being an end in itself, is at best metaphorical; it must be interpreted on the basis of the projection of the actor into his field of action. Acting "according to nature" then means orientating one's actions with respect for the fact that many things (including ourselves) have come into being without our active help.[15]

These explanations may suffice to make clear that "views of nature" commonly assumed to be unproblematic and accepted as such can carry normative settings standing in the way of a solution of conflicts relevant to the environment.

1.1.2
Setting Limit Values as a Rational Process

In order to avoid setting limit values without any rational basis, the question has to be asked what the criteria of rationality are in setting limit values. Of particular importance is the question, following the discussion above, of which kind the rationality is in setting a limit value that is not or only indirectly based on threshold values. Environmental quality targets play a crucial role in justifying limit values. In the interest of a consensual environmental policy, it is mandatory that the setting of environmental quality targets and the establishment of limit values guided by these targets is shaped as a rational process.[16] Threshold values and risk assessments must be considered as the principal instruments of making rational decisions.

The rationality of an act is measured against its ends. Environmental standards ought to fulfil the function of preventing damage to humans and nature, i.e. of prescribing risk limits for activities burdened with risks. Hence the concept of risk is, as it were, the latent core concept of the discussion on environmental standards. Concerning the rationality of laying down environmental standards, one must focus on the question if and, should the occasion arise, how risks can be weighed in a rational manner.

In the discussion about risks of the modern, technological civilization, the notion of *risk* is used with a variety of meanings.[17] Communication on coping with environmental risks requires a conceptual reconstruction of the term, making action under risk-distribution aspects accessible in the sense that a distribution takes place according to criteria justifiable transsubjectively. A reconstruction is expedient, if it enables us to talk about risks in such way that risks are comparable. Hence, "distributability" recurs to *comparability*. If the comparison between risky actions is to lead to results – as demanded – that are valid not just subjectively (individual or group-specific), it must be possible to perform a discussion on risks that can be measured against criteria of generalizability. Thus, distributability also recurs to *generalizability*.

---

[15] See Gethmann (1992a).

[16] Rationality is the term describing the ability to develop procedures for the discursive redemption of validity claims, to follow these procedures and to dispose of them (see Gethmann 1995). This expressly includes practical rationality, meaning that one can decide rationally on requests and norms, too. This general concept of rationality comprises more specialized notions of rationality like the rationality of objectives or purposes.

[17] On the philosophical reconstruction of the risk concept, see e.g. Gethmann (1993), Nida-Rümelin (1996), Ott (1998) and Rescher (1983).

Within the scope of this reconstruction, generalizability and comparability are necessary prerequisites of a justified distribution. The understanding of the term "risk", which formulates the requirements of generalizability and comparability with regard to the "distributability" of relevant actions as a requirement on the concept of risk, shall be referred to as the *systematic* notion of risk. In this chapter, risk always means risk in the sense of the systematic notion of risk. By this we are not assuming that there is just *one* risk concept under this notion. However, a number of meanings occasionally attached to the word "risk" must be discarded as inexpedient for our purpose. Factors like the fears or aversions experienced due to a risky activity do not flow into the concept of risk. The fear occasioned by an action (the idea of an action) says something about the subjective perception of the action, but it cannot serve as a generalizable indicator of the risk. This means that such a concept of risk cannot be used in a controlled manner in situations of *transsubjective consultation* either.

Recurring to the elementary pragmatics helps grasping the methodological implications of this risk concept: Ideally, we act in such a way as to bring about results (of first to n-th order) up to a goal at which the intended objectives are realized. Our experience in the lifeworld however tells us that this ideal is constantly in danger of "disruptions". Such disruptions can be of various types:

- Since we do not know if the consequences of our actions, especially the higher-order results will actually, our doing is dominated by uncertainties.
- Even if the planned results come about, it cannot be excluded that unwelcome statess, too, will result from a course of action. The higher-order the result, the more it has to be feared that such *side effects* do occur, which, however, are not unwelcome in every case. Hence, the possibility of welcome side effects should also play a role when weighing risks and chances.[18]
- Sometimes, our actions are followed by events that we cannot at all understand as consequences of our actions – for instance because there is no causal relationship between the result of the action and the event, or (which is the same in pragmatic terms) because we do not dispose of the causal knowledge. Both events, the action and the other event, belong to another type of events, making us consider the event as a *coincidence*.
- Finally we execute actions that are prerequisites but not the sole cause, i.e. necessary but not sufficient to ensure the occurrence of certain results. That is the case, when the intended conditions come about only if further preconditions are met. In this context, one talks of *chance* ("Geschick"), which may turn out as *luck* or *misfortune*, depending on whether the outcome is welcome or unwelcome.

Towards chance, we can take one of two ideal-typical attitudes: *resignative*, as one does not dispose of the additional "coincidental" preconditions, or *confident*, because one sees a possibility that the preconditions can be met in part. The last attitude is characterized by the confidence that it may be possible, under certain circumstances, to *manage* the uncertainty of chance i.e. to *avoid*, to *remove* or, in case

---

[18] Low doses of certain air pollutants, for example, have a fertilizing effect on plants.

of misfortune, to (fully or partly) *compensate* the uncertainty. Hence, the confident attitude towards chance expresses itself in the readiness to enter a venture that can carry a chance or a risk. Considering life from the perspective of managing chance, on the other hand, is an element of the modern self-image of (western) man. The humans of the modern age and of enlightenment, who are distinguished by a confident attitude to life, try to pursue their risk management by taking precautions. Among the paradigms showing the non-resignative attitude towards the vagaries of chance are the insurance against misfortunes (fire, illness, death etc.) and the rational betting behaviour in gambling. Both examples show that the genesis of risk assessment lies in the human-cultural contexts, and that it does not primarily represent a category for the assessment of appliances, machines and installations. The origin of the modern risk concept is by no means technomorphic.

Insurance and betting were the social need conditions triggering the development of probability theory. Since we have an apparatus for calculating probabilities at out disposal, we can specify the risk concept by establishing a numerical degree of risk: The degree of a risk (chance) equals the product of the – numerically expressed – harm (benefit) and the – numerically expressed – probability for the occurrence of the harmful (beneficial) event. This rational risk concept is, so to speak, a *stylization onto a higher level of our managing the chance related to our actions in the lifeworld.*[19]

Risk assessment must not be confused with the subjective (individual or collective) *perception of danger*. In contrast to the perception of a danger, risk assessment endeavours to determine the danger incurred by a *type* of action, independent of the respective situation. While danger is an element of the concrete event faced by an individual or a collective, risk characterizes a type of situation in relation to a typical participant of the situation. Hence, we have danger connected with a particular situation, on the one hand, the risk as the typified misfortune and chance as the typified luck, on the other.

The distinction between risk assessment and the perception of danger enables us to understand that, for instance, the gambler may believe to be one step away from the lucky win, while the chances, as viewed by the bank, are always distributed in the same way. Equally, the perception of danger (the fear of flying) has no deterministic influence on the actual risk (of a plane crash). Thus, the insurance premium is not set according to the individual perception of danger, but to the risk. Precaution against danger by assessing the risk does not replace the aversion of hazards (accident insurance does not replace the safety belt), and vice versa (the safety belt does not replace the accident insurance).

In principle, where a risk concept is available for qualifying actions with regard to their potential for danger, they can also be compared rationally. For the compari-

---

[19] At this point, we have to guard against a misunderstanding arising from the mix-up between the *meaning of* a concept and the *subsumption under* a concept. The fact alone that risks can be perceived differently should not lead to the introduction of accordingly different risk concepts. For the reference point for our speaking of the perception of a risk must be independent of this perception itself, if one wants to avoid the disintegration of the concept of risk as the core concept of health and environmental policy. It is exactly the situative transsubjectivity what distinguishes the systematic concept of risk as a reference point. However, the subsumption thereunder does not mean a reduction.

son, Bayes' law advises us to choose the option carrying the lower risk when choosing between two options of action with identical benefits and costs.[20] Several objections have been raised against the procedure of risk-risk-comparisons as a measure of rationality:[21]

1. There is the thesis that risks are absolutely incommensurable. The advocates of this thesis assume commensurability being a kind of natural property of actions. In reality, however, the establishment of a principle for comparison is a matter of forming rational conventions to achieve an objective, making different things comparable. Concerning terminology, there must be a sharp distinction between "comparing" and "equating". Comparability does not exclude (it actually includes) that classes can be formed regarding the risk of actions, within which one considers risk-risk-comparisons as a plausible reference frame (risk classes), whereas comparisons beyond these boundaries of such classes are viewed as implausible.
2. There is the objection that risk-risk-comparison leads to an additional, unacceptable subjection to risk. For instance, individuals already taking general life risks could not be expected to accept the additional burden of living close to large-scale technological installations. This objection confuses the accumulation of component risks in the context of risk balances with a comparison between options, the latter of which assumes that one *or* the other option is realized.
3. Furthermore, it is asserted that risks are acceptable only if they are taken voluntarily. However, there are numerous examples for the functionality of a measure restricting action being based exactly on its non-voluntariness, e.g. regulations of road traffic, the obligation to employ a notary for closing certain contracts, the liability for maintenance and the duty of chance children to school. In these contexts, however, one should not talk of voluntariness, but of approving acceptance. In many areas of life, voluntariness is not a necessary criterion of acceptance.
4. It is argued that natural risks must not be included in a risk-risk-comparison. The natural risk as such is indeed bare of any normative content. The acceptance of natural risks, however, is a different matter. For instance, an individual who considers a natural risk as tolerable in principle must also be prepared to tolerate this risk himself.

To compare risks in a rational manner, one requires three types of knowledge and skill:

- One must know that certain actions result, with a certain probability, in certain effects or – if applying an ordinal scale – how the relative probabilities of different effects of the action are to be assessed. Hence, one must have some knowledge on causal or at least conditional relationships. Scientific research concern-

---

[20] Bayes' law means that in any decision the option should be preferred that carries the largest expectation value. What we discuss here is a special case of this rule. It should be noted that the application of the rule to decisions in which several individuals are involved requires a common and unique scale of benefits and costs. On decision processes in general, see Bitz (1981), Stegmüller (1973) and Zimmermann and Gutsche (1991).

[21] Especially see Beck (1993), Douglas and Wildavsky (1982), Jungermann and Slovic (1993), Rescher (1983). On arguments for the risk-risk-comparison in such contexts, see section 1.3.1 and *Akademie der Wissenschaften zu Berlin* (1992).

ing environmental issues is pursued in order to establish such causal relationships. Scientific findings are therefore an indispensable element of a rational setting of limit values.

- One must also have some knowledge on what actors as well as the individuals or groups affected by their actions regard as harmful. Concerning elementary effects, this question may be trivial, but for chain effects and complex, communal systems of preference, for competing systems of preference, for latently tabooed or otherwise complicated preference situations, it is an issue of socio-logical (including psychological) research.
- Finally one needs the general know-how of risk comparisons, meaning not the skill of calculating risk values, but the knowledge of how to assess the results of such calculations, i.e. one must know about the scale against which comparisons are performed. This involves questions like how to determine variation margins, how to take various environmental media into account, the question of distribution problems and compensation, etc. Comparability is an essential, normative element of setting limit values. These questions are, at their core, questions to be put to philosophical ethics.

Hence it is crucial for the structurally equivalent treatment of combined exposures to work out a solid foundation of knowledge on how to determine risks related to such exposures. This is the endeavour of chapter 2, in particular, where the problem will be approached by categorizing the combinations according to interaction principles[22].

1.1.3
Pragmatic Consistency and Practical Generalizability

When setting limit values, not only certain risks are excluded but also, in many cases especially for stochastic effects, certain risks are accepted or allowed. Therefore all risk comparisons, which are at the core of setting limit values, contain a normative element, insofar as a limit value implies that anyone, who is subject to this limit value, may have to tolerate certain harmful consequences and may be permitted to subject others to the same consequences, i.e. such that correspond to a lower degree of risk than that stipulated in the limit value concerned. However, the political debate makes it abundantly clear that such toleration cannot be taken for granted. The normative problem therefore boils down to the question as to which risks can be accepted at all, and by what right. The reason for the controversial discussion in this context lies in the varying assessment of commensurability and risk acceptability.

The question of by what right others can be expected to accept anything has always been a fundamental problem of societies. However, at the moment, in the context of establishing limit values, this is an even more controversial topic. Firstly, it must be remembered that the question of de facto acceptance does not provide, by principal, any reliable guidance for establishing a legitimate set of limit values. De facto acceptance is generally vague in content and fluctuates over time. The particular interests of individuals and groups often lead to diverging expectations. How-

---

[22] See section 2.1.

ever as a principle, limit values should be acceptable to anyone and be of long-term validity. Hence, they have to be the instrumental result of mutual objectives. To resolve disagreements with regard to set purposes it is necessary to reach an understanding on the ends pursued, which in many cases may be realized through various, different purposes. However, discursive conflict solving may be easier, if common objectives are already agreed upon explicitly or implicitly. Conclusions can therefore be drawn from discovering which purposes and objectives actors in society, by virtue of their own actions, already accept implicitly ("revealed preferences"). On the basis of such a set of purposes, an actor can be required to accept that a certain degree of risk may be involved in a given action if he otherwise accepts that degree of risk on the basis of the same benefit/harm assessment. A rule of this kind would constitute a criterion of acceptability for risk toleration. Therefore, not *acceptance* but *acceptability* is the criterion for justifying validity claims expressed in the form of environmental standards.

The differentiation between acceptance and acceptability essentially marks the difference between philosophical and sociological treatments of this problem area. Psychological and sociological research[23] has taught us about regularities in the perception of dangers by individuals and collectives. These regularities concern the de facto *acceptance* behaviour without, however, carrying any information about the acceptability of a risky action. *Acceptability* is a normative concept measuring the acceptance of options, each carrying risks, against rational criteria of action under conditions of risk. An acceptable attitude to risk would be the attitude shown by a cognitively and operatively perfect decision-maker faced with several options for action. However, this does not imply the possibility of formulating *categorical* imperatives in a way that everyone could be required to take a certain risk (e.g. by using airliners). On the other hand it is possible to formulate *hypothetical* imperatives for acting under risk, which relate the degree of risks already accepted to options of action to be decided on. Such a hypothetical imperative could be for example: "Anyone accepting the risk of driving a car should also be prepared to risk using a train!" Hypothetical imperatives could also take care of subjective risk tolerances. Even if nothing definite can be said on the assessment of harm and the probability of the harmful event, one still demands that an individual or a group behaves towards a certain risky situation in the same way as they already have behaved in a situation carrying a comparable degree of risk. This demand constitutes a postulate of reliability supposedly ensured by the "internal rationality" of the individual. The *principle of pragmatic consistency* generalizes the consistency requirement into a collective rule.[24]

The question of acceptability therefore raises the problem of justifying norms i.e. universal requests (requests applying to everyone). Justification problems of this kind are the topic of the philosophical discipline, *ethics*, while the factually propagated or executed imperatives represent elements of morals currently in force. Ethics concerns itself in more detail with certain normative questions, namely the

---

[23] See Jungermann a. Slovic (1993), Renn (1984).

[24] First formulated in Gethmann (1991). On the discussion of the objections, see Gethmann (1993). If someone refuses driving a car because of the risk attached to this activity, and if she acts accordingly, i.e. does not drive, the argument brought forward here is irrelevant to her. The reverse conclusion that she should not use trains only because she does not drive cars is, however, not applicable. – On the relevance of this principle for environmental policy, see section 1.3.2.

questions relevant to conflict. Where the resolution of conflicts is attempted by means of arguments in conversation, we will refer to discursive conflict resolution. More precisely, ethics is the art (and the science of mastering it) of leading such conflict-related discourses (justification discourses). The description and explanation of morals, on the other hand, is a matter of sociological research.

Party-invariant orientations can be found by asking which presuppositions are common to the parties in conflict, although such presuppositions depend on the basis of experience and action in the lifeworld. This lifeworld approach does not lead to theoretical imperatives, but it may well lead to obligations in some cases.[25] Hence, the obligatory is a formularization of what we are practically guided by. Not the normativeness of the factual rules but the normativeness of that which is presupposed in the de facto normative.[26]

The principle of pragmatic consistency is by no means trivial; in some circumstances it implies a strong intervention in the de facto acceptance behaviour. In order to justify it, we would have to embark on a comprehensive discussion of the traditional strategies for treating ethical problems. Instead, we merely point out that the principle can be derived both from Kant's categorical imperative and from Hares principle of generalizability.[27]

The principal of pragmatic consistency assumes that from the actions of the agents it is possible to infer their willingness to accept risks. This process can be reconstructed as a fictitious discourse. In this, a statement can be incorporated which, though never actually expressed, still represents an "eloquent testimony".

Complications in this approach arise from, among others, the facts that

- an individual's way of life is generally not of his or her own choice,
- the use of the term "decision" is problematic for ways of life solely defined by traditions, and
- situations of radical reversal can make the reference to past actions appear implausible.

In such cases, the recourse to a fictitious discourse and to an approval given within the bounds of that discourse is not immediately possible. However, a methodical anchor is at hand, when actors expect from others a de facto risk acceptance. The principle of pragmatic consistency demands one's willingness to accept a risk, the acceptance of which one expects from others. This applies both to small groups and to collective, social relationships.

The fall-back on de facto risk toleration, which is necessary for the functioning of the principle of pragmatic consistency, appears to be partly affected by objections to revealed-preference arguments. These can be discussed using the example of the dilemmata of risk management formulated by Kristin Schrader-Frechette[28]:

---

[25] See Gethmann (1993).
[26] See Gethmann (1992b).
[27] See Kant (1785, p. 421): „*Handle nur nach derjenigen Maxime, durch die du zugleich wollen kannst, daß sie ein allgemeines Gesetz werde.*" ("Always act according to the maxim which also enables you to wish for it become a general rule") and Hare (1952, p. 168f.): "I propose to say that the test, whether someone is using the judgement 'I ought to do X' as a value-judgement or not is, 'Does he or does he not recognize that if he assents to the judgement, he must also assent to the command "Let me do X"?'"
[28] Schrader-Frechette (1991, p. 68ff.)

1. The *facts-and-values dilemma* supposedly lies in the problem that risk assessments can neither be based only on scientific procedures nor only on democratic processes. The dilemma is rooted in an ethical position absolutely denying the rationality of judgements on prescriptions. In principle, however, prescriptions, too, can meet criteria of generalizability (e.g. in the context of the rationality of ends and means). In the lifeworld as well as in sciences (e.g. in economy, jurisprudence and ethics) the possibility of judgement is always assumed. In the same way as assertive statements can be true or wrong, imperative statements can be right or misguided. Only if normative irrationality were insurmountable, the dilemma could not be resolved.

2. The *standardization dilemma* states that a standardization of risk assessment procedures does not take into account contextual factors. This dilemma can be resolved by the consideration that it is possible to standardize the way of taking into account those factors. Apart from that, scientific findings are defined by contexts too (e.g. by an experimental set-up). The art of generalizability is exactly to make a finding independent of those factors ("context-invariant").

3. The *Sorites dilemma* (contributors' dilemma) is based on the fact that many risks, while being unproblematic on their own, can be unacceptable in combination. A certain accumulation leads to a change in their prescriptive quality (like one grain of sand does not make a sandpile, neither do two or three – and still there are sandpiles). This dilemma can be resolved by reconstructing the edge of quality change. Where this is not possible pragmatically, one has to look for models allowing an alternative (or even provisional) assessment. In this study, this particular dilemma will be important with regard to the vast multitude of combined effects.[29]

3. The *de minimis dilemma* goes back to the observation that, at a certain degree of improbability of an event, one has to equate pragmatically this very improbable occurrence of the event with its non-occurrence (pragmatic zeroness). Thus we must expect that, for pragmatic reasons, a limit value has to be given, below which a risk is negligible, for instance 10^{-6} of the annual probability of an event. In this case, we cannot recognize a dilemma at all. It is rather recommendable to assume such limits.

4. The *consent dilemma* comes from the fact that the individuals or groups affected are generally not qualified, with regard to their knowledge, as legitimate decision makers, and that legitimate decision makers tend to avoid being affected. The problem touched upon here cannot be denied. One can even use it in the opposite direction in order to qualify the rationality of a discourse. Generalizability is a result of discourse only if the author of a requirement, at least, can put himself into the position of the addressee.[30] In this case, too, we are therefore not dealing with a dilemma but with a pointer to an important criterion for decisions.

---

[29] See chapter 2.

[30] This requirement corresponds to an explanation of categorical imperative formulated by Kant: *„Es gehört aber ein vernünftiges Wesen als Glied zum Reiche der Zwecke, wenn es darin zwar allgemein gesetzgebend, aber auch diesen Gesetzen selbst unterworfen ist. Es gehört dazu als Oberhaupt, wenn es als gesetzgebend keinem Willen eines anderen unterworfen ist."* ("A reasonable being belongs as a member to the realm of ends, if he is generally legislating therein, but at the same time is subject to these rules. He belongs as a prince, if he as legislator is not subject to any other's will.") (Kant 1785, p. 433).

Thus it turns out that none of these "dilemmata" stands in the way of applying the principle of pragmatic consistency in the context of risk-risk-comparisons. Such comparisons therefore allow the rational evaluation of options, taking recourse to indifferences regarding the relevant reference situations. The aspects characterizing risks can be taken account of without surrendering to them in the sense that comparisons are regarded as impossible, or in Rescher's words: "Its size or magnitude is not something a negativity *has*, it is something it *gets*."[31]

1.2
Methodological Problems of Establishing Limit Values

The establishment of limit values to fulfil normative prescriptions by environmental standards, on the one hand, and with reference to threshold values or risk-risk-comparisons (e.g. with regard to natural exposures) on the other, has to be based on methodologically secure knowledge, due to the multitude of problems of defining and laying down concrete values. This applies *a fortiori* to the treatment of combined effects, which, in order to get closer to realistic conditions, are to supplement or replace single-substance investigations.

1.2.1
The Effect Principle

In the context of environmental standards, conditions are formulated to guide human action directly or indirectly. On the one hand, courses of action are set limits to; on the other, however, environmental standards can create a need for action. Hence, limit values firstly are conventional, legal regulations of activities, using statutory or economic instruments. Still, limit values shall not be arbitrary but related to demonstrable effects, for which e.g. (in the case of non-stochastic effects) threshold values have been established.

However, a cause-effect relationship[32] cannot always be interpreted in a simple mechanistic context.[33] In principle, one has to accept a broad effect mechanism consisting of many steps of mediation. Moreover, effects cannot always be defined as isolated effects, e.g. phenomenal health damage. Even the accumulation in the human body of a chemical that in itself is not harmful may produce a damaging effect. Thus certain substances having accumulated in the body's fatty tissue may find their way into breast milk and initiate a harmful effect on the infant.

In many cases, effects can only be clearly demonstrated in animals or cell cultures, because corresponding experiments with humans are widely regarded as

---

[31] See Rescher (1983, p. 27).
[32] The concept of cause is understood here in the wider sense commonly used in the philosophy of science. Consequently, administering a certain dose is deemed to be a cause too.
[33] Pharmacological (and toxic) effects are changes in biological functions and/or structures initiated by substances (Forth et al. 1996, p. 4).

morally reprehensible.[34] The particular problems of transferring such findings to the human case i.e. of extrapolating the results are discussed in the next chapter.

For a long time, limit values were only laid down with a view to human health, meaning that the medical effect on humans was the only effect accepted. This approach turned out to be too narrow. Instead of only accepting effects on humans, the effect concept has to be understood in a wider sense. Where necessary, other living systems must be included too. A prominent example is the effect on plants of sulphur dioxide, which is generally harmless to humans in usual concentrations.

However, the wide interpretation of the effect concept carries the risk that the term might lose its semantic contours, so that a unique subsumption of candidate cases becomes impossible. Ultimately it has to be avoided that vague speculations lead to momentous limit values. Risk aversion, after all, mostly incurs considerable (monetary and/or social) costs. Consequently, one always has to consider whether the funds available would not be better used for other purposes – e.g. to avert other risks. This question affects not only the decision if or if not a limit value should be laid down, but also the issue of how far risk precaution should be driven by the standard concerned. In the case of stochastic effects, there is no threshold to support a specific limit value; natural exposure can serve as a reference here.[35]

These considerations result in the need for a *relevance criterion* to measure effects against. For instance, the question arises whether the widespread fear of being harmed by "environmental poisons" should be addressed, too, in order to arrive at a "broad" cause-effect concept. Many supporters of non-standard theories like e.g. the psychosomatic medicine advocate an interpretation of this kind. Here we will have to point out that not every relationship is a cause-effect relationship. The fear of an effective agent belongs to another context than the agent itself, whereas an organic phenomenon (like a skin allergy) can, in principal, belong to the same context. This statement does not conflict with the fact that fears can be psychologically examined by methodologically valid means.[36] If we dropped this distinction, the fear of noxae would lead to their regulation through a limit value without the occurrence of a reproducible harmful effect.[37] The examples show that, in individual cases, considerable effort has to go into differentiation, which can only be done on the basis of scientific and medical research.[38]

In this context it also becomes clear that to establish effects and causes implies considerable requirements on the measurability of dependencies. Only where a cer-

---

[34] Apparent exceptions like clinical tests of pharmaceuticals do not contradict this statement. The fact that such tests are subject to stringent licensing laws and must be preceeded by extensive preparatory research rather highlights its truth. Clinical studies are therefore not experiments in the narrower sense. This also holds for other clinical data obtained not through experiments but through clinical practice.

[35] See section 1.2.2.

[36] The view that psychology, too, investigates cause-effect relationships in the scientific sense can only be defended on the basis of a rigorous behaviourism, which is in disaccord with the view of this study, if only because the "effect" of good arguments is not interpreted in the context of stimulus-reaction schemes.

[37] This does not mean that fears should not be taken seriously; one just has to address them in different ways than through laying down limit values.

[38] Still, psychosomatics is of importance for the treatment of combined effects on the basis of separation of cause-effect spheres described here, too. This link will be followed in chapter 3.

tain exposure to a harmful substance is accessible to experimental investigation and if an effect can be observed, one can speak of cause and effect in a *meaningful* way.[39]

A limit value not accompanied by rules of measuring is ultimately meaningless in operational terms, since measurability has to be assumed for requirements and restrictions to prevent systematic performance deficits. Still, the requirement of measurability is by no means easy to meet in all cases. There are areas where relatively good and reliable methods of dosimetry are available, e.g. for ionizing radiation (especially external radiation). For many chemical agents, however, the issue of measurability is quite problematic, and unfortunately there is no correlation of such kind that a higher harmful potential means better measurability. In some areas, the correlation rather seems to be inversely proportional. Particularly difficult with regard to measurability are the combined effects e.g. of chemical agents, especially in cases where a super-additive effect is suspected.

Consequently, one must not succumb to the temptation of mistaking good measurability for a sufficient reason to set over-restrictive limit values. Environmental policy makers finding themselves under sweeping political pressures to act are particularly "seducible" in this respect. Moreover, the economic context has to be kept in mind: Manufacturers of measuring equipment would quickly see the opportunities arising from an increasing refinement of the measuring accuracy, if measurability were the decisive for setting limit values.Despite of the difficulties, one must not depart from the regulative, methodological principle that limit values should be based on experimentally established cause-effect relationships. If cause-effect relationships are not available yet, the principle implies an imperative to research in order to establish them. The difficulties mentioned above give reason to explain in more detail the conceptual relationship between cause and effect in the context of the concept of explanation and of experiment and measurement.[40]

According to Hempel and Oppenheim[41], a general scheme of scientific explanation can be formulated as follows:

S1, N | S2,

or in plain words: a situation S2 is explained through the natural law N and the situation S1. S1 is the cause for the occurrence of the effect S2. In logic: S2 follows from S1 and N. The logical pattern in itself is however not sufficient to characterize a causal relationship of this kind. The statement that all ravens are black, for instance, may explain that a particular raven is black. The fact that an animal is a raven would then be the cause for its being black.[42] Such kind of "explanation" can of course not be easily accepted as a scientific one. This would be the case, for example, if the general statement were a result of the definition of the term "raven".

---

[39] "Meaningful" refers to the pragmatic reconstruction of the concepts of cause and effect, as lined out in the following.

[40] On the discussion of the topic of explanation and causality in analytical philosophy, see the comprehensive treatment in Stegmüller (1983).

[41] See Hempel and Oppenheim (1948).

[42] The example was taken from von Wright (1971), where it is discussed in detail; see p. 19ff.

When choosing a pragmatic approach, the key to an adequate reconstruction of the concepts of natural law, cause and effect is found at the core of scientific method: the experiment.[43]

By carrying out experiments, one ensures the reproducibility of certain courses ("Verläufe"), which are transitions from one situation to another without further intervention. The experimentalist prepares the initial situation and observes the transition to the final situation.[44] If one feeds a certain number of mice over a certain number of days in a certain way with a certain quantity of a certain substance before examining their liver cells after a certain period, then one might find changes compared to animals which had been kept under identical conditions, apart from having not been given the substance concerned.[45]

When formulating a connection by natural laws, one has to vary the experimental conditions in order to ensure that all parameters considered – i.e. all "definite" parameters – are relevant, and that no relevant parameter was left out of consideration.[46] In the example above, for instance, one can vary the method of administering the substance in order to find out if this is relevant for transformations in the liver cells.

On the basis of a scenario such validated by variation, a *natural law* can be formulated: N: $S_1 \rightarrow S_2$. This describes the scenario created by the experiment and can therefore be expressed in the above scheme of scientific explanation:

S1, N: S1 → S2 | S2

On the descriptive level, we can now speak of a *causal connection* between situations of the type S1 and situations of the type S2.

A causal connection like this can in turn be explained by means of *constructs*. A certain natural law, for instance, could be described by the attachment of the smallest particles of a substance to a receptor. The receptor would be, in this construct, e.g. a macromolecule containing a certain functional group in a certain steric constellation and bonding to the smallest particles of the substance.

By formulating – proceeding from the representation of the natural law in an appropriate construct – intermediate steps, S_{1k}-S_{z1k}-S_{z2k}-...-S_{zik}-S_{2k}, each compatible with the data, one arrives at a m*echanism*. Accordingly, the receptor-agent attachment represents a partial mechanism. In principle, however, the natural law is compatible with a multitude of mechanisms.[47] Still, this condition of the law being under-determined does not affect the "quality" of the scientific product, since any decision, whether a mechanism is adequate with regard to the aim of the investiga-

---

[43] See e.g. Hartmann (1993), Janich (1997) and Tetens (1987).

[44] These situations can be very complex. Moreover, such a simple description by two situations is not sufficient for most experiments. In principle, however, nothing changes when several courses have to be considered. – A special case, which does not need to be further discussed here, is given, when the final state occurs immediately at the time of preparation of the initial state. Then, one refers to laws of states, not of courses.

[45] We do not need to go into the necessary statistical evaluation here, as it does not change the structure of the experimental context.

[46] See Tetens (1987, p. 20ff.) and Hartmann (1993, p. 128ff).

[47] Hence, the categorization of substance combinations on the basis of mechanistic assignations, as proposed in section 2.1, would represent a metalaw, through which, based on a multitude of experimental results, a connection is formulated that is accessible to further experimental examination.

tion, is supported by the experimentally secured phenomena and hence is not left to arbitrariness.[48] Based on this explanation, the problem of the precise definition of the effect concept can be specified. With regard to the common graphical representation, this problem could as well be put as the *ordinate-problem:* Which effect is examined with respect to which result?

But first we shall construct, in an exemplary manner, levels of argument in the context of setting limit values – levels that can mark the way from an experimental, toxicological end point to the effect on human health or the human environment[49]:

- The experimental clarification of the cause-effect connection
- The transition to effects and effect contexts different from the experimental context, which is possible on the basis of other cause-effect connections. Thus, this step does not involve extrapolations.
- The extrapolation test animal to human, cell culture to human, higher dose range to lower dose range[50] etc.
- The establishment of the health or environmental effects

The dose-response curves presented in the following sections usually show effects concerning the first level of argument. Therefore, different end points are marked on the ordinate (transformations in tissues, numbers of mutations, survival rates etc.[51]). Although the statements of the fourth step of argument are the ultimate aim of the scientific effort, these results constitute the rational fundament of the argument as a whole.

Apart from the establishment of health or environmental effects in this way, two other aspects are important, which we mention at this point for reasons of completeness:

- The analysis of the emission, immission and exposure contexts, and
- taking into account individually higher sensitivities of humans (risk populations).

The view exists that one could investigate the health effects directly, since, although experiments on and with humans cannot be performed, epidemiological results are obtainable. The problems with this approach become clear as soon as one considers the structure of epidemiological arguments: They are founded on statistical correlations of lifestyle or experiences between larger groups of individuals. A study on workers in a certain industry, for example, who had to carry out a certain activity in a certain environment, e.g. at a furnace, in comparison with individuals not having been exposed to the respective influences, may find that the furnace workers fall ill with a certain lung disease in significantly more cases than the "control" individuals.

---

[48] This precondition of being "founded in the phenomena" can be read as a criterion for the relevance of the precautionary principle (see below). The required, scientifically plausible conjecture would therefore be equivalent to an appropriate foundation of the relevant constructs or models.

[49] For establishing a limit value, further levels must be considered; these are discussed further below.

[50] The transition from the higher to the lower dose range represents an extrapolation because effects cannot be established experimentally in the lower dose range.

[51] On the various end points, see chapter 2.

The result appears to be an appraisal along the lines of the fourth point on the list above. Still, the causality of the work conditions is by no means proven in this way. Even if an effect like this were postulated, the relevant influences could not be established (the abscissa problem of epidemiology); to say nothing of the fact that one is not dealing with standardized populations. Hence, investigations of that kind cannot be regarded as arguments within the scope of the scheme outlined above; they represent as-if arguments of the kind as if a certain situation were a cause: Cause-effect connections (effect constructs) are postulated, according to which the statistical material is surveyed or produced. More cannot be done on the basis of the epidemiological method. Establishing causal connections remains the task of the experimental disciplines[52].

1.2.2
Natural Exposures

On the basis of the causality-theoretical discussion, the relationship between stochastic and non-stochastic effects can be clarified, too. The etymology of the word stochastic leads back to the Greek "stochasticos": skilled in supposing/guessing, shrewd, depending on chance. The science of stochastic phenomena now includes the fields of probability theory, descriptive statistics and inference statistics.[53] For an adequate understanding of the distinction stochastic vs. non-stochastic it is essential not to confuse it with the epistemological distinction deterministic vs. probabilistic. Every scientific, hence also toxicological and radiobiological experiment results in a probabilistic statement. Statements of necessity, in the stringent sense, are possible only in the realm of non-empirical (e.g. analytical) truths.[54] The distinction between stochastic and non-stochastic refers to a differentiation between cases within the class of probabilistic statements. For these statements, either an effect threshold (with the probability p) i.e. an interval within which the effect sets in and below which no effect can be detected (non-stochastic effect) can be established, or there is no effect threshold, so that even for the smallest dose an effect must be assumed with some probability (stochastic effect).[55] The absence of an effect threshold is explained by the stochastic nature of molecular trigger events.

Threshold values, i.e. the corresponding threshold intervals are found by scientific experimentation and can be read, from a graphical representation, as properties

---

[52] It is of course possible to broaden the concept of causality to the extent that one can speak of epidemiologically established causality (see e.g. Beaglehole et al. 1997). The hard concept of causality introduced here is however justified by the objective to provide a solid experimental fundament for effects of combinations of noxae – a fundament that that ultimately allows to link-up experimental effect research and epidemiological work.

[53] A constructive realization of these fields can be found in Hartmann (1993).

[54] On this discussion in the philosophy of science, see e.g. Lorenzen (1987, p. 177ff.) and Stegmüller (1973, S. 65ff).

[55] On the distinction stochastic vs. non-stochastic, see the more detailed treatment in section 2.2, especially 2.2.1.6. The difficulties of extrapolation to very low dose ranges are discussed in chapter 2. On the discussion concerning a threshold dose applicable to stochastic effects too, also see Streffer (1996), according to which there is no evidence justifying the assumption of a threshold.

of curves. However, as just discussed, not all effects can be attributed with a threshold value. Limit values, on the other hand, are set according to certain objectives, as explained above. They may possibly correspond or be orientated to threshold values. But even the use of safety margins, for example, already leads to a difference between threshold value and limit value. By the very fact that stochastic effects are characterized by the lack of thresholds, they present a twofold problem. First because of the effect even at lowest doses, second due to the absence of a threshold for setting a limit value within a dose-response diagram. Hence other indicators have to be found. *When apportioning the risk acceptability of stochastic effects, one has to proceed from the mean bandwidth of the background exposure – presuming pragmatic, "normative" indifference towards such exposure.*

The recourse to natural exposure must not lead to falling into the trap of the naturalistic fallacy. The existence of an exposure does not imply its acceptability. It is rather the case that only from the de facto acceptance as "revealed preference" one can draw a conclusion by way of the prescriptive premise of pragmatic consistency: Who through his actions conclusively announces his accepting, for himself and for others, of a natural background exposure, must be expected to accept the same or a lower exposure, too.[56]

Where there is a natural exposure to a certain noxa, as in the case of ionizing radiation, exposure to heavy metals and the ubiquitous polycyclic aromatics, it is pragmatically unreasonable – assuming pragmatic, "normative" indifference towards those exposures – to orientate environmental standards to limit values below the mean bandwidth of the natural exposure. It is a matter of discussion whether the mean bandwidth itself is accepted as a reference value or if one keeps a certain "safety distance" from it. In any case, even where a safety margin is included in the limit value, an exposure below this value is by no means without risk. Even for the most stringent limit value, mortality, morbidity or some other dimension of harm can be calculated. Thus, every limit value for stochastic effects implies an expected risk acceptance. The question is under which conditions such an expectation can be justified pragmatically. It is surely unreasonable, pragmatically, to require a lower risk acceptance e.g. for technological installations than given by the bandwidth of the corresponding natural exposure.[57] A background exposure is therefore a valid reference point for setting a limit value.

For stochastic effects of chemical agents, there is, in many cases, no or no relevant natural background exposure with regard to its level or distribution. In these cases, the existence of background exposures to other agents may be of help, insofar as a comparable hazard e.g. through ionizing radiation and certain chemical agents can be established. It needs to be examined whether one can set certain limit values for chemical noxae in analogy to the practice in radiation protection.[58]

---

[56] See above, section 1.1.3.
[57] For details, see *Akademie der Wissenschaften zu Berlin* (1992).
[58] This topic has not been treated yet in sufficient depth; also see Gethmann (1998).

1.2.3
Combined Effects

Combined effects occur between physical (e.g. ionizing radiation) and chemical agents as well as between physical and chemical agents, respectively, among each other.[59] Especially the last – combinations between chemical agents – give rise to considerable problems concerning their methodical treatability, due to the immense number of possible combinations. Therefore, a perspective for the pragmatic management of the issue of environmental standards can only be gained through finding appropriate paradigms and models.

Concerning terminology, one has to define what we understand under the toxicity of a substance. As soon as a substance is called toxic, it has to be indicated which species or collective one refers to, which harmful or transforming effect one speaks of, and over which period and at which concentration the substance was administered, and how. In addition, the effect context needs further specification etc.

The concept of "substance" is fundamental for chemistry.[60] It can be reconstructed as an abstractor, for which the respective, relevant substantial properties are constitutive, i.e. one disregards all other properties of a thing when speaking of the substance of which it consists. The relevance is context-variant, meaning that the relevant substantial properties vary depending on the application. For a building material, for instance, other properties are important than for an active substance. In analogy to the distinction between intended effects and side effects, one could differentiate between relevant properties and secondary properties, be they welcome or unwelcome. One relevant property of a building material would be e.g. its insulating effect; a toxic effect would be an unwelcome secondary property. In order to avoid negative experiences of the kind that a widely used building material turns out to be toxic, substances are routinely examined for certain types of harmful effects. The classes of the relevant properties, on the one hand, and of the secondary ones, on the other, are however by no means disjunct. Therefore, dilemmatic situations can occur, of the kind that a relevant property is an unwelcome secondary one at the same time. Many pesticides, for example, do not dissolve in water. By adding dissolving mediators one arrives at the homogenous phase, on the one hand, which can be easily spread; on the other hand, however, the pesticide is easier taken up through the skin. In this semantics, the consideration of combined effects merely represents another aspect – another index or appreciator of "toxic" – that has to be taken into account explicitly.

There is no standard terminology to be found in the relevant literature on combined effects.[61] Indeed, one speaks of a "confusion of terms"[62]. Therefore, one of the aims of this study is to formulate a standard terminology in order to facilitate or enable communication within and between the individual disciplines concerned.[63]

---

[59] The interactions with physical influences, which are categorically completely different from these combinations, are treated in chapter 3.

[60] For details on the philosophy of chemical science, see Hanekamp (1997).

[61] See e.g. Kordell and Pounds (1991).

[62] See Könemann and Pieters (1996).

[63] On the terminology, also see section 2.2.2.1.

A *phenomenological terminology of combined effects* for cases where several substances also individually applied are combined could present itself as follows. For the example of two substances, A and B, it reads (x = A or x = B):

EA+B>EX   Synergism rel. to x   S(A,B) or S(B,A) resp.

EA+B < EX   Antagonism rel. to x   A(A,B) or A(B,A) resp.

Hence, synergism and antagonism have to be qualified i.e. a reference substance x has to be indicated. The choice of a reference substance x must however not be interpreted in a mechanistic fashion, e.g. that an antagonism in relation to x would mean that the other substance or substances would have an effect on x. Effect in a chemical reaction has to be established as a symmetrical relationship. Furthermore, it must be taken into account that a synergism or an antagonism of two substances does not have to be effective over the entire range of concentrations. Consequently, antagonism and synergism in relation to another substance are not substantial properties.

This introduction of the terms synergism and antagonism will be particularized in the following chapters, where, if the reference substance is not explicitly indicated, the substance that singularly shows the strongest effect will be taken as the reference point.

Additivity Models

As an extension of the phenomenological terminology, additivity models may be used as comparative tools. However, the description of the deviations from such "zero-point" models has to be clearly differentiated, terminologically, from the phenomenological terminology. Furthermore, additivity models can serve as zeroth-order approximations in cases of information deficit.[64]

The Model of Dose Additivity

The starting point for the formulation of the dose additivity model is the observation of equal slopes of the dose-response curves of individual substances, for it is assumed that the doses of the single substances, in some cases lowered or increased by a constant factor, can be added up in order to determine the effect of the total dose on the basis of a common dose-response relationship. Thus, with regard to its effect, one substance is treated like another substance in the mixture or like a dilution of the same. One simply adds up the single doses and reads the effect on the graph of the common dose-response-curve.[65]

The Model of Effect Additivity

In the effect additivity model, one merely adds up the single effects in order to determine the combined effect. The model of effect additivity can therefore be applied to any combination of the same effect.

---

[64] See section 2.2.6.
[65] Disregarding its heuristic roots, the dose additivity model can also be used for substances with dose-response relationships of different slopes. In the lower dose range, for instance, one can assume a linear relationship, so that the model can be applied without any problem (see e.g. section 2.1.3).

A synergism can be described more detailed by putting it into relation to a model. The synergism can either be in agreement with the model, or it leads to stronger or weaker effects than the model predicts. Using effect additivity as an example, a further development of the phenomenological terminology could look like this:

$$E_{A+B} > E_A + E_B \qquad \text{Effect super-additive syn.} \qquad S_{sup}(A,B)$$

$$E_{A+B} = E_A + E_B \qquad \text{Effect additive syn.} \qquad S_{add}(A,B)$$

$$\begin{aligned} &E_{A+B} > E_X \text{ and} \\ &E_{A+B} < E_A + E_B \end{aligned} \qquad \text{Effect sub-additive syn. rel. x} \qquad S_{sub}(A,B) \text{ or } S_{sub}(B,A)$$

Other than in the case of sub-additive synergisms, additive synergisms and super-additive synergisms are uniquely determined without a reference substance being given. Otherwise, for effects without an index, the singularly most effective component is taken as the reference point.[66]

Even if the terminology just introduced is cumbersome and will probably not be used in specialist circles, it may serve to illustrate an important achievement of scientific terminology, here for the example of synergisms: When calling a synergism sub-additive, it must follow from the context that one refers to a model and to which model one refers. The terminology above decontextualizes this semantic extension and hence is understandable without further information. This is of benefit especially in interdisciplinary contexts.

An important difference, with regard to establishing limit values, between dose additivity and effect additivity exists in their application to combinations of substances in doses that do not show any effect when given separately. Under the assumption of effect additivity, a combination of any number of substances in individually ineffective doses will not show any effect either, whereas dose additivity can lead, for the same combination, to a relative dose that is effective.[67] In current environmental law, the "more cautious" dose additivity model is applied – subject to the emergence of more precise scientific results.[68]

The Model of Independence (Relative-Effect Additivity)

The independent model – or model of relative-effect additivity – rests on the consideration that the combined effect of a mixture results from a fictitiously assumed, consecutive exposure to single doses of the combination partners. One further assumes that the single doses are related to a total effect calculated in a way that the single doses cause a fraction of the residual effect (total effect minus actual effect) proportional to their relative single effect; or simply:

---

[66] It should be noted that the effect super-additive synergism made explicit here is often equated to synergism as such. This would however mean that every combined effect stronger than the sum of all single effects had to be called a synergism. The scientific terminology used here does not agree with that view.

[67] By adding the doses, one arrives at a region of the dose-response curve, where the effect is greater than zero (pragmatic zero).

[68] See section 5.1.3.4.

A has the single effect E_A

B has the single effect E_B

$$E_{A+B} = E_A + (E_B/E_{max}) * (E_{max} - E_A)$$
$$= E_A + E_B - E_A * E_B/E_{max}$$
$$E_{A+B} = E_A + E_B - E_A * E_B \qquad \text{for } E_{max} = 1$$
$$= E_A + E_B (1 - E_A)$$

This formula is nothing else than the standard form of statistical independence without correlation. Independence does, however, not refer to independent events.

As already mentioned, terms for combined effects deviating from the respective models disagreement should not belong to the same terminology as those for combination phenomena. Therefore, an effect should not be called [model]antagonistic or [model-]synergistic but rather [model-]sub-additive or [model-]super-additive, respectively.

1.3
Environmental Policy

1.3.1
The Risk Concept as a Fundamental Concept of Environmental Policy

The principal, basic concept of environmental policy is the risk concept, which is understood in this study – as already explained in detail – in the sense of a systematic notion of risk.

There are a number of objections against this risk concept. However, the alternative risk concepts that have actually been proposed are either quantitative variations of the concept as used by us, or they are based on a mix-up between risk and risk perception.

Jurisprudence differentiates gradually, dependent on to the probability of the undesirable effect, between danger, risk and residual risk. Accordingly, there will be a danger when the probability is very high, a risk when the probability falls into a medium range, and a residual risk in cases where the probability is low enough that it can be called "general, civilizational burden"[69]. This differentiation is practical for cases that are supposed to be covered by e.g. police law, i.e. for cases of medium damage. For issues of environmental law, however, such differentiation is fraught with problems, since the continuous transition from danger to risk does not allow the desired, clear-cut distinction between cases of averting a danger and taking precautions [against a risk]. Hence it is possible that the classification has to be performed for individual cases.

Psychological risk research[70] provides important insights concerning the various behaviours of individuals in risk situations, insights that are important both for the

---

[69] See Rehbinder (1991).
[70] See e.g. Jungermann and Slovic (1993).

normative (ethical) aspects of acting under risk and for the political "risk assessment". Nevertheless, immediate guidelines for action cannot be gained from individual risk perceptions.

Hence, with regard to normative contexts, the notions of risk flowing into sociological and psychological research[71] share a common deficit: They cannot provide a foundation for generalizable, prescriptive requirements, since they, without a single exception, register de facto perceptions and acceptances of dangers and risks. The attempt to make judgements about acceptability on this basis would lead to a naturalistic fallacy.

The risk notions common in sociology do not lend themselves to a reduction towards a common concept. For instance, one can make out a sociopsychological access (Renn), a cultural-anthropological access (Douglas, Wildavsky) and a system-theoretical approach (Luhmann), all distinct from each other – but the aspect of autonomy and heteronomy, respectively, is a central one. When speaking of risk, all harm arising from it is ascribed to an autonomous decision; when dealing with danger, it is a matter of the environment. Risk is explicitly defined as the opposite concept to danger, but not as the opposite of safety.[72] The notion of risk thus defined is prerequisite for the possibility of risk management, in the sense that dangers can be avoided only if risks can be influenced, in the sense of the confident attitude towards chance explained above.[73]

It appears to be an uncontroversial assumption that one speaks of possible harm when using the notion of risk.[74] However, proceeding from this sparing introduction, a plethora of different approaches are described.[75] Both relevant factors, possibility and harm, can be characterized within the scope of the dichotomy subjective/transsubjective.

As extreme cases, one can cite a "technical" and a "cultural-sociological" concept of risk. The first assumes universal, cardinal scales for harm and probability. The risk is represented by the arithmetic product of the harm and its corresponding probability. A risk in the cultural-sociological sense varies depending on the cultural context. Consequently, a context-transcending ordinal or cardinal scaling is not possible.[76] The psychometric approach, for instance, is somewhere in the middle; it regards subjective appraisals as relevant, but it also believes in its ability to establish them in an empirical manner.[77]

Under the label "social amplification of risk", the effort was made of unifying the aspects emphasized by the different approaches into one concept.[78] According to this [unifying] approach, a multitude of factors of influence corresponding to the respective aspects must be considered within a broadly constructed model of risk perception. In that context, the usual dichotomy between objective and socially constructed risk then surely falls down. Following a recent extension of this model,[79] harm is particularized

[71] In the following, the literature relevant to this matter will be indicated in a cursory manner.
[72] See Luhmann (1990, 1991, 1993).
[73] See Renn (1992, 1995) and Luhmann (1993).
[74] See e.g. Renn (1995).
[75] See Renn (1992).
[76] See Douglas and Wildavsky (1982).
[77] See Jungermann and Slovic (1993).
[78] See Kasperson and Renn et al. (1988).
[79] See Renn (1998).

according to categories arising from an aggregation of types of harm, and to categories of harm, like distribution, persistence, reversibility, quantity etc. The probability of the harmful event is either cardinally quantifiable, by the corresponding confidence intervals, or at least it can be arranged ordinally. Beyond that, purely subjective aspects have to be formulated, aspects defying the transsubjectification demanded below. Apart from this principal "limit of transsubjectifiability", this categorization points to factual limits. The aggregation of types of harm, for instance, or the comparability of harm characteristics often are presuppositions difficult to achieve.

For the following, let us suppose the aspect pluralism pointed out above must be taken into account for successful risk management. But even then, that plurality would not justify our dropping the distinction between risk and risk perception in the context of the questions of regulation at issue here. For this distinction is at the core of risk management.

A risk concept introduces as a systematic notion of risk – as a mathematical product of numerically expressed harm and a numerically expressed probability – thus takes us beyond a definition of risk as possible damage, to its quantifiability.[80] It defines the reference point for the assessment of risk perceptions.

This risk concept is applied in an explicitly normative context, i.e. it requires us to treat risks in a rational manner in order to be able to operationalize the distinction between risk and risk perception and to arrive at decisions accordingly.

One has to bear in mind that this introduction [of the risk concept] is not the same as the technical variant mentioned above, for it does not postulate a universal scale of harm.[81] It rather requires us to constitute risk classes, within which comparisons *shall* be allowed. The demand of consistent action within these classes, concerning the risk accepted, corresponds to the principle of pragmatic consistency.[82] "Objective" entities are not called upon either. Despite the qualifiability demanded, subjective perceptions and subjective preferences are used, which however have to meet the relevant criteria of rationality i.e. transsubjectivity. Hence in this context subjectivity does not at all mean irrationality.

A classification of the particularizations of the notion of risk as the possibility of harm can be achieved as shown in figure 1-1. The transsubjectification of the risk notion in the context of environmental policy – which has been formulated as a desideratum here – is represented by the horizontal arrows. The vertical arrows lead from the nominal, through the ordinal, to the cardinal scale level. If a nominal scale means the mere designation of different types of risk, the ordinal scale allows the comparison between degrees of risk. Cardinal scaling finally enables us to indicate numerical ratios. A monetarization, for instance, represents a cardinal scaling.

---

[80] See Gethmann (1987, 1991).
[81] See Gethmann (1993).
[82] See Gethmann (1993). Beyond that, risk classes could be constituted on the basis of a 2nd-order principle of pragmatic consistency: the consistency of the classification, meaning that the refusal to make risk comparisons across the class boundaries has to be as consistent as the comparison within a class.

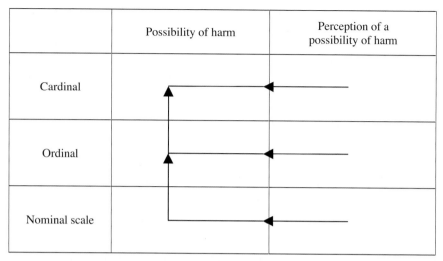

Fig. 1.1: Transsubjectification if risk perceptions and raising the scale level as principles of risk assessment

1.3.2
Principles of Environmental Policy

In the environmental legislation in Germany (and other European countries) there is no valid, uniform conception of how to set environmental standards. Current environmental legislation is indeed guided by a variety of very different principles[83]:

- The *principle of protection* is the principle of averting or preventing danger, where "danger" stands for a factual situation that, without intervention, would most probably lead to harm. The level of probability to be considered as too high depends on the type of harm. Hence, the principle of protection sets in at a relatively high level of rational risk.
- The *precautionary principle* asserts that whenever there is hazard, the state must intervene and prevent the hazard in a precautionary manner.[84] As distinct from danger, hazard means that the probability is low or, in analogy to the terminology developed in the previous section, the rational risk is relatively low. The precautionary principle can be interpreted or elaborated in different ways:
  - According to a categorical *principle of minimization* as a principle of exclusion or avoidance, an exposure would have to be minimized disregarding any considerations of proportionality. Firstly this principle cannot be applied without operationalization; secondly it breaks the weighing principle, which requires considerations of proportionality.

---

[83] Summary treatments of these principles can be found in Breuer (1989), Kloepfer (1998, p. 161 ff.) and Rehbinder (1998). Also see the compilation of tools and criteria of the decision process in *Akademie der Wissenschaften zu Berlin* (1992, p. 345 ff.).

[84] See Rehbinder (1991). On the distinction between the principles of protection and precaution, respectively, see Rehbinder (1998, margin nos. 20–26).

– The *alara principle* (alara = "as low as reasonably achievable") demands that limit values shall be set as low as reasonably achievable. At present it is the common form of the (non-categorical) minimization principle. Obviously, this principle on its own is of little help in many cases, because what should be considered as reasonably achievable is often debatable. The reference to "reasonability" implies, by the way, not only the weighing of risks, but also the reference to the expectable benefit of an exposure. For instance, the assessment of a radiation dose used for fitting a shoe (which can as well be done by other means) can be different from the assessment of the same dose when used for a vital diagnostic measure.

The principle of orientation to the *state of the art*, meaning state of technology, is problematic because it is not explicitly guided by aims. It links precaution to a development contingent of the aims of the precaution.

Further principles are discussed, which cross the distinction between precautionary and protection principles:

- The *principle of total-cost minimization (efficiency principle)*[85] puts costs and benefits or risks and chances, respectively, in relation to each other. In order to carry out such a comparison, cost and benefits have to be given on a single scale. The budgets for different areas are established through efficiency considerations. Hence, there is no fixed budget for environmental protection, on which decisions are based. Instead, a decision on raising/reducing this budget is made possible for instance by comparing the limit costs/limit benefits of all budgets considered. This comparison can however only be made with regard to an overriding objective of environmental policies. One particular difficulty appears to be the single (usually monetary) scale mentioned above.

- The *principle of cost-efficiency (principle of economic efficiency)*[86] is called upon especially in situations where the question of the optimal allocation of instruments of environmental policy is debated.[87] According to this principle, such instruments should be allocated where they achieve most with the resources available. Hence, in contrast to the previous principle of total-cost minimization, there is a fixed budget. First the marginal costs of each risk to be considered relative to a unit of benefit are determined, and then the available budget is allocated to optimal benefit. However, the budget principle can be applied only if several suppositions are actually met. For example, it has to be assumed that benefits and costs can be expressed at least on an ordinal scale or preferably on a cardinal one, although there does not have to be a single, unique scale. There are however aims and values, the numerical expression of which is not accepted, for moral or legal reasons. This is essentially the case, in Germany and in many other countries, with phenomena linked to the area of health. In the same way as organ sales are prohibited (although, of course, one could attach prices to organs, from a purely economic

---

[85] See Cansier (1996a, p. 25ff.)

[86] Another term used for this is *budget principle*. One could also call it the principle of effectivity, which differs from the efficiency principle in the sense that the latter represents a simultaneous consideration of costs and benefits.

[87] See *Akademie der Wissenschaften zu Berlin* (1992, p. 363ff.).

point of view), an employee, for example, is not supposed to strain his health beyond a certain degree and, in a sense, "sell" it. In the context of measures to reduce emissions, as an example, the benefit may well be expressed in terms of saved human lives, saved man-hours etc., so that a direct monetary scale is not required. One can see that the budget principle will not be particularly helpful, especially not for those environmental areas under controversial discussion.

- The *principle of sustainability*[88] is characterized by an emphasis on the long-term perspective and the consideration of resources. It is based on the concept of "sustainable development". A development is sustainable if it allows meeting current needs without impairing decisively the fulfilment of the needs of future generations (WCED 1987). This approach is more far-reaching than the ones discussed so far, since both social/economic and cultural changes become relevant. The long-term nature of the perspective makes it difficult to grasp methodically. For instance, the principle of sustainability can mean precaution concerning resources as well as "options precaution". The exclusive focus on resources precaution may miss the mark in the sense that the needs of future generations may have been changed by innovation processes; options precaution, which allows the consumption of certain non-renewable resources, has to rely on long-term prognoses for, among other things, technological progress in order to enable itself to appraise substitutabilities.

This kind of plurality of principles is unsatisfactory for methodological reasons. Simple examples will help to explain that the different principles will lead to different results. One such example is the alternative use of the precautionary principle and the cost-efficiency principle. Methodological order could be achieved by establishing a selection criterion by which one could choose among the "principles".

For every principle, areas can be cited where its application is plausible. The principle of precaution for example is meaningful if one suspects harmfulness on the basis of scientific criteria, but where little is known about the degrees of cause and effect. The more knowledge is gained about the quantitative relationships, the less plausible becomes the application of the precautionary principle, since more knowledge enables more targeted regulations. Under certain circumstances, precise knowledge of the quantitative relationship will even allow precautionary measures to be entirely abandoned in favour of aftercare procedures (e.g. repair of material damage). Ultimately one should not lose sight of the fact that it is nonsensical from a pragmatic point of view and legally not permissible, in view of the principle of proportionality, to undertake extensive – and expensive – precautionary measures concerning extremely small risks.

If sufficient knowledge is available, targeted measures can be introduced under balanced consideration of the relevant risks. This material "weighing" of risks takes place according to a certain principle of environmental policy, the application of which can in turn be reconstructed as the result of a weighing process. Hence one must always distinguish between two levels on which a weighing has to take place.

---

[88] See Cansier (1995, 1996b), Kloepfer (1998, p. 167), Rehbinder (1998, margin nos. 5766) and Schröder (1995).

Balanced considerations with the objective of choosing an environmental policy principle, performed within boundary conditions such as the quality of existing knowledge relevant to environmental policy or the type of risks involved, constitute the selection criterion we were looking for. As a meta-principle we shall call this the *weighing principle*. It will be the principle according to which the application of individual environmental policy principles will be decided on. Regarding the relevant knowledge, the weighing principle will also help deriving the most urgent research imperatives.

The principles of transsubjectivity and consistency discussed above are subsidiary to the weighing principle, since the result of a weighing process should be based on consistent conditions and should not depend on the individual conducting the weighing process or on the context in which the process takes place.

1.3.3
Instruments of Environmental Policy

In principle, environmental policy can call upon statutory/regulatory, economic, or so-called "soft" instruments to meet environmental standards. With respect to economic *instruments*[89] (licences, certificates, duties, environmental liability), environmental standards prescribe nominal targets or limits to be met or observed by the markets, respectively. The choice of the economic instruments will be guided by their respective effectiveness but also by fundamental attitudes of society towards environmental policy. Economic instruments will be met with disapproval for instance when measures of environmental policy are aiming at a rapid minimization of an emission or even at zero-emission.[90] If threshold values are taken seriously, on the other hand, there will be no objections to e.g. selling pollution licences, i.e. to making the adherence to environmental standards accessible to weighing according to cost-benefit aspects.

Regulatory instruments[91] are essentially characterized by the feature that they do not leave space for market reactions. Instead, they prescribe fixed cardinal (or at least ordinal) quantities. Regulations can even be of a private or semi-governmental nature, such as standards (e.g. DIN standards), recommendations for action (e.g. the MAK Values Commission of the Deutsche Forschungsgemeinschaft (DFG) or of business associations) etc. In contrast to regulatory and economic instruments, *soft (flexible) instruments*[92], such as voluntary agreements of business and industry, or information or organizational duties agreed on by companies, are non-interventionist tools. They aim at avoiding disruptions to the activities of economic actors, e.g. efficiency losses, which can occur due to the application of regulatory as well as economic instruments. As a precise targeting is not achievable by soft instruments,

---

[89] On economical instruments, also see Bonus (1990), Cansier (1996a), Siebert (1998) and Steinmann and Wagner (1998).
[90] A gradual reduction of emissions possible by economic instruments too, e.g. through continuous downgrading of certificates.
[91] See Breuer (1989), Kloepfer (1998), Rehbinder (1998).
[92] See Rehbinder (1998, margin nos. 221ff.); *Der Rat von Sachverständigen für Umweltfragen* (1998, margin nos. 266ff).

they are unsuitable for averting danger; they are, however, found in the area of prevention.

The question of which instruments to use in environmental policy is often treated as a strategic choice come to a head between market economy and a regulatory approach, as if the issue were to prove the power of the respective instruments *per se*. That the instrument question can be brought to such a point is understandable for instance against the background of contradictory ideas of government – liberalist, on one side, and etatist, on the other. According to liberalist understanding, the state ought to guarantee the framework for the market-economic settlement of interests without itself giving any material prescriptions, whereas in the etatist understanding, it is a genuine task of the state to realize material prescriptions by way of positive legislation. What both positions have in common is a hardly adequate vision of modern societies owing to an overestimation of the functionality of markets, on one side, and an overestimation of the capacity of state regulation, o the other. Ultimately, however, nobody would want to advocate neither of those extreme positions anymore, by which the corresponding polarization of the instrument question would now be obsolete, too.

By resolving the contrarious relationship of liberalistic and etatist positions into a polar-contrarious one, hence by formulating a continuum between the two extreme positions, the instrument question can be depolemicized and put into concrete environmental-political contexts. Then, the question to be answered is which instruments are adequate with regard to the objective of achieving a given environmental quality target.

1.4 Literature

Akademie der Wissenschaften zu Berlin (1992) Umweltstandards. Grundlagen, Tatsachen und Bewertungen am Beispiel des Strahlenrisikos. Springer, Berlin

Beaglehole R, Bonita R, Kjellström T (1997) Einführung in die Epidemiologie. Bern

Beck U (1993) Politische Wissenstheorie der Risikogesellschaft. In: Bechmann (1993), pp. 305–326

Bechmann G (ed.) (1993) Risiko und Gesellschaft. Westdeutscher Verlag, Opladen

Bitz M (1981) Entscheidungstheorie. Vahlen, München

Bonus H (1990) Preis- und Mengenlösung in der Umweltpolitik. In: Jahrbuch für Sozialwissenschaft 41, pp. 343–358

Breuer R (1989) Verwaltungsrechtliche Prinzipien und Instrumente des Umweltschutzes. Bestandsaufnahme und Entwicklungstendenzen

Cansier D (1995) Nachhaltige Umweltnutzung als neues Leitbild der Umweltpolitik. In: Hamburger Jahrbuch für Wirtschafts- und Gesellschaftspolitik 40, pp. 129–149

Cansier D (1996a) Umweltökonomie. Lucius & Lucius, Stuttgart

Cansier D (1996b) Ökonomische Indikatoren für eine nachhaltige Entwicklung. In: Kastenholz HG, Erdmann KH, Wolf M Nachhaltige Entwicklung. Zukunftschancen für Mensch und Umwelt. Springer, Berlin

Colburn Th, Dumanoski D, Myers JP (1996) Our Stolen Future. New York

Douglas M, Wildavsky A (1982) Risk and Culture. Berkeley

Forth W, Henschler D, Rummel W, Starke K (1996) Allgemeine und spezielle Pharmakologie und Toxikologie. Spektrum, Weinheim

Gaido KW et. al. (1997) Estrogenic Activity of Chemical Mixtures: Is There Synergism? CIIT Activities. 17, pp. 1–7

Gethmann CF (1987) Ethische Aspekte des Handelns unter Risiko. VGB Kraftwerkstechnik 12, pp. 1130–1135

Gethmann CF (1991) Ethische Aspekte des Handelns unter Risiko. In: Lutz-Bachmann M (ed.) Freiheit und Verantwortung. Berlin, pp. 152–169

Gethmann CF (1992a) Das Setzen von Umweltstandards als Ausdruck gesellschaftlicher Risikobewältigung. In: Wagner G (ed.) Ökonomische Risiken und Umweltschutz. Vahlen, München, pp. 11–26

Gethmann CF (1992b) Universelle praktische Geltungsansprüche. Zur philosophischen Bedeutung der kulturellen Genese moralischer Überzeugungen. In: Janich P (ed.) Entwicklungen der methodischen Philosophie. Suhrkamp, Frankfurt a.M., pp. 148–175

Gethmann CF (1993) Zur Ethik des Handelns unter Risiko im Umweltstaat. In: Gethmann CF, Klöpfer M (eds.) Handeln unter Risiko im Umweltstaat. Springer, Berlin, pp. 1–54

Gethmann CF (1995) Rationalität. In: Mittelstraß J (ed.) Enzyklopädie Philosophie und Wissenschaftstheorie, Vol. 3. Metzler, Stuttgart

Gethmann CF (1996) Zur Ethik des umsichtigen Naturumgangs. In: Janich P, Rüchardt CH (eds.) Natürlich, technisch, chemisch. Verhältnisse zur Natur am Beispiel der Chemie. De Gruyter, Berlin, pp. 27–46

Gethmann CF (1997) Praktische Subjektivität und Spezies. In: Hogrebe W (ed.) Subjektivität. Fink, München. 1998, pp. 125–145

Gethmann CF (1998) Umweltstandards. Grundlegungs- und Umsetzungsprobleme. In: Janich et al. (eds.), pp. 25–36

Hanekamp G (1997) Protochemie. Vom Stoff zur Valenz. Königshausen & Neumann, Würzburg.

Hare RM (1952) The Language of Morals. Oxford

Hartmann D (1993) Naturwissenschaftliche Theorien. Wissenschaftstheoretische Grundlagen am Beispiel der Psychologie. Bibliographisches Institut, Mannheim

Hartmann D, Janich P (1996) Methodischer Kulturalismus. In: ibid (ed.) Methodischer Kulturalismus. Zwischen Naturalismus und Postmoderne. Suhrkamp, Frankfurt a.M., pp. 9–69

Hempel Carl Gustav, Oppenheim P (1948) Studies in the Logic of Explanation. In: Philosophy of Science 15

Janich P (1996) Natürlich künstlich. Philosophische Reflexionen zum Naturbegriff der Chemie. In: Janich P, Rüchardt CH (eds.) Natürlich, technisch, chemisch. Verhältnisse zur Natur am Beispiel der Chemie. De Gruyter, Berlin, pp. 53–79

Janich P (1997) Kleine Philosophie der Naturwissenschaften. Beck, München

Janich P, Thieme PC, Psarros N (1998) Chemische Grenzwerte. Eine Standortbestimmung von Chemikern, Juristen, Soziologen und Philosophen. Wiley-VCH, Weinheim

Jungermann H, Slovic P (1993) Die Psychologie der Kognition und Evaluation von Risiko. In: Bechmann (1993), pp. 167–208

Kant I (1785) Grundlegung zur Metaphysik der Sitten. In: Werke (Akademieausgabe). De Gruyter, Berlin, Vol. 4: 385–464

Kaspersen RE, Renn O, Slovic P, Kasperson JX, Emani S (1988) The Social Amplification of Risk: A Conceptual Framework. Risk Analysis, 8: 177–187

Kern L, Nida-Rümelin J (1994) Logik kollektiver Entscheidungen. Oldenbourg, München

Kloepfer M (1990) Rechtsstaatliche Probleme ökonomischer Instrumente im Umweltschutz. In: Wagner (1990), S 241–261

Kloepfer M (1998) Umweltrecht. Beck, München

Kordell RL, Pounds JG (1991) Assessing the toxicity of mixtures of chemicals. In: Krewski D, Franklin C (eds.) Statistics in Toxicology. New York

Könemann WH, Pieters MN (1996) Confusion of Concepts in Mixture Toxicology. Food and Chemical Toxicology. 34, pp. 1025–1031

Lorenzen P (1987) Lehrbuch der konstruktiven Wissenschaftstheorie. Bibliographisches Institut, Mannheim

Luhmann N (1990) Risiko und Gefahr. In: ibid. Soziologische Aufklärung 5. Westdeutscher Verlag, Opladen, 131–169. Also in: Krohn W, Krücken G (eds.) Riskante Technologien: Reflexion und Regulation. Suhrkamp, Frankfurt a.M. 1993, pp. 138–185

Luhmann N (1991) Soziologie des Risikos. De Gruyter, Berlin

Luhmann N (1993) Die Moral des Risikos und das Risiko der Moral. In: Bechmann G (ed.) Risiko und Gesellschaft. Westdeutscher Verlag, Opladen. 1997

Mittelstraß J (1995) Artikel 'Naturalismus'. In: ibid. Enzyklopädie Philosophie und Wissenschaftstheorie. Vol. 2. Metzler, Stuttgart

Nida-Rümelin J (1996) Ethik des Risikos. In: ibid. Angewandte Ethik. Die Bereichsethiken und ihre theoretische Fundierung. Kröner, Stuttgart

Ott K (1998) Ethik und Wahrscheinlichkeit: Zum Problem der Verantwortbarkeit von Risiken unter Bedingungen wissenschaftlicher Ungewißheit. Nova Acta Leopoldina NF 77, 304, pp. 111–128

Der Rat von Sachverständigen für Umweltfragen (1996) Umweltgutachten 1996. Metzler-Poeschel, Stuttgart

Der Rat von Sachverständigen für Umweltfragen (1998) Umweltgutachten 1998. Metzler-Poeschel, Stuttgart

Rehbinder E (1991) Das Vorsorgeprinzip im internationalen Vergleich. Werner, Düsseldorf

Rehbinder E (1998) Ziele, Grundsätze, Strategien, Instrumente. In: Salzwedel et al. (eds.) Grundzüge des Umweltrechts. Schmidt, Berlin

Renn O (1984) Risikowahrnehmung der Kernenergie. Campus, Frankfurt a. M.

Renn O (1992) Concepts of Risk: A Classification. In: Krimsky S, Golding D (eds.) Social Theories of Risk. Westport

Renn O (1995) Risikobewertung aus Sicht der Soziologie In: Renn O et al. Risikobewertung im Energiebereich. Zürich

Renn O (1998) WBGU-Gutachten 1998

Rescher N (1983) Risk. A Philosophical Introduction to the Theory of Risk Evaluation and Management, Lanham

Schäfer L, Ströker E (1993–96) Naturauffassungen in Philosophie, Wissenschaft und Technik (4 vol.). Alber, Freiburg

Schäfer L (1993) Herrschaft der Vernunft und Naturordnung in Platons Timaios. In: Schäfer L, Ströker E, Naturauffassungen in Philosophie, Wissenschaft, Technik. Vol. 1, Antike und Mittelalter. Alber, Freiburg

Schröder M (1995) „Nachhaltigkeit" als Ziel und Maßstab des deutschen Umweltrechts. In: Wirtschaft und Verwaltung. 2/1995. 65–79

Shrader-Frechette KS (1991) Risk and Rationality Philosophical Foundations for Populist Reforms. Berkeley

Siebert H (1998) Economics of the Environment. Springer, Berlin

Stegmüller W (1973) Probleme und Resultate der Wissenschaftstheorie und Analytischen Philosophie. Vol. IV. Personelle und Statistische Wahrscheinlichkeit. Springer, Berlin

Stegmüller W (1983) Erklärung Begründung Kausalität. Springer, Berlin

Steinmann H, Wagner GR (eds.) (1998) Umwelt und Wirtschaftsethik. Schäffer-Pöschel, Stuttgart.

Streffer C (1996) Threshold Dose for Carcinogenesis: What is the Evidence? 12th Symposium of Microdosimetry. Oxford

Tetens H (1987) Experimentelle Erfahrung. Eine wissenschaftstheoretische Studie über die Rolle des Experiments in der Begriffs- und Theoriebildung der Physik. Meiner, Hamburg

Wagner GR (1990) Unternehmung und ökologische Umwelt. Vahlen, München

WCED (Weltkommission für Umwelt und Entwicklung) (1987) Unsere gemeinsame Zukunft. Greven: Eggenkamp (engl. Original: Our Common Future. Oxford 1987)

Wright GH von (1971) Explanation and Understanding. Ithaka

Zimmermann H-J, Gutsche L (1991) Multi-Criteria Analyse. Springer, Berlin

2 Scientific and Medical Foundations

A great number of limit values has been introduced through statutory regulations in recent decades, in order to put constraints on the exposure to toxic agents at the workplace and in the environment. As far as these limits have been determined on the basis of scientific data especially from experimental studies, they usually are derived from investigations on exposures to single agents. For the simultaneous exposure to several noxae, on the other hand, limit values have rarely been regulated, and only in few cases such limits are supported by systematic studies of combined exposures. One of the reasons for this is the immense complexity presented by combined exposures, resulting from the following facts:

- There are manifold physical, chemical, and biological noxae potentially affecting human beings and other organisms in the environment. Two noxae may be dominant among a much larger number of harmful agents actually being effective.
- The chronological sequence of exposures can vary greatly (continuously simultaneous, successive with varying intervals, different sequences for different exposures, etc.)
- The concentrations of individual noxae can vary considerably, thus leading to a large variety of effects.

Boundary conditions as varied as these can significantly influence the behaviour of the individual noxae, i.e. they can determine if the agents behave like additive, sub-additive, or super-additive noxae. Hence, in an assessment of combined exposures it is not possible to reach a clear judgement in every detail, taking into account every one of those boundary conditions. This is particularly true for agents with interacting effect chains. However, a practicable solution could follow from approaching the effect mechanisms of the individual noxae with regard to their harm to human health or the development of environmental damage. In the first instance, one will assume that for combined exposures the single effects behave additively. This assumption has indeed been made in regulating for such scenarios. A departure from additivity can occur predominantly in situations where the single noxae or the processes in the development of the damage events interact with each other. Damage events following exposure to noxae, especially events leading to cancer develop in a chain of successive steps. Thus, e.g. for a substance to act super-additively with ionizing radiation, it is necessary that the substance interferes in one or more steps in this chain of development. An appropriate assessment of combined exposures could therefore be established by considering the effect mechanisms of the individual noxae with regard to their harm to health or the environment and by trying to find out if these effect mechanisms present ways

Table 2.1-1 Possible combinations of agents with regard to their mechanisms and their impact on the development of stochastic effects (The details will be explained later.)

Agent combination	Expected combined effect	Conditions for the combined effect	Determination of health risks	Examples of already examined combinations of an agent with ionizing radiation
I) Genotoxic agents affecting the same phase of tumour development	Addition/ intensification/ attenuation	Low specificity of the agents	Linear extrapolation from high to low doses	Benzo[a]pyrene Alkylating agents
II) Genotoxic agent/agent inhibiting mechanisms for damage repair	Super-additivity	Successive order of exposure	Generally not by linear extrapolation	Metals Caffeine Chemotherapeutics
III) Genotoxic agent/agent inducing mechanisms for damage repair	Antagonism	Successive order of exposure; half-life of the adaptive processes	Linear extrapolation from high to low doses may overestimate the risk	Radioprotectors Certain food components
IV) Genotoxic agents affecting different phases of tumour development	Departure from addition	Extent dependent on the specificity of the agents and the order of exposure	Dependent on the degree of departure from additivity	Viral genetic sequences
V) Genotoxic agents/ non-genotoxic agent (initiator + promoter)	Super-additivity	Successive order of exposure	Generally not by linear extrapolation	12-O-Tetradecanoylphorbol-13-acetate (TPA); some hormones; certain food components; cigarette smoke; carbon tetrachloride; asbestos

of interaction between the individual development chains. The overall effect may be synergetic, perhaps super-additive or possibly antagonistic. Considerations like this could make it possible to arrive at general principles for more precise standards of assessment. Hence, in this chapter we will try to crystallize these mechanisms and principles as far as they are proven by experimental research and, where possible, by epidemiological studies.

2.1
Possible Principles of Interaction between two Noxae, especially with the Involvement of Ionizing Radiation

On the basis described above, we will try to establish general mechanisms with regard to possible interactions, which in turn can be applied, by analogical inference, to situations not examined yet. Table 2.1-1 shows a selection of such possible principles with the expected interactions derivable from them, ideas on how to determine health risks, and examples of agents, the investigation of which in combination with other noxae has led to the formulation of the principles quoted.

2.2
Dose-Response Relationships and Models for Combined Effects

2.2.1
Dose-Response Curves and Mechanisms

Effects and combined effects of chemical substances and physical factors can be calculated and graphically displayed as dose-response curves by plotting the effect of an agent over its dose or concentration (fig. 2.2-1). In the context of environmental standards, effects in the so-called lower dose range are of special interest (fig. 2.2-1). This is the region where small effects just occur or where no effects can be found yet.

Available data on effects in this lower dose range are very limited indeed. Considering the importance of carcinogenic agents, some experimental studies with large numbers of test animals have been performed in order to obtain more data points in the lower dose range (Zeise et al. 1987). In the excellent ED_{01}-study with 24,192 mice, by Littlefield et al. (1979), an increase in the tumour incidence by just 1% was still statistically significant. An increase of the tumour rate by one case in 100, as observed in that study, is, however, a lot higher than incidence rates of 1 in 100,000 or 1 in 1,000,000 humans, which is the range environmental standards for carcinogenic agents are dealing with. Still, from a known dose-response relationship in the medium dose range, a tentative extrapolation to the lower dose range is possible. Dose-response relationships are calculated by means of corresponding, suitable models and represented graphically as (linear) curves.

The main point of this procedure is the graphical representation of a dose-response relationship between the individual data points and the effort to extrapolate this relationship to the lower dose range.

In principle, the effects as such or the incidence rate of a certain effect can be plotted along the y-axis over the doses (or concentrations) on the x-axis. Then the effect or the effect incidence as such, or the change in a control value can be used to characterize the dose response. Effects or incidence rates, respectively, and doses are plotted either linearly or on a logarithmic scale, as illustrated in fig. 2.2-1.

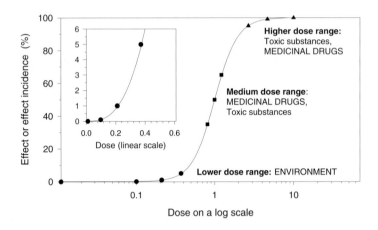

Fig. 2.2-1 The dose-response relationship and the dose ranges in a simplified representation. The dose-response curve is S-shaped when plotted over a logarithmic dose axis. The dose ranges are marked by different symbols. The lower dose range is highlighted in the insert, where a linear dose axis is used.

In practice, of course, an assessment of physical factors or of chemical substances by appropriate dose-response curves cannot be done for every conceivable factor or chemical. We know, however, that substances similar in chemical terms or with similar features behave similarly, too, with regard to their desired and undesired effects, so that analogical conclusions can be drawn from results with "model agents".

Among the well-researched agents are, above all, radiation and medicinal drugs (pharmaceuticals), for which we also know something about the mechanisms leading to effects. For certain effect mechanisms, mathematical models were developed, which is of advantage compared to a purely phenomenological description, because in this way an effect of an unknown mechanism can be assigned to a known effect mechanism, albeit with reservations. The problem can be explained by the fact that certain mechanisms must follow equally certain dose-response relationships. In turn, phenomenologically similar dose-effect relationships can also arise from other mechanisms which happen to appear similar from a phenomenological point of view. Such superficial similarities are to be expected, especially where a dose-response relationship was analyzed only for a limited dose range.

In order to understand biological effects and their dose-response relationships, it appears useful to consider the basic mechanisms leading to effects. These are referred to as reversible and irreversible mechanisms, which should be seen, in the first instance, as mechanisms that do or do not operate in a backward as well as in a forward direction respectively.

2.2.1.1
Reversible Effect Mechanisms

We find reversible effect mechanisms in the very most reactions occurring in the human body, their characteristic feature being their rapid subsidence after exposure to an agent. Examples are sensory stimuli and the workings of messenger substances and hormones, which bind only weakly and therefore reversibly to their corresponding receptor molecules. Insofar as reversible mechanisms involve substance transformation, enzymes are the mediators.

As a rule, effects brought about by reversible mechanisms can be computed and represented well by S-shaped dose-response curves (see fig. 2.2-1). Curves of this type are often referred to as sigmoid curves, and the model they are based on is called the sigmoid model. Irreversible effects, which will be discussed later, are also described well by the sigmoid model.

Sigmoid Curves

Many physiological and pharmacological effects can be represented by sigmoid curves. In the typical logarithmic-linear plot of effect (incidence) over dose, they appear as S-shaped curves (fig. 2.2-1), which can be computed as a four-parameter logistic function through the following mathematical equation (De Lean et al. 1978).

$$Y = ((a-d)/(1+(x/c)^b)+d$$

Y:	effect
x:	dose
a:	minimum
b:	slope
c:	half-maximum effective dose or concentration (ED_{50})
d:	maximum

Interestingly, not all such curves feature the same slope i.e. the same increase in effect with increasing dose (fig. 2.2-2a). A slope of 1 corresponds to processes following the law of mass action for substances attaching to a molecular binding site like a receptor for messenger substances, and thereby triggering a reaction sequence that ends in the biological effect. Steeper dose-response curves e.g. with a slope of 2 can sometimes be interpreted in terms of so-called allosteric mechanisms where, for instance through a change of conformation by binding of molecules at a different site, binding e.g. to a classic receptor is favoured. In other cases, which are the rule, one assumes "functionally allosteric" mechanisms, where an effect promotes further triggering of that same effect in the sense, for example, of a hidden previous damage promoting some manifest damage. Therefore, it is no surprise that many toxic-effect curves are steep, with slope values of 1 to 12 (fig. 2.2-3a). Receptor-mediated reactions are often characterized by slope values of (about) 1.0, carcinogens by steeper slopes between 1.3 and 2.0 (fig. 2.2-3b).

With another choice of scales than in this log-linear plot, the sigmoid model not always results in S-shaped curves, as illustrated in fig. 2.2-2b. Therefore, such curves are also referred to as supra-linear (slope = 1), linear (slope about 1.5), and sub-linear (slope > ca. 1.5).

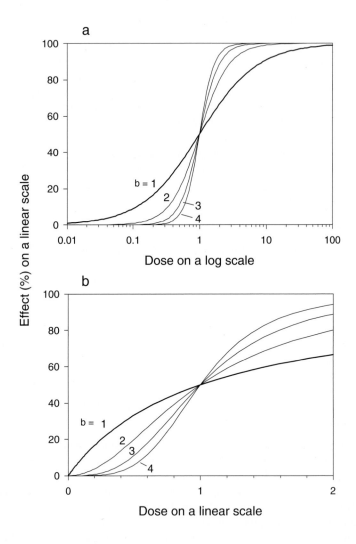

Fig. 2.2-2 Sigmoid dose-response curves of various slopes, a) with a logarithmic x-axis, b) with a linear x-axis. All curves in a) turn out S-shaped, in b) supra-linear (b = 1), linear (b = 1.4 in fig. 2.2-4b), or sub-linear for higher slope values (in this case b = 2 or higher).

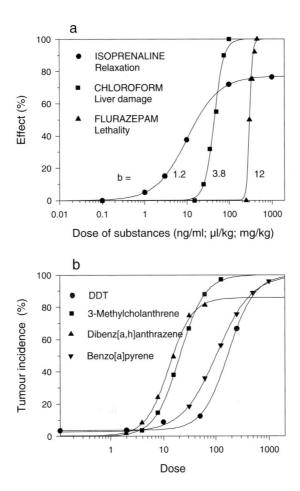

Fig. 2.2-3 Experimental examples a) of sigmoid curves of various slopes, from own data (Isoprenaline), Plaa and Hewitt (1982: Chloroform) and Pöch et al. (1990b: Flurazepam), b) of curves with similar slopes, from Grieve (1985: DDT, in: Food Safety Council 1980) and Chou (1980: other carcinogenic substances).

2.2.1.2
Irreversible Mechanisms

Irreversible mechanisms are characterized by the fact that the resulting effect does not simply abate after exposure to a certain dose, but typically becomes stronger over the period of exposure. It has to be noted, however, that such effects can often be repaired, either fully or to some extent. In contrast to the weak binding forces between a molecule and its receptor, e.g. between a substrate and an active centre of an enzyme, in reversible mechanisms, irreversible mechanisms following an exposure to chemical substances involve covalent bonds between molecules (or radicals).

Stochastic Effects

Exposure to radiation also leads to chemical reactions e.g. at the DNA, although these reactions do not primarily arise from chemical processes, but from the absorption of radiation energy, which causes randomly distributed molecular transformations in the DNA. Hence, these mechanisms are referred to as stochastic in nature.

Exponential Curves

At present, it is generally assumed that genotoxic effects occur randomly and rise with increasing doses. It is further assumed that any given dose always harms the same percentage of cells. The mathematical representation of such a mechanism is an exponential function:

$$Y = a * e^{(-b*x)} + y_0 \qquad \text{exponential decrease}$$
$$Y = a * (1 - e^{(-b*x)}) + y_0 \qquad \text{exponential increase}$$

a: coefficient
b: coefficient
y_0:   y-value (effect) at dose 0

According to this type of dose-effect relationship, damage to the DNA or to cells by a unit dose always leads to the same fraction of yet unharmed DNA molecules or cells becoming affected. Because this dose-effect relationship is represented by a straight line, it generally is referred to as a linear relationship (also see section 2.3.1). Corresponding to two kinds of graphical representation, mutagenic effects and cell death will be treated separately, before we discuss dose-response curves for tumour development.

2.2.1.3
Mutagenic Effects

A number of experimental studies have shown that exposure to radiation frequently leads, depending on the dose, to a linear increase of mutations. Similar findings exist for chemical mutagens (see e.g. Zeise et al. 1987). Fig. 2.2-4a shows this at the example of the well-known mutagen benzo(a)pyrene and for a nitro compound. Fig. 2.2-4b explains this linearity as an expression of an exponential effect. The points follow an exponentially rising curve (not shown here). We notice that, while the points fit a straight line in the lower dose range, an exponential function like this looks very similar to a sigmoid dose-response curve with a slope of b = 1.4. Indeed, some results are represented just as well or even better by sigmoid curves, as an analysis (Pöch unpublished) of the data by George et al. (1991) and Nestmann et al. (1987) has shown.

Cell Death Caused by Mutagens

Cell death by mutagens or other agents is usually represented as in fig. 2.2-5, with the survival rate on a logarithmic scale over the dose on a linear scale. Such curves are referred to as „survival curves“, with the exponential function displayed as a

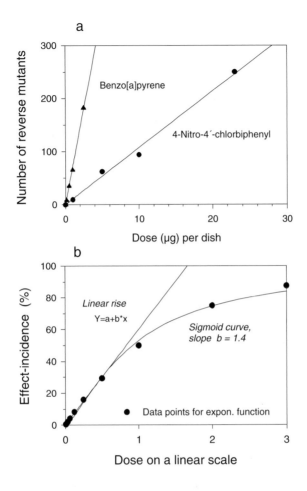

Fig. 2.2-4 a) Linear dose-effect function, illustrated by experimental results with mutagens (Donnelly et al. 1988), b) theoretical, exponential dose-effect function (full circles). Linear fit to the dose range 0–29% (linear rise), and a sigmoid curve fit with a slope of b = 1.4. Both in a) and in b) the dose scale is linear.

straight line not only in the lower dose range of a cytotoxic effect. In experimental findings, such linearity is rather the exception than the rule. It is observed for cells without any significant repair function (fig. 2.2-5a), while cells with repair capacity often show a so-called shouldered (survival) curve (Kumazawa 1994). Such curves consist of two components, the shoulder and an adjoining straight line. The shoulder can be computed as a sigmoid curve segment, the straight line as an exponential one (fig. 2.2-5a). In other cases, the entire curve can be approximated by a sigmoid curve, as shown in the insert in fig. 2.2-5b.

The results in fig. 2.2-5 and a host of similar results reported in the literature touch on the question if pragmatic zero doses can be assumed not only for non-genotoxic

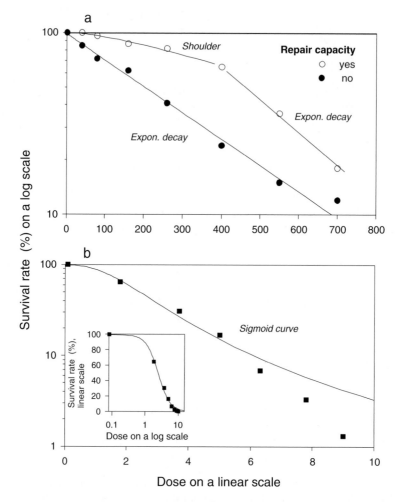

Fig. 2.2-5 Experimental survival rates following exposure to radiation a) in rad, b) in Gy. a) Various curves for CHO wild type cells (O) and camptothecine-resistant cells (●) according to Mattern et al. (1991: fig. 3), b) from data by Kumazawa (1994). A sigmoid curve, shown in the insert on a logarithmic scale, was fitted to all experimental data points.

chemicals, but also for radiation and chemical agents (see e.g. Purchase and Auton 1995). Unfortunately, this question cannot be answered (on the basis of present knowledge) with a positive yes or no.

2.2.1.4
Tumours

In the section on sigmoid curves, we already referred to curves for tumour incidence. Fig. 2.2-3b shows that the experimental tumour incidence due to chemical substances as well as other effects (fig. 2.2-3a) can be represented by sigmoid curves. Interestingly, most sigmoid tumour incidence curves feature slope values between about 1.3 and 2.0 (fig. 2.2-3b).

Unfortunately, only a small number of experimental studies with animals are available to supply us with sufficiently large numbers over a wide range of doses to make a comparative analysis of curve fits statistically meaningful. Even the initially mentioned, large ED_{01}-study (Littlefield et al. 1979) is less than ideal for such a comparison of curves, because of the scarcity of data points in the medium and higher dose range. However, the much better data situation in the lower dose range gives reason to show the results in the form of sigmoid curves assuming a tumour incidence of 100% in the high dose range (fig. 2.2-6a).

One should also mention the experimental investigations on rats with diethyl and dimethyl nitrosamine (Peto et al. 1982). The tumour rates from both studies are well represented by sigmoid curves. A comparative analysis of the published incidence rates of liver tumours in male and female rats, caused by diethyl and dimethyl nitrosamine respectively, produced significantly better fits with the sigmoid model than with the exponential model for three of the four experimental set-ups. In the fourth case, the difference between the fits with the two models was of no statistical significance (p = 0.38) (Dittrich and Pöch unpublished). The dose-effect relationship for male rats treated with dimethyl nitrosamine (Zeise et al. 1987: table 18) is well fitted by a sigmoid dose-response curve with a slope of 2.74. With a value of $r^2 = 0.986$, the regression coefficient, a measure of the quality of the fit, is very close to the ideal value of $r^2 = 1$. The fit of an exponential curve to the same data points is distinctly worse, which is also confirmed by a comparative F-test analysis with p = 0.0001.

Numerous studies on tumour experiments have shown that linear, supra-linear, or sub-linear curves are suitable representations of the occurrence of tumours following exposure to chemical substances and radiation. Sub-linear dose-response functions are most frequent, followed by linear functions. As already shown in fig. 2.2-4b, the linear part of an exponential curve can also be well described as a sigmoid curve with a slope of b = 1.4. Fig. 2.2-2b shows that a supra-linear function corresponds to a sigmoid curve with a slope of b = 1. Sub-linear curves match sigmoid curves with slope values of b = 2 or higher. As we will show later, the linear-quadratic curve is equivalent to a sigmoid curve with a slope of b = 2.5 (fig. 2.2-6b).

The comparison between the linear, supra-linear and sub-linear polynomial functions, on the one hand, and the sigmoid curves, on the other hand, shows that in the lower dose range every polynomial function seems to be describable by a sigmoid curve. For experimental data, it usually is impossible to decide which dose-response function produces a better fit. This observation also applies to the comparison between exponential and sigmoid curves, although one finds exceptions here too, as shown by the result with rats pre-treated with nitrosamine, which we described above and where, in three out of four cases, the sigmoid model produced significantly better fits, illustrated at an example in fig. 2.2-6a. Also, epidemiological studies show at least equally good fits by sigmoid curves as by a linear-quadratic or a linear function (fig. 2.2-7), which we will look into later.

2.2.1.5
Dose-Response Curves in the Lower Dose Range

As we have mentioned at the beginning, the lower dose range is of particular importance in the context of environmental standards, especially for the assessment of carcinogenic agents. This area presents an exceptional challenge to science, because experimental effects are far more difficult to detect for low, hardly effective doses than for the medium dose range. With the same number of experiments, the half-maximum effective dose, ED_{50}, can be determined with much more certainty than for instance an ED_1. This is textbook knowledge. Fig. 2.2-1 shows clearly that the effect only increases slowly with rising doses in the lower dose range, quite contrary to the situation in the medium dose range. In the higher dose range, we see effects slowly rising again with increasing doses.

Hence, in order to determine the dose-response function in the lower dose range it would appear reasonable to examine effects in the medium dose range and extrapolate from there to the dose-response curves in the lower dose range. Against this consideration stands the fact, however, that one cannot decide in all cases or with appropriate certainty, which dose-response function should be applied for the extrapolation, especially since the differences between the diverse curve models are extremely slight in the medium dose region, while they are more pronounced in the lower dose range. This problem especially affects extrapolations to the extremely low dose range for carcinogenic agents.

For this purpose, a number of models were developed and suggested (see Zeise et al. 1987). The majority of these models are based on polynomial functions (linear and sub-linear function) and it appears doubtful, at least, if these can offer a better description of the corresponding dose-response functions, especially in the lower dose range, than the sigmoid model. Perhaps one should also note that a linear or sub-linear dose-response function, unlike the sigmoid curve function, does not describe experimental effects at the maximum.

For practical purposes, a dual approach is often used. On one hand, a polynomial function is fitted to all data points in the lower and medium dose range or above and, on the other hand, a dose-response function is established to determine the dose-effect relationship in the lower dose range. The fact that this procedure is applied at all (see Zeise et al. 1987) already indicates that it is not always possible to find a polynomial function representing both the "full range" and the lower dose range.

However, finding a common function, which would adequately describe effects both over full range and in the lower dose range, appears to be a desirable goal. To this end, it should be reasonable to examine the suitability of sigmoid curves, particularly since such curves can also be derived as sums over Gaussian bell-shaped curves representing statistical incidence distributions.

Epidemiological Studies: Smoking and Radiation

Tobacco smoke and radiation are, without any doubt, the best-researched carcinogenic agents, on which there exist good epidemiological studies with human beings (see table 2.3-13 and section 2.4.4.3). We will now attempt using the sigmoid model to describe the tumour incidence in dependence of the dose of the carcinogenic agent.

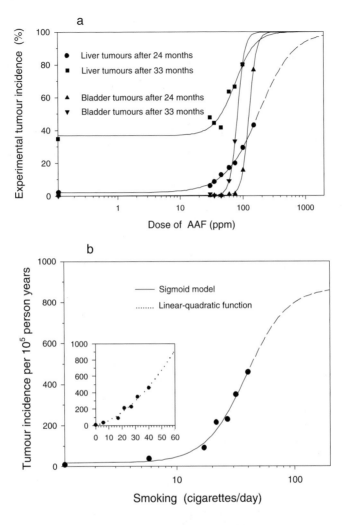

Fig. 2.2-6 a) Experimental tumour incidence in mice following continuous treatment with 2-acetylaminofluorene (AAF). Curve fitted to data points, using the sigmoid model with d = 100%, from results of the ED_{01}-study by Littlefield et al. (1979); b) epidemiological investigations on smoking and lung cancer; fitted with the sigmoid model and with a linear-quadratic function (insert).

Smoking

Zeise et al. (1987, pp. 275-276) discussed a very careful study performed on British physicians, the results of which were published by Doll and Peto (1978). While the first analysts described as linear the relationship between the number of cigarettes smoked and the occurrence of lung cancer, later analyses have shown that a linear-quadratic function represents this relationship significantly better (p < 0.01). As fig. 2.2-6b shows, the dose-response function can be described equally well by the sigmoid model, with a slope of b = 2.5, as by a linear-quadratic function (insert in fig. 2.2-6b).

Radiation

Because of their exposure to radiation, many survivors of Hiroshima and Nagasaki developed leukemias and cancers years after the nuclear bomb explosions. Studies on this subject were carried out with a great number of victims. The analysis by Pierce et al. (1996) is based on a total of 86,572 individuals, about half of which had been exposed to radiation doses of more than 0.005 Sv, of which the majority, again, of 32,915 people was exposed to 0.005–0.1 Sv, and 1,914 subjects, still, had suffered exposure to 1.0–2.0 Sv of radiation. Fig. 2.2-7a shows the incidence of leukemia among the survivors of the nuclear catastrophe over the years from 1950 to 1990, the doses corresponding to the averaged dose ranges. The data points fit almost perfectly to the sigmoid model with a slope value of b = 1.3. The deviation of the points from this sigmoid curve, yielding p > 0.8 in a chi-squared test of the fit, is not significant. The linear fit shown in the insert in fig. 2.2-7a performs slightly worse and, with p = 0.05 in the chi-squared test, is close to the border of significance.

Fig. 2.2-7b shows the relative risk for the survivors of Hiroshima and Nagasaki to fall ill with cancer (UNSCEAR 1994). Concerning the curve fit, we see a picture quite similar again to the situation in fig. 2.2-7a, namely an almost perfect fit with the sigmoid model (slope b = 1.3) and a slightly worse fit to a straight line (insert in fig. 2.2-7b).

The maximum in the incidence of leukemia and cancer appears to be around 15% or about seven times as high as for individuals not damaged by radiation. However, these maxima should be interpreted with great reservations, since no data are available in this dose region. The best basis on which to determine the cancer risk induced by radiation are epidemiological studies, first of all on the said consequences of the nuclear blasts of Hiroshima and Nagasaki, where a statistically significant increase of cancer was only observed for radiation doses of 0.2 Gy and above. This dose range is beyond those exposures occurring in the environment and requiring limitation through environmental standards. Nevertheless, according to present understanding radiation risks from low, environmentally relevant doses cannot be excluded entirely. The number of cases, however, becomes so small that other variable factors – lifestyle, genetic disposition, etc. – having an influence on the risk of cancer and leukemia completely supersede and hence conceal the radiation-induced risk.

Radiation risks in dose ranges relevant for occupational radiation exposure (a few mSv to some 10 mSv/year) and, above all, in yet lower ranges, for instance in the vicinity of nuclear installations (some 10 to 100 µSv/year), have thus to be derived by extrapolation from the empirically established risks at higher radiation doses. These extrapolations extend over wide dose ranges and are accordingly afflicted with uncertainties. One should also bear in mind that the epidemiological data on the increased incidence of leukemia and cancer were collected after exposures not only to high radiation doses, but also to a high dose rate (distribution of dose over time). The radiation impact in the case of the nuclear bomb explosions was characterized by an extremely high dose rate, whilst exposures at the workplace

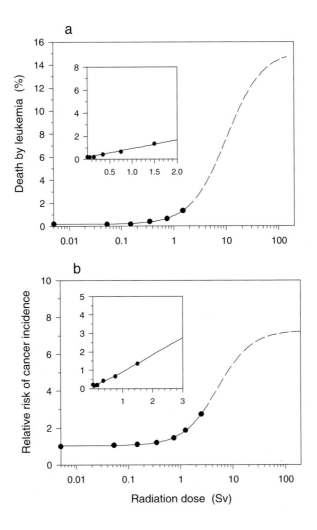

Fig. 2.2-7 Epidemiological studies on leukemia and cancer incidence following a nonrecurring radiation exposure due to nuclear bombs in Hiroshima and Nagasaki. Data points fitted with the sigmoid model and with a straight-line function (insert). a) Deaths by leukemia (%) according to Pierce et al. (1996), b) cancer incidence expressed as relative risk (UNSCEAR 1994).

and even more markedly in the environment occur not only at much lower doses, but also at extremely low dose rates.

In this range, there is no empirical knowledge on the shape of the dose-effect relationship. For the medium and high dose range, there exist various types of dose-response curves based on different theoretical assumptions about the mechanism of radiation effects. Risk estimations in the lower dose range, on the other hand, generally start from two postulates: It is assumed, firstly, that the dose-response relationship for stochastic effects is a linear function of the radiation dose in this dose range and, secondly, that there is no threshold dose (below which no effect would occur). This assumption has its origin in the biological mechanism of a monoclonal growth of tumours (see section 2.3.2). Accepting these postulates rather leads to an overestimation than to an underestimation of the risk, which is in the interest of the conservative risk estimation common in radiation protection and hence takes into consideration the idea of prevention.

The postulated linear shape of epidemiological dose-response curves in the lower dose range and the assumption of no threshold dose being applicable have not been proven and are virtually improvable (see section 2.1.1). The results shown in fig. 2.2-7 make it appear utterly possible that the postulate of a missing pragmatic threshold dose is indeed misguided, since the dose-response relationship is well described by a sigmoid curve, and because one does find effects with threshold doses in curves of this type. There seems to be an urgent need for further studies on the mechanisms of tumour development, on the one hand, and the dose-effect relationships, on the other hand. For the latter issue, review-type studies in terms of a meta-analysis could be considered too.

2.2.1.6
Differentiation between Stochastic and Non-stochastic Effects

It is frequently assumed that stochastic and non-stochastic effects can be differentiated between by their different dose-effect curves (see e.g. Streffer 1991), as in the example in fig. 2.2-8a. In graphs with linear x- and y-axes, an exponential dose-effect relationship in the lower dose range appears as a straight line as illustrated in fig. 2.2-4. A linear relationship like that is regarded as characteristic for stochastic (unicellular) effects. In the same double-linear notation, non-stochastic (multicellular) effects, on the other hand, are described by S-shaped curves that do not rise from zero-dose but, quite visibly, from a threshold dose. Fig. 2.2-8a seems to be a good example for this observation. For a certain strain of mice, irradiation causes the development of malformations, with unicellular embryos behaving differently from multicellular ones, as shown by the curves in fig. 2.2-8a. Without any doubt, mutations as a direct result of stochastic events feature this linear dose-effect relationship, which can be expressed as an exponential relationship between radiation dose and mutagenic effects. Similar exponential effects one also observes for mutagenic chemicals, as illustrated in fig. 2.2-4a. In cases of malformations that can be triggered by mutations and of radiation-induced cancer, mutations are the initial, stochastic event, followed by non-stochastic processes resulting in malformations and tumours. Dose-effect relationships for such processes are well represented by sigmoid dose-response curves over a logarithmic dose axis, as the examples above have shown. This also applies to the malformation experiments in fig. 2.2-8. Both dose-effect functions in fig. 2.2-8a can be described just as well or better by sigmoid curves of various slopes (fig. 2.2-8b). These appear as S-shaped lines in the double-linear plot of fig. 2.2-8a only if the slope value is above ca. $b = 2$ (fig. 2.2-2b).

With this, we touch on a point that is regrettably still under debate, the issue of the threshold dose. The sigmoid model can be regarded as a curve model for non-stochastic effects. According to theory and experience, the curves of the sigmoid model can have virtually any slope value between about b = 1 and b = 10 or higher (see the examples in fig. 2.2-3). In a double-linear plot of effects over doses, only at slope values >> 2 threshold doses are clearly noticeable. There is, however, an abundance of unequivocally non-stochastic processes showing slope values < 2, e.g. typical, physiological effects of messenger substances and hormones.

2.2.1.7
Conclusions

As the many examples in this chapter have shown, the dose-effect relationship can be described adequately by sigmoid curves, on the one hand, and by polyno-mial functions, on the other hand, as well as by exponential curves in a small number of cases. Independent of the mechanisms of the agents examined, many

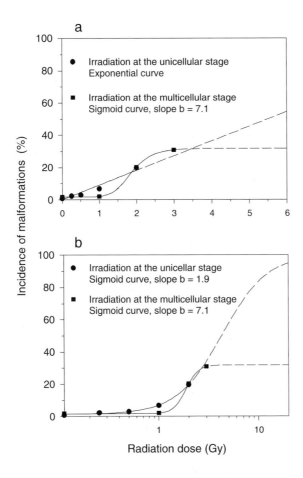

Fig. 2.2-8 Malformations caused by radiation. Different dose-response curves for the irradia-tion of unicellular and multicellular embryos, from Müller et al. (1994), a) with a linear dose axis, b) with a log-scale.

dose-effect relationships can be represented by sigmoid curves. They appear to be suitable for reversible as well as for irreversible effects, at least in many cases. Hence, there appears to be some justification for describing and interpreting combined effects, too, in terms of sigmoid terms, with the exception of direct, stochastic events. Where stochastic events trigger a reaction chain, however partly followed by nonstochastic processes (see section 2.3.1.2), dose-effect curves for cancer and malformations (Pöch et al. 1997) appear to be well described by sigmoid curves too.

Above all, using the sigmoid model offers the particular advantage that the curves of this model fit well almost every experimental or epidemiological dose-effect relationships and offer a good representation of them, a representation that is in fact independent of the slope of the curve, which emerges as a result of the fitting process itself. For fits using polynomial functions, on the other hand, any particular function, e.g. a linear or linear-quadratic equation, must be defined beforehand.

2.2.2
Combined Effects

While the combined action of agents can lead to milder as well as to more severe effects than an exposure to single agents, stronger effects in the lower dose range are of special relevance in the context of environmental standards. In a combination of agents, the same dose of an agent may induce equal or stronger effects already at low doses. In this section, we will discuss and illustrate, by means of graphs and tables, basic aspects of combined effects, especially with regard to the lower dose range. A more detailed analysis of empirical effects of combined exposures, in particular to medicinal drugs, will be the subject of the later section 2.2.3.

2.2.2.1
Terminology

Combined effects are frequently described by the terms synergism and antagonism. Unfortunately, so far there is no agreement about when these expressions should be used. The situation may be explained, in a simplified fashion, like this: Combined effects are compared with the corresponding effects of the individual agents in effect (single effects) and, alternatively, with model effects in combined application. A deviation of observed combined effects from the single effects or the model effects, respectively, is often judged to be due to synergism or antagonism. The diversity of interpretations of the terms synergism and antagonism forces each researcher to determine at first which terminology is used in any piece of literature.

2.2.2.2
Comparison with Single Effects: Synergism and Antagonism

For the reason given above, the terms synergism and antagonism are largely avoided in this paper, or they are used in the comparison between combined and single effects, particularly since the combined-effect models still to be described already assume a synergism of the individual components, meaning that each of the model effects is stronger in combination with others than as a single effect. The term synergism characterizes the combined effect being stronger in comparison with the effects of the components (single effects) or, more precisely, with the effects of the more effective component. For a combination of the agents A and B, the combined effect E_{A+B} exceeds the effect of the stronger component, E_A:

$$E_{A+B} > E_A$$

Synergism thus describes the direction, but not the extent of the effect amplification. The term antagonism marks an effect lessened in combination with others as compared to the single effects:

$$E_{A+B} < E_A$$

When dealing with agents causing identical or similar effects, we can generally expect amplification i.e. synergism. The opposite applies to agents with effects counteracting each other. In combinations of such effects, we can expect a mutual diminution or attenuation, an antagonism. Deviations from this rule occur especially with chemical substances of natural as well as synthetic provenances, if toxicokinetic interactions take place between them (see section 2.4.1.3), because under such conditions the concentration of a substance or effective metabolite in the blood and thus at the site of the effect can change and counteract the effect. These effects are particularly pronounced where a change in the concentrations due to toxicokinetic interaction acts in the same direction as a toxicodynamic amplification or attenuation.

2.2.2.3
Comparison with Models

A comparison of observed combined effects with model effects is often interesting, not least because one hopes to gain information about the way individual components work or on how they combine in their effect. Beyond that, models are also called on to predict combined effects. The models to be described yet are pharmaco- and toxicodynamic models respectively. This means that these models describe only certain possibilities of a pharmaco- or toxicodynamic combination at the site of the effect, provided there are no toxicokinetic changes in concentrations.

In the context of environmental standards, we are interested, above all, in synergism models, which is why only these will be discussed here. The models dose additivity, effect additivity and independence deserve our special attention. At this point, we only give a short characterization of these models, which will be discussed in more detail later. In the dose additivity model, the agents behave like dilutions of one and the same substance. Effect additivity means that the effects of the individual components add together. Independent combined effects are characterized by unchanged relative effects of the components in combination.

A comparison between observed combined effects with computed model effects at given doses can result in an agreement or a discrepancy with the model. The same is true for the comparison between observed combined doses with doses calculated using the model for a given effect.

In this paper, any effects stronger than the corresponding model effects are marked by the prefix "supra", while the prefix "sub" indicates combined effects falling short of the computed model effects.

In the interest of a consistent terminology, the customary identification of "supra" with synergism and "sub" with antagonism shall be avoided. For quite rarely one observes sub-additive combined effects exceeding the single effects. The equation sub-additive = antagonism would lead to the contradictory statement of antagonism with regard to a model and synergism in respect to single effects. Let us assume, for example, each of the agents A and B on its own causes 60% of the possible maximum effect. Even with the strongest possible amplification, i.e. to 100% in combination, it would still fall short of the strength predicted by the model of effect additivity, which is 120%.

2.2.2.4
Toxicokinetic and Toxicodynamic Interactions

Amplification as well as attenuation effects can have various causes. They can arise or result from "kinetic" interactions or from "dynamic" ones. The term interaction is therefore used in the context of mechanisms, although this does not always imply a *mutual* action, as the term suggests. In toxicological interactions, the influence can be one-sided, as a substance A may e.g. inhibit the breakdown of a substance B, but not the other way round. In general, toxicokinetic interactions can be explained as changes of the kinetics of a substance A by another substance, where the change often affects the metabolism of A (see section 2.4.1.3).

Models can describe combined effects based on kinetic interactions only with great difficulty, especially in a situation, where effects of other kinds may occasionally occur as well. The latter is only possible under particular conditions, if an insignificant metabolic reaction sequence of a substance A becomes an important one through the influence of a substance B and the metabolite formed in the process has a toxic effect different from the effect of A or a biological effective metabolite of A. Such a mechanism can explain the toxic effect of cocaine in the presence of alcohol (Odeleye et al. 1993), even if a toxicodynamic interaction cannot be excluded too.

In contrast to the toxicokinetic combined effects, the additive i.e. toxicodynamic effects of combined impacts, which are of interest in this context, can be computed quantitatively.

2.2.2.5
Representation of Combined Effects

Dose-Response Curves

Depending on the purpose of the investigation, the researcher can choose one of two experimental set-ups and ways of evaluation:

a) One dose-response curve is produced for an agent on its own and one for the agent in the presence of one or more fixed doses of a second. This is a standard procedure in pharmacology.

b) Dose-response curves are produced for (both of) the combination partners on their own and for a mixture with a fixed dose ratio.

Depending on the question in hand, combined effects are represented and interpreted following either the standard procedure or the mixture approach. In any case, observed combined effects, too, can now be compared with theoretically calculated curves (points) of model effects. The standard procedure is particularly well suited for clarifying pharmaco- and toxicodynamic mechanisms, while the mixture procedure has its special strength in viewing combined effects in the lower dose range.

Isobolograms

Another widely used representation of combined effects is the isobologram (see section 2.3.3), in which equieffective doses and the corresponding dose contributions of two agents, A and B, are plotted against each other, both on a linear scale. This procedure will be discussed and illustrated in more detail in the context of dose additivity.

Dose-Response Surfaces

Perhaps the most comprehensive representation of a combined effect of two agents is the three-dimensional dose-response surface. It requires, however, a high "data density", which is why this representation, in contrast to dose-response curves and isobolograms, is less likely to be found in literature (e.g. Sühnel 1992). The latter two representations may be visualized as contour plots of cuts through the dose-response surface. Vertical cuts produce dose-response curves; horizontal cuts allow the derivation of isoboles of effects of equal strength. It should be noted here that dose-response curves and isobolograms can also be arrived at, when (complete) dose-response surfaces are not available. The latter can be deduced from dose-response curves.

Advantages and Drawbacks of Dose-Response Curves and Isobolograms in Representing Combined Effects

The basic advantage of dose-response curves is that they allow comparing combined effects both with single effects and with model effects. For the standard method as well as for the mixture method, curves of the model effects can be calculated. The fact that there is no computer software commercially available for the latter method, however, must be considered a drawback of the representation as dose-response curves, when such a comparison of observed combined effects and the corresponding model effects is desired, although there are a number of not too difficult – including computer-aided – ways of calculating theoretical curves.

The essential advantage of the isobologram representation lies in the fact that it directly shows the doses of the combination partners, single or in combination, required to give rise to a certain effect. It may be possible to take these doses from dose-response curves, too, but not "at a glance". One disadvantage of the isobolic representation is the large experimental effort involved, because each mixture experiment produces only one data point from a given mixture. Another drawback is that an isobologram does show neither the single effects nor the dose-effect relationships, although dose-effect relationships can be called on to produce isobolograms. Additional, numerical information about the effects may provide some remedy here.

A non-specialist may be confused by finding that the sub-additive regime is located above the additivity isobole, with the supra-additive region below it. However, one should bear in mind that isobolograms tell us which doses are required in order to produce a given effect, which explains why sub-additive combinations just involve higher doses, while supra-additive combinations require smaller doses for the same combined effect than additive combinations.

Both representations, dose-response curves and isobolograms, enable us to compare effects or doses, respectively, with any models. Even the simultaneous comparison with more than one model is possible. Thus, in dose-effect analyses, combined effects can be compared with the corresponding effects in the dose additivity model as well as in the independence model (Pöch et al. 1990a). This is possible, too, in isobolograms concerning the corresponding doses (Steel and Peckham 1979, Pöch et al. 1990b). The procedure used by Steel and Peckham (1979) will be described and discussed in more detail in section 2.3.3.1.

2.2.2.6
Models for Combined Effects

Models are suitable for the more detailed description of combined effects that are synergistic through pharmacodynamic and toxicodynamic interactions. Two models of synergistic combination, "dose additivity" and "independence", are based on certain mechanisms, in contrast to the model of "effect additivity", which does not involve a mechanism. That non-mechanistic model's special features will be discussed later.

Pharmacodynamic/Toxicodynamic Models Based on Mechanistic Concepts
The models discussed here may be interpreted as mechanisms of a synergistic combination at the site of action (Pöch 1993a, Pöch et al. 1996), i.e. as pharmacodynamic or toxicodynamic models. We are interested in greater effects in a combination as well as in the dose contributions of the components which in combination produce an effect of the same strength (as the single components). Examples for underlying mechanisms of the models will be given.

Dose Additivity: Combined Effects Stronger than the Single Effects
This model is based on the assumption of the same effect mechanism for all combination partners. For chemical substances, it can be explained in terms of combination partners binding to the same molecular site, e.g. to a receptor, an enzyme or an ion channel. The molecules compete for this binding site. This mechanism of a biological interaction is therefore referred to as competitive interaction. Here we are interested in that type of competitive interaction, in which the partners show the same dose-response relationship, namely a maximum effect of identical strength and the same curve slope. Due to the identical "site of action", the partners must feature the same slope of the dose-response curve, since an identical site of action absolutely requires an identical effect mechanism. But as the two dose-response curves (of the two combination partners) can be steeper in some cases and less

steep in others, the effect of a dose-additive combination cannot be directly computed from the effects of the components as such, with the exception of a linear dose-response curve, in which case dose additivity and effect additivity coincide. This will be discussed again later. At this point we will address, because of their practical importance, different survival curves following stochastic processes manifesting themselves as mutations. Even with the same mechanism at work, different deviations from exponential curves can occur in the lower dose range, which can be explained by repair processes of differing efficiency. Although these differences are not to be interpreted as a sign of different mechanisms, they indicate differences between agents, if the damages done by them are repaired with different efficiency. In the higher dose range, such agents behave very similarly, notably in the corresponding, exponential sections of the curves.

In any case, the effect of a dose-additive combination can be derived by adding up equally effective (equieffective) doses. Let us assume, for example, that each of the combination partners A and B on its own causes an effect of 50% at a dose of 1. In the case of dose-additivity, the effect of A + B equals the effect of the dose 2 of A or B, respectively. With an S-shaped dose-response curve of a sigmoid curve slope = 1, this results in an effect of 67%. For steeper curves of the combination partners, one arrives at a combined effect > 67%. The larger the slope value of the curves, the stronger is the corresponding combined effect.

For the above example, the additivity model means, with equieffective doses of 1A and 1B:

Effect of 1A + 1B = Effect of 2A = Effect of 2B.

Of particular practical importance is the fact that dose-additive combined effects can be computed not only for half-maximum effective "single doses", but also for threshold doses or doses of a very slight effect. It should also be noted that this model – as well as the others – is not limited to combinations of only two components.

Dose Additivity: Equal Effect in Reduced Doses of the Single Components
Of equal practical importance and easier to grasp for the non-specialist is the calculation of the doses to be contributed by partners in combination in order to cause an effect of equal strength as the full doses of the partners as single agents. As an example, let us assume that the relevant effect level is achieved with a dose 1 of A or B. In the case of dose additivity, we can therefore expect that the effect of a single dose of A or B will be achieved with half the doses of A and B in combination, but also with any other dose fractions adding up to 1, for instance with 0.25 A plus 0.75 B.

Again, the mathematical relationship becomes clear from the example of equieffective doses 1A and 1B:

Effect of 0.5 A + 0.5 B = Effect of 1A = Effect of 1B

These are the concepts behind the well-known *isobologram representation*. If the combination partners have the same effects, observed combined effects must agree with dose additivity in an isobologram, provided that kinetic interaction does not take place. Conversely, combinations appearing dose additive can arise through other mechanisms, too, especially where only a narrow dose range is included in investigations of this type.

Fig. 2.2-9c shows an isobol representation derived from the dose-response curves in figs. 2.2-9a and 2.2-9b. Fig. 2.2-9a demonstrates that for linear dose-response curves, a dose-additive combination can be computed and displayed simply as the sum of the single effects. In the example shown in fig. 2.2-9 it is further assumed that the single effects of A and B are identical i.e. that the combination partners A and B are equieffective. This is generally not the case, although it is useful for illustrating the relationships.

Fig. 2.2-9a makes it perfectly clear that the 1:1-mixture of A and B is arrived at by forming the sum of the single effects and that for the combined effect shown, half doses of the combination partners are sufficient. The latter is particularly evident from the isobologram in fig. 2.2-9c.

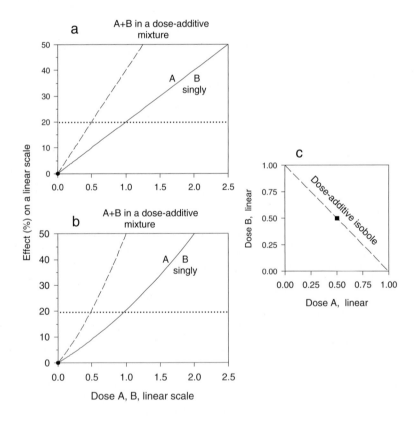

Fig. 2.2-9 a) Linear and b) non-linear dose-response curves for A and B as single agents and in a 1:1-mixture. The curves for A and B on their own coincide, since equieffectivity of A and B was assumed. c) Isobologram representation of the mixture effects in terms of the dose contributions resulting in a 20%-effect in a) and b).

From fig. 2.2-9b it becomes apparent that a dose-additive combined effect can be computed or determined from the graph, also for non-linear dose-response curves of the same type, albeit not by simply adding up the effects, but by first totalling the doses ($1 + 1 = 2$ in this example) and then finding the corresponding effect ($2A = 2B$).

Examples of Equal, Non-stochastic Effects

Obviously, effects of combination partners with the same site of action and therefore with the same mechanisms are rarely observed. They are rather the exception than the rule. There are, however, examples of practical importance, for instance cholinesterase-inhibitors like the phosphoric acid esters malathion and parathion, which bind to the same point at the enzyme cholinesterase. Other important examples are the "dioxins", which bind to the so-called Ah-receptor. Certain hormonally effective substances, too, feature the same molecular binding sites, as e.g. various estrogens bind to a certain receptor type, while androgens bind to another one.

A striking example of a competitive interaction is shown in fig. 2.2-10a, in the standard experimental set-up, namely sulphate inhibiting the enzyme activity of transketolase in the presence of different, fixed concentrations of phenyl phosphate. The results shown stem from work by Kremer et al. (1980), in which various, additional tests have shown that both enzyme inhibitors, sulphate and phenyl phosphate, most likely compete for the same binding site at the enzyme. Accordingly, the combined effects observed are not significantly different from the computed effects of a dose-additive, competitive effect. The isobologram in the insert in fig. 2.2-10a, too, shows a good agreement with dose additivity.

Combined effects resembling the model of dose additivity

There are numerous combined effects behaving nearly, but not quite as predicted by the dose-additivity model, and which, at the same time, can hardly be explained by the mechanisms discussed above. In individual cases, it is often difficult to decide on an explanation for the similarities between the combined effects and dose additivity. One possibility is that substances bind to different receptors before triggering the same sequence of reactions. Leff (1987) described a dedicated model for such combined effects, the two-receptor one-transducer model, the effects of which, in many cases, closely resemble the dose-additivity model, which can be regarded as a one-receptor one-transducer model.

We are justified in referring to "similar" effects of the combination partners here, because the partners trigger the same mechanisms. The question if the narcotic effects e.g. of solvents can be described by this model cannot be answered yet. However, the term "similar effects" is also used in a much more general way, in the sense that the effects are not necessarily based on the same mechanisms.

These considerations also touch upon the question of the importance of the additivity model especially for the calculation and estimation of combined effects which, in phenomenological terms, are roughly in accordance with this model.

Independent Effects

The independence model starts from the argument that the agents cause their effects via different sites of action *and* that these effects contribute to the combined effect

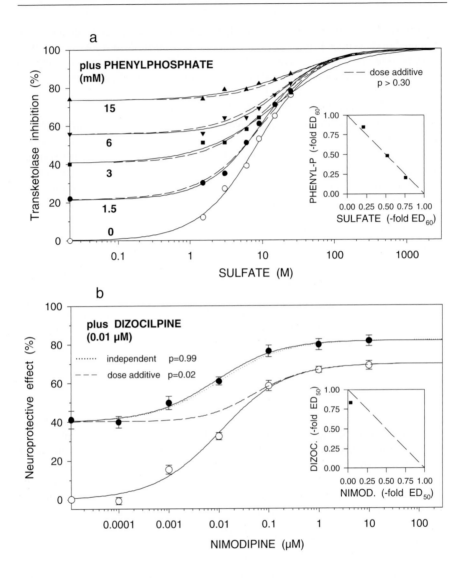

Fig. 2.2-10 Examples of combined effects a) according to the dose-additivity model, b) according to the independence model; a) from converted data by Kremer et al. (1980: table 2.2-1) for an enzyme inhibition, b) for a neuroprotective effect (from Krieglstein et al. 1996: fig. 3a). Insert in a): Experimental data points in the isobologram for a 60%-effect. Insert in b): Experimental data point in the isobologram for a 50%-effect. The broken, diagonal lines in the inserts mark a dose additive combination.

independently of each other. This model is not limited to chemical substances with different binding sites; it does not require any binding to molecular structures like receptors. Hence, its applications are virtually universal, particularly in assessing toxic effects of chemicals and physical factors. Like the other combination models, it describes a particular if very general type of synergistic combinations, where the combined effect can be directly computed from the single effects. We present here just two examples of the various formulae for calculating the combined effect (see Pöch 1993a, b).

For effects, E, which can be expressed as a *fraction of a possible maximum effect*, the independent total effect of A + B equals the effect of A plus the effect of B minus the effect of A multiplied by the effect of B:

$$E_{A+B} = E_A + E_B - (E_A * E_B)$$

$$E_{A+B} = 0.5 + 0.5 - (0.5 * 0.5) = 0.75$$

This result can also be interpreted in the way that A in the presence of B causes, in relative terms, the same effect as on its own, in this example 50% of the maximum effect. Naturally, this applies to B as well (also see chapter 1.2.3).

For changes in the control values, e.g. survival rates, the calculation is even more straightforward, as the fraction values under A are simply multiplied with those under B. The corresponding formula, where FC stands for the *Fraction of Control*, reads for the above example:

$$FC_{A+B} = FC_A * FC_B$$

$$FC_{A+B} = 0.5 * 0.5 = 0.25$$

Hence, we arrive at virtually the same result as with the formula given above, the sole difference being that the effects relate to the control values, in this case, and not to the maximum achievable effect. In this example, A or B on their own both reduce the control values 1 to 0.5, while both together reduce them to 0.25.

The "multiplication formula" can also be applied to (1-E), where E stands for the maximum possible effect. This enables us to calculate the independent effect of multiple combinations, too, in a very simple way, e.g.

$$(1\text{-}E)_{A+B+C} = (1\text{-}E)_A * (1\text{-}E)_B * (1\text{-}E)_C.$$

2.2.2.7
The Importance of the Independence Model

In the same-kind combined effects we are dealing with, the combination partners mostly cause the same kind of effect through different sites of action. Hence, one could feel tempted to assume that such effects in combination are generally described by the independence model. Still, this would mean overlooking the fact that this model describes the special condition of an independent interplay of effects. Chemical substances and/or physical factors having "independent" sites of action does not automatically imply that they combine independently of each other. Therefore, the question arises if combined effects in our environment follow this model at all, or if we are dealing with a mere theoretical artefact. We will go further into this question elsewhere. At this point, we confine ourselves to the observation that there are studies in existence, according to which the combined effects are well described by the independence model. One example is the combined effect of differently acting substances shown in fig. 2.2-10b, where the neuroprotective effect of the calcium antagonist nimodipin in the presence of the MNDA-antagonist dizocilpine in nerve cell cultures is virtually identical to the computed, independent combined effect, but in significant disagreement with the additivity model (Krieglstein et al. 1996). This discrepancy with the dose-additive effect, which is also noticeable in the isobologram (insert in fig. 2.2-10b), is explained by the combination partners using different sites of action. Combined effects often exceed the effects of an independent combination, *but not in the lower dose range*, a fact we will further explore at a later stage.

2.2.2.8
Independent Effects of Different Kinds

One has to distinguish between independent effects of the same kind, described above by the independence model, on the one hand, and independent effects of different kinds. In the latter, agents in „combination" cause an effect in the same way, to the same extent, or with the same incidence rate as if they were single agents, i.e. acting singly. Let us assume a certain dose of an agent A on its own causing skin irritation in 10% of all cases, while an agent B gives rise to liver damage in 10% of the cases. Since the effects of the two agents are of different kinds, we may expect, as independent effects of A + B, 10% skin irritation *and* 10% liver damage. If A never caused skin irritation and B never led to liver damage, we would expect from A and B in an independent combination, 0% skin irritations and 0% liver damages. Without any doubt, different-kind effects of possible combination partners are much more common than same-kind effects. Insofar as there is no reason to expect any interaction between agents with effects of different kinds, it is only logical to assume independent effects in such cases.

2.2.2.9
Combinations of Stochastic Effects

Stochastic effects of different doses can be pictured as independent of each other, since a certain dose always harms or kills the same fraction of cells still alive. Hence, it does not surprise at all that combined effects of agents acting stochastically, like radiation or certain chemicals, can be described by the independence model, too, at least in theory. Fig. 2.2-11a shows this behaviour for the standard experimental approach. In this plot, the dose-response lines of the agent A on its own and in the presence of a fixed dose of B are parallel, straight lines. The picture for mixtures, fig. 2.2-11b, does not show this parallelism.

We can further say that the theoretical combination effects in fig. 2.2-11 are in agreement with the dose-additivity model as well as with the independence model, because of the exponential dose-response function for stochastic effects like cell death. In the example in fig. 2.2-11a, for instance, we see the dose 1 of A reducing the control values from 100% to 10% i.e. to the same degree as the fixed dose of B. A dose-additive combination of 1 A in the presence of B therefore equals the effect of the dose 2 of A, which reduces the control value from 100% to 1%. As mentioned

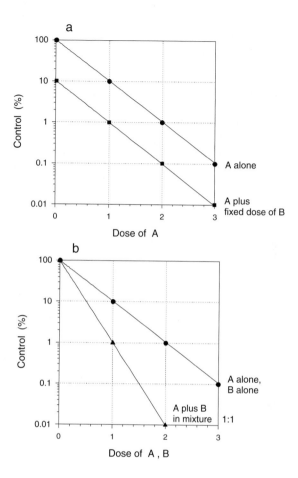

Fig. 2.2-11 Stochastic effect of A on its own and in the presence of B, plotted in analogy to fig. 2.2-1a. a) Computed combined effects for A in the presence of a fixed dose of B, b) in a fixed dose ratio. The combined effects follow the dose-additivity model as well as the independence model.

before, this combined effect also corresponds to the computed effect for an independent combination. The dose 1 of A causes a reduction of the control value down to 1/10. The same dose with B present leads to a decrease to 1/10 as well, that is from 10% to 1%, and hence shows the same effect as the dose 1 of A in the absence of B.

We already pointed out that fig. 2.2-11 also shows that the curve A in the presence of a fixed dose B is parallel to the single-agent curve, but shifted to lower control values, if the combined effect behaves according to the independence model and to the dose-additivity model as well. The same is true for the far more common "shouldered curve", where the curves for A and for A in the presence of a fixed dose of B have to be parallel too.

2.2.2.10
Combinations of Non-stochastic Effects: Dose Additivity and Independence Compared

In contrast to stochastic effects, non-stochastic effects do not yield a general agreement between the two mechanistic models. In the (extreme) low dose range, though, we notice a broad agreement for binary combinations, whereas differences are more likely for multiple combinations. We will look into two-partner combinations in the lower dose range and the differences between the model effects for multiple combinations at a later stage. Multiple combinations are of course relevant because exposure to more than two substances or physical factors is the rule rather than the exception in the environment.

A Point-by-Point Comparison between the Model Effects

The most convenient way of comparing the models of combined effects discussed so far is a "point-by-point" approach, as demonstrated in the following tables 2.2-1 –2.2-3, in which the effects of A and B, single and in combination, are listed. The combined effects are the computed effects for the three models.

Table 2.2-1 Computed combined effects (%) of A + B for single effects of A and B = 30% and for different slope values of the sigmoid curve for A

Slope of curve A	Single effect of A or B	Model effect of A + B		
		independent	effect additive	dose additive
1				46
2				63
3				77
4				87
5	30	51	60	93
6				96
7				98
8				99
9				100
10				100

Table 2.2-1 shows the situation at the example of two agents, A and B, each causing a single effect of 30%. The computed effect for an independent combination effect amounts to 51%; the sum of the effects is of course 60%. The effects of the dose-additivity model have been calculated for dose-response relationships represented by *sigmoid curves* of various slopes and naturally show a strong dependence on the slope of the curve. (It was assumed here that the slope of the curve for A is known and that B behaves like a fixed dose of A). We notice that the combined effects with dose additivity exceed the results for both other models, except for the curve with slope = 1. For stochastic dose-response relationships, on

the other hand, there is no difference between independent and dose-additive effects. Tables 2.2-2 and 2.2-3 show a comparison between the model combined effects for small single effects of A and B, 1% in table 2.2-2 and 0.1% in table 2.2-3, from which we can draw two important conclusions. Firstly, the differences between the independence model and the effect-additivity model disappear in this "lower dose range". Secondly, the combined effects of the dose-additivity model must not be neglected in this dose range either, especially for steeper curves. This observation is also well reflected in the dose-response curves of A in the presence "threshold doses" of B.

Table 2.2-2 Computed combined effects (%) of A + B as in table 2.2-1, but for single effects of both A and B = 1%

Slope of curve A	Single effect of A or B	Model effect of A + B		
		independent	effect additive	dose additive
1				2
2				4
3				8
4				14
5	1	2	2	26
6				40
7				57
8				71
9				84
10				91

Table 2.2-3 Computed combined effects (%) of A + B, as in table 2.2-1, but for single effects of both A and B = 0.1%

Slope of curve A	Single effect of A or B	Model effect of A + B		
		independent	effect additive	dose additive
1				0,2
2				0,4
3				1
4				2
5	0,1	0,1	0,1	3
6				6
7				10
8				20
9				34
10				51

Comparison of Dose-Response Curves:
A in the Presence of a Fixed Dose of B

Testing and representing a dose-response relationship of a substance A on its own and in the presence of one or several fixed doses of B is a standard procedure in pharmacology. This approach is also suitable for the analysis of toxic effects i.e. for testing on toxicokinetic and toxicodynamic interactions. For the latter, experimental effects can be compared with model effects, too, as pointed out earlier. We notice that the curves according to the models of dose additivity and independence behave differently to each other for different slopes of the agent A and for different effect levels of B. The ED_{50} of the combination curve for dose additivity compared to the ED_{50} for independent effects is of particular interest here. The latter is always equivalent to the ED_{50} of A.

The case that A alone already can cause the maximum possible effect yields the following exemplary results (Pöch and Pancheva 1995):

Slope of A	Effect of B (%)	ED_{50} of A: Dose additive
1	1	1.01
1	10	1.11
1	50	2.00
2	1	0.91
2	10	0.78
2	50	0.75
6	1	0.54
6	10	0.34
6	50	0.20

At higher slope values, B causes a left-shift of the dose-response curve of A, even for low effects of B. Such a shift is usually referred to as potentiation.

A and B in Mixture

The graphical representation of the single and mixed effects of A and B in a fixed dose ratio is extremely well suited for analyzing combination effects in the lower dose range. A and B require separate dose scales according to the dose ratio in their combination. (For a 1:1-mixture of equieffective doses of A and B, *one* dose scale for both is sufficient, as e.g. in figs. 2.2-9a and b). This representation allows the comparison not only between the experimental (observed) combined effects and the corresponding single effects of the combination partners, but also between the experimental (observed) effects in combination and the corresponding model effects. Beyond that, it makes possible a comparison between a threshold dose of the stronger component A on its own and in the presence of B.

Fig. 2.2-12 shows two theoretical examples for approximately equieffective components on their own and in combination. A is, with $ED_{50} = 1$, the agent with the stronger effect, B is slightly less effective, with $ED_{50} = 1.2$. The curves for A and B in figs. 2.2-12a and 2.2-12b clearly differ with regard to their slopes, $b = 1.5$ in fig. 2.2-12a and $b = 6$ in fig. 2.2-12b. The effect of the 1:1.2-mixture is represented by the computed dose-additive and independent combination curves.

B Increasing the Effect of A

Above a threshold, all combined effects are noticeably more marked than the corresponding single effects. In fig. 2.2-12a, the model combined effects in the lower dose range are only slightly stronger than the single effects. At a dose of 0.047 of A, for instance, we see an effect of 1%, at the dose of 0,039 of B an effect of 0.8%, and for a combined action of the same doses an effect of 1.8% for an independent combination and of 2.5% for dose additivity.

Fig. 2.2-12b shows a similar relationship between single effects of A and independently combined effects, although dose-additive effects are clearly stronger, due to the steeper slope of the curves for A and B. The "threshold doses" of A and B, which on their own cause effects of 1% and 0.4%, respectively, lead to a combined effect of 1.4% in an independent combination compared to 28% with dose additivity.

Hence it emerges that the effect of an independent combination in the region of "threshold doses" roughly equals the sum of the single effects – independent of the slope of the dose-response curves. The effects of dose-additive combinations, on the other hand, are markedly stronger. This difference is more pronounced for steeper curves (fig. 2.2-12b) than for curves with a lesser slope (fig. 2.2-12a).

B Shifting the Threshold Dose of A

A representation of mixture effects as in fig. 2.2-12 also holds information concerning the question if or to which degree a threshold dose of the stronger component A is shifted to lower doses by the component B. Since the described models of combined effects can be interpreted as models of a synergistic interplay, the model effects, too, have to be subject to a shift of threshold doses of A by B. The lower one assumes the threshold effect to be, the less distinct one expects the shift to be. For the dose-additivity model, the shift will be more pronounced for steeper curves of A.

Assuming a threshold of 1%, fig. 2.2-12 shows the following *threshold doses* of A on its own and in combination with B (fractions of ED_{50}):

Fig. 2.2-12a: A = 0.047; A + B: independent = 0.032; dose additive = 0.026
Fig. 2.2-12b: A = 0.47;   A + B: independent = 0.440; dose additive = 0.25

We should note at this point that evaluations of different experimental effects of agents acting in different ways did *not* produce any evidence of a lowering of the threshold dose of "A" by "B". If these results are confirmed by further investigations, this would establish something like a threshold dose for the model of independent combined effects.

2.2.2.11
Non-mechanistic Models

Apart from the models described so far, researchers can and do apply purely mathematical (input – output) models, too, for *describing combined effects*, for instance

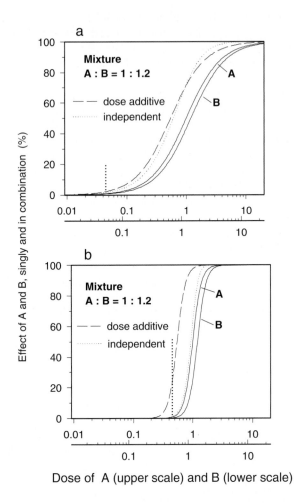

Fig. 2.2-12 Combined effects for approximately equieffective doses of A and B on their own (full lines) and in dose-additive and independent mixtures, as indicated. a) Slope of the dose-response curve b = 1.5, b) slope b = 6. The 1%-threshold doses are shown as vertical, dotted lines. The dose 1 corresponds to the ED_{50} of A and B on the upper or lower x-axis respectively.

the model of effect additivity. The comparison between observed effects with such model effects is suitable for a phenomenological description, but (generally) not for clarifying equal or different mechanisms of the combination partners.

Effect Additivity

This is an often used, easy to grasp model, as its combination effects can be computed simply by adding up the single effects. The model is not based on any pharmacodynamic/toxicodynamic interaction mechanism – with one important exception. The exception (heteroadditivity, in Steel and Peckham 1979) is explained by the addition of reductions in the survival rate, which is performed graphically on a logarithmic scale (see fig. 2.3-3). In mathematical terms, of course, this is multiplication, not addition. Independent reductions of control values can be calculated by multiplying the single values expressed in terms of their contribution to the control value (also see the preceding section, "Independent Effects").

Although effect additivity is an essentially non-mechanistic model, there are conditions where the sum of the single effects equal the effects computed according to the independence model. In the context of environmental standards, this is of particular interest, because the effects of the two models are virtually identical in the lower dose range and hence the lower effect range. The following summary table demonstrates this fact for percentage effects in combinations of equieffective agents.

Another point of practical importance is that one can always add the single effects, while an independent effect can be calculated only if the effects can be expressed as fractions (or percentages) of the maximum possible effect or of the

Single effects of A, B	Effect-additive combination	Independent combination
1	2	2
10	20	19
20	40	36
30	60	51
40	80	64
50	100	75

control values. Hence, the model of effect additivity can be applied where independent effects cannot be calculated for the reason just given. For very small effects, the model of effect additivity can literally "stand in" for the independence model. This also implies that, where observed combined effects exceed the sum of the single effects, they have to exceed independent effects too.

Applying the model of effect additivity leads to impossible and therefore meaningless results when the sums exceed the absolute maximum of 100%. Any comparison between the observed values and the sum of the single effects should hence be restricted to those effect regions, where effect additivity roughly equals independent joint action i.e. where the single effects do not exceed about 10% (see the table above). In comparisons of stronger effects with the sum of the single effects, a considerable amplification can still pose as sub-additive and be interpreted as antagonism (as against the model effects).

2.2.2.12
Dose Additivity for Agents Behaving Differently

The mechanistic model of dose additivity assumes that two agents, A and B, behave in equal ways. For substances with the same molecular site of action, this same behaviour can now be explained in terms of these substances' competition for a common binding site. A bound molecule of A triggers the same sequence of effects as a bound molecule of B. Consequently, the dose-response curves of such substances have the same slope. Such combinations are, without any doubt, the exception, but not the rule. Substances with dose-response curves of differing slopes cannot induce the same reaction chain through a common "receptor". They behave differently with regard to their effect mechanisms, even if their behaviour appears similar from a phenomenological point of view. Nevertheless, even in these cases one can calculate "dose additivity", for instance: Effect of $1A + 1B$ = Effect of $(2A + 2B)/2$, if the doses $1A$ and $1B$ are equieffective. An agreement between the combined effects of agents acting in different ways and this generalized additivity model must therefore be interpreted in phenomenological terms only, since it is, for all practical purposes, a non-mechanistic model.

It is, however, conceivable that this model mathematically describes a mechanism similar to the "two-receptor one-transducer" model, i.e. a similar action. The "stringent", mechanistically justifiable additivity model, on the other hand, describes a situation where the combination partners have the same action, making it a model of the "one-receptor one-transducer" type. With regard to similar, but not identical effects, there is not much more than speculation at present. The central, depressant effect of alcohol, on the one hand, and of benzodiazepines, on the other hand, may perhaps be regarded as an example of a "similar" effect. As it is shown later, the dose-response curves for these substances show different slopes, whereas the combined effect of their mixture is approximately dose additive.

2.2.2.13
Stochastic and Non-stochastic Effects in Combination

Combined effects of agents acting stochastically and non-stochastically, like combinations of stochastic radiation effects and chemicals having a non-stochastic effect, are of great practical importance for environmental standards. Naturally, such combined effects can also occur between chemical substances only. In their analysis, we must keep in mind that stochastic and non-stochastic effects can only be based on different mechanisms.

For the assessment of synergistic combination effects of this type, the models of isoadditivity and heteroadditivity are often called upon. These models are comparable, to a large degree, to the model of dose additivity and the independence model respectively.

Isoadditivity and Heteroadditivity
Since the publication of work by Steel (Steel and Peckham 1979), the models of isoadditivity and heteroadditivity are often used for the isobologram representation of stochastic and non-stochastic agents in combination (see section 2.3.3.1).

2.2.2.14
Non-stochastic Multiple Combinations

The model effects described here are usually calculated or explained for binary combinations i.e. combinations of two partners, although they can also be applied to multiple combinations with more than two components. Table 2.2-4 shows the effects of 10 components. Again, not only can we compute the effects for the different models; comparisons between the model effects are possible too. For multiple combinations, the discrepancies between the effects calculated according to the different models are, of course, more pronounced than for binary combinations. The decision, which model to use for predicting effects of certain combinations, is of even greater importance here.

Table 2.2-4 Computed combination effects (%) of 10 components, each with a single effect of 0.1%, acting in the same direction

Slope of curve A	Single effect of the components	Model effect of the 10 components		
		independent	effect additive	dose additive
1				1
2				10
3				51
4				91
5	0,1	1	1	99
6				100
7				100
8				100
9				100
10				100

Fig. 2.2-13 illustrates an effort to represent theoretical combined effects of a mixture of 16 components as dose-response curves with "medium" slope values of 2 to 4. The assumptions were that the 16 components are present in approximately equieffective doses (fig. 2.2-13a) or in non-equieffective doses (fig. 2.2-13b). In reality, the first case will rarely occur and therefore reflects a rather pessimistic scenario; the second case is much more likely and realistic. Accordingly, the dose-additive combination appears to have a much more dramatic effect in fig. 2.2-13a than in fig. 2.2-13b, where it is still a lot stronger than the effect of the strongest substance in the mixture.

Another interesting point is the independent effect of this multiple combination, which, for approximately equieffective components, is markedly stronger than the single strongest component (fig. 2.2-13a), even if it is still vastly weaker than that of a dose-additive combination. Under the more realistic assumption of non-equieffective components, the computed independent combination effect in the lower dose range hardly exceeds the effect of the strongest component (fig. 2.2-13b).

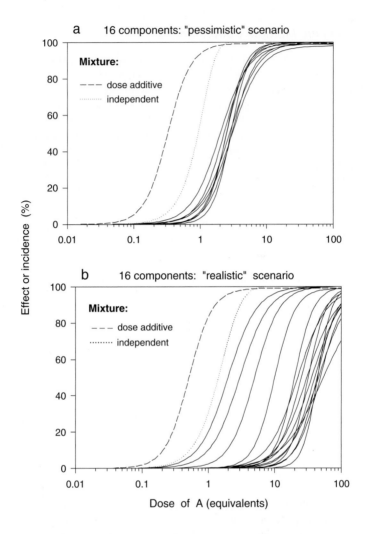

Fig. 2.2-13 Theoretical dose-response curves for 16 agents on their own and in a mixture, computed for dose additivity and independence. The 16 components are assumed to be (a) approximately equieffective or (b) non-equieffective, respectively. Each of the solid lines in (a) represents two equal curves.

In experiments with multiple mixtures of 14 to 16 single substances acting *differently* with toxic effects on algae, Backhaus et al. (1998) showed recently that computations based on the independence model provide a very good estimate for the effect of the mixture. This finding is of particular interest, not least because these experimental investigations also included the lower dose range.

Studies on multiple combinations acting *"similarly"*, on the other hand, produces the appearance of dose additivity in the combined effect (Scholze et al.

1998). 18 substances (s-triazines) specifically binding to the D1-protein of the photosystem II of the algae – and hence inhibiting the electron transport in photosynthesis in the same, specific way – were classed as acting "similarly" in those studies.

2.2.2.15
Stochastic Multiple Combinations

As already discussed and demonstrated elsewhere (fig. 2.2-11), combinations of stochastic effects are characterized by the effects of the independence model coinciding with dose-additive effects. The independence model can therefore be used for computing the overall effects of a stochastic multiple combination, or the dose-additivity model can be applied in calculating the dose contributions summing up to the same effect.

In the lower dose range, we can also equate the effects of the independence model with the effect-additivity model and thereby reduce the computation of stochastic total effects to adding up the single effects.

2.2.2.16
Stochastic and Non-stochastic Multiple Combinations

The effects of agents acting stochastically with chemicals acting non-stochastically appear to be of particular interest with regard to environmental standards, especially if one considers an increased risk of cancer. Understandably, combined effects of this type cannot be calculated as easily as the effects of agents behaving in a purely stochastic manner, although the mechanisms of the combination partners offer a clue to such combined effects. Assuming that all components promote the development of cancer via different mechanisms, without any special interaction taking place, this risk can be estimated through the independence model.

Where special interactions do occur, as e.g. between agents initiating cancer development and inhibitors in DNA repair processes, we must expect potentiation effects, meaning a higher risk than for an independent combination. The precise extent of such potentiation effects can only be approximated. Experimental studies with two-component mixtures show an amplification effect equivalent to the dose of the initiating agent increased by a factor of 2 to 4.

The effects of multiple exposures to stochastic agents combined with chemicals acting non-stochastically are therefore difficult to assess, which is, without any doubt, a very unsatisfactory situation. We desperately need targeted studies in this direction in order to obtain empirical estimations, where the emphasis should be on the mechanisms of the combination partners. One possible strategy could be to calculate the risk posed by multiple combinations by summing up the single effects, as in the case of agents acting in a "purely" stochastic manner, and to apply a safety factor accounting for possible potentiation effects.

In the following section, empirical effects of substances behaving non-stochastically, especially medicinal drugs, will be discussed, comparing combined effects with single effects as well as with model effects. Toxic effects will be treated on a broad basis and illustrated by examples in further chapters (see chapters 2.3, 2.4 and 2.5).

2.2.3
Examples of Empirical Combined Effects

Combined effects have always raised interest, and the literature on this subject reports numerous experimental studies. Regrettably, however, many of these papers are of little general interest or are accessible to specialists only. Some publications do not even show if the effects are stronger or weaker in combination than as single effects, because the combined effects were compared with models or model parameter values, but not with single effects.

When studying publications on combined effects with regard to environmental standards, one has also to bear in mind that the incidence rate and the extent of reported, synergistic combination effects most likely leads to an overestimation of the problem with regard to environmental agents. The reasons for this are twofold. Firstly, the majority of publications in all sciences deal with "positive" or "exceptional" results. Secondly, investigations on this subject are performed almost exclusively with well detectable i.e medium or higher doses of agents.

In the field of drug-drug interactions, the situation is slightly better, since negative findings, i.e. non-existing interactions are of equal practical importance for the safety of a therapy as positive ones. An interaction may be defined as an interplay of agents, in which the effects of one agent are changed by another one. In the literature, the term interaction is often used for undesirable cross-effects, while desirable interactions are referred to as drug combinations.

Obviously, one cannot identically apply interactions or combined effects of medicinal drugs to the conditions (doses) of toxic agents in the environment. What is definitely called for here is an integrated approach which also takes into account mechanisms of environmental agents/pollutants.

2.2.3.1
Interactions between Medicinal Drugs: Mechanisms, Extent and Incidence

Undesirable Interactions

Everybody is familiar with undesirable interactions in therapeutic applications, but the mechanisms, the extent and the incidence of such interactions are less well known. A recent study (Pöch 1995) arrived at the following result: Of the reported undesirable interactions, 60% are kinetic interactions and about 33% dynamic ones. The remainder is explained by kinetic-dynamic hybrids or by unknown mechanisms. The pharmacokinetic interactions, which represent the majority of cases (see section 2.4.1.3), are mostly caused by a changed metabolism of the drugs in the liver. Pharmacodynamic interactions manifest themselves mainly as undesirable amplification effects. Desirable effects of this kind are usually referred to as drug combinations, meaning desirable interactions, which will be discussed later.

The incidence rate of undesirable interactions is difficult to quantify, not least because undesired effects can also be brought about by side effects of a medicinal drug administered on its own. Unwanted interactions produce increased side effects of the drugs, if they do not impress by therapeutic effects being too strong or too feeble. Data will vary depending on the criteria, by which unwanted interactions are defined, and on the methods applied for recording cases. Estimates range from about 0.1% to 20% of all cases, where more than one drug was taken at the same time, increasing near exponentially with the number of drugs taken.

For some years now, undesirable interactions have also been assessed with regard to their seriousness (see e.g. Hempel and Zagermann 1993). Of the 278 interactions reported, 75 (27%) are classed as minor, 148 (53%) as of medium severity, and 55 (20%) as clinically grave. Degrees of seriousness of these unwanted interactions can be expressed or estimated in terms of dose factors. Generally, a dose factor of 2 to 4 can be assumed for interactions of any clinical relevance (Pöch 1989).

Desirable Interactions: Medicinal Drug Combinations

Desirable interactions, too, are relevant in the context of environmental standards, since an assessment in terms of "desirable" and "undesirable" reflects a wish rather than some biologically different mechanisms. Any effect (e.g. a decrease in the blood pressure) may be welcome at one time and unwanted at another time. Furthermore, pharmacokinetic and pharmacodynamic interactions can have essentially the same causes and mechanisms.

Combined effects of medicinal drugs can be found in countless publications, the majority of which describe effects of combined single doses, partly compared with the effects of single doses of the components (see Pöch 1993b). A comparison between combination effects with the corresponding effect of the stronger component shows that in clinical application, combined effects exceed the corresponding effects of the single drugs (see Pöch 1993b, table 3-5). According to the definition of synergism and antagonism used in this paper, these are, without exception, synergistic drug combinations. However, we must not conclude from this finding that there could never be an antagonism in the combined effect of drugs with additive effects, e.g. because of a particular pharmacokinetic interaction, although such combinations should be the proverbial exception that proves the rule.

Since the combined effects described above can be regarded as synergistic in most cases, an additional comparison between the observed combined effect and the computed effects of independent action appears to be of fundamental interest. The observed clinical combination effects were less homogeneous in comparison with the independent effects than with regard to the effects of the stronger component on its own. The majority of combinations may have shown stronger or equally strong effects as predicted by the independence model, but markedly lesser effects occurred too, even for equieffective single effects. This result showed clearly that not all synergistic amplification effects necessarily equal or exceed independent effects.

2.2.3.2
Dose-Response Analyses

In this section, we describe experimental and some clinical studies on combined medicinal drug effects, based on the analysis of dose-response curves. We further discuss results with teratogens, with special emphasis on the lower dose range.

Drugs with Effects on Smooth Muscle

In recent years, numerous investigations concerning the question of dose-response analysis of combined effects were carried out – especially in the framework of dissertations and theses at the university of Graz (Austria) – with pharmaceuticals and isolated muscle strips. We cannot go into this work in detail, but we will try to summarize the results, as far as they are relevant for the issue of effect amplification. The studies were mostly performed with drugs acting similarly with regard to smooth-muscle relaxation or contraction, taking the following approach.

1. Dose-response curves were determined for a given substance, A, on its own and with one or several doses of a second substance, B, in presence.
2. Dose-response curves were determined both for A and for B each on its own and in a certain mixture i.e. a fixed dose ratio.

Countless experiments with various substances always showed the same picture for substances causing the same effect by different mechanisms. The combined effects exceeded the single effects in virtually every case, above all in the medium and higher dose range, as shown by the examples in fig. 2.2-14. Furthermore, the result in fig. 2.2-14 is interesting with regard to a comparison with independent model effects, for the vast majority of the said experiments with smooth-muscle relaxing drugs showed that combined effects in the medium dose range equalled or just exceeded independent effects. Fig. 2.2-14 shows that the relaxing effect of the active metabolite of molsidomin, SIN-1, in the presence of the selected three fixed concentrations of nicorandil can be adequately explained through an independent combination of SIN-1 and nicorandil (Holzmann et al. 1992).

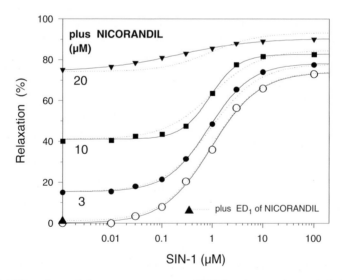

Fig. 2.2-14 Experimental dose-response curves of SIN-1 on its own (open circles) and in the presence of fixed concentrations of 3, 10 and 20 µM nicorandil at isolated strips of bovine coronary artery (full symbols). The dotted curves represent the computed independent effects (from Holzmann et al. 1992), also calculated for the ED1 of nicorandil.

These results justify the assumption that SIN-1 in lower concentrations of nicorandil, for instance the ED_1, will also be in agreement with an independent effect. The computed independent effect is shown in fig. 2.2-14, too. It is all but identical to the effect of SIN-1 as a single agent.

Hence, we could come to the generalized conclusion that this relationship between the majority of the experimental combined effects and the theoretical independent effect should also hold in the regime of extremely low doses. However, as the effect of a substance A is virtually identical to the independent effect of A in the presence of B, this means that there is no amplification of effects in the lower dose range, where substance B itself only shows very little effect or no effect at all.

In order to illustrate this finding, let us assume, for example, that the fixed dose of B on its own causes a 0.1%-effect. The independent effect of A with B present is virtually identical to the sum of the single effects. Then the combined effects of A in the presence of B exceed the effects of A on its own by 0.1%.

Chemotherapeutics for Tumour Treatment

In recent times, many dose-response analyses were performed on data from in-vitro tests with antitumoural substances on their own and in combination, including the cytostatics cyclophosphamide, 5-fluorouracil, paclitaxel, and cisplatin. The mixture effects were partly in accordance with an independent effect, being slightly stronger or just weaker than predicted by the model (Vychodil-Kahr 1997).

Tests with "individual" tumour cells from patients brought a variety of results, while the mean values formed from six test series showed an unequivocal trend, as shown in fig. 2.2-15. On average, a stronger effect of A + B compared with A was found in the medium and higher dose range, where the combination of their effects somewhat exceed the computed effects of an independent combination as well (fig. 2.2-15a). In the lowest dose range, on the other hand, all curves appear to meet in one point, while B does not seem to cause any lowering of a threshold dose of A (fig. 2.2-15b).

For antiviral combinations, too, effect amplifications can be found in the literature, some of which can be represented as dose-response curves (see e.g. Pancheva 1991). In some cases, the combined effects in the medium and lower dose ranges significantly exceed independent effects (Pöch and Pancheva 1995, Pöch et al. 1995). Fig. 2.2-16 demonstrates this finding for the combination of acyclovir and BVDU. We immediately recognize that the effect of acyclovir can be markedly enhanced by fixed doses of BVDU (fig. 2.2-16a) and vice versa (fig. 2.2-16b). Only when these components have an effect of less than 0.1%, no effect amplification is detectable.

Apart from the independent effects of antiviral substances (see e.g. Bryson and Kronenberg 1977), plotted as dotted curves in fig. 2.2-16, and from stronger effects, there is also the case of the combined effect of phosphonoacetate (PAA) and phosphonoformate (PFA), which act in a largely dose-additive fashion (Johnson and Attanasio 1987, Pöch and Pancheva 1995: fig. 4). These substances compete for the same binding site at the virus-induced DNA-polymerase (see Johnson and Attanasio 1987).

Sedative and Anesthetizing Medicines

Some interesting clinical studies on centrally sedating and sleep inducing drugs have produced dose-response curves e.g. for midazolam and propofol (Short and

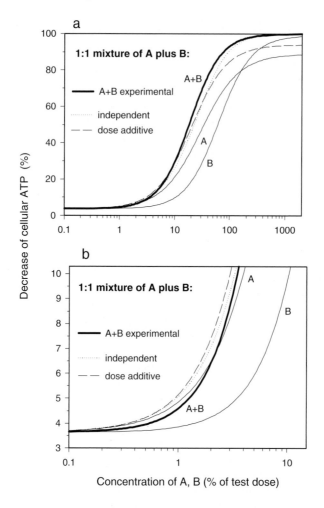

Fig. 2.2-15 Experimental dose-response curves of tumour-inhibiting substances on their own (A, B) and in combination (mixture A + B), compared with the model curves for dose-additive and independent combinations. The experimental curves represent the mean values derived from six test series. From investigations by Vychodil-Kahr (1997), a) over the full dose-response range, b) in the lowest dose range.

Chui 1991). The mixture of both substances is dramatically more effective than the equieffective components. At a 1%-effect of propofol, the effect of midazolam increases from about 10% to nearly 100% (see Pöch 1993b).

Similar results were found with regard to anesthetic effects of midazolam with thiopental (Tverskoy et al. 1988), methohexitone (Tverskoy et al. 1989), and fentanyl (Ben-Shlomo et al. 1990). 10%-effective doses of midazolam are amplified to ca. 70-80% by virtually ineffective doses (effect of the order of 1%) of the respective combination partner.

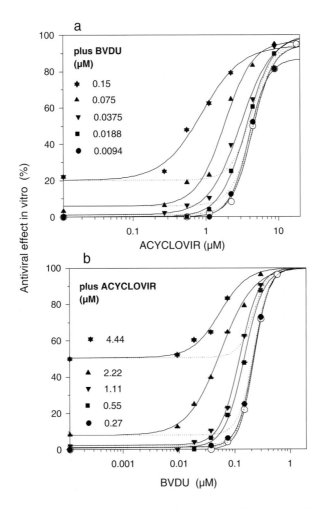

Fig. 2.2-16 Experimental dose-response curves for antiviral substances (Pöch and Pancheva 1995), a) for acyclovir on its own (○) and in the presence of fixed concentrations of BVDU (full symbols), b) for BVDU on its own and in the presence of fixed concentrations of acyclovir, as indicated. The dotted curves represent independent effects in combination.

Sedatives and Alcohol

Similar, but even more conspicuous are amplification effects between centrally sedating substances and alcohol, as fig. 2.2-17 demonstrates with dose-response curves for mild anesthetic effects (Pöch et al. 1996). Doses of flurazepam, which on their own cause an effect of only a few percent, already yield a maximum effect when practically ineffective doses of alcohol are present (fig. 2.2-17a). A similar amplification was observed for the effects of alcohol in combination with flurazepam (fig. 2.2-17b).

Notwithstanding the many similarities in the effects of the two agents, ethanol does not behave like flurazepam. This we can conclude from the deviation of the experimental curves for flurazepam in the presence of 2,500 mg/kg ethanol from the corresponding dose-additive curve (fig. 2.2-17a). Neither is the experimental effect of ethanol in the presence of 150 mg/kg flurazepam in agreement with the computed dose-additive effect (fig. 2.2-17b).

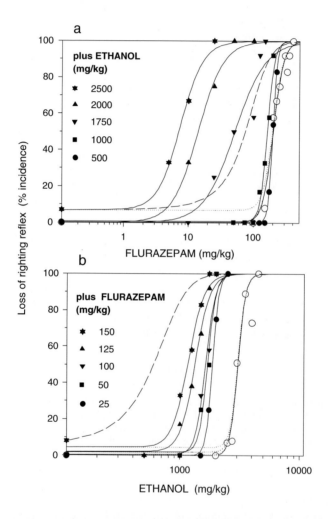

Fig. 2.2-17 Experimental dose-response curves for a light narcosis with (a) flurazepam on its own (O) and in the presence of fixed doses of ethanol (full symbols), (b) with ethanol on its own (O) and in the presence of fixed doses of flurazepam (full symbols), from investigations by Pöch et al. (1996). Computed curves for dose additivity (broken line) (a) in the presence of 2500 mg/kg ethanol, (b) in the presence of 150 mg/kg flurazepam.

2.2.3.3
Effect Amplification in the Lower Dose Range?

Naturally, combined effects in the lower dose range are of special importance for issues of environmental standards. Combined effects in the lowest dose range can be derived, with reservations, from dose-response curves, primarily from mixture curves for A + B compared to the curves for A and B as single agents. The following analysis looks at experiments with chemotherapeutics, alcohol, and teratogens.

Chemotherapeutics for Tumour Treatment

Fig. 2.2-15 shows that the mixture of the two chemotherapeutics, A + B, is by no means stronger in its effect in the lower dose range than the components of the mixture as single agents. The averaged curve of the mixture, A + B, even crosses the curve of the stronger component, A; in the lower dose range, it runs below the curves for dose additivity and independence (fig. 2.2-15b).

Sedatives and Alcohol

Fig. 2.2-18 shows the combination curve for flurazepam/alcohol mixtures of 1:10 and 1:20 in comparison with the curves for the components. At a mixture ratio of 1:10, only flurazepam has an anesthetising effect in the lower dose range, while alcohol appears to have no effect. Therefore, it is no surprise that the mixture of the two substances is not more effective than flurazepam on its own in the lower dose range.

However, there might be a surprising aspect to this result, insofar as the effect of flurazepam in the presence of a fixed, yet ineffective dose of 1,750 mg/kg of alcohol was already markedly intensified (fig. 2.2-17a). Lower doses than that, on the other hand, did not give rise to a combined effect stronger than the effect of flurazepam itself, with a flurazepam dose of about 100 mg/kg i.e. just above the effectivity threshold (fig. 2.2-17a). This again explains why the combination of 100 mg/kg flurazepam and 1,000 mg/kg alcohol did not show an effect any stronger than 100 mg/kg flurazepam on its own (fig. 2.2-18a). In this lower dose range, we are dealing with a combination of, as it were, one effective and one "wholly ineffective" component.

Even for a dose-additive combination of a mixture ratio of 1:10, any dramatic effect in the lower dose range would have been a surprise (fig. 2.2-18a). Remarkably, however, a maximum combined anesthetic-narcotic effect was observed with a mixture of 1:20 (100 mg/kg flurazepam plus 2,000 mg/kg ethanol), which is equally (in)effective in the lower dose range (fig. 2.2-18b).

This result already showed up in a closer analysis of the curves in fig. 2.2-17, namely for flurazepam plus 2,000 mg/kg ethanol (fig. 2.2-17a) and for ethanol plus 100 mg/kg flurazepam (fig. 2.2-17b). Hence, the combined effect at about 100 mg/kg flurazepam plus 2,000 mg/kg ethanol in fig. 2.2-17 seems to support the results in fig. 2.2-18b.

A similar result, if not quite as spectacular as the last, was observed for sleep induction with the two medicinal drugs midazolam and propofol in a 1:7.1-mixture. At doses of 0.1 mg/kg midazolam and 0.71 mg/kg propofol, the combined effect was 100% opposed to 10–20% after administration of midazolam or propofol each

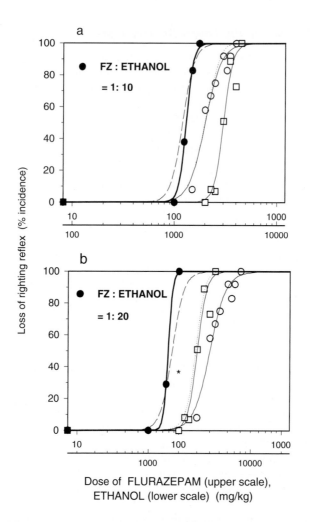

Dose of FLURAZEPAM (upper scale),
ETHANOL (lower scale) (mg/kg)

Fig. 2.2-18 Results of the same experiments as in fig. 2.2-17, but with the combined effect of the mixture 1:10 (a) and 1:20 (b) of flurazepam (O) and alcohol (□), from Pöch et al. (1996). The combined effect marked with a star, at 1.500 mg/kg ethanol, was interpolated to 75 mg/kg from the effects at 50 and 100 mg/kg flurazepam. Theoretical combined effects: Dose additive (broken line) and independent (dotted line).

on its own. At threshold doses with effects of less than 1%, the combined effect was still 60–70% (see Pöch 1993b: fig. 2.2-6).

2.2.3.4
Reduction of a Threshold Dose of A in the Presence of B?

The important issue, in the context of environmental standards, of effect amplification in the lower dose range leads to the question if a threshold dose of an agent A –

where such a threshold is manifest – is shifted towards lower doses by another agent. Because of the "uncertainties" in the lower dose range, we will call upon two examples including dose-response curves gained from the average of a larger number of experiments.

Example 1: Chemotherapeutics for Tumour Treatment

Going back to the result shown in fig. 2.2-15, particularly with regard to the possible lowering of a threshold dose of A through the influence of B, we find that an assumed threshold for A, which is 0.1% above the original dose, is not at all lowered, but even (insignificantly) raised by a factor of 3.66. The obvious absence of any lowering of the threshold dose has its explanation in the significantly steeper slope of the curve for A + B as compared to the curve for A. The slope of the combination curve for A + B is 1.56, the slope value for A is just 1.25.

Example 2: Teratogens

In experiments with frog embryos it could be shown that the combined exposure to teratogenic substances (agents causing malformations) acting in different ways, in an approximately equieffective dose ratio, leads to effects which on average exceed the independent effects – but not in the lower dose range. Combinations with only one effective component further showed an effect not exceeding that of the stronger component on its own (Pöch and Dawson 1996).

Results with roughly equieffective mixture components appeared to be particularly interesting with regard to environmental standards and were therefore re-evaluated. The sigmoidal dose-response curves of the individual substances (A and B) were constructed with identical minima for a series of experiments, since we can assume that the spontaneous rate of malformation in a series are the same for A and B each on its own and in combination. The spontaneous rate of malformations corresponds to the minimum of the curves.

The results of this study with 15 pairs of teratogens acting differently (Pöch and Dawson 1988) are illustrated in fig. 2.2-19. The approximate equieffectivity of the mixture components is easily noticeable (fig. 2.2-19a). The ED_{50} for A or B, respectively, was defined as 1 (in terms of multiples of the ED_{50}). B, with $ED_{50} = 1.1$, was marginally less effective. For almost the entire dose range, the combination of A and B, at a mixture ratio of 1:0.909, showed stronger effects than predicted by the independence model. Accordingly, the dose-response curve for the mixture showed, with b = 6.54, a significantly ($p < 0.02$) steeper slope than A (b = 5.02).

In the controls, the minimum incidence came to 5.06%. The arbitrarily assumed threshold at 0.1% above the minimum, i.e. at 5.16%, corresponded to a concentration of 0.255 for A on its own and to 0.259 for A with B present, meaning that the threshold concentration of A was not lowered by B. The actual rise of the threshold dose was insignificant too (p = 0.07). In fig. 2.2-19b, these threshold doses are represented by an open (A on its own) and by a closed circle (A in the presence of B), respectively, which practically coincide in the plot.

The results of this study are remarkable, especially because they are based on a large number of embryos (ca. 450 per curve in each series). The malformations

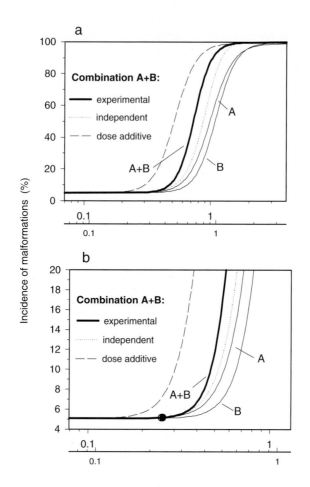

Concentration of A (upper scale) and B (lower scale)

Fig. 2.2-19 Dose-response curves for teratogenic effects of 15 pairs of substances, A and B, each with a different site of action, single and in combination. The curves were constructed on the basis of the averaged parameter values of sigmoid curves, a) for the entire effect range, b) for (malformation) incidence rates between 4% and 20%. O marks the threshold dose for A on its own, □ A with B present.

were mostly osteolathyrism and microcephaly. Combinations of components causing the same malformation resulted in the same picture that emerged from the study as a whole (Pöch et al. 1997).

B does not lower the threshold for A, because the mixture curve for A + B is steeper than the curve for A. This appears to be true for dose-response curves of agents with slope values not much above 1 (chemotherapeutics for tumours) as well as for agents with slope values much higher than 1 (teratogens), e.g. for the combined effect of flurazepam and alcohol discussed earlier (fig. 2.2-18).

2.2.3.5
Slope Values of Dose-Response Curves and their Impact on Amplification Effects in Combinations

Beyond the points already discussed, this and other studies to be found in the literature (see Pöch 1993a) show that amplification effects occur much more frequently in combinations of components with similar effects and with very steep dose-response curves. A steeper slope of the curves is, as such, obviously not the reason for amplification effects, although it seems to be something like a factor in favour of amplification and hence a risk factor.

For toxicokinetic interactions, this is easily explained. A slight increase in the concentration of a substance in the blood through a kinetic interaction causes a disproportionately larger increase in the effect than for substances with dose-response curves of lesser slope. This connection becomes clear immediately, when one considers the following situation: A on its own causes an effect of 1%, and B increases the concentration of A in the blood by, say, a factor of 2. This results in an effect of 2% for a slope of $b = 1$, 8% for $b = 3$, but 91% for $b = 10$.

Somewhat more difficult to explain is a slope-dependent, *toxicodynamic* amplification effect, except for dose-additive combinations. However, one can imagine that dynamic amplification effects may not arise from changed concentrations of substances in the blood, but are determined by mechanisms behaving *as if* there had been a change in those concentrations or an increase in doses.

2.2.3.6
Conclusions

The studies discussed here – studies on single doses in combination, on mixture curves and on effects of A in the presence of a fixed dose of B – produced results relating to substances acting in the same direction, through pharmacodynamic or toxicodynamic interaction. For a comparison with model effects, the "standard model" (see fig. 2.2-14, but also fig. 2.2-10) proved to be very useful.

For the comparison of observed combination effects with same-kind effects of the individual components, the "mixture model" (see fig. 2.2-12) seems to be more suitable than any other experimental approach or model, including the standard model (see fig. 2.2-15, -18, -19). The mixture model allows not only the direct comparison between the effects of the agents, but also a comparison with theoretical combined effects. And, last not least, it offers a good representation of the effect in the lower dose range, too.

From the given examples with same-kind effects, in connection with the theoretical foundations, we can draw the following conclusions:

1. Combined effects in the medium and higher dose ranges are often stronger than the effects of the single strongest component (A). In the lower dose range, we generally find *no* amplification effects.
2. For antitumoural substances and teratogens with different molecular sites of action, the threshold doses of a substance A are *not* lowered in the presence of B.
3. Agents acting in different ways usually show combination effects in agreement with the independence model. In the medium dose range, the combined effects

may be stronger than the model predicts, but not in the lower dose range. Hence even if such combinations, especially equieffective ones, produce effects stronger than the independent model effects in the medium dose range, it is still to be expected that the combined effects in the lower dose range are in agreement with the computed independent effects or with the effects of the dominant component alone (see the mixture of antitumoural substances and teratogens).

4. Substances with the same site of action show dose-additive combined effects. These cases are very rare. For such combinations, there is no reason to expect effects stronger than predicted by the dose-additivity model, neither in the medium nor in the lower dose range. This might also apply to substances acting "similarly" with different sites of action, but the same reaction mechanism (see combinations with alcohol).

5. For dose-additive combinations, the slope of the curves determines the extent of the amplification effect compared to the single effects. Equieffective combinations further show stronger effects than non-equieffective mixtures. This follows from the model of dose additivity.

6. Of all combinations discussed here, the combined effect of alcohol and flurazepam is the most spectacular, even or especially in the lower dose range. Similar amplification effects were reported for other combinations of centrally sedating, narcotic substances with alcohol.

7. *Judging by the results available so far, the present standards for single substances do not have to be reduced for such substances in combination, if the substances cause, through different mechanisms, a certain effect of the same kind and if no toxicokinetic or special, toxicodynamic interactions have to be considered.*

The few results with agents that act in the same way, the combined effects of which can be explained and described mechanistically as a dose-additive combination, show stronger theoretical effects in the lower dose range and a lowering of the threshold dose of A in the presence of B.

In equieffective mixtures, the threshold dose of A is reduced by half through the presence of B. For equieffective multiple combinations, one has to expect a lowering of the threshold by a factor dependent on the number of components. In such cases, a correction to the threshold doses for the single agents must be considered, if these agents are effective simultaneously. This correction relates to equally effective agents. For non-equieffective substances, the shift of the threshold dose of A through the presence of B is smaller the smaller the effect of B in comparison to A is. Such dose-additive combinations will usually contain chemically similar substances that, due to the characteristics of their chemical structure, compete for a common receptor or a common binding site, for instance cholinesterase inhibitors or dioxins.

What is true for different mechanisms of same-kind effects, probably also holds for different effects i.e. effects of different kinds. However, this subchapter cannot contribute any more recent experimental observations concerning this point (also see section 2.2.2.8).

2.2.4
Literature

Backhaus T, Altenburger R, Boedeker W, et al. (1998) Predictability of the aquatic toxicity of multiple mixtures of dissimilarly acting chemicals. SETAC-Europe 8th Annual Meeting 14–18 April 1998, Bordeaux France, Abstract 4A/015

Ben-Shlomo I, Abdel-El-Khalim H, Ezry J et al. (1990) Midazolam acts synergistically with fentanyl for induction of anaesthesia. Br J Anaesth 64: 45–47

Bryson YJ, Kronenberg LH (1977) Combined antiviral effects of interferon, adenine arabinoside, hypoxanthine arabinoside, and adenine arabinoside-5´-monophosphate in human fibroblast cultures. Antimicrob Agents Chemother 11: 299–306

Chou T-C (1980) Comparison of dose-effect relationships of carcinogens following low-dose chronic exposure and high-dose single injection: an analysis by the median-effect principle. Carcinogenesis 1: 203–213

De Lean A, Munson PJ, Rodbard D (1978) Simultaneous analysis of families of sigmoidal curves: application to bioassay, radioligand assay, and physiological dose-response curves. Amer J Physiol 235: E97–E102

Doll R, Peto R (1978) Cigarette smoking and bronchial carcinoma: dose and time relationship among regular smokers and lifelong non-smokers. J Epidemiol Commun. Health 22: 303–313.

Donnelly KC, Brown KW, Estiri M et al. (1988) Mutagenic potential of binary mixtures of nitropolychlorinated dibenzo-p-dioxins and related compounds. J Toxicol Environ Health 24: 345–356

Food Safety Council (1980) Quantitative risk assessment. Food Cosmet Toxicol 18: 711–734

George SE, Chadwick RW, Creason JP et al. (1991) Effect of pentachlorophenol on the activation of 2,6-dinitrotoluene to genotoxic urinary metabolites in CD-1 mice: a comparison of GI enzyme activities and urine mutagenicity. Environ Mol Mutagen 18: 92–101

Grieve AP (1985) Risk extrapolation in carcinogenicity studies. 31. Biometr. Kolloquium, pp. 66–72, Biometr. Ges. Bad Nauheim, Germany

Hempel L, Zagermann P (1993) Arzneimittel-Interaktionen erkennen und richtig interpretieren, eds. API, ABDA, ÖAK, SAV, Schweiz Apoth.verein, Bern-Liebefeld, Switzerland

Holzmann S, Kukovetz WR, Braida Ch, Pöch G (1992) Pharmacological interaction experiments differentiate between glibenclamide-sensitive K+ channels and cyclic GMP as components of vasodilation by nicorandil. Eur J Pharmacol 215: 1–7

Johnson JC, Attanasio R (1987) Synergistic inhibition of anatid herpesvirus replication by acyclovir and phophonocompounds. Intervirology 28: 89–99

Kremer AB, Egan RM, Sable HZ (1980) The active site of transketolase. Two arginine residues are essential for activity. J Biol Chem 255: 2405–2410

Krieglstein J, Lippert K, Pöch G (1996) Apparent independent action of nimodipine and glutamate antagonists to protect cultured neurons against glutamate-induced damage. Neuropharmacology 35: 1737–1742

Kumazawa S (1994) A new model of shouldered survival curves. Environ Health Persp 102, suppl. 1: 131–133, Table. 1

Leff P (1987) An analysis of amplifying and potentiating interactions between agonists. J Pharmacol Exp Ther 243: 1035–1042

Littlefield NA, Farmer JH, Gaylor DW (1979) Effects of dose and time in a long-term, low-dose carcinogenic study. J Environ Pathol Toxicol 3: 17–34

Mattern MR, Hofmann GA, McCabe FL, Johnson RK (1991) Synergistic cell killing by ionizing radiation and topoisomerase I inhibitor topotecan (SK&F 104864). Cancer Res 51: 5813–5816.

Müller WU, Streffer C, Pampfer S (1994) The question of threshold doses for radiation damage: malformations induced by radiation exposure of unicellular or multicellular preimplantation stages of the mouse. Radiat Environ Biophys 33: 63–68

Nestmannn ER, Brillinger RL, McPherson MF, Maus KL (1987) The SIMULTEST: a new approach to screening chemicals with the Salmonella reversion assay. Environ Mol Mutagen 10: 169–181

Odeleye OE, Watson RR, Eskelson CD, Earnest D (1993) Enhancement of cocaine-induced hepatotoxicity by ethanol. Drug Alcohol Dependence 31: 253–263

Pancheva SN (1991) Potentiating effect of ribavirin on the antiherpes activity of acyclovir. Antiviral Res 16: 151–161

Peto R, Gray R, Branton P, Grasso P (1982) Effects on two tonnes of inbred rats of chronic inges-
tion of diethyl- or dimethylnitrosamine: An unusually detailed dose-response study. Imperial
Cancer Research Fund, Cancer Studies Unit, Nuffield Dept Clin Med, Radcliffe Infirmary,
Oxford, UK

Pierce DA, Shimizu Y, Preston DL, Vaeth M, Mabuchi K (1996) Studies of the mortality of atomic
bomb survivors. Report 12, Part I. Cancer: 1950-1990. Radiat Res 146: 1–27

Plaa GL, Hewitt WR (1982) Potentiation of liver and kidney injury by ketones and ketogenic substances, in: Yoshida H, Hagihara Y, Ebashi S (eds) Advances in pharmacology and therapeutics
II, vol 5, Pergamon Press, Oxford – New York, pp. 65–75, Fig. 1

Pöch G (1989) Pharmakodynamische Interaktionen. Öst Apoth Ztg 43: 661–665 (Austria).

Pöch G (1993a) Combined effects of drugs and toxic agents. Modern evaluation in theory and
practice. Springer-Verlag, Wien – New York

Pöch G (1993b) Die klinische Prüfung von Arzneimittel-Kombinationen. In: Kuemmerle H-P,
Hitzenberger G, Spitzy KH (eds) Klinische Pharmakologie, 4. edn. 39. suppl. del.3/93.
Ecomed Verlag, Landsberg, III-2.19, pp. 1–18

Pöch G (1995) Arzneimittel-Interaktionen: Unerwünschte Wirkungen. In: Kuemmerle H-P,
Hitzenberger G, Spitzy KH (eds) Klinische Pharmakologie, 4. edn. 46. suppl. del.9/95.
Ecomed Verlag, Landsberg, II-2.12, pp. 1–14

Pöch G, Dawson DA (1998) Threshold doses in combination. A dose-response-curve study. Int.
Conf. on Complex Environ. Factors, Baden b. Wien (Abstract)

Pöch G, Pancheva SN (1995) Calculating slope and ED50 of additive dose-response curves, and
application of these tabulated parameter values. J Pharmacol Toxicol Methods 33: 137–145

Pöch G, Dawson DA, Dittrich P (1997) Teratogenic mixtures: Analysis of experimental dose-
response curves and statistical comparison with theoretical effects. Arch Complex Environ
Studies 9: 23–33

Pöch G, Dawson DA, Reiffenstein RJ (1996) Model usage in evaluation of combined effects of
toxicants. Toxicol Ecotoxicol News (TEN) 3: 51–59

Pöch G, Dittrich P, Holzmann S (1990a) Evaluation of combined effects in dose-response studies
by statistical comparison with additive and independent interactions. J Pharmacol Methods 24:
311–325

Pöch G, Dittrich P, Reiffenstein RJ et al. (1990) Evaluation of experimental combined toxicity by
use of dose-frequency curves: comparison with theoretical additivity as well as independence.
Can J Physiol Pharmacol 68: 1338–1345

Pöch G, Reiffenstein RJ, Unkelbach HD (1990b) Application of the isobologram technique for the
analysis of combined effects with respect to additivity as well as independence. Can J Physiol
Pharmacol 68: 682–688

Pöch G, Reiffenstein RJ, Köck P, Pancheva SN (1995) Uniform characterization of potentiation in
simple and complex situations when agents bind to different molecular sites. Can J Physiol
Pharmacol 73: 1574–1581

Purchase IFH, Auton TR (1995) Thresholds in chemical carcinogenesis. Regul Toxicol Pharmacol
22: 199–205

Scholze M, Faust M, Altenburger R, Backhaus T et al. (1998) Predictability of the aquatic toxicity
of multiple mixtures of similarly acting chemicals. SETAC-Europe 8th Annual Meeting 14–18
April 1998, Bordeaux France, Abstract 4A/016

Short TG, Chui PT (1991) Propofol and midazolam act synergistically in combination. Br J
Anaesth 67: 539–545

Steel GG, Peckham MJ (1979) Exploitable mechanisms in combined radiotherapy-chemotherapy:
The concept of additivity. Int J Radiol Oncol Biol Physiol 5: 85–91

Streffer C (1991) Stochastische und nichtstochastische Strahlenwirkungen. Nucl-Med 30:
198–205

Sühnel J (1992) Zero interaction response surfaces, interaction functions and difference response
surfaces for combinations of biologically active agents. Arzneim-Forsch (Drug Res) 42:
1251–1258

Tverskoy M, Fleyshman G, Bradley EL, Kissin I, Jr. (1988) Midazolam-thiopental anesthetic
interaction in patients. Anesth Analg 67: 342–345

Tverskoy M, Ben-Shlomo I, et al. (1989) Midazolam acts synergistically with methohexitone for
induction of anaesthesia. Br J Anaesth 63: 109–112

UNSCEAR (1994) United Nations Scientific Committee on the Effects of Atomic Radiation.
Sources and Effects of Ionizing Radiation. United Nations, New York

Verna L, Whysner J, Williams GM (1996) 2-Acetylaminofluorene mechanistic data and risk assessment: DNA reactivity, enhanced cell proliferation and tumor initiation. Pharmacol Ther 71 83–105

Vychodil-Kahr S (1997) Chemosensitivitätstestung für Cytostatica in vitro. Computerisierte Auswertung der Einzel- and Kombinationswirkungen und ihre Bedeutung. Diss. University Graz, pp. 1–137

Zeise L, Wilson R, Crouch EAC (1987) Dose-response relationships for carcinogens: a review. Environ Health Perspect 73: 259–308

2.3
Combined Exposure to Radiation and Substances

2.3.1
Biological Effects of Ionizing Radiation

2.3.1.1
Overview

Exposure to radiation can harm cells of all tissues, i.e. soma cells (Greek, soma = body) as well as germ cells. In the first case, somatic radiation effects occur in the exposed individual; in the latter, the genetic material of a germ cell can be affected and changed, leading to mutations in the descendants of the individual exposed.

In order to assess radiation risks, the dose-effect relationship i.e. the functional dependence between the radiation effect and the radiation dose has to be known. In radiation protection, one discerns between two fundamentally different types of radiation effect (Streffer, 1997c): For a *non-stochastic (deterministic)* radiation effect to occur, a threshold dose must be exceeded before certain effects can be induced. Only when this threshold dose is surpassed, the number of effects (number of affected individuals) and the severity of the effect will rise (ICRP, 1977). The development of such radiation damage happens through a multicellular mechanism: A large number of cells must be affected before defects become manifest. This class of radiation effects includes, for instance, all acute radiation effects, developmental anomalies, eye cataracts and a number of other delayed effects.

For the second type of radiation effect, the *stochastic effects*, it is assumed that there is no threshold dose and that the probability for such effects to occur rises with the radiation dose. Because of the absence of a threshold dose, stochastic radiation effects are of crucial importance for radiation protection, especially in the lower dose range. "Stochastic" is a term used in statistics, where it denotes the branch of this mathematical science concerned with random events. The primary [radiation-induced] ionization events in the DNA are indeed random events governed by statistical distributions. However, the "processing" of the radiation damage, e.g. the preferential repair of certain locations in the DNA, may then lead to specific damage patterns. Nevertheless, when a group of people suffers exposure, one can only state the probability for any individual of the group to experience stochastic radiation effects. This class of radiation effects includes the induction of inheritable defects and malignant illnesses. A statistically significant increase of genetic defects through radiation has only been found in animal experiments so far. There is no epidemiological evidence for such an increase among humans, for the obvious reason that large enough groups of individuals have never been exposed to doses sufficient to produce measurable genetic effects (Streffer 1991, UNSCEAR 1993). Epidemiological studies found a significantly increased incidence of malignant diseases only for radiation doses in the region of one tenths to one Sievert (Preston et al. 1994, Thompson et al. 1994). Based on considerations concerning the mechanism and on experimental observations, one now assumes those effects to be of a unicellular nature. In the context of genetic mutations, this argument seems quite plausible. For inheritable defects to occur, just one germ cell needs to be damaged, which leads, after its fertilization, to a mutation in the following generation. The situation involving the development of leukemia and solid cancer is considerably more complex, since these are processes involving several steps of mutations. The suspicion is that the malignant transformation of *one* cell is enough to cause an illness – with a certain probability (according to the concept of monoclonal tumour growth) (UNSCEAR 1986, 1993). Each transformed, malignant cell has a potential to develop a tumour, as is suggested by the important observation that cancerous cells often show chromosomal transformations which then appear in every cell of an individual tumour (Streffer 1997 a). Cell clones emerge and cause a clonal growth of the tumour.

The most important study to record radiation-induced cancer risks is the survey of the survivors of the A-bomb explosions over Hiroshima and Nagasaki. A significant rise in cancer mortality was found for the dose range of about 100 mSv and above. In one of the more recent papers it was reported that a radiation dose of 50 mSv already led to a significantly higher cancer mortality in Hiroshima and Nagasaki (Pierce et al. 1996). This finding, however, has been received with some scepticism. Below that range (10-50 mSv), deviations from the "natural" incidence rate of cancer may have been observed, but these turned out to be of no statistical significance when considering a larger group of individuals of all age groups. Other studies confirmed this result: Wei et al. (1990) for instance examined the cancer death rate in regions with high (2.2 mSv per year) and low (0.8 mSv per year) exposure to external radiation from natural sources. No increased cancer induction was found for exposures to radiation doses of 10-50 mSv. However, investigations following prenatal, diagnostic radiation exposures of fetuses in utero showed that a dose in the range from 10mSv leads to an increased risk of leukemia and solid tumours developing in early childhood (Doll and Wakeford, 1997).

Again, it comes as no surprise that the radiation effect following the exposure of adult humans to doses that low cannot be detected. The main reason is that it is not possible so far to distinguish between cancer caused by radiation and "spontaneous" cancer i.e. cancer not induced by ionizing radiation. The "spontaneous" cancer rate does however show considerable variations mainly due to genetic pre-dispositions, endogenous influences (e.g. of a hormonal nature) and exogenous factors (e.g. lifestyle, nutrition). Considering the variation margin of the "spontaneous" cancer rate, the effect of a radiation dose < 100 mSv is indiscernible. Consequently, epidemiological studies cannot answer the question if small doses have an effect or if there is a threshold dose, respectively. According to current knowledge, risks in the lower dose range can be quantified only through extrapolation from medium to higher doses. For cancer risk estimates, a linear dose-response relationship is assumed as a matter of precaution, in order to avoid underestimating the radiation risk (Streffer, 1997 a).

2.3.1.2
The Induction of Malignant Diseases

An exposure to carcinogenic agents, ionizing radiation or carcinogenic chemicals, can lead to genetic changes in a cell and result – if the damage is not or not successfully repaired – in the *initiation of a tumour development*. Transformations of the

DNA obviously play an essential role in this process; they are the primary starting point of the cancer development. As for chemicals, the substances themselves or their metabolites enter direct reactions with the DNA. The same is true for ionizing radiation, where such mutations occur either through a direct modification of the DNA or via radicals. Radicals – molecules that contain unpaired electrons and therefore are very aggressive chemically – are created in the radiation-induced ionization processes in the target molecule itself (direct effect) as well as in molecules close to the target, especially in water (indirect effect). Even radiation doses smaller than 0.01 Sv can still cause mutagenic transformations (UNSCEAR, 1986, 1993). According to our present understanding, for each cell transformed in this way there is a certain probability that a solid tumour or leukemia develops. This is one of the reasons why the assumption of a dose-response relationship without a threshold dose is plausible. Proliferation-increasing agents can promote the malignant cell transformation by multiplying pre-neoplastic cells and by raising the mutation rate.

Only in a later, enormous cell proliferation, an initiated cell can give rise to a clinically manifest tumour *(promotion of tumour development)*. The affected cells „transform", meaning they undergo changes causing them to show characteristics of tumour cells. The processes leading to that state can develop over many years or some decades. Apart from genotoxic factors (e.g. activated proto-oncogenes or deactivated tumour suppressor genes), non-genotoxic factors, the effects of which however depend on surpassing a threshold value, are of importance here, too. The following processes can be relevant for the promotion of the tumour:

– Stimulation of cell proliferation
– Inhibition of DNA repair
– Suppression of the immune system
– Induction of a radiosensitizing hypoxia

In contrast to the initiator, the effect of which is irreversible, promoters are often effective repeatedly or over an extended period, because otherwise their effects could be reversible. The tumour develops through further cell proliferation, mutation and other processes like vascularization; the tumour cells multiply and the malignancy increases *(progression of tumour development)*.

Fig. 2.3-1 shows a schematic representation of the "multistage model" described so far.

Fearon and Vogelstein (1990) have described such a model by showing the succession of mutation steps leading to the induction of the colorectal carcinoma.

Ionizing radiation initiates tumour development, as Berenblum and Shubik have shown through their pioneering work. Shubik et al. (1953) proved that the exposure of skin to beta radiation – which, applied on its own, did not induce tumours during the observation period of the experiment – indeed had a carcinogenic effect when followed by a treatment with the phorbol ester TPA. Berenblum and Trainin (1960) showed that treatment with urethane after X-ray exposure causes leukemia, whereas the radiation dose alone did not have any effect. Further results have proved that ionizing radiation initiates transformations that survive in viable cells for a long time and find expression only if an additional, "promoting" transformation takes place (e.g. Hoshino and Tanooka 1975). Such carcinogenic effects on tumour

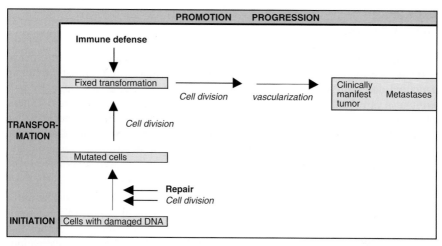

Fig. 2.3-1 The multistage model of tumor development

induction (especially for skin tumours) were found not only in animal experiments, but also for humans, when carcinogenic doses of various chemicals were combined with ionizing radiation (e.g. Cloudman et al. 1955, Canellos et al. 1975). Under certain conditions, evidence was found for ionizing radiation itself showing characteristics of a promoter (Tanooka and Ootsuyama 1993) and an amplifying effect on the progression of tumours (Jaffe et al. 1987).

On the whole, we can say that chemicals modulate the radiation damage by either interfering in the processes leading to the damage or by inhibiting the developments that should restore the original state. It was recognized that certain mechanisms are of special importance for interactions, which are influenced by

1. forming of and reactions of radicals,
2. structural or numerical transformations of the chromosomes,
3. DNA repair,
4. stimulation of cell proliferation.

"Super-additive" effects occur when interactions take place between the processes of the damage development induced by radiation, on the one hand, and by chemical substances, on the other hand.

This means that agents, of which we know that they can influence one of the said processes, demand our special attention with regard to their possible influence on the radiation risk. Risk-reducing and risk-increasing mechanisms can occur in the early as well as in the later molecular and cellular processes leading to tumour development, but experience so far shows that most substances protecting from radiation damage must be present during exposure, whereas many risk-increasing substances are effective also when they are added (even a long time) after the radiation exposure.

2.3.1.3
Extrapolations

The assessment of radiation effects often involves two extrapolations: In order to estimate effects in the lower dose range, one extrapolates from high and medium radiation doses to the lower dose range (see 2.3.1.1, linear dose-response curve without a threshold value). Apart from that, results gained from animal experiments are often applied to humans. Investigations with different species often yield results that are markedly different from each other. Naturally, data gained from research on humans ensure better accuracy, but in many cases – where quantitative results from humans are not available – animal experiments have to be drawn upon for investigations.

Any quantitative estimation of the stochastic *risk* due to radiation must be based on studies comparing the leukemia and cancer incidence in groups of individuals having been exposed to radiation with the corresponding number for groups that have not suffered exposure. Such comparative epidemiological investigations enable us to assess if and to which degree the incidence of malignant diseases has risen in the group having experienced radiation exposure. The principal findings in this respect stem from studies on the survivors of the nuclear explosions of Hiroshima and Nagasaki (UNSCEAR 1988).

A statistical analysis (Bayes Model) by DuMouchel and Groer (1989) performed to check the agreement between risk appraisals for radionuclide-induced bone tumours included 13 studies with results from rats, beagles and humans. These results showed that the slope of the dose-response relationship can vary by the factor 4, leading to the conclusion that an extrapolation for this parameter between these species of mammals appears inappropriate. An empirical comparison between the carcinogenic potentials of some chemicals detected in animal experiments, on the one hand, and for humans, on the other hand, also warns against transferring risk appraisals (Meijers et al. 1997). Another study, however, which compared the results for 23 chemicals, found – for risk assessment too – a strong correlation with the tendency to overestimate the risk for humans (Allen et al. 1988).

Concerning the *mechanisms,* we usually have to depend on animal experiments or experiments in vitro with established cell strains. The results are applied to humans, under the assumption that the processes here and in a suitable animal species are similar qualitatively. Experimental data support this thesis. For instance, the present teaching of genetics is based, in large parts, on results from the animal model, although in this case, too, there are counterexamples: In the case of Thalidomide, the transfer of results gained with various species (certain strains of rats, mice, guinea pigs, rabbits, hamsters, cats and dogs) proved to be misleading. Only after administering Thalidomide to a certain rabbit strain (the "New Zealand White Rabbit") the teratogenic effect of the substance could be recognized. The malformations of the extremities caused by Thalidomide, typical for affected humans, could only be found in primates (Hendrickx et al. 1966).

For chemical agents, the application of results from animal experiments to human beings seems to be more problematic than for the noxa "radiation", because in the first case, metabolic processes, which may influence the carcinogenic effect of the chemical substance, can differ quantitatively and/or qualitatively between animal and human.

Results on combination effects between radiation and chemical agents, gained in animal experiments, were confirmed by clinical or epidemiological observations on humans in many cases. Therefore, it appears to be justified to consider a substance showing a super-additive effect in animal experiments as a potential amplifier for effects in humans, too, and to regard such noxae as particularly interesting in our context.

2.3.2
Radiation Protection

2.3.2.1
Radiation Dose Units

The first quantitative measurements of X-ray radiation were carried out with the ion dose i.e. the number of ionizations (ion pairs) created in air. The dose unit was the Roentgen. The ion dose (nowadays measured in units of Coulomb per kg of air) was then replaced by the energy dose as the most widely used unit. This dose quantity represents the energy absorbed in a volume or mass of tissue. The dose unit is the Gray [Gy]. One Gray corresponds to one Joule of energy absorbed in a mass of one kilogram (1 J/kg). The rad is an older unit for the same quantity (100 rad = 1 Gy).

Studies on the biological effects of different types of radiation and different radiation energies have shown that the extent of the radiation effects depends on these radiation qualities (type and energy). To draw an exact comparison between the dose effects of different-quality radiation exposures, the relative biologic effectiveness (RBE) of each kind of radiation is determined. The RBE is the ratio of the radiation doses of two different qualities of radiation to be compared with each other. The doses entered in this ratio are the doses of the two kinds of radiation leading to radiation effects of the same strength. The reference radiation is, by convention, X-ray of 200 kV energy. A multitude of experimental and clinical studies led to the conclusion that the RBE-values depend on the dose level, the tissue or cell type, the biological end point and on other characteristics of the exposure. It was therefore agreed to determine rating factors, or quality factors, on the basis of RBE-values measured under many different conditions. The energy dose is multiplied by these quality factors, the result being the dose equivalent, given in units of Sievert [Sv]. The older unit for the same quantity is the rem (100 rem = 1 Sv).

Today, the effective dose equivalent is often used to determine radiation risks. However, this kind of risk assessment can be performed for stochastic effects only (cancer and mutations). To this end, the various organs and tissues are assigned statistical weight factors according to their respective sensitivity for radiation. The organ-specific weight factors are multiplied with the dose equivalents in the respective organs. The sum of the products arrived at in this way is the effective dose (ICRP 1977, ICRP 1991). The effective dose is only applicable for risk assessments in the lower dose range.

When ionizing radiation and radioactivity is quantified, one often comes across the unit "Becquerel" [Bq]. This unit is not a dose unit; it merely quantifies the amount of decays emerging from a radioactive substance. To determine the dose from the amount of radiation, the route of exposure (external exposure, internal exposure) must be known. When a radioactive substance has been incorporated, it has to be taken into account how the substance has been resorbed and distributed in the body. From these biokinetic data and from the physical data concerning the type of radiation, its energy and its half-life, one can determine a dose factor. Multiplying the dose factor with the amount of radioactivity (in Becquerel) results in the corresponding dose equivalent.

2.3.2.2
Radiation Exposure to the Population

As everybody knows, all life on earth has always been exposed to ionizing radiation. To this natural exposure – ca. 2.4 mSv effective dose equivalent on average, per head of population per year – we have to add a component caused by man i.e. the application of ionizing radiation in medicine and technology (ca. 1.56 mSv per head of population per year). The relative contributions, into which these radiation exposures can be subdivided, are shown in fig. 2.3-2.

The natural radiation exposure comprises external as well as internal radiation following the incorporation of radioactive substances contained in food or in air. The external exposure consists of *cosmic radiation* (ca. 0.3 mSv/year at sea level) and *terrestrial radiation exposure* (about 0.6 mSv/year on average). Cosmic radiation mainly

Table 2.3-1 Notes on some radiation dose units

Quantity	Symbol	Unit	Definition
Energy dose	D	Gray [Gy] 1 Gy = 1 J/kg	The energy dose is the ratio of the radiation energy transferred to the material exposed and the mass of the material.
Dose equivalent	H	Sievert [Sv] 1 Sv = 1 J/kg	The dose equivalent is the product of the energy dose in tissue and a dimensionless rating factor q (for photons and electrons: q=1; for neutrons: q=10; for alpha particles: q=20).
Effective dose equivalent	HE	Sievert [Sv] 1 Sv = 1 J/kg	The effective dose equivalent is the sum of the dose equivalents, each multiplied by the corresponding weight factor wT, for the relevant organs or tissues. (Not the doses in different organs are summed up, but the risk contributions of the organs, e.g. wT for the gonads: 0,25; the weight factors for the individual organs sum up to 1.)

arises from nuclear processes on the sun's surface. The magnitude of this radiation dose depends on the height above sea level and therefore differs considerably depending on the region considered (Streffer 1997 b). Terrestrial radiation is a result of the fact that we are surrounded by radioactive substances, e.g. in the ground and in building materials, which emit ionizing radiation in the course of their radioactive decay. Depending on the characteristics of rock and soils in a region, this part of the exposure is extremely variable. Apart from these sources of radiation, there are *internal exposures* (ca. 0.3 mSv/year) due to the incorporation of radioactive substances contained in food, mainly the radionuclide potassium-40, which is a natural ingredient of vegetables, potatoes and meat, and which becomes distributed homogeneously in the whole organism, just as the stable, non-radioactive potassium. In the lung, there is the particular radiation exposure due to inhaled, radioactive radon and its decay products (effective dose ca. 1.2 mSv/year). Radon itself is a decay product of uranium or radium, respectively, which are natural trace elements in soil and rock. It diffuses from the ground to the surface, into mineshafts and into the air inside buildings. Being an inert gas, radon is exhaled quite quickly. But its decay products are metal **ions** (mainly polonium-218, lead-214, bismuth-214 and polonium-214), which are attached to aerosols, like dust particles, and hence are inhaled with air and stay in the respiratory system for some time. Because of their short physical half-lives, a major fraction of these radionuclides on aerosols decay in the respiratory system, causing a stronger exposure of the lung by the alpha radiation emitted. Here, too, different regions and even individual buildings can lead to very different doses.

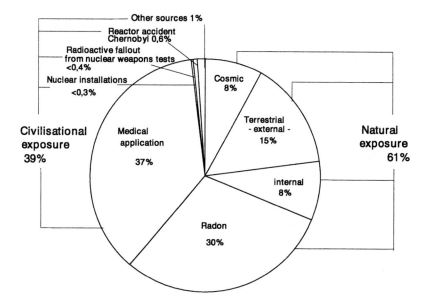

Fig. 2.3-2 Individual sources of exposure and their relative contributions to the average annual radiation exposure (effective dose) of the population of Germany in 1990 (from the Annual Report of the Federal Minister for the Environment, Conservation and Reactor Safety [Bundesminister für Umwelt, Naturschutz und Reaktorsicherheit], 1992).

This section illustrates that the natural radiation exposure can vary considerably, depending on the region, the living conditions and individual human behaviour. Studies on terrestrial radiation and radon concentrations in buildings in Germany, in particular, have shown that this radiation component is subject to large variations. In buildings, for instance, indoor concentrations of radon between 5 Bq/m^3 and some 10 kBq/m^3 were measured, with an average value of 40-50 Bq/m^3.

Apart from exposures to radiation from natural sources, the population is exposed to radiation from medical and other man-made radiation sources. The medical exposure is predominant here, especially due to X-ray diagnostics and nuclear medicine. Nuclear installations and radioactive fallout from the nuclear weapons tests performed in the atmosphere in the 1950s and 1960s contribute relatively little. Beyond that, there is exposure to radioactive substances – radionuclides of caesium in particular – going back to the reactor accident in Chernobyl in 1986. The relative contributions of the average exposures of people in Germany are given in fig. 2.3-2.

2.3.2.3
Dose Limits

The dose limits of legal regulations in Germany and other countries are designed so that, if these limits are not exceeded, non-stochastic radiation effects do not occur, because their threshold doses are not exceeded. On the other hand, it is assumed that even in the region of the dose limit and below stochastic effects can be induced, because there is no threshold dose for these radiation effects. With approximately 1 mSv/year, the dose limit for the general population is, however, far below the dose range for which any measurable harm to human health can be registered. Although the ICRP (International Commission on Radiological Protection) recommends a limit of 1 mSv per year on non-medical radiation exposures for the general population, German regulations prescribe lower limit values for individuals living in the neighbourhood of nuclear installations. These limits are guided by the natural radiation exposure and its margin of variation, which exceeds the limit value. Hence, when an individual experiences a change in his or her living conditions, including a change of residence within Germany, this can lead to changes in the individual's natural radiation exposure that exceed the margin for the dose limit in the vicinity of nuclear installations. In this way the dose limit is guided by living conditions involving cancer risks, which our population is subjected to or subjects itself to all the time.

This limit as well as other standards set for agents in the environment and at the workplace was determined predominantly for conditions of exposures to single agents, though the majority of exposures are combinations of several toxic agents. In the following, we will endeavour, by reporting the present state of scientific knowledge, to provide the basis for assessing the adequacy of the existing limit values in situations with combined exposures.

2.3.3
Interactions between Ionizing Radiation and Chemicals

2.3.3.1
Basic Considerations on Estimating Combined Effects

The fundamental question in treating combined effects is the following: Is there a discrepancy between the combined effect and the effect to be expected if the combination partners just behaved additively? Furthermore, it needs to be clarified how "additivity" can be established as the reference point for possible deviations caused by interactions between the combination partners. By answering these questions, we will be able to detect super-additive or sub-additive effects.

The determination of the additive reference effect of two combination partners becomes most simple, when they, each on its own, follow a linear dose-response relationship, in which case the additive effect emerges as the sum of the single effects.

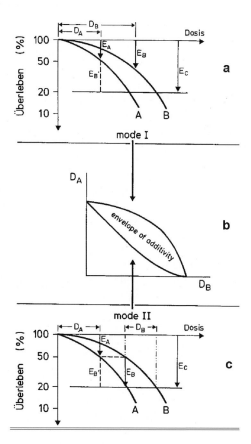

Fig. 2.3-3 An isobologram analysis using the concept of the "envelope of additivity", by Steel and Peckham (1979); explanations in the text.

Where at least one of the dose-response relationships is not linear, there are two mathematical approaches to establishing additivity (Frei 1913, Loewe 1953). In slightly simplified terms, we are dealing with an addition either of the *effects* of the single agents (heteroaddition) or with of the *doses* of the single agents (isoaddition)[1]. An addition of the effects is appropriate if the agents involved in the combination act through different (hetero) mechanisms. Dose addition should be applied, if the partners act through the same (iso) or at least similar mechanisms. In the latter case, one assumes that agent A behaves like agent B, with regard to their effects, and could therefore replace a certain dose of agent B. Since the effect mechanisms are unknown in many cases, one cannot decide which expected value constitutes the basis for the comparison with the combination value actually observed. Consequently, both procedures (heteroaddition and isoaddition) can be taken into account.

The theory required for this approach was developed by Loewe (1953). Steel and Peckham (1979) made this very complex model more accessible through their work on the „envelope of additivity", where the problem caused by the non-linearity of the dose-response curves is solved by isobologram analyses (see 2.2.2.5). The term "isobologram" refers to the fact that one obtains curves of "equal effect", with the shape of these curves depending on the dose-response curves of the single agents. First, the isoboles are calculated through heteroaddition or isoaddition. Then, from the coordinates of the observed effect following combined exposures in relation to the "additivity area" enclosed by the two isoboles, the quality of the combined effect is derived. If the observed value lies in this "envelope", the effect is judged to be additive. The area to the left of the envelope is the region of super-additivity; the area to the right represents sub-additivity.

In the following, we will explain how an isobologram analysis is performed using Steel's and Peckham's (1979) approach. Fig. 2.3-3 shows the procedure of such an analysis established for cell survival curves. The first step is a heteroaddition (fig. 2.3-3 a): An arbitrary dose D_A of the agent A causes an effect E_A. Before the observed combined effect E_C can be achieved, another effect $E_{B'}$ is needed, which can be supplied by the dose D_B. In this way, one arrives at a dose pair, D_A and D_B, that can give rise to the combined effect E_C – provided the mechanisms of the two agents are independent of each other. The resulting dose pair is plotted as in fig. 2.3-3b. This procedure is then repeated for other doses of agent A, until a curve can be constructed from such doses of the agents A and B which each on their own cause the combined effect E_C (= intersections of curve with the x- and y-axes) (upper isobole).

The isoaddition (fig. 2.3-3c) starts in the same way, by determining the effect E_A following the application of the dose D_A. However, in this case we assume, for the further construction of the isobole, that the dose D_A is equivalent to a certain dose of the agent B, which is the dose value of agent B that would cause the effect E_A on the dose-response curve of agent B. To achieve the combined effect E_C, we then only need the dose D_B, since it is assumed that agent A has replaced part of agent B. Again, this procedure is repeated for the construction of a curve as in fig. 2.3-3b (lower isobole).

According to Steel and Peckham (1979), an effect is called super-additive or sub-additive only if the dose pair actually used in the combination is represented by a

---

[1]   See section 2.2.2.12.

point left or right of the additivity area, respectively. Every dose pair within the "envelope of additivity" corresponds to the value expected for additivity in this model. It should be noted that this analysis does not provide any information about the mechanisms on which the effects of the agents are based.

Essentially, this method of analyzing combined effects – which requires the knowledge or the existence of does-response relationships – is a tool for evaluating scientific experiments. The epidemiological approach, on the other hand, which is concerned with issues of risk assessment in cases of combined exposure, calls upon statistical methods, which will be explained at a later stage, in the context of a relevant example (see section 2.3.3.2, "Cigarettes/Radon").

2.3.3.2
Exemplary Discussion of Combined Effects

Many experimental and epidemiological studies have been looking at situations following combined exposures with ionizing radiation as one of the components of the exposure. Radiation offers two advantages. First, the exposure dose at the target location can be measured very well, and the harmful agent has a direct impact on the DNA or other cellular structures. Second, there are no transport and distribution steps or any metabolic processes involved, which would influence the intracellular concentration or the activation or deactivation of the noxa. Therefore, we will first discuss examples of combined harmful effects involving ionizing radiation.

Considering harmful effects on human health of noxae at the workplace and in the environment, exposures causing cancer are very much at the centre of attention. Hence, we will focus our comments on such diseases following exposures, especially combined exposures to harmful agents, based on the current state of knowledge.

Proceeding from the principles introduced in table 2.1-1, we will present examples of proven interactions between radiation and chemical agents. We must point out, however, that the considerations behind table 2.1-1 are rooted largely in laboratory studies and experience from clinical medicine, mainly tumour therapy. In general, the boundary conditions of the experiments do not correspond to conditions in the environment. The emphasis of the following descriptions will be on pointing out existing principles, the further development of which should enable us to predict interactions. The categories of table 2.1-1 will be explained with some examples, even if the differentiation between the mechanisms is often not that clear.

Genotoxic Agents Affecting the Same Phase of Tumour Development

In theory, the combination of genotoxic agents affecting the genome in a stochastic way and influencing the same tumour state leads to an additive total effect. For a combined exposure to e.g. ionizing radiation and an alkylating agent, we therefore expect an additive effect, comparable to a corresponding increase of the dose of the single agent.

Benzo[a]pyrene (BP) (see table 2.1-1) is a polycyclic hydrocarbon common in the environment and known as a constituent of tobacco tar. It is a chemical carcinogenic acting as a secondary genotoxic, meaning that an intracellular, enzymatic modification of the substance gives rise to highly reactive products damaging the DNA in a way similar to the action of ionizing radiation. Little et al. (1978) examined the effect of a simultaneous exposure of hamsters to Benzo[a]pyrene (15 weekly instillations, each of 0.3 mg BP/animal) and to the radioactive alpha emitter ^{210}Polonium (Po) on the induction of lung tumours. Benzo[a]pyrene alone led to a tumour incidence of 11%. A high Po-dose (3 Sv lifetime lung dose) induced lung tumours in 44% of all animals. The combined exposure resulted in a tumour rate of 49%. Experiments with lower Po-doses (0.15-0.75 Sv lifetime lung dose) induced lung tumours in 16% of the animals; the combination treatment gave rise to an incidence of 27%. These results clearly reflect an additive effect of this combined exposure.

One has to expect an additive total effect, too, from a combination of ionizing radiation and alkylating substances. Following an exposure to alkylating agents, reactions with the cellular macromolecules can occur in every organ. The covalent reaction products are referred to as "adducts"; their fragments can be detected e.g. after the breakdown of the alkylated proteins or nucleic acids in the body fluids. Alkylated DNA molecules can be repaired; other alkylated macromolecules are broken down and replaced. Remaining, latent damage to the DNA can become expressed by an exposure to another genotoxic agent – e.g. to ionizing radiation.

Nevertheless, the investigations so far performed following irradiation and application of alkylating substances (see table 2.1-1) often showed deviations from the postulated additivity towards sub-additive effects. In the following, we will point out, by means of examples, some of the processes responsible for this discrepancy.

The carcinogenic ethylnitrosourea (ENU) has been studied intensively. It develops its effect by alkylating DNA bases. Products of methylating and alkylating at the DNA lead to changes in the genetic code and hence to the induction of mutagenic and carcinogenic effects. A single in-utero exposure of prenatal rats as well as a neonatal exposure to this substance, which acts diaplacentally, causes a high incidence of brain tumours in the offspring. The adult animal, on the other hand, is less susceptible to this effect, by a factor of about 50. Various authors showed that a combined exposure of fetuses (Kalter et al. 1980, Knowles 1984, Stammberger et al. 1990) or newborn animals (Knowles 1985, Hasgekar et al. 1986) to radiation and ethylnitrosourea reduces the incidence of induced tumours. An increased mortality of tumour-sensitive animals could be excluded as a possible cause for this occurrence. Quite frequently, the plausible possibility was discussed that the killing of the target cells of the carcinogenic by the radiation could explain the phenomenon of the reduced tumour rate. In the fetal period, when the irradiation takes place, these cells (glioblasts) are in a phase of active proliferation and hence are more sensitive to radiation. However, the question of the extent of the cell recovery in the period between the irradiation and the ENU exposure remained unanswered.

Investigations by Stammberger et al. (1990) produced data on a biochemical explanation: There is a repair mechanism that can correct the ENU-induced damage to the DNA. Involved in this process is the protein O^6-alkylguanine-DNA-alkyltransferase (AT), which can be induced by radiation. Therefore, the relationship between the radiation-induced repair capacity and the ENU-induced tumour incidence was examined. Rat fetuses were subjected to radiation on day 16 post conceptionem (1 and 2 Gy X-ray) and exposed to ENU (50 mg/kg) two hours later. In a long-term study (observation period 26 month, latency time 335-369 days), the AT activity measured in the fetal brain 24 hours after the exposure was compared with the postnatal brain tumour rate of the test animals (table 2.3-2). In animals treated with ENU, a decrease in the AT activity was found, compared to untreated control animals, as well as a correlated increase in the number of brain tumours. The inducing effect of the radiation on the repair activity correlated with a reduction of the brain tumour incidence following a combined treatment. Hence, a comparison between the biochemical and the morphological results leads to the conclusion that the antagonistic effect is a result of the AT induction by the radiation. Schmahl and Kriegel (1978) treated pregnant mice with X-rays in their fetal period (day 11-13 post conceptionem, 1 Gy on each day) and ENU (day 17 post conceptionem, 0.5 mM/kg). During an observation period of 18 months, the offspring showed a spectrum of tumourous and degenerative diseases (table 2.33). Considering the total rate of induced tumours, the differences between the effect of the combined treatment and the single effect of radiation or ENU, respectively, are of minor statistical significance (p < 0,05). Still, if one differentiates between specific tumour entities, one finds a significant increase in the leukemia rate compared to animals treated either with X-ray or with ENU alone. In contrast to this, the incidence of lung and liver tumours (as well as ovary cysts) is significantly lower after a combined treatment. Extensive liver necroses and cystic kidney degenerations presented particularly noticeable pathological transformations. The described quantitative shift in the tumour spectrum was attributed to differently modified proliferation capacities of the individual fetal tissues following the radiation insult. No influence of ENU on the radiation-induced incidence of malformations was observed.

Yokoro et al. (1987) found, after a combination of a whole-body irradiation (4 Gy X-ray) followed by an ENU exposure (5 mg per animal), synergistic effects with regard to the induction of T-cell lymphomas in mice. While the application of ENU alone proved to be not more than moderately carcinogenic, with a lymphoma inci-

Tab. 2.3-2 Relative AT activity in fetal rat brains (in relation to the value 1.00 for un-treated control animals) and the fraction of test animals developing brain tumours after exposures to radiation and ethylnitrosourea (ENU), from Stammberger et al., 1990. The numbers showing statistically significant differences to a treatment with ENU alone are highlighted by a grey background (*p<0,05; **p<0,01; ***p<0,001; ap=0,06)

Mode of treatment	AT activity		Brain tumour incidence		Mortality
	Decrease to (relative values)	Increase	Decrease to [%]	Increase	[%]
ENU treatment [50 mg/kg]	0.61			44.1	2.9
X-ray treatment [1 Gy]		1.31**	-	-	5.9
X-ray treatment [2 Gy]		2.02***	-	-	28.6
1 Gy + ENU		1.03*	26.8[a]		4.7

Tab. 2.3-3 The effect of a prenatal exposure to 3 x 1 Gy X-radiation and ethylnitrosourea [0.5 mM/kg] on the tumour development in mice (from Schmahl and Kriegel, 1978); the numbers showing a statistically significant discrepancy (p < 0,05) between combined and single treatment are highlighted by a grey background

| Post mortem diagnosis | Number of animals developing illnesses following | | | |
	Irradiation	ENU	Irradiation + ENU	Control
Brain tumour	0 (0%)	0 (0%)	0 (0%)	0 (0%)
Leukemia	3 (5.3%)	3 (2.4%)	10 (12.6%)	2 (2.3%)
Lung tumour	8 (14.3%)	22 (17.8%)	6 (7.6%)	11 (12.8%)
Hepatoma	2 (3.5%)	6 (4.9%)	2 (2.5%)	1 (1.1%)
Pancreas adenoma	0 (0%)	1 (0.8%)	2 (2.5%)	0 (0%)
Intestinal tumour	0 (0%)	2 (1.6%)	0 (0%)	0 (0%)
Ovarial tumour	6 (7.0%)	0 (0%)	10 (11.2%)	0 (0%)

dence of 20%, the combined treatment induced this kind of tumour in 92% (time interval of 5 days between irradiation and ENU) and 40% (30 days between irradiation and ENU) of the animals respectively. Apart from the amplification of the ENU effect by the preceding irradiation, a shortening of the latency time was observed for the combined exposure. It was shown that the supposed synergism could be attributed to cellular kinetics: As a reaction to the 4 Gy X-ray treatment and the cell killing induced thereby, the bone marrow and the thymus recover quickly through regenerative repopulation. Hence, the principal effect of the radiation treatment is that a sufficiently large population of sensitive cells becomes available for the transforming effect of the chemical carcinogenic. This explanation is supported by an experimental approach in which the field of irradiation is limited to the thymus, while the dose is left unchanged. In this case, the rate of induced lymphomas is considerably lower than after whole-body radiation treatment and correlated with the proportion of sensitive cells. The irradiation can also be replaced by an ENU treatment (4 mg per animal). When second ENU doses (1 mg pro Tier) are administered after the time intervals quoted above, the tumour rates become comparable with those induced in the radiation/ENU set-up.

The reported shift in the tumour spectrum was also observed for a combination of ionizing radiation and N,N´-2,7-Fluorenylenbisacetamide (FAA, Nagayo et al. 1970, Vogel and Zaldivar 1971). This chemical – in contrast to ionizing radiation – has a carcinogenic effect especially on the liver. Combinations with X-ray or neutron radiation increased the hepatoma incidence in a synergistic fashion. For other types or locations of cancer, however, antagonistic effects were detected. These effects also occurred for malignancies "typical for radiation", like tumours of the small intestine and the chest, lymphomas and leukemia. The considerably shorter life-span of the test animals due to the application of the physical and the chemical noxa (273 days average survival time following the combined treatment, compared to 669 days for untreated control animals) is a possible reason for this occurrence, because the manifestations of various tumour types could not be registered any-more. This means that in the dose range of radiation and chemicals, where the induction of cell death has to be expected, an effect reduction may arise ("overkill"), which cannot necessarily be attributed to an interaction between the agents.

There are plentiful other studies on genotoxic, predominantly alkylating sub-stances showing deviations from additivity when combined with radiation. How-ever, the numerous cases of synergism and antagonism standing against additivity rather appear to be the result of the processes described in this section than of a spe-cific genotoxicity. The investigations with alkylating substances described here exemplify the essential processes: The non-additivity of the observed combination effects could be attributed to radiation-induced modifications of repair processes, to changes in the proliferation capacity of certain tissues or cell types, and to the induction of cell death ("overkill") by one of the noxae. These mechanisms are of importance mainly in the higher dose range, where the coincidence of different damage events overshadows the actual interactions. In the lower dose range, on the other hand, it is very unlikely that two damage events induced by individual noxae occur sufficiently close together in the DNA to give rise to interactions

Genotoxic Agent Combined with an Agent Inhibiting Mechanisms for Damage Repair

In many cases, damage to the DNA polynucleotide chain can be cured through effi-cient repair systems. The break-up of one of the DNA polynucleotide chains can occur in such way that a single enzymatic step (by the enzyme ligase) leads to the complete repair of radiation damage. In other cases, however, several enzymatic steps are necessary for repairing the damage (excision repair): First, an enzyme of the endonuclease class cuts the affected DNA strand close to the damage. Then the dam-aged components of the DNA must be extracted from the polynucleotide chain (by an enzyme of the polymerase class) and replaced by the original base sequence, based on the information contained in the complementary, undamaged DNA strand. Finally, the enzyme ligase closes the polynucleotide chain again. Despite of the high effi-ciency of this repair system, residual damage can remain due to incomplete or faulty processes. If a noxa inhibits the repair, a significant strengthening of the radiation effect can occur, where the combined effect exceeds additivity of the single effects by far.

Interference with the DNA repair represents an important mechanism for poten-tiating the damage from combined exposures. The organism, however, also disposes of damage-recovery mechanisms on other levels of regulation (e.g. elimination of toxic substances and immunological defence processes), the impairment of which can lead to effects of similar importance.

The assumption is that there is causal connection between mutagenicity and oncogenicity. However, the examination of mutagenic activity by means of the Ames test (also see chapter 2.4) yielded negative results for metals which in many cases have established, carcinogenic effects (e.g. Chambers et al. 1989). It is consid-ered as proven that metals (see table 2.1-1) do not directly harm the DNA. Although little is known so far about the exact mechanisms of the oncogenic effect of these inorganic chemicals, there are numerous clues that various metals act as inhibitors of the biosynthesis of DNA and therewith of the DNA repair capacity. This mechanism, which damages the DNA indirectly, could explain the carcinogenic potential of metals. For some metals, evidence was found that they contribute, through a variety of mechanisms, to the impairment of recovery processes. Among the mechanisms reported are for instance the modulation of enzyme activities (Vallee and Ulmer 1972), the direct inhibition of excision repair (Rossman 1981), the bonding to the DNA-synthesis and repair enzyme, DNA polymerase (Jung and Trachsel 1970), interference in the cell cycle (Brooks et al. 1989) and the depletion of the energy – as ATP – that is essential for the repair process (Vallee and Ulmer 1972).

Metals cause biological effects through a multitude of mechanisms, for instance through interactions with cell components and by influencing enzyme activities (e.g. Hartwig 1995, 1996). They show a high bonding affinity to receptors involved in the transduction of signals. Any damage through low concentrations of metals is probably limited by the cystine-rich protein metallothionein (MT). Since metals show a high affinity to molecules carrying sulfhydryl groups, they bind to MT, the production of which in the tissue is increased as a reaction to an exposure to heavy metals and some other noxae. In this way, the exposed tissue can be detoxified.

Various interactions with chromosomal material were found for beryllium. These mechanisms could well be the cause of the tumour-inducing properties of the metal. Brooks et al. (1989) examined, in in-vitro experiments, the interaction between X-radiation and beryllium sulphate with regard to the development of chromosome

Table 2.3-4 The induction of chromosome aberrations by irradiation and beryllium (Be) exposure (*mean value with standard error); the numbers showing statistically significant differences (p<0,05) between combined and single treatments are highlighted by a grey background (from Brooks et al. 1989)

Treatment	Number of normal cells / number of examined cells	Total rate Aberrations/cell*
Control	192/200	0.05 ± 0.015
0.2 mM Be	194/200	0.03 ± 0.003
1.0 mM Be	182/200	0.09 ± 0.02
1.0 Gy	166/200	0.21 ± 0.03
2.0 Gy	150/200	0.34 ± 0.03
1.0 Gy + 0.2 mM Be	163/200	0.22 ± 0.04
1.0 Gy + 1.0 mM Be	134/200	0.45 ± 0.04
2.0 Gy + 0.2 mM Be	144/200	0.34 ± 0.03
2.0 Gy + 1.0 mM Be	122/255	0.85 ± 0.03

damage (aberrations). The exposure of the CHO cells as such to relatively high concentrations of beryllium (1mM) on its own only induced a low rate of aberrations, but a combined exposure including an X-ray treatment with 1 Gy or 2 Gy during the last two hours of the 20-hour beryllium exposure led to a synergistic effect (table 2.3-4). 90% of the lesions detected were chromatid aberrations, meaning that only one of the two "arms" of the chromosomes was affected. The synergistic effect was limited to cells in particular phases of cell proliferation (in the S- and G_2-

phases of the cell cycle). These observations point to an S-phase dependence as described for many other metals: In order to produce chromosome aberrations, the affected cells must still be in a DNA synthesis (S) phase. Obviously, higher doses and concentrations of beryllium strongly reduce the cells' capacity for repairing radiation-induced damage.

As described for an impaired DNA repair, we would expect greater harm following the impairment of any other recovery mechanism. Combination experiments with beryllium and plutonium dioxide, for instance, led to the following observations: Inhaled beryllium oxide prevents the release of plutonium from the lung, if the plutonium is inhaled after the beryllium. This retardation leads to an increased radiation dose for the lung tissue. Still, despite of this connection, Sanders et al. (1978) did not find any incidence of lung tumours exceeding the rate induced by plutonium alone. Experiments with 5456 rats (Finch et al. 1991, 1994 a) however produced some evidence for a combined effect beyond additivity. 65% of the animals developed lung tumours after inhaling 50 µg of beryllium (lung dose). This high rate is probably caused by beryllium delaying the release of other carcinogenic substances from the lung, as described above. A lung exposure of up to 170 Bq of ^{239}PuO$_2$ induced a lung tumour incidence of 10%. The combination of these exposures resulted in tumour rates of 90%. Still, these studies leave open the question if the rise of the tumour incidence in the lung can be fully attributed to the increased lung dose, or if it is the result of interactions on the molecular or cellular level. Arsenic has a high affinity to proteins containing a sulfhydryl (SH) group. One suspects that the arsenic-induced mutagenic effects are partly caused by this chemical feature, which blocks the activity of the repair enzyme ligase. If the final step of the DNA repair is prevented due to an inhib-

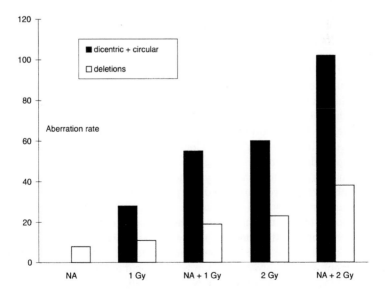

Fig. 2.3-4 Aberration rate (chromosome aberrations/200 cells) in human lymphocytes following X-ray treatment and exposure to sodium arsenite (NA); results from one donator. (The absolute values in this graph were taken from fig. 4 in Jha et al., 1992.)

ited ligation, more chromosomal damage will arise. This biological effect is documented, for instance, in the work of Jha et al. (1992). Human lymphocytes exposed to 1 and 2 Gy in vitro and treated with sodium arsenite (5 μM) two hours later. Then the incidence of unstable chromosome aberrations (dicentric chromosomes + circular chromosomes; deletions) was analyzed. (Fig. 2.3-4 shows the results of a representative experiment). Considering all the experiments, the combined exposure led to a significant increase in the radiation-induced frequency of dicentric chromosomes and circular chromosomes, by the factor 2, and for acentric fragments by a factor of 1.3 to 2.5.

The combined effect of arsenic and radiation was investigated in epidemiological studies on occupationally exposed mineworkers. Taylor et al. (1989) and Xuan et al. (1993) looked at workers at a tin mine in China (Yunnan province). In the process of mining as well as during the smelting of ore, arsenic dust is released, which was shown to be a strong risk factor for lung cancer. Individuals exposed to arsenic showed an up to 20-fold increase in the risk of lung cancer, with a clear dose-response relationship compared to groups of people who had not been in contact with this noxa. The duration of the exposure seemed to be of greater importance than its intensity. The aim of the study was to determine possible interactions between arsenic and radon (see 2.4). No evidence was found to support the assumption that there is such an interaction with regard to the induction of lung tumours. At the same time, however, the study emphasized the difficulty of differentiating and identifying the single effects of radon and arsenic. A study on workers at a gold mine in Ontario (Kusiak et al., 1991) postulates independent effects of radon and arsenic i.e. an additive total effect concerning the induction of lung tumours.

Müller and Streffer (1987) used a number of parameters in their study on the risk to pre-implantation mouse embryos following their combined exposure to X-rays and various metallic salts. Details of the morphological embryo development (formation and development of the blastocysts) can point to interferences in the course of development processes (table 2.3-5). In principle, this examination demonstrates the importance of the shape of the dose-response curves of the single agents for determining the combined effect. For none of the conditions examined, any significant influence of arsenic on the radiation risk was observed. The effects simply added up. A result for cadmium showed a significant tendency (p < 0.05) to the risk-increasing effect with regard to the morphological development of the embryos. For lead, an additive effect on the morphological development and on cell proliferation was found, but also a super-additive increase in the radiation risk regarding chromosomal damage (micronucleus induction). Mercury turned out to increase clearly the radiation risk for morphological development and cell proliferation (p < 0.05).

Since neither mercury and cadmium nor X-radiation follows a linear dose-response relationship for the induction of the end point examined, the discrepancy between the combined effect and the sum of the single effects could simply be a result of the shape of the dose-response curves. Hence, in order to determine the area of deviation from additivity that can be attributed to non-linear dose-response relationships, the "envelope of additivity" was computed (see fig. 2.3-3). To this end, more detailed checks on the dose-response relationships and isobologram analyzes (Steel and Peckham 1979, Streffer and Müller 1984, Burkart et al. 1997) had to be carried out. These investigations showed that the observed effect following the

Table 2.3-5 Observed and expected effects on the morphological development of pre-implantation mouse embryos, following the combined exposure to heavy metals and X-ray [1 Gy]. In control cultures, 85-95% two-cell embryos developed into blastocysts, and in 80-90% the blastocysts hatched from the zona pellucida. The numbers showing statistically significant differences (p<0.05) between the sum of the single effects and the observed effect are highlighted by a grey background (from Müller and Streffer 1987)

Metal		Formation of blastocysts observed/expected [%]	Hatching of blastocysts observed/expected [%]
Arsenic	0.3 µM	67/70	18/27
Cadmium	0.3 µM	93/92	46/69
	3.0 µM	18/35	0/2
Lead	0.4 µM	56/60	12/15
	3.6 µM	39/47	10/10
Mercury inorganic	3 µM	84/76	34/52
	10 µM	69/81	14/29
organically bonded	0.2 µM	75/89	26/54

combined exposure did lie outside the "envelope of additivity", in the area of super-additivity. The difference to the additive region of the envelope is, however, too slight to be of statistical significance (fig. 2.3-5a).

Hence, the assumption of an interaction between the two agents is not necessarily justified, although the combined effect exceeds the sum of the single *effects*. For the effect to be expected based on the sum of the single *doses* is greater than the observed combined effect. For mercury, on the other hand, the findings turned out to be a result of an interaction, the effect of which was actually super-additive (fig. 2.3-5b; Müller et al. 1987).

For caffeine (see table 2.1-1), the information gained over the last decades appear to be confusing at first. In the midst of all these contradictions, the following picture takes shape:

The conclusion seems to be safe that caffeine shortens the duration of the radiation-induced G_2-block (e.g. Lücke-Huhle 1982). The extension of the G_2-phase prior to the phase in which the chromosomal material is distributed to two daughter nuclei, which is normally observed following radiation treatment, is used for repairing damage at the genome. Any shortening of this "block", by high concentrations of caffeine, can promote either the death of damaged cells or, possibly, the proliferation of genetically damaged cells. Through isobologram analyses ("envelope of additivity", Steel and Peckham 1979), Müller et al. (1983) found a clearly super-additive combined effect (fig. 2.3-5c). For mouse embryos treated with caffeine after their radiation exposure, the risk for the embryonal development doubled compared to the single exposures. Müller et al. (1985) further described a super-additive effect for the induction of chromosomal damage (micronuclei) in mouse embryos treated with radiation and caffeine. However, all these clearly super-addi-

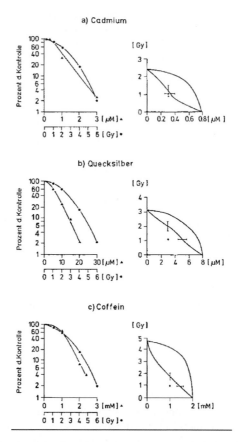

Fig. 2.3-5 Isobologram analysis (envelope of additivity) for a) cadmium sulfate, b) mercury chloride and c) caffeine. Left: The dose-response curves for the chemical and for X-radiation respectively; end point: hatching of the blastocysts 144 h p.c. Right: the corresponding isobolograms; the point in the graph represents the dose pair actually used in combination; the 95%-confidence intervals are shown as well (from Müller et al. 1983, 1987).

tive effects were only achieved with extremely high doses of caffeine (> 1mM, corresponding to about 800 cups of tea or coffee per day).

There are indications that caffeine – which features a structural similarity with the purine bases of DNA – can be built into the damaged DNA in a covalent bond. Domon et al. (1970) demonstrated this finding for bacteria. Through changing the DNA conformation, an increased number of start signals can be made available to DNA synthesis, leading to the cancellation of the temporary inhibition of the synthesis after the irradiation. In this way, cells with faults that have not been repaired are "forced" into replication (Painter 1980). Furthermore, caffeine modifies phosphorylating processes, which influence the regulation of the cell cycle (Jung and Streffer 1991).

Due to the established role of mercury and caffeine in combined effects with ionizing radiation, a multiple exposure to mercury (0.5-5 µM), caffeine (0.52 mM) and X-rays (0.25-2 Gy) in combination have been studied (Müller, 1989). As a result, a marked increase in the risk for the morphological development, the proliferation capacity and the occurrence of cytogenetic damage in pre-implantation mouse embryos was observed. A three-dimensional isobologram analysis allowed attributing this increase to interactions between radiation and caffeine, although no risk increase specific for the threefold combination was found.

The herbicide, paraquat (a dipyridine salt), slows down the progression of C3H10T1/2 cells through the cell cycle. Dependent on the paraquat concentration, one finds cell populations where the cell-division activity comes to a complete halt. Following a combined exposure to 3 Gy of gamma rays, the radiation effect with regard to this induction of cell death was found to increase by the factor 1.2. For the induction of oncogenic transformations, too, a combined effect exceeding the sum of the single effects was detected (Geard et al. 1984).

Some of the biological interactions described here are used in clinical medicine for combining radiation therapy with chemotherapeutics (see table 2.1-1). Any further clarification of the patterns of interaction between therapeutic radiation treatment and accompanying chemical agents could provide us with insights into analogue mechanisms under conditions found in the environment and at the workplace.

One essential mechanism for a cytotoxic agent sensitizing tissue is the inhibition of damage repair induced by radiotherapy.

For the combined effect of cisplatin and irradiation, for instance, it is assumed that induced breaks in one strand of the DNA double helix are not repaired as usual, but interact with another single-strand break. As a result, it comes to double-strand breaks, which are lethal events for the affected (tumour) cells.

Other agents, e.g. actinomycin D, bind to the DNA by intercalation, thereby preventing DNA synthesis and repair. Since the repair following an exposure to weakly ionizing radiation is particularly effective in the lower dose range, the effects of a repair inhibitor will be especially distinct there. Streffer (1982) showed for instance that actinomycin D increases the risk in the lower dose range to a markedly greater extent than in the higher dose range. The synergistic effect of actinomycin D was demonstrated clearly for the two-cell state of the mouse embryo: While 80 µCi of Tritium per ml of the culture medium was needed for a 50% reduction of development, the activity required went down to 35µCi/ml in the presence of actinomycin D (10^{-4} µg/ml). This dose modifying effect can be expressed as a factor of 2.3 (the ratio 80 µCiml^{-1}/35 µCiml^{-1}).

The inhibition of specific enzymes, too, can cause cytotoxic effects. Hydroxyurea, for instance, leads to a depopulation of the pool of the basic DNA building blocks (nucleotides) through this mechanism, so that the DNA repair enzymes are lacking a substrate. However, there are molecules that are incorporated into the DNA in place of the pyrimidine base thymine and in place of the uracil into the RNA, respectively (pyrimidine analogue), whereby the DNA transcription, synthesis and repair are hampered.

The combination of radio- and chemotherapy achieves cell-killing effects – which can lead to the healing or delayed growth of a tumour – in excess of the sum of the single effects. Still, this therapy can give rise to long-term complications, e.g. the induction of secondary tumours (see Tucker 1993 for an overview).

The examples quoted so far clearly demonstrate that a reduced repair capacity intensifies the damage to the DNA. This damage can be made permanent as a transformed cell.

The immune system represents a second defence facility by which the body can eliminate such cells. This factor and its potential impairment therefore deserve our special attention, when we consider risks.

The role of the immune system in defending against radiation damage was examined as long as several decades ago (Ainsworth and Chase 1959). After giving the immunostimulant lipopolysaccharide (LPS) to mice prior to exposing them to doses of radiation normally inducing the lethal haematopoietic syndrome, an increased recovery of the blood producing system was detected. The recovery pattern was very similar to that observed after transplanting isogenic (genetically identical to the receiver's) bone marrow (van Bekkum 1969), leading to the conclusion that the effect mediated by LPS can probably be attributed to a restoration of the haematopoietic cells and the lymphocytes. Today we know that the protective effects in vivo induced by LPS and other anti-inflammatory agents can be attributed to the activity of a spectrum of cytokines. The last are endogenously produced, hormone-like polypeptides. Interleukin-1 and tumour necrosis factor (TNF) alpha showed in animal experiments that they mediate protection against radiation doses that would normally be lethal, if they are given before the radiation exposure. In the lower dose range, other cytokines, too, showed effects promoting recovery. Cytokines are involved in the early steps leading to the initiation of the repair of potentially lethal damage. They were found to have a strongly inducing effect on substances neutralizing oxygen radicals (for an overview, see Neta and Oppenheim 1991).

The role of the immune system in the expression of radiation damage can also be "modelled" by looking at animal strains with defined genetic defects of the defence function. In a study by Kobayashi et al. (1996), mice lacking the thymus function (nu/+) were treated with 1-4 Gy of ^{137}Cs gamma radiation and 0.5 mg of urethane per gram of body weight. The radiation treatment caused lung tumours in only about 10% of the mice, while urethane caused this development in 70–80% of the animals, with the tumour incidence being quite similar for the two phenotypically different groups of test animals. The combined effect proved to be synergistic for both groups. On the other hand, a tendency to higher tumour rates was registered in nu/+ mice – an indication that the impeded T-cell immune function of the nu/nu mice does not promote the development of lung tumours in this system. For an assessment, however, one should keep in mind that urethane on its own already causes a high incidence of lung tumours.

Ataxia telangiectasia (AT) is a genetic disease in humans, which affects the sensitivity for radiation as well as the immune system. The disease is inherited in an autosomal recessive fashion and is characterized by a defective DNA synthesis. This disorder particularly affects the cells' capacity for repairing X-ray-induced DNA damage, leading to a repair deficiency accompanied by a defective cellular immune defence. Another feature is that some classes of antibodies (IgA, IgE, IgG$_2$, in some cases IgG$_4$) are completely missing in the individuals affected. The majority of chromosome breaks in the patients' cells occur on chromosome 14, which encodes the heavy chain of the immunoglobulins. It is still not clear if there is a genetic connection between the increased sensitivity to radiation and the immune deficiency.

Genotoxic Agent Combined with an Agent Inducing Mechanisms for Damage Repair

Apart from the processes described in the previous section, which potentiate the damage, mechanisms protecting the cell or the organism against radiation have been reported too. The principal effect mechanisms are – as far as we know – similar to the ones described above, but with the opposite sign. These substances are of special interest for general radiation protection, but also for purposes of tumour therapy, where practitioners would use combinations of radiation and chemicals in their endeavour to achieve an optimum in tumour control while, at the same time, providing the best possible protection of normal tissue against the cytotoxic effect of the therapy.

In this respect, cytosine arabinoside is a very effective agent for use on the bone marrow. The results by Millar et al. (1978) demonstrate, among other things, how varied and even conflicting the effects of a substance can be: While cytosine arabinoside is known for its repair-inhibiting effect, this special investigation showed that the substance also stimulates an increased repopulation of the surviving fraction of cells.

Chemical radiation protectors (see table 2.1-1) (Fritz-Niggli 1991) actually influencing the radiation sensitivity of cells, generally act through the detoxicating mechanism intrinsic to the cell, i.e. through antioxidative enzymes. The latter remove or detoxify any reactive oxygen compounds and their products induced by radiation. One class of radioprotectors studied quite intensively are the thiols. Glutathione is the most important of them and is present in every cell. Due to the effectiveness of this substance class, particularly sulphurous molecules, which in most cases also contain an amino group, were modified by attaching a thiophosphate group, when chemical radiation protection was developed. Results from these studies usually refer to the protection against radiation of a low linear energy transfer (LET). Table 2.3-6 gives an overview of the magnitude of the dose reduction through some of these substances. The data were taken from Langendorff (1971) and Messerschmidt (1979). The dose reduction factor was calculated as a ratio of quantities representing the lethal dose for 50% of the protected animals and for 50% of the unprotected control animals respectively (both 30 days after the radiation treatment). To obtain any

Table 2.3-6 Some substances offering radiation protection, and their dose reduction factors (DRF) for mice (from Langendorff 1971 and Messerschmidt 1979)

Radiation protection substance	DRF
Cysteamine	1.39-1.42
Cystamine	1.68
S-Aminoethylisothiuronium (AET)	1.45
Para-aminopropiophenol (PAPP)	1.56-1.63
Serotonin	1.9
AET+serotonin+cysteine+glutathion+cystamine	3
WR-2721, WR-638	2.5-3

radiation protection, the substances must however be administered in almost toxic doses, and they have no therapeutic effect when applied after the radiation treatment. Furthermore, the duration of their effect, between 30 minutes and four hours, is relatively short. Most of the substances offer no or only slight protection against neutrons. In most experiments, the survival of mice or rats was tested. There is only little experience concerning the causation of cancer.

The thiols dominated the research and development of protective substances for about three decades. More recent studies in the field of radiation protection concentrated on cytokines (Murray 1996), polysaccharides (Maisin et al. 1980) and prostaglandins (Hanson and Thomas 1983). In contrast to the chemical substances for radiation protection, some of these endogenous protectors can modulate the damage induced by radiation even when given after the exposure, which is why they could be of importance, too, for therapy following radiation accidents. The dose reduction factors for these agents, though, are relatively low (generally < 1.5).

Table 2.3-7 The relative risk of lung cancer in dependence of smoking and drinking habits and of food supplements with alpha-tocopherol or beta-carotene (from Albanes et al., 1996)

Treatment group	Relative risk	95%-confidence interval
alpha-tocopherol	0.99	0.87-1.13
beta-carotene	1.16	1.02-1.33
beta-carotene; 5-19 cigarettes/day	0.97	0.76-1.23
beta-carotene; > 19 cigarettes/day	1.25	1.07-1.46
beta-carotene; <10 g Ethanol/day	1.03	0.85-1.24
beta-carotene; > 10 g Ethanol/day	1.35	1.01-1.81

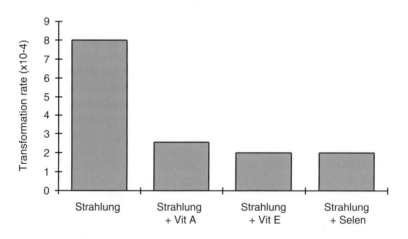

Fig. 2.3-6 The transformation rate of surviving C3H10T1/2 cells following the combined exposure to gamma rays (4 Gy) and nutrition factors. (The absolute values in this plot were taken from fig. 2 in Hall and Hei, 1990.)

Consequently, combinations of them with each other and with chemical radioprotectors were tested (Maisin et al. 1993, Murray 1996). It was shown in experiments
with rodents that the interplay between different mechanisms strengthens the protection against radiation effects while maintaining a tolerable toxicity. Unfortunately, this approach so far turned out to be less successful for higher mammals,
where the protective effect proved to be weaker and the toxic side effects stronger
than in previous experiments.

Table 2.3-8 The relative risk of lung cancer for the colletive treated with retinol and beta-
carotene compared to the placebo collective (from Omenn et al., 1996)

Treatment group	Relative risk	95%-confidence interval
Placebo; lung cancer	1.28	1.04-1.57
Placebo; other causes of death	1.17	1.03-1.33
Placebo; lung cancer	1.46	1.07-2.00
Placebo; cardiovascular diseases	1.26	0.99-1.61

There are examples for food ingredients (see table 2.1-1) combined with radiation having an antagonistic effect on the induced tumour rate. Some vitamins (E
and K) act as antioxidants. They can bind and render harmless the radiation-
induced, highly reactive radicals and oxygen compounds. Vitamins C, E and K
further possess the capacity to prevent the formation of the carcinogenic
nitrosamines in the body. Beta-carotene (the precursor of vitamin A) and selenium seem to have a positive effect on the cell-cell communication, the distur-
bance of which is of importance in carcinogenesis. The radioprotective powers of
vitamin A, vitamin E and selenium were demonstrated in a study by Hall and Hei
(1990). C3H10T1/2 cells were incubated in vitro with these substances and
exposed to 4 Gy of gamma radiation. Fig. 2.3-6 shows the results concerning the
change in the transformation rate. The reduced incidence of transformed cells
should lead to a reduced cancer incidence. In clinical-epidemiological studies, the
question was investigated if one could lower the lung cancer incidence for smokers by supplementing their food with vitamins. Albanes et al. (1996) gave daily
doses of alpha-tocopherol (vitamin E analogue; 50 mg), beta-carotene (vitamin A
precursor; 20 mg) or of a placebo to 29,133 men (50-69 years old) for 6.1 years
on average. Table 2.3-7 shows the relative risks found in this way. (For an explanation of the term relative risk, see section "Cigarette Smoke and Radon".)
Alpha-tocopherol supplements did not influence the lung cancer rate. The beta-
carotene intake led to a higher relative risk of lung cancer. When the same group
of individuals was divided into two subgroups according to their smoking behaviour, a dependence of the relative risk on the number of cigarettes smoked was
observed. Another finding was that the relative risk of lung cancer depends on the
intake of alcohol per day.

In another study (Omenn et al., 1996), the daily diet of 18,314 individuals was
supplemented, over a period of 4.0 years on average, with a combination of retinol
(vitamin A; 25,000 IU) and beta-carotene (30 mg) or with a placebo. The relative

lung cancer risks experienced by the subjects treated with retinol and beta-carotene compared the placebo subcollectives are listed in table 2.3-8.

The authors of those two studies concluded from their results that food supplements of beta-carotene and vitamin A given to smokers do not reduce the relative lung cancer risk; such supplements, given in pharmacological doses, rather seem to cause a slightly increased risk, which is associated to the individuals' behaviour with regard to smoking and alcohol consumption.

Genotoxic Agents Affecting Different Phases of Tumour Development

In general, the primary biophysical damage events initiated by genotoxic agents are distributed randomly over the genome. In other words, the specificity of many genotoxic agents, with regard to reactions at different gene locations, seems to be low. It is therefore relatively unlikely, in statistical terms, that one genotoxic agent causes damage to the DNA that initiates a tumour and then another genotoxic agent mediates a promoting activity. It will become clear from the following section that most promoting substances do not act through genotoxic mechanisms.

However, the combination of radiation and viruses – or viral genetic sequences integrating themselves into the host DNA (see table 2.1-1) – represent an exception from this rule. Astier-Gin et al. (1986) investigated the effect of radiation combined with the infection with a retrovirus on the induction of leukemia in mice. The animals received an injection with non-pathogenic retroviruses (T1223/B virus) and were exposed to radiation doses (2 x 1.75 Gy) demonstrably not inducing leukemia. After injecting the virus before or after the radiation treatment, 31% and 19% of the animals, respectively, fell ill with leukemia. The active contribution of the retrovirus was proved by identifying a recombining provirus in each of the tumours induced by the combined treatment. These studies, as well as earlier ones e.g. by Yokoro et al. (1969), describe a carcinogenic effect observed *only* when using the combination. The data indicate that a synergism takes place, in the literal meaning of the word ("acting together"). It is conceivable that the integration of the virus genome into the host DNA causes genetic transformations, which in turn enable the tumour development.

Genotoxic Agents Combined with a Non-genotoxic Agent (Initiator/Promoter)

It is generally accepted that carcinogenesis is a multistep process that can be modelled as being subdivided into the following three phases: Initiation of the genetic damage, tumour promotion, and malignant progression (fig. 2.3-1). This notion has developed over many years of tumour research (e.g. Boveri 1914, Mottram 1935, Berenblum 1941, Bishop 1991). Proceeding from experiments on the skin of mice, Berenblum showed in the 1940s that by giving a sub-effective dose of a carcinogenic followed by several exposures to a promoting agent one can induce multiple tumours. Although this model was examined most intensively by experiments on mouse skin, we have plenty of evidence today for the existence of initiating and promoting phases in other tissues too.

Initiating agents such as ionizing radiation usually act as genotoxics. However, many chemicals in our environment do not specifically attack the DNA, but influence the cell proliferation or differentiation e.g. by modifying the processes of intercellular communication or signal transduction. Substances not acting in a genotoxic fashion usually feature a concentration threshold, below which the effect does not occur. By definition, tumour promoters are tissue-specific substances that induce the proliferation in the respective target tissue and hence potentially induce the clonal expansion of malignant, transformed cells.

The tumour promoter 12-O-Tetradecanoylphorbol-13-Acetat (TPA, see table 2.1-1) can drastically raise the incidence of radiation-induced tumours. TPA is a phorbol ester found in oil extracts from the seed of the croton plant. With regard to combined effects, it is one of the best-studied substances. It is considered the strongest tumour promoter for skin neoplasia.

Several authors have shown that applying TPA to the skin of radiation-treated mice leads to a synergistic increase in the skin tumour rate as well as to the increased occurrence of some other tumour entities (e.g. Shubik et al., 1953; Jaffe and Bowden, 1987; Vorobtsova et al., 1993). Table 2.3-9 shows the results published by Jaffe et al. (1987) as a representation of the general picture.

Table 2.3-9 The tumour incidence in mice treated with various doses of radiation (initiation) and, two weeks later, with TPA applied to their skin twice a week, 8 nmol each (promotion); the numbers showing statistically significant differences ($p < 0.01$) between combination and single treatment are highlighted by a grey background (from Jaffe et al. 1987)

Ray treatment [Gy]	TPA promotion [weeks]	Animals with tumours	Tumours/mouse
0	10	0/20 (0%)	0
0	60	6/23 (26%)	0.26
0.5	0	0/25 (0%)	0
0.5	10	1/23 (4%)	0.09
0.5	60	7/22 (32%)	0.36
11.25	0	3/22 (14%)	0.14
11.25	10	2/25 (8%)	0.08
11.25	60	11/25 (44%)	0.44

The tumour rate found in animals after local irradiation with 0.5 Gy or 11.25 Gy succeeded by a promotion period of 60 days for both doses (0.36 and 0.44 tumours/ mouse respectively) did not differ significantly from each other ($p > 0.05$; chi-squared test). However, animals only treated with TPA showed a significantly lower tumour rate than mice which had also been treated with radiation ($p < 0.01$; chi-squared test). The duration of the TPA promotion clearly influenced the tumour incidence, though this can be attributed mainly to an increase in the papilloma rate, independent of the radiation dose. These data demonstrate that TPA, as a promoting agent can increase the radiation-induced carcinogenesis in vivo.

The results of Kennedy et al. (1978) and of Han and Elkind (1982) confirm this in-vivo finding through in-vitro experiments performed on a strain of mouse embryos (C3H10T1/2): Transformations induced by X-rays *in vitro*, too, were amplified by TPA (fig. 2.3-7). Kennedy et al. (1978) found that radiation doses of 0.25 Gy to 1 Gy triggered transformation rates that were hardly detectable or very low, although the promoting activity of TPA proved to be particularly distinct in the dose range from 0.5 Gy. After a treatment with 0.5 Gy, cell transformations increased 26-fold, 1 Gy led to an increase by the factor 19 and for 4 Gy, the increase was threefold. These results were obtained, however, only when certain conditions concerning the cell culture were strictly observed. The cell density, in particular, proved to be a crucial factor.

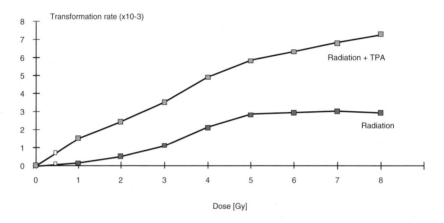

Fig. 2.3-7 The transformation rate of surviving C3H10T1/2 cells after X-ray exposure, with and without incubation in TPA (0.1 µg/ml) thereafter. (The absolute values in this plot were taken from fig. 2 in Han and Elkind, 1982.)

The dose-response curve representing the effect following radiation treatment (fig. 2.3-7, from Han and Elkind, 1982) again shows a negligible transformation rate after the exposure to 0.5 Gy and a strong amplification by a TPA treatment after the irradiation. A radiation dose of 4 Gy led to a considerably weaker combined effect with regard to the relative values.

Kennedy et al. (1980) did not detect any TPA-mediated promotion for initiated cells in the phase of confluence. Hence we can conclude that proliferating cells are the primary target for TPA. When confluent cells were re-spread at a low density, however, their susceptibility to TPA-induced promotion proved to be preserved. This result confirms the finding from in vivo experiments that the TPA effect is independent of the time passed since the exposure to the initiating agent. Jaffe et al. (1987) also studied the effect of a TPA *pre-treatment* on the initiating potential of irradiation in vivo. The results are summarized in table 2.3-10. A TPA treatment *prior to* the initiation pattern irradiation/promotion by TPA (following the treatment protocol described above) led to a small increase of the total tumour rate for animals treated with 0.5 Gy of radiation. However, neither this effect nor the reduction of the tumour incidence through the pre-treatment in the 11.25 Gy group were statistically significant ($p > 0.05$; chi-squared test).

Table 2.3-10 The effect of TPA pre-treatment (17 nmol, 24 hours prior to irradiation) on mice, in which tumours were induced by irradiation alone and by irradiation followed by TPA treatment; tumour-positive animals/total number of animals (from Jaffe et al. 1987)

Pre-treatment	Irradiation [Gy]	TPA promotion [weeks]	Animals with tumours	Tumours per mouse
-	0.5	-	0/25 (0%)	0
-	11.25	-	3/22 (14%)	0.14
-	-	60	6/23 (26%)	0.26
TPA	0.5	60	9/24 (38%)	0.5
-	0.5	60	7/22 (32%)	0.36
TPA	11.25	60	7/25 (28%)	0.44
-	11.25	60	11/25 (44%)	0.44

Han and Elkind (1982), too, found a slight increase in the transformation rate following the incubation in vitro with TPA (0.1 µg/ml) of C3H/10T1/2 cells untreated with radiation. Due to these in-vitro effects and the in-vivo findings on the TPA pre-treatment (Jaffe et al. 1987), it cannot be excluded that TPA can also act as an initiator. Another finding of that study confirms this supposition indirectly: TPA entered synergistic interactions with X-rays as well as with neutron radiation. In radiobiological terms, neutrons are the more effective type of radiation. Hence one should expect the net effects of the interactions X-ray + TPA and neutrons + TPA will reflect this difference too – if TPA only acts as a promoter. Still, there are experimental findings pointing to an independent TPA effect.

Investigations by Hill et al. (1987, 1989) showed a promoting effect of TPA, also after an exposure to gamma radiations at the extremely low dose rate (distribution of the dose over time) of 10 cGy/day. This protracted radiation treatment on its own remained sub-effective with regard to the transforming effect, in vitro, on C3H10T1/2 cells. Hence, the effect of a TPA treatment following the irradiation shows that a damage sector – and therefore an attack point for the effect of a promoting substance – remains even after an obviously unhampered progress of the DNA repair process.

Also in the context of the combined treatment with X-ray and TPA, an interesting effect on germ cells was described. Vorobtsova et al. (1993) subjected two successive generations of mice – the descendants of irradiated (whole body exposure of 4.2 Gy X-ray/animal) and of untreated animals respectively – to TPA treatment (twice weekly application to the skin, 6.2 µg TPA/application, over a period of 24 weeks). It was shown that the descendants of the radiation-treated animals developed more skin tumours after the TPA exposure than the descendants of the untreated animals. In the first generation after the radiation treatment, 20.1% of the male and 36.6% of the female offspring of untreated fathers had developed skin tumours. The skin tumour rates among the offspring of the radiation-treated animals were 75.0% and 67.5% respectively. The difference between the tumour rates among descendants of untreated animals as compared to those of radiation-treated animals was statistically significant both for male (p < 0.001) and for female animals (p < 0.01). Some of the F1-descendants (first generation offspring) were allowed to mate prior to the TPA treatment, and the F2 generation was treated with TPA in the same way as the F1 generation. The incidence of skin papilloma amounted to 57.8% and 40.0%, respectively, for the F2 generation. Again, the differences in the tumour rates (p < 0.001 for male, p < 0.01 for female animals) were statistically significant. For both generations, the tumour rates for male and for female descendants were markedly different from each other (p < 0.05). Another remarkable finding was that there were more animals with multiple (> 4) skin papillomata among the descendants of radiation-treated male mice. These results support the hypothesis that the radiation treatment of male animals prior to mating raises the susceptibility to the promoting stimulus in the carcinogenesis among the offspring. The suspected mechanism is the induction of a persistent genetic instability in the germ cells.

Another study (Nomura et al. 1990) was concerned with the effect of a postnatal TPA exposure (36x10 µg/animal, 6 weeks after birth) on *in utero* irradiated mice (0.3-1.03 Gy X-ray, dose rate 0.54 Gy/min, exposure 11 days after fertilization). The radiation doses alone did induce somatic mutations in the embryos, which however turned out not to be carcinogenic. The postnatal administration of TPA, on the other hand, gave rise to skin tumours and hepatomas in the animals radiation-treated in utero, at incidence rates rising with the radiation dose. This result is remarkable also with regard to the outcome of an epidemiological study (Kato et al. 1989), in which the cancer incidence among those survivors of Hiroshima and Nagasaki who had been exposed to radiation in utero was investigated. Here, too, the risk of falling ill with solid tumours at an adult age proved to be higher in this group of individuals. The results from animal experiments and epidemiological studies suggest that in-utero radiation exposure induces neoplastic transformations attributable to somatic mutations in the embryo. Through the contact of the affected individuals with promoting agents e.g. in the environment, at the workplace or in food, these mutations can develop into tumours. Interestingly, Nomura et al. (1990) found that the rates of mutations and tumours following an exposure to TPA and radiation at a low dose rate (0.043 Gy/min with the same total dose as for the exposure to a high dose rate, see above) were reduced by 75% and. 80% respectively.

Various studies have shown that TPA interacts synergistically with X-rays as well as with neutron radiation. TPA is effective even in nanomolar concentrations, after very short periods of exposure and long after an irradiation. As the specific cellular receptor for TPA, the protein kinase C, an enzyme mediating the promoting activity of various hormones to the respective target tissue, was identified. Due to this conspicuous relationship between the biochemical structure and the biological activity, on the one hand, and between the established pleiotropic effects on the other, TPA is thought to act in a hormone-like fashion (Blumberg 1981). The supposed, principal effect mechanism is that TPA interferes in the cell-cell communication and stimulates the cell proliferation.

Hormones (see table 2.1-1) and substances acting hormone-like might well be among the most important chemical substances with regard to their influence on environmental radiation risks. This observation is connected to the finding that a number of hormones and similar substances stimulate the cell proliferation because an increased cell proliferation is one of the essential mechanisms during the promoting phase of tumour development. Numerous studies have established that hormones in combination with ionizing radiation increase the tumour risk.

In animal experiments, an increased concentration of the TSH hormone (stimulating the thyroid gland) was shown to be a risk factor for the development tumours in the thyroid gland (Doniach 1963). Experiments with rats proved that tumours of the thyroid gland can emerge as a result of a hyperplasia. A hyperplasia is an organ enlargement caused by a supernormal cell proliferation. In the thyroid gland, this is induced by an increased secretion of the TSH hormone from the frontal pituitary gland. The triggering factor for this development could be e.g. a fall in the iodine level in the blood caused by nutritional deficiency, partial thyroidectomy or radiation. A radiation exposure of the thyroid gland initially leads to the cells losing their capacity to divide. The hyperplasia, which then ensues, represents a compensatory reaction of the organ to the radiation damage. Therefore, the exposure of the thyroid gland to a low dose of radiation can lead to the manifestation of a malignant tumour, if the latent, malignant transformations initiated by the radiation are promoted by the stimulation of a hyperplasia.

During puberty and pregnancy, the presence of the TSH hormone in humans is naturally increased. Various epidemiological studies indicate that women who become pregnant after having received therapeutic radiation treatment run a higher risk of developing a tumour of the thyroid gland than women who do not become pregnant at a time after the treatment (Conard 1984, Ron et al. 1987, Wingren et al. 1993). For the study by Ron et al. (1987), a medical anamnesis of 159 thyroid tumour patients was carried out. A radiation exposure of the head and neck, especially during childhood, was shown to be a risk factor for tumours of the thyroid gland; 9% of all cases could be attributed to this cause. Among the group of individuals who had developed such tumours and who had had radiation therapy in their childhood, the subgroup of young women who had given birth or suffered miscarriages was noticeable. It seems fair to assume that endogenous hormones, including TSH, should be considered as a promoting factor for the malignant transformations induced by the radiation treatment in childhood. In the Swedish study by Wingren et al. (1993), the number of pregnancies was postulated as an additional, risk-increasing factor.

In animal experiments on the induction of breast cancer, the hormone prolactin proved to be a strong promoter of cells, the transformation of which had been initiated by radiation or other agents (Yokoro et al. 1987). In fact, prolactin is such a strong promoter that it can be used to detect the carcinogenicity of even small doses of an agent. For instance, female rats were irradiated with fission neutrons (1.1-16.6 cGy) and supplied with the hormone by implanting a prolactin-secreting tumour into the animals, either a few days or 12 months after the radiation treatment. The results are summarized in table 2.3-11.

Just 2 out of 62 rats (3.2%) of the group only exposed to radiation developed breast tumours (one adenocarcinoma, one fibroadenoma). In strong contrast to these numbers, such tumours developed in 21 out of 63 animals (33.3%) which were treated with prolactin immediately after the irradiation (mainly adenocarcinomas). The combined treatment also shortened the latency period following the radiation treatment. In cases where the prolactin was given 12 months after the irradiation, breast tumours were induced in 12 out of 51 animals (23.7%). This incidence, which is still relatively high, points to a long survival time of the potentially malignant cells induced by the radiation treatment. Effects similar to the ones described were observed for the combination of slow neutrons and prolactin. Following an exposure to ^{60}Co gamma radiation (10 cGy, observation period 12 months), no breast tumours were found, neither after a single exposure nor after combined exposure. As a result, one can largely exclude any influence of the gamma component on the induction, as described, of breast tumours by neutron radiation.

The study demonstrates that the carcinogenic effect on the breast tissue of rats, which is induced by apparently sub-effective neutron doses, can be promoted strongly by supplying prolactin to the tissue.

Shellabarger et al. (1983) investigated the influence of the length of the time interval between the neutron irradiation and the treatment with diethylstilbestrol (DES, a synthetic estrogen) on the induction of breast cancer in female rats. The synergistic effect was strongest after a simultaneous exposure. It should be noted that (as described for prolactin too, see table 2.3-11) the extent of the synergistic effect did not decrease when there were longer time intervals (up to one year) between the irradiation and the DES exposure, meaning that the DES promotion is still effective

Table 2.3-11 The induction of breast tumours in female rats through irradiation with fission neutrons and treatment with prolactin (P). The differences between the tumour incidences in the groups 4 and 8*, 5 and 9**, and 6 and 10*** were significant (*$p < 0.05$; **$p < 0.025$; ***$p < 0.005$; from Yokoro et al., 1985)

Group	Treatment	Animals with tumours	Latency period p.r. [days]
1	none	2/24 (8%)	597, 843
2	Prolactin only	0/14 (0%)	
Irradiation only			
3	1.1 cGy	1/14 (7%)	304
4	4.1 cGy	1/16 (6%)	205
5	7.3 cGy	0/16 (0%)	
6	16.6 cGy	0/16 (0%)	
		2/62 (3%)	
Prolactin given shortly after irradiation			
7	1.1 cGy + Prolactin	1/15 (7%)	237
8	4.1 cGy + Prolactin	6/16 (38%)	191-334
9	7.3 cGy + Prolactin	5/15 (33%)	165-365
10	16.6 cGy + Prolactin	9/17 (53%)	163-365
		21/63 (33%)	
Prolactin given 12 months after irradiation			
11	1.1 cGy + Prolactin	1/ 6 (17%)	502
12	4.1 cGy + Prolactin	4/15 (27%)	463-567
13	7.3 cGy + Prolactin	3/15 (20%)	504-549
14	16.6 cGy + Prolactin	4/15 (27%)	375-526
		12/51 (24%)	

even after long periods of time. This result makes us conclude that the neutron irradiation initiates a breast cancer oncogenesis which repair mechanisms cannot cure.

The male sex hormones, the androgens, also lead to an increased tumour rate, when they are combined with ionizing radiation. This was demonstrated in experiments by Hofmann et al. (1986): Male rats were treated with 40µCi of Na^{131}J, which induced, at a high incidence of 94%, follicular adenomas and carcinomas of the thyroid gland.

For animals that were castrated prior to the radiation treatment, a reduction in the incidence of thyroid gland tumours by 34%, compared to irradiated but not castrated animals, was observed. The exogenous administration of testosterone to the castrated and radiation-treated animals led to an increase in the tumour frequency depending on the time interval between the irradiation and the hormone treatment: Testosterone replacements in physiological concentrations within a year resulted in tumour rates of 100% (hormone replacement within 6 months post radiation treatment, p.r.), 82% (6-12 months p.r.), 70% (12-18 months p.r.) and 73% (18-24 months p.r.) respectively. It was shown that the concentration of serum TSH in irradiated rats was higher than in animals not subjected to radiation treatment, and that the TSH concentration was higher in castrated, irradiated and testosterone-replaced animals than in castrated, irradiated, but not testosterone-replaced animals. Therefore, we can assume that a testosterone-mediated THS stimulation is involved in the radiation-induced oncogenesis of thyroidal cancer in rats.

It is a well-known fact that eating habits can modify the effect of chemical carcinogenics. Silverman et al. (1980) studied the influence of the fat content in foods on the radiation-induced incidence of breast cancer (table 2.3-12). Rats aged 50 days were subjected to a whole-body exposure to X-rays (3.5 Gy) and given, from the age of 30 days (group 4) and 50 days (all other groups), food rich (20%) or poor in fat (5%) respectively. 40% of the irradiated animals died before the experiment was concluded. Compared to the animals in group 6 (kept on a diet low in fat), the radiation-treated animals in groups 4 and 5 (high fat content in the food from 30 or 50 days of age respectively) suffered an overall higher incidence of tumours overall,

Table 2.3-12 The effect of the fat content in the diet given to rats on the radiation-induced (3.5 Gy) incidence of mamma carcinomata in the animals. The difference between group 4 and group 6 is statistically significant ($p < 0.01$; chi-squared test)

Treatment	Group no.	Number of animals	Age of the animals at the start of the diet [days]	Mamma carcinomata		Average latency period [days]
				Number of rats with tumours	Total number of tumours	
Control group	1	9		0	0	
20% fat (HF)	2	27	50	0	0	
5% fat (LF)	3	29	50	0	0	
Radiation+HF	4	49	30	21	28	185
Radiation+HF	5	47	50	12	20	188
Radiation+LF	6	50	50	8	9	210

both in terms of the tumour frequency per animal and of a shorter latency period. However, the results carried statistical significance ($p < 0.01$; chi-squared test) for the comparison between group 4 and group 6 only.

For humans, too, a connection was identified between the spontaneous rate of breast cancer and the amount and type of fats taken in with food (Welsch 1987). There are indications that the mechanism driving these effects involves changed balances between certain hormones, especially between prolactin and estrogen.

In recent years, the role of a diet (see table 2.1-1) with regard to its modifying potential for the development of experimental as well as human tumours has been emphasized worldwide (e.g. U.S. Department of Health and Human Services 1988, Weisburger 1991). A study by Doll and Peto (1981) attributes the factor food/nutrition with a 35% causal contribution to the total mortality due to cancer. It is considered a proven fact, for instance, that different fats are differently strong promoters of the cancer development. Cancer advice organizations suggest dietetic cancer prophylaxes warning of ω-6-multisaturated fatty acids and recommending oils, like cold-pressed olive oil, containing mono-unsaturated fatty acids. From the studies so far (see above, alpha-tocopherol and beta-carotene) we can conclude, however, that not a single substance is counteracting the emergence of a tumour, but that a healthy lifestyle and eating habits are the important factors.

Cigarette smoke (see table 2.1-1), even as a single agent, is considered extremely relevant with regard to public health issues. Beyond that, smoking tobacco probably represents the most important case of an activity increasing risks posed by noxae in the human environment. An increased risk of death from lung cancer was found for the lung carcinogenics arsenic, asbestos, dichloromethyl ether and radon (Steenland and Thun, 1986; also compare chapter 2.4).

In this section we will report the results on how smoking cigarettes increases radiation risks. Because of the public-health relevance of this combination, it appears to be appropriate at this point to look into the fundamental question of the epidemiological proof for any agent modifying the risk posed by an(other) environmental noxa, too.

In a paper by Steenland (1994), the incidence of death by lung cancer among 3,568 American uranium miners, who were occupationally exposed to radon, are compared to the corresponding rate for US veterans in order to determine the influence of tobacco smoking on the number of lung tumours. The data collected refer to the years from about 1950 to 1980. Table 2.3-13a shows the respective lung cancer mortality for different age groups. The relative risk posed by smoking cigarettes was determined by calculating the ratio of the lung cancer death rates among smokers not exposed to radon and the corresponding rate among non-smokers not exposed to radon (RR_1); in the same way, the relative risk solely due to radon resulted by dividing the lung cancer death rates among the non-smoking uranium miners by the rate for the non-smoking US veterans (RR_2). Hence one obtains a relative death risk of 1 for the group "non-smokers without radon". For all exposure groups, an age dependence was detected in the risk of death by lung cancer. Above the age of 60 years, a plateau or a decrease of the effect was observed. This could be explained by the overall effect of the two strong risk factors being so high that, due to the limited population of sensitive individuals, it cannot come to its full expression for the specific causation.

Table 2.3-13a The relative risk of dying from lung cancer for uranium miners of different age groups, dependent on their status as smokers or non-smokers (Steenland 1994). The calculation used the death rates for exposed and non-exposed individuals; further explanations in the text.

	Relative risk of dying from lung cancer			
Age	Non-smokers without radon (n=55,049)	Smokers without radon (n=142,518)	Non-smokers with radon (n=516)	Smokers with radon (n=3,052)
30-50 years	1	4.6	7.9	28.8
50-59 years	1	9.1	21.9	72.3
60-69 years	1	9.0	13.4	47.5
≥ 70 years	1	8.0	6.2	36.6

Moolgavkar et al. (1993) and Leenhouts and Chadwick (1997), too, found a clearly dependence between the lung tumour risk and the age of the individuals at the time of the exposure.

In scientific experiments, the qualified identification of combined effects was arrived at through an extensive assessment of the dose-response relationship and by isobologram analyses based on the dose-response relationships of the noxae and their combinations (see 2.2.2.5). This approach offered insights into the interactive mechanisms, which is not possible, as a rule, in epidemiological analyses. The purpose of the statistical models applied here is to develop equations that describe the combination results already known and to estimate outcomes that are not known yet. Epidemiological data are compared with statistical values calculated according to an additive or multiplicative model and used as the additive basis. Equality of the observed and the computed data means that no interaction between the combination partners has taken place. Any deviation from the computed values, on the other hand, is as sign of interaction. One should note the semantic problem of "interaction" being used as a statistical term. The statement of a statistical interaction does not necessarily mean that there was an interaction between biological processes and mechanisms.

The additive model is defined as $\qquad$ $RR_{12} = RR_1 + RR_2 - 1$
or, alternatively, as $\qquad$ $R_{12} = R_1 + R_2 - R_0.$

The multiplicative model is defined as $\qquad$ $R_{12} = RR_1 \times RR_2$
Or, alternatively, as $\qquad$ $R_{12} = R_1 \times R_2/R_0,$

where RR represents the relative risks and R the effect rates for the agents 1 and 2; the relative risks are calculated as the ratio of the effect rate following an exposure to the agents 1 or 2 and the spontaneous rate R_0.

Results in agreement with the expectations according to the additive model indicate independent effects of the individual agents. The agents cause their effects through different attack points; the respective single effects contribute independently to the effect of the combination (see 2.2). If the observed data reflect stronger

Table 2.3-13b The risk of dying from lung cancer following the combined exposure to radon and cigarette smoke; comparison of the relative risks calculated through the additive or the multiplicative model with the relative risk* reflected in the observed rate (from Steenland 1994)

Age	Expected relative risk according to the		Relative risk "observed"*
	additive model	multiplicative model	
30-50	11.5	36.3	28.8
50-59	30.0	199.3	72.3
60-69	21.4	120.6	47.5
≥ 70	13.2	49.6	36.6

effects than predicted by the additive model, this could be interpreted as an indication that the combined effect of the exposure exceeds the sum of the single effects. A deviation from the additive model is equated to the phenomenon of effect modification, meaning the relative risk is modified by the effect of the second agent. In both cases, any deviation from the respective model "prediction" means that individuals exposed to the agents concerned face a higher health risk. Avoiding or reducing one of the two noxae would therefore help to protect one's health.

For all age groups, the relative risk arrived at by Steenland (1994) for the combination of the two lung carcinogenics turned out to range between the values to be expected according to computations with the additive and the multiplicative model respectively (see table 2.3-13b).

The multiplicative model best describes the results for the youngest and oldest age groups, with major deviations in between these two extremes. The varying, age-specific discrepancies between the observed data and the risk model might point to unknown mechanism underlying the effect. The exact mechanisms, by which radon or cigarette smoke induce lung tumours, are in fact not fully established. It is still under discussion if these agents act as initiators, as promoters or as complete carcinogenics.

Scores of other epidemiological studies on mineworkers (e.g. Thomas et al. 1985, Burkart 1989, Lubin et al. 1994, Yao et al. 1994) concerned the combined exposure to radon and cigarette smoke. The study by Lubin et al. (1994) is one of the most extensive investigations on this matter, combining the results from 11 groups of mineworkers. This work, too, demonstrates that the combination of ionizing radiation and cigarette smoke leads to stronger than additive effects with regard to the lung tumour rate. Again, the combined risk often ranges between additive and multiplicative.

Considering those epidemiological studies, one should always bear in mind that the risk factors arrived at involve exposures accompanying radon in a mining environment (e.g. the exposure to arsenic). Furthermore, the relatively small number of "control subjects" reduces the statistical significance of the results (non-smokers with lung tumours). The radiation doses involved are often relatively high, while the degree of the smoking habit has not been determined with a sufficient accuracy in most cases.

Pershagen et al. (1994) performed a case-control study in Sweden. They investigated the natural background radiation – i.e. radon at low concentrations – in connection with smoking cigarettes. 586 women and 774 men were included as "cases" and 1,380 women and 1,467 men as "controls" for this analysis. All lung tumours diagnosed between 1980 and 1984 were registered. For the single exposure to radon, the researchers arrived at relative risks (as related to radon concentrations up to 50 Bq/m^3) of 1.3 (for 140-400 Bq/m^3) and 1.8 (at concentrations > 400 Bq/m^3). For the combined exposure to radon and cigarette smoke, the authors found an effect stronger than additive, approaching a multiplicative effect. These risk assessments are in good agreement with those derived from the mineworker data.

Another group of individuals, for whom cigarette consumption presents an increased risk of radiation-induced lung tumours, are radiotherapy patients with breast cancer (Neugut et al. 1994) or with morbus Hodgkin's lymphoma (van Leeuwen et al. 1995). In the study by Neugut et al. (1994) the question was investigated if smoking cigarettes and radiotherapy have a multiplicative effect on the risk of a developing cancer thereafter (secondary tumour). In this case, the control subjects were patients who had developed secondary tumours, too, the genesis of which was however demonstrably not attributable to radiation or smoking. Both for smokers and for non-smokers, who had developed a secondary cancerous disease more than ten years after being diagnosed with breast cancer, a threefold, radiotherapy-associated increase in the risk of lung cancer was demonstrated. A synergistic effect was detected in women who had been exposed both to therapeutic radiation and to cigarette smoke. It was assumed in this study that radiotherapy protocols common before 1980 were applied; it is not clear yet if today's therapies, which involve lower radiation doses for the lungs, have similar effects.

The finding with rats that a synergistic effect can be achieved experimentally only if the radon exposure takes place prior to the inhalation of cigarette smoke or if the exposures occur simultaneously, suggests that one or several component(s) of cigarette smoke act as promoters for lung tissue (Chameaud et al. 1982, Gray et al. 1986).

There are indications that a gene of the cytochrome P450 family is of importance for the super-additive combination effect of tobacco smoke and radon (Douriez et al. 1994). CYP 1A1 is involved in the metabolic activation of polycyclic hydrocarbons into reactive, mutagenic substances. To a large degree, its expression is also induced by cigarette smoke. In the interaction between cigarette smoke and the alpha radiation of plutonium, another mechanism was identified: Cigarette smoke delays the removal from the lung of inhaled plutonium oxide (Talbot et al. 1987, Finch et al. 1994 b). It is, however, still unknown if the synergistic effect in this is induced by interactions between the radiation and the cigarette smoke on the molecular or cellular level, or if it is attributable to the locally increased lung dose due to the retention of the plutonium.

A subcutaneous injection with tetrachlorocarbon (CCl_4, see table 2.1-1) following a radiation treatment with fast neutrons showed a synergistic effect on the incidence of radiation-induced liver carcinomas in mice (Cole and Howell 1964, Benson et al. 1994). The results by Cole and Nowell documented a threefold increase in the hepatoma rate for mice that were not only irradiated (1.65-3.06 Gy neutrons) but also injected with CCl_4 following the radiation exposure. The neutron radiation treatment induced hepatomas in 19% of the animals, CCl_4 on its own proved to be ineffective in the observation period of 22 months, and the combined exposure resulted in a tumour rate of 61%.

A strong carcinogenicity was demonstrated for asbestos fibres (see table 2.1-1). Even small amounts of asbestos fibres can induce bronchial tumours and malignant pleuramesotheliomas. The effect mechanism differs greatly from that of genotoxic substances: Asbestos fibres do not react with the DNA. They do not act like direct mutagenics, but they induce chromosomal aberrations leading to the formation of cells with a chromosome set other than normal (aneuploidy), which is correlated with the transformation of cells. Phagocytosis leads to the accumulation of the fibres in the cell close to the nucleus, where they impede the distribution of the chromosomes between the daughter cells. Asbestos fibres also cause an increased cell proliferation, through mechanical irritation and cell killing (Barrett 1992). The discovery of the super-additive effect of the exposure to asbestos and tobacco on the induction of lung tumours is one of the first examples of a true, synergistic combination effect relevant for the protection of workers (Selikoff et al. 1968, also see chapter 2.4). For this combined exposure, a shift of the tumour spectrum from mesotheliomas to broncho-pulmonary carcinomas was observed. Similar results (including the shift of the tumour spectrum) of less statistical significance were found with rats exposed to radon (1600 WLM) and injected subcutaneously with chrysotile fibres (20 mg). While the control groups registered a lung tumour incidence of 1% (asbestos exposure only) and. 49% (radon inhalation only) respectively, the combined treatment resulted in a lung tumour rate of 55%, which was considered significantly higher than the result for radon exposure alone (Bignon et al. 1983).

In-vitro exposure to asbestos fibres and alpha emitting helium-3 ions, which stimulate the radon emission, leads to a risk markedly increased compared to the single effects (Hall and Hei, 1990). C3H10T1/2 cells were exposed to crocidolite fibres (5 μg/ml), helium-3 ions (0.66 Gy) and to a combination of the two noxae (24-hours pre-treatment with crocidolite). Fig. 2.3-8 shows the results, which indicate that the combined exposure induces a transformation rate in vitro about four times higher than the value expected for additivity.

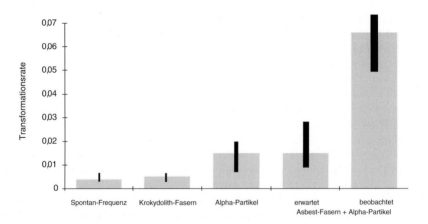

Fig. 2.3-8 The transformation rate in C3H10T1/2 cells following an exposure to alpha particles and asbestos fibres. (The absolute values in this plot were taken from fig. 4 in Hall and Hei, 1990; the dark bars represent the 95%-confidence intervals.)

Many agents not acting as direct genotoxics evidently cause clearly synergistic effects when combined with ionizing radiation. However, for some of the substances studied experimentally (e.g. TPA), we do not know of any functional analogues in the environment, though some of these agents are of a high environmental relevance; they just do not occur in sufficiently critical concentrations in the environment, or there is a science deficit with regard to the impact of relatively low concentrations of these chemicals. Nevertheless, the examples presented here demonstrate some basic mechanisms that could underlie the interactions between radiation and other noxae. In this sense, they contribute to our basic understanding of these interactions.

2.3.3.3
Summary Assessment and Recommendations for Action

The DNA in the human body (consisting of 10^{13}–10^{14} cells, to give the order of magnitude) undergoes, according to biochemical estimations, about 10^3–10^4 oxidative damage events per cell. Apart from these oxidative effects of radicals, the DNA suffers direct damage e.g. through natural radiation exposures. Oxidative reactions with radicals are statistically distributed over the genome of mammalian cells, while this is not the case for direct damage caused by ionizing particles. Along the passage of an ionizing particle through the chromatin or the double helix, clusters of ionization and therewith radiochemical transformations occur. The clusters consist of accumulations of radiation damage (multiply damaged sites) reflecting accumulations of ionizations along the trace of such a particle through the chromatin or the DNA. The extent of these clusters depends on the quality of the radiation, since different qualities of radiation feature different ionization densities in the respective material. The natural exposure to endogenous and exogenous processes relates to the impact of toxic agents in the environment (or at the workplace). According to an estimation by Doll and Peto (1981), 1.4% of all cases of death by cancer occurring in the USA every year can be attributed to the exposure to background radiation (see 2.4.2).

Due to the natural sources of ionizing radiation in our environment and because of the stochastic nature of radiation effects, it would be unpractical to aim for a complete suppression of radiation effects. The existence of a threshold dose for such probability-related effects can be largely excluded. The basic assumption of the concept of risk in radiation protection, which was introduced by the International Commission on Radiological Protection in 1977, therefore proceeds from the thesis that the average dose equivalent in a tissue or organ rises proportionally with the probability that a tumour occurs in this tissue or organ. By setting dose limits for the general population and for persons occupationally exposed to radiation, one limits the individuals' risk of falling ill with a radiation-induced tumour. The ALARA (As Low As Reasonably Achievable) rule aims at optimizing radiation protection against doses below the respective limit values too. For safety reasons, any estimate of the effects of exposures to low doses of radiation is always based on a linear dose-response relationship – even if this probably leads to an overestimation in many cases. Extrapolations to the lower dose range start from doses in the range of 0.2-2 Sv, where the effects are well known from observations in humans (not just from animal experiments).

As a matter of fact, each substance possibly damaging human health always occurs in combination with our exposure to natural radiation. Due to this naturally combined exposure, but also to anthropogenic influences, mutually amplifying or attenuating effects are conceivable in the lower, environmentally relevant dose range, too, as factors in the observed risks of e.g. cancerous illnesses. Hence arises the issue of an appropriate regulatory protection against combined exposures.

Against this background, the effort was made in the preceding section to draw a picture, based on a multitude of data from literature, of our present knowledge on how chemicals influence the radiation risk. Because of the vast number of chemicals – about 14 million compounds altogether, according to present estimates, a variable number of which, depending on the country concerned, is present in the environment (100,000–200,000) – and the variety of possible combinations between them, it makes sense to accept that a systematic analysis of combination effects for every single agent is utterly unfeasible. As an analytical instrument, a grid of "interaction principles" was drawn up (see table 2.1-1). The individual principles are guided by basic mechanistic features shared by two noxae with regard to their interaction patterns. Since stochastic effects are at the centre of risk considerations for the lower dose range, the development of cancer was investigated as the main biological end point of the interactions.

In the assessment of combined effects, one assumes a relatively homogenous, representative radiation sensitivity among the population, although we know that some individuals are much more sensitive to radiation than the vast majority. Such individuals fall ill with cancer more often, too. Frequently, during and after radiotherapy acute and delayed reactions in the normal tissue of these patients are observed; these events are the first manifestation of the primary illness in many cases. Quite often such patients show a genetic predisposition, e.g. patients with ataxia telangiectasia or Fanconi's anaemia. Above all, disorders of the DNA repair process and changes in the regulation of the cell cycle occur, causing an increased sensitivity to radiation, accompanied by a stronger genetic instability and cell death (e.g. German 1983, Murnane and Kapp 1993, Streffer 1997d). This is true not only for the homozygous carriers of heritable information of this syndrome, but also for the heterozygous carriers, which cannot be easily recognized by phenotypic transformations and could be more prevalent among cancer patients (Swift et al. 1991). It is a matter of discussion if for this "subgroup" of the population the combination effects must be assessed in a different way from that applied for the "general" population in order to achieve a better health protection for this group. However, there are no studies on the issue of combined exposures for individuals with the said genetic pre-dispositions.

In the following section, we will try to derive some general statements and principles concerning the appraisal of the influence of chemicals on the radiation risk for "the" population. These "rules" are continuously checked against new discoveries and are subject to change if necessary.

2.3.3.3.1 Detection of Combined Effects: Animal experiments as well as cell-biological and epidemiological studies have shown that, in the vast majority of cases, the result of a combination of ionizing radiation and toxic substances is either the addition of the single effects or an effect weaker than the sum of the single effects. Under certain conditions, synergistic or super-additive effects have been observed after exposures to some single substances combined with ionizing radiation. Essentially, these are combinations of ionizing radiation with:

- Smoking tobacco: Numerous studies especially on uranium miners have provided proof that stronger than additive effects can result; the significance of these results for the environmentally relevant lower dose range is still unclear.
- Hormones: There are many indications that hormones or substances acting hormone-like can raise the risk from radiation. The question, however, if the observed increases are applicable not only to the high radiation doses studied (1 Sv and above) but also for the small doses (some mSv) occurring in the environment or in occupational dealings with ionizing radiation, has yet to be settled. For food components (e.g. fats) that can influence the hormonal status, effects on the radiation risk were actually shown. Typically, these substances stay effective over prolonged periods after the radiation exposure.
- Viruses: The few data available point to a possible synergistic effect.

2.3.3.3.2 The Condition of Specificity: Radiation injury, e.g. the causation of cancer, develops through several steps. Extensive radiobiological studies show that for an interaction to take place, the chemical agent has to intervene in events in the development chain after the irradiation, in order to increase the radiation effect.

In principle, we would expect an amplification of the malignant cell transformation through interactions between ionizing radiation and genotoxic chemical substances. In practice, however, the effect of the agents investigated proved to be of insufficient specificity; in many cases, sub-additive effects were observed instead. One possible reason for this could be that the damage events following the exposure to ionizing radiation and genotoxic substances happened so far apart in the genome that an interaction could not take place.

A number of non-genotoxic substances, on the other hand, which often act as promoters increasing the cell proliferation or hampering the DNA repair following radiation exposure, emerged as remarkably effective. The increased cell proliferation leads to a multiplication of radiation-induced tumour cells and thereby to a rise in the cancer rate. Through inhibition of DNA repair processes, more mutations occur than after the course of an unhampered repair of radiation damage, resulting in genetic effects or increased transformation rates of normal cells into tumour cells. The specificity of these mechanisms can be explained as follows: For the cell proliferation to be influenced, interactions between the substances and specific receptors on the cell membrane have to take place; these interactions then trigger signal transduction processes. The inhibition of DNA repair requires an equally specific inactivation of such enzymes that are involved in the repair processes. Both processes, the modified cell proliferation and the suppressed DNA repair, can also combine or interplay.

The importance of the increased cell proliferation for e.g. the induction of breast tumours in rats through radiation treatment combined with doses of estrogen has been shown experimentally. In epidemiological studies, it was observed that a hormone-induced rise of the cell division rate in the thyroid gland leads to a super-additive increase in the tumour incidence after the radiation exposure. A synergistic effect was also found in the development of lung tumours initiated by combinations of radiation and cigarette smoke. Here, too, the underlying mechanism is thought to be an increased cell proliferation caused by substances contained in tobacco smoke.

The inhibition of DNA repair can be just as effective in causing super-additive combination effects between ionizing radiation and chemical substances. This has been demonstrated e.g. for heavy metals or caffeine in high concentrations. The amplification to be expected, from the mechanistic point of view, through damage to the immune system is still under discussion.

2.3.3.3.3 The Time Interval between Radiation and Substance Exposure: Some studies have demonstrated quite impressively that amplifying effects can also occur when the exposure to the substance takes place long after the radiation exposure. This applies mainly to the effects of certain promoters e.g. some hormones. In many cases, however, the interval between the substance intake and the radiation exposure or between the irradiation and the substance intake, respectively, must not be too long, as the results in table 2.3-14 show in a representative manner. This is particularly true for interactions on the DNA-repair level.

In the study on which table 2.3-14 is based, the development (morphology, proliferation) of pre-implantation mouse embryos was investigated. The results show that, for the induction of a super-additive total effect to occur, the mercury exposure has to take place a few minutes after or even before or during the irradiation, although the presence of mercury during the radiation exposure does not lead to any further amplification beyond the effect achieved through incubation immediately after the radiation exposure. Furthermore, the mercury exposure must extend over a relatively long time interval.

For chemical substances acting as radioprotective agents, it has been found that they must be taken immediately prior to irradiation.

2.3.3.3.4 The Relevant Concentration/Dose Range: In most cases, super-additive results were obtained when the substance was given in medium or high doses. Due to the involvement of signal transduction processes or reactions with enzymes described above, very small exposures are not expected to have an effect, since a

Table 2.3-14 The impact of the time factor on combination effects, demonstrated for the exposure of pre-implantation mouse embryos to X-rays and mercury (from Müller and Streffer, 1988); x = additive effect, xx = effect stronger than additive, xxx = effect markedly stronger than additive

Radiation ↓ exposure

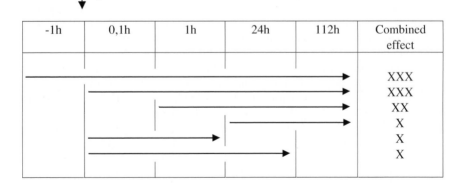

-1h	0,1h	1h	24h	112h	Combined effect
					XXX
					XXX
					XX
					X
					X
					X

sufficiently high number of substance molecules must bind to receptors in order to cause a signal transduction or to perform an enzymatic step quickly enough. It is extraordinarily difficult to arrive at any statement on the respective substance concentration leading to a combined effect with ionizing radiation. In contrast to the situation for ionizing radiation, for which a definite estimation can be made concerning the dose in tissues and cells, substance exposures "on target" (the target organ or molecule) cannot be determined with the same certainty. For the knowledge of the concentration applied does not imply sufficient information about the quantity of the substance that actually interacts with the radiation-induced damage at the target organ after pharmacokinetic processes have taken place.

For the combination partner "ionizing radiation", too, any effect amplification observed mostly related to medium and higher dose ranges, with only few exceptions.

When analyzing the existing literature, one notices such an exception being the finding that some substances cause a markedly stronger increase in the relative radiation risk in the lower dose range than in the higher range (e.g. Han and Elkind 1982, Kennedy et al. 1978, Streffer 1982). This observation applies, above all, to substances promoting the development of tumours or inhibiting DNA repair. The effect does not surprise, since promoters predominantly act by inhibiting the DNA repair or by stimulating the proliferation of cells. As DNA repair is particularly effective in the lower dose range, the effects of the repair inhibitor, too, should be more distinct in that range. The stimulation of cell division is a process relevant in the lower and medium dose range as well; higher radiation doses inhibit the capacity of cell division or lead to cell death, so that a growth stimulation of malignant, transformed cells cannot take place.

Experimental radiation doses, for which an increased risk of stochastic effects in combination with chemicals was found in animal experiments, typically ranged between 1 and 10 Gy. Only such studies that were concerned exclusively with the extremely strong promoter TPA found combination effects following an irradiation with 0.5 Gy. Obviously, any proof for an increased risk in the lower dose range can only be obtained with systems more sensitive than the ones described here.

Table 2.3-15 reflects quite "typical" results from experiments in which only high doses (> 0.5-1 Gy) and substance concentrations (1-2 mM in the case of caffeine,

Table 2.3-15 The dose and concentration dependence of combined effects in pre-implantation embryos of mice (from Müller et al., 1985; Müller and Streffer, 1987). End points for a): morphologic development, proliferation, cytogenetic damage; end points for b): morphologic development, proliferation; x = additive effect, xxx = effect stronger than additive

a) Radiation + caffeine

[mM] \ [Gy]	0.1	0.5	1	2
0.25			x	
0.5		x	xxx	
1			xxx	xxx
2				xxx

b) Radiation + mercury chloride

[mM] \ [Gy]	0.5	3	5	10
0.25			x	
0.5	x		x	
1		xxx		xxx
2		xxx		

see table 2.3-15a, about 3 μM for mercury chloride, see table 2.3-15b) led to a super-additive effect of the combined exposure. The caffeine concentration required, in particular, was extremely high – like several hundred cups of coffee –, so that the observed risk increase will certainly not occur in practice. Still, the result is of scientific interest in the sense that it helps clarify the question of mechanisms. The mercury concentration, too, corresponds to about 100 times the average mercury concentration found in the blood of pregnant women.

2.3.3.3.5 The Influence of the Radiation Quality: Effects caused by densely ionizing radiation – i.e. radiation that is biologically more effective at a given energy dose, notably neutrons – can hardly be influenced by chemicals interfering in early events of the damage development e.g. the DNA repair. This fact is connected to special characteristics of the biological effect of this radiation type. Compared to processes relevant for the development of tumours (see 2.3.2), neutrons are distinguished by the following features: Hardly any repair processes take place after an exposure to neutrons. Therefore, a substance attributed with repair-inhibiting properties can raise the risk at most very slightly when combined with neutron radiation. It is not possible, in this case, that chemicals would achieve a synchrony in a certain phase and keep the cells in either a radiation-sensitive or a radiation-insensitive phase. Hence, the effect of densely ionizing radiation is largely independent of the cell cycle. The oxygen effect does not play a significant role either; substances will hardly be able to act as protectors by reducing the oxygen concentration. "Radical catchers" will also fail to contribute to a protection against neutron radiation, since the direct radiation effect (excitation and ionization of molecules) dominates for this radiation quality, resulting in relatively few radicals being created.

Another picture emerges for effects following combinations of neutrons and certain hormones. In such combinations, interactions with the densely ionizing radiation could be demonstrated. Prolactin (Yokoro et al. 1987) and diethylstilbestrol (DES, Shellabarger et al. 1983), in particular, were shown to contribute to combination effects with neutron radiation. Effects on the cell-proliferation level probably play a role here.

For most tissues, the contribution of densely ionizing radiation from natural sources amounts to not more than a few per cent. The lung, on the other hand, is an example where the radon isotopes and their decay products change this picture. There is no evidence for chemicals influencing the radiation damage as such on the cellular or molecular level. Instead, it was suggested in several papers (e.g. Sanders et al. 1978, Talbot et al. 1987, Finch et al. 1991, Finch et al. 1994b) that chemicals have an impact on the biokinetics of isotopes that emit alpha particles in their decay, leading to a delayed elimination of the radiation source from the lung and hence to an increased effect. More information concerning the processes is needed here, especially in the context of the everyday exposure to the combination of radon and cigarette smoke.

2.3.3.3.6 Risk Appraisal for Stochastic Effects Following a Combined Exposure in the Environmentally Relevant Dose Range: In principal, any influence of a chemical agent on the radiation risk should find expression in the dose-response curve, which reflects the underlying mechanism (Streffer and Müller 1984). Where a chemical inhibits the damage repair, one expects that the normal, sigmoid shape of the dose-response curve is modulated towards a linear shape. This implies that the change is particularly manifest in the lower dose range, since repair processes are most important there. Such an effect was, however, found only after the administration of high doses of the chemical agents researched so far (see table 2.3-15). If the substance is a promoting agent, we expect not only that a proliferation stimulus causes an increase in the tumour risk for the lower dose range, but also that the dose-response curve takes an overall steeper shape. This observation is related to the promoting agents' capacity to stimulate "sleeping" tumour cells into proliferation, thereby increasing the tumour incidence. The frequently observed shortening of the latency period by promoting agents contributes to this shape, too. For most other risk-increasing mechanisms, the main expectation would be a steepening of the dose-response curve, while most risk-decreasing mechanisms should be expressed in a flattening of the slope.

Unfortunately, the consequences for the dose-response curve rarely are unambiguous enough to allow the identification of the underlying mechanism. Furthermore, most studies do not show dose-response curves for the single noxae, but only comparisons at certain points. In this way, selective influences on the risk can be found, expressed in deviations between the combined effect and the sum of the single effects. However, even a significant disagreement with the expected value does not necessarily indicate a true interaction between the noxae; for further clarification, an isobologram analysis (see 2.1) would have to be performed, which in turn would require the knowledge of the respective dose-response curves. Dose-response curves can also be evaluated directly. One approach of appraising the risk from combined effects without having to take into account the dose-response curves of the single agents is the risk model according to Rothman (1976) (also see section 2.2). This model is based on the consideration that, for an assessment of the risk following a combined exposure to two agents, the crucial question is if the combined effect exceeds the sum of the single effects or otherwise:

$$P(exp.) \ = \ P(A) + P(B) - P(A) \times P(B),$$

where P(exp.) is the expected probability for the combination causing the effect investigated; P(A), P(B) are the probabilities for an effect occurring after an application of agent A or agent B. The equation, which stems from probability theory, applies to cases where two causes act independently from each other [also see chapter Pöch]. If the observed effect is in agreement with the combined effect expected according to the equation above, then Rothman calls it an additive combined effect. If the observed value exceeds the expected value, the finding, in terms of this model, is synergism; is the observed effect weaker than the expected one, it is antagonism. This model suffers the drawback, however, that it only allows statements for the conditions of combined exposures actually investigated, since the shape of the dose-response curves has a considerable influence on the result of a combination.

Fig. 2.3-9 shows relative-risk coefficients (for different end points and experimental conditions) calculated with this model and some data from literature.Fig. 2.3-9 demonstrates that the influence on the risk is small in many cases. The risk modification rarely exceeds doubling or halving the risk. Even these effects are only achieved with high doses of the substances used or with high radiation doses respectively. The spectacular cases with high risk coefficients usually represent very special

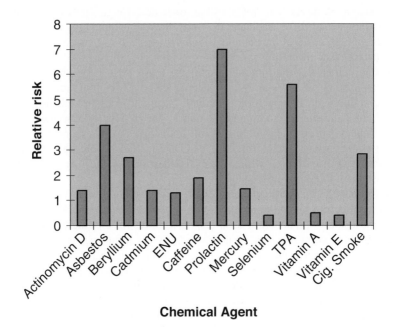

Chemical Agent

Fig. 2.3-9 Some relative-risk coefficients following combined exposures to radiation and chemicals, calculated with data by Streffer, for actinomycin D (1982), Hall and Hei for asbestos, selenium, vitamin A and vitamin E (1990), Brooks et al. for beryllium (1989), Müller and Streffer for Cadmium (1987), Schmahl and Kriegel for ethylnitrosourea ENU (1978), Müller et al. for caffeine (1985), Yokoro et al. for prolactin (1987), Müller and Streffer for mercury (1987), Kennedy for the phorbol ester TPA (1978), and Finch et al. for cigarette smoke (1994 b). Please note that the coefficients of the relative risk (modified factors) relate to different end points and to experimental conditions with maximum effects, because high doses untypical for the environment were used. Data from low exposures much more common in the environment, are scarce or not existing. (An additive effect corresponds to a relative risk of 1.)

experimental conditions that are not or hardly ever found in everyday life. (The effect of TPA, for instance, could only be achieved if a relatively low cell density in the culture was not exceeded; the effect of asbestos is limited to in-vitro conditions.)

Statements on the influence of genotoxic chemicals on the radiation risk in the lower dose range, which is the relevant range in the environment and at the workplace, are arrived at through mathematical extrapolations from the higher doses to the lower dose range. For several reasons, this approach is however inappropriate for combinations with non-genotoxic agents. For some of these substances, threshold values apply, below which an effect even in combination with ionizing radiation can be excluded. For other substances it was shown that the relative-risk increase in combination with low doses of radiation is considerably larger, giving rise to cases where a linear extrapolation would even lead to an underestimation of the risk.

Based on present knowledge, the situation concerning a regulatory recommendation for action presents itself as follows: Under conditions *in the environment* near nuclear installations operating according to their specifications, a detectable modification of the health risk for the population caused by a combined exposure to ionizing radiation and a chemical agent, compared to the corresponding single exposure, appears unlikely. However, exposures due to breakdown or accidents can lead to different situations.

One has to bear in mind that the data on and the knowledge about the mechanisms of interaction are insufficient for the relevant dose range. If combination effects occur in this low dose range, we can only assume that these effects are hidden among variations of the natural radiation exposure and in the variation of the effects due to it; hence the combination effects would be "unknowingly" covered by the conventional regulations concerning single noxae. Still, investigations into biological effects following combined exposures with ionizing radiation as a combination partner are of scientific interest in the sense that they help us gaining a better understanding of the mechanisms, on which basis we could draw conclusions with regard to the lower dose range too.

Super-additive effects from noxious combinations acting through certain mechanisms cannot be excluded. For some agents, a basic potential to change the radiation risk was found, which however unfolds under very special conditions only. Smoking probably poses a super-additive risk, although it is not clear yet if the observed combination effects are relevant for the lower radiation dose range. In many cases, the substance concentrations needed for achieving a risk modification are relatively high and hence largely excluded by the existing legislation.

Those substances that were shown to cause synergistic combined effects have in common that they influence certain processes in a specific way. For this reason, these agents and the respective mechanisms should remain at the centre of scientific interest. The processes concerned mainly affect the creation of radicals, DNA repair and cell proliferation.

2.3.4
Literature

Ainsworth EJ and Chase HB (1959) Effect of microbial antigens on irradiation mortality in mice. Proc. Soc. Exp. Biol. Med. 102, 483–485

Albanes D, Heinonen OP, Taylor PR, Virtamo J, Edwards BK, Rautalahti M, Hartman AM, Palmgren J, Freedman LS, Haapakosi J, Barrett MJ, Pietinen P, Malila N, Tala E, Liipo K, Salomaa E-R, Tangrea J-A, Teppo L, Askin FB, Taskinen E, Erozan Y, Greenwald P, Huttunen JK (1996) Alpha-Tocopherol and Beta-carotene supplements and lung cancer incidence in the alpha-tocopherol. beta-carotene cancer prevention study: effects of base-line characteristics and study compliance. J. Natl. Canc. Inst. 88 (21), 1560–1570

Allen B, Crump K and Shipp A (1988) Correlation between carcinogenic potency of chemicals in animals and humans. Risk Analysis, 8, 531–544

Ames B (1989) Mutagenesis and carcinogenesis: endogenous and exogenous factors. Environ. Mol. Mutagen. 14, 66–77

Astier-Gin T, Galiay T, Legrand E and et al. (1986) Murine thymic lymphomas after infection with a B-ecotropic murine leukemia virus and/or X-irradiation: proviral organization and RNA expression. Leukemia Research, 10, 809–817

Barrett JC (1992) Mechanisms of action of known human carcinogens. In: Vainio H, Magee PN, McGregor DB, McMichael AJ (eds.) Mechanisms of Carcinogenesis in Risk Identification, International Agency for Research on Cancer (IARC). Lyon, pp. 115–134

Benson JM, Barr EB, Lundgren DL et al. (1994) Pulmonary retention and tissue distribution of 239Pu nitrate in F344 rats and syrian hamsters inhaling carbon tetrachloride. ITRI-144, 146–148

Berenblum I (1941) The mechanism of carcinogenic action related phenomena. Cancer Res. 1, 807–814

Berenblum I, Trainin N (1960) Possible two-stage mechanism in experimental leukemogenesis. Science 132, 40–41

Bignon J, Monchaux G, Chameaud J et al. (1983) Incidence of various types of thoracic malignancy induced in rats by intrapleural injection of 2 mg of various mineral dusts after inhalation of Rn-222. Carcinogenesis 4, 621–628

Bishop JM (1991) Molecular themes in oncogenesis. Cell 64, 235–248

Blumberg PM (1981) In vitro studies on the mode of action of the phorbol esters, potent tumour promotors. CRC Crit. Rev. Toxicol. 3, 152

Boveri T (1914) Zur Frage der Entstehung maligner Tumoren. Gustav Fischer, Jena

Brooks AL, Griffith WC, Johnson NF, Finch GL, Cuddihy RG (1989) The induction of chromosome damage in CHO cells by beryllium and radiation given alone and in combination. Radiat. Res. 120, 494–507

Burkart W (1989) Radiation biology of the lung. The Science of the Total Environment 89, 1–230

Burkart W, Finch GL, Jung T (1997) Quantifying health effects from the combined action of low-level radiation and other environmental agents: can new approaches solve the enigma? The Science of the Total Environment 205, 51–70

Canellos GP, Arseneau JC, DeVita VT, Wlang-Peng J, Johnson REC (1975) Second malignancies complicating Hodgkin's disease in remission. Lancet 1, 947–949

Chambers PL, Chambers CM, Greim H (1989) Biological monitoring of exposure and the response at the subcellular level to toxic substances. Arch. Toxicol., suppl. 13

Chameaud J, Perraud R, Chretien J et al. (1982) Lung carcinogenesis during in vivo cigarette smoking and radon daughter exposure in rats. Cancer Res. 82, 11–20

Cloudman AM, Hamilton KA, Clayton RS, Brues AM (1955) Effects of combined local treatment with radioactive and chemical carcinogens. J. Nat. Cancer Inst. 15, 1077–1083

Cole LJ, Howell PC (1964) Accelerated induction of hepatomas in fast neutron-irradiated mice injected with carbon tetrachloride, Ann. N. Y. Acad. Sci. 114, 259–267

Conard RA (1984) Late radiation effects in Marshall Islanders exposed to fallout 28 years ago. In: Boice JD Jr., Fraumeni JR Jr. (eds.), Radiation Carcinogenesis: Epidemiology and Biological Significance. Raven Press, New York, pp. 57–71

Doll R, Wakeford R (1997) Risk of childhood cancer from fetal irradiation. Brit. J. Radiol. 70, 130–139

Doll R, Peto R (1981) The Causes of Cancer: Quantitative estimates of avoidable risks of cancer in the United States today. J. Natl. Cancer Inst. 66 (6), 1192–1315

Domon M, Barton B, Forte A, Rauth AM (1970) The interaction of caffeine with ultraviolet-light-irradiated DNA. Int. J. Radiat. Biol. 17, 395–399

Doniach I (1963) Effects including carcinogenesis of I-131 and X-rays on the thyroid of experimental animals – a review. Health Phys. 9, 1357–1362

Douriez E, Kermanach P, Fritsch P et al. (1994) Cocarcinogenic effect of cytochrome P-450 1A1 inducers for epidermoid lung tumour induction in rats previously exposed to radon. Radiat. Prot. Dosim. 56 (1-4), 105–108

DuMouchel W, Groer PG (1989) A Bayesian methodology for scaling radiation studies from animals to man. Health Phys. 57 Suppl 1, 411–418

Fearon ER, Vogelstein B (1990) A genetic model for colorectal tumorigenesis. Cell 61 (5), 759–767

Finch GL, Haley PJ, Hoover MD, Griffith WC, Boecker BB, Mewhinney JA, Cuddihy RG (1991) Interactions between inhaled beryllium metal and plutonium dioxide in rats: effects on lung clearance. 49–51

Finch GL, Hahn FF, Carlton WW et al. (1994 a) Combined exposure of F344 rats to beryllium metal and 239PuO2 aerosols. ITRI-144, 81–84

Finch GL, Nikula KJ, Barr EB et al. (1994 b) Exposure of F344 rats to aerosols of 239PuO2 and chronically inhaled cigarette smoke. ITRI-144, 75–77

Frei W (1913) Versuche über Kombination mit Desinfektionsmittel. Z. Hyg. 75, 433–496

Fritz-Niggli H (1991) Biologisch-chemischer Strahlenschutz. In: Strahlengefährdung/Strahlenschutz. 3. edn., Verlag Hans Huber, Bern, Stuttgart, Toronto, 126–131

Geard CR, Shea CM, Georgsson MA (1984) Paraquat and radiation effects on mouse C3H10T1/2 cells. Int. J. Radiat. Oncol. Biol. Phys. 10, 1407–1410

German J (1983) Patterns of neoplasia associated with the chromosome breakage syndromes. In: German J Chromosome mutation and neoplasia. Alan R. Liss, New York, 97–134

Gray R, Lafuma J, Parish SE et al. (1986) Lung tumors and radon inhalation in over 2000 rats: Approximate linearity across a wide range of doses and potentiation by tobacco smoke. in: Thompson RC and Mahaffey JA (eds.) Life-span radiation effects in animals: What can they tell us? Department of Energy, Washington, D.C., 592–607

Hall EJ, Hei TK (1990) Modulation factors in the expression of radiation-induced oncogenetic transformation. Environ. Health Perspect. 88, 149–155

Han A, Elkind MM (1982) Enhanced transformation of mouse 10T1/2 cells by 12-O-tetradecanoylphorbol-13-acetate following exposure to X-rays or to fission-spectrum neutrons. Cancer Res. 42, 477–483

Hanson WR, Thomas C (1983) 16.16-Dimethyl prostaglandin E2 increases survival of murine intestinal stem cells when given before photon radiation. Rad. Res. 96, 393–398

Hartwig A (1995) Current aspects in metal genotoxicity. Biometals 8, 3–11

Hartwig A (1996) Interaction of carcinogenic metal compounds with Deoxyribonucleic Acid Repair Processes. Ann. Clin. Lab. Sci. 26 (1), 31–38

Hasgekar NN, Pendse AM, Lalitha VS (1986) Effect of irradiation on ethyl nitrosourea induced neural tumours in Wistar rats. Cancer Lett. 30, 85–90

Heinonen OP, Albanes D, Taylor PR et al. (1996) Alpha-tocopherol and beta-carotene supplements and lung-cancer incidence in the alpha-tocopherol, beta-carotene cancer prevention study – effects of base-line characteristics and study compliance. J. Natl. Cancer Inst. 88 (N21), 1560–1570

Hendrickx AG, Axelrod LR, Clayborn LD (1966) 'Thalidomide' Syndrome in Baboons. Nature 210, 958–959

Hill CK, Han A, Elkind MM (1987) Promotion, dose rate and repair processes in radiation-induced neoplastic transformation. Radiat. Res. 109, 347–351

Hill CK, Han A, Elkind MM (1989) Promotor-enhanced neoplastic transformation after gamma ray exposure at 10 cGy/day. Radiat. Res. 119, 348–355

Hofmann C, Oslapas R, Nayyar R, Paloyan E (1986) Androgen-mediated development of irradiation-induced thyroid tumors in rats: dependence on animal age during interval of androgen replacement in castrated males. J. Nat. Canc. Inst. 77, 253–260

Hoshino H, Tanooka H (1975) Interval effect of beta-irradiation and subsequent 4-nitroquinoline-1-oxide painting on skin tumour induction in mice. Cancer Res. 35, 3663–3666

ICRP (1977) Recommendations of the International Commission on Radiological Protection. ICRP Publication 26, Pergamon Press Oxford

ICRP (1991) Recommendations of the International Commission on Radiological Protection. ICRP Publication, Pergamon Press Oxford

Jaffe DR, GT Bowden GT (1987) Ionizing radiation as an initiator: effects of proliferation and promotion time on tumour incidence in mice. Canc. Res. 47, 6692–6696

Jaffe DR, Williamson JF, Bowden GT (1987) Ionizing radiation enhances malignant progression of mouse skin tumors. Carcinog. 8, 1753–1755

Jha AN, Noditi M, Nilsson R, Natarajan AT (1992) Genotoxic effects of sodium arsenite on human cells. Mutat. Res. 284, 215–221

Jung EG, Trachsel B (1970) Molekularbiologische Untersuchungen zur Arsencarcinogenese. Arch. klin. exp. Derm. 237, 819–826

Jung T, Streffer C (1991) Association of protein phosphorylation and cell cycle progression after X-irradiation of two-cell mouse embryos. Int. J. Radiat. Biol. 60 (3), 511–523

Kalter H, Mandybur TI, Ormsby I, Warkany J (1980) Dose-related reduction by prenatal X-irradiation of the transplacental neurocarcinogenicity of ethylnitrosourea in rats. Canc. Res. 40, 3973–3976

Kato H, Yoshimoto Y, Schull WJ (1989) Risk of cancer among children exposed to A-bomb radiation in utero: a review. IARC Sci. Publ. 96, 365–374

Kennedy AR, Mondal S, Heidelberger C, Little JB (1978) Enhancement of x-ray transformation by 12-O-tetradecanoyl-phorbol-13-acetate in a cloned line of C3H mouse embryo cells. Canc. Res. 38, 439–443

Kennedy AR, Murphy G, Little JB (1980) The effect of time and duration of exposure to 12-O-tetradecanoyl-phorbol-13-acetate (TPA) on x-ray transformation of C3H 10T1/2 cells. Cancer Res. 40, 1915–1920

Knowles JF (1984) Reduction of N-nitroso-N-ethylurea-induced neurogenic tumors by X-radiation: A life-span study in rats. J. Nat. Canc. Inst. 72, 133–137

Knowles JF (1985) Changing sensitivity of neonatal rats to tumorigenic effects of N-nitroso-N-ethylurea and X-radiation, given singly or combined. J. Nat. Canc. Inst. 74, 853–857

Kobayashi S, Otsu H, Noda Y et al. (1996) Comparison of dose-dependent enhancing effects of gamma-ray irradiation on urethan-induced lung tumorigenesis in athymic nude (nu/nu) mice and euthymic (nu/+) littermates. J. Cancer Res. Clin. Oncol. 122, 231–236

Kusiak RA, Springer J, Ritchie AC et al. (1991) Carcinoma of the lung in Ontario gold miners: possible aetiological factors. Br. J. Ind. Med. 48, 808–817

Langendorff H (1971) Der gegenwärtige Stand der biologisch-chemischen Strahlenschutzforschung. In: Der Strahlenunfall und seine Behandlung, Vol. XI, Strahlenschutz in Forschung und Praxis. Thieme, Stuttgart, p. 90

Leenhouts HP, Chadwick KH (1997) Use of a two-mutation carcinogenesis model for analysis of epidemiological data. In: Health Effects of Low Dose Radiation: Challenges of the 21st Century. British Nuclear Energy Society, London; pp. 145–149

Little JB, McGandy RB, Kennedy AR (1978) Interactions between Polonium 210 irradiation, benzo[a]pyrene, and 0.9% NaCl solution instillations in the induction of experimental lung cancer. Canc. Res. 38, 1929–1935

Loewe S (1953) The problem of synergism and antagonism of combined drugs. Arzneim. Forsch. 3, 285–290

Lubin JH, Boice JD Jr., Edling C et al. (1994) Radon and lung cancer risk: A joint analysis of 11 underground miners studies. 94–3644

Lücke-Huhle C (1982) Alpha-irradiation-induced G2-delay: A period of cell recovery. Radiat. Res. 89, 298–308

Maisin JR, Gerber GB, Lambiet-Collier M, Mattelin G (1980) Chemical protection against the long-term effect of whole-body exposure of mice to ionizing radiation III. The effects of fractionated exposure to C57BL mice. Radiat. Res. 82, 487–497

Maisin JR, Albert C, Henry (1993) Reduction of short-term radiation lethality by biological response modifiers given alone or in association with other chemical protectors. Radiat. Res. 135, 332–337

Meijers JMM, Swaen GMH, Bloemen LJN (1997) The predictive value of animal data in human cancer risk assessment. Regulatory toxicology and pharmacology 25, 94–102

Messerschmidt O (1979) Über den chemischen Strahlenschutz in seiner Bedeutung für die Katastrophen- und Wehrmedizin. Wehrmed. Mschr. 23, 193

Millar JL, Blackett NM, Hudspith BN (1978) Enhanced post-irradiation recovery of the haemopoietic system in animals pretreated with a variety of cytotoxic agents. Cell Tissue Kinet. 11, 543–553

Moolgavkar SH, Luebeck EG, Krewski D et al. (1993) Radon, cigarette smoke and lung cancer: a re-analysis of the Colorado Plateau uranium miners'data. Epidemiology 4 (3), 204–217

Mottram JC (1935) The origin of tar tumours in mice, whether from single cells or many cells. J. Pathol.40, 407–409

Müller W-U, Streffer C, Fischer-Lahdo C (1983) Effects of a combination of X-rays and caffeine on preimplantation mouse embryos in vitro. Radiat. Environ. Biophys. 22, 85–93

Müller W-U, Streffer C, Wurm R (1985) Supraadditive formation of micronuclei in preimplantation mouse embryos in vitro after combined treatment with X-rays and caffeine. Teratog. Carcinog. Mutagen. 5, 123–131

Müller W-U, Streffer C (1987) Risk to preimplantation mouse embryos of combinations of heavy metals and radiation. Int. J. Radiat. Biol. 51, 997–1006

Müller W-U, Streffer C (1988) Time factors in combined exposures of mouse embryos to radiation and mercury. Radiat. Environ. Biophys. 27, 115–121

Müller W-U (1989) Toxicity of various combinations of X-rays, caffeine, and mercury in mouse embryos. Int. J. Radiat. Biol. 56, 315–323

Murray D (1996) Radioprotective agents in Kirk-Othmer, Encyclopedie of chemical technology, fourth edition, Volume 20, ISBN 0-471-52689-4, John Wiley & sons inc. (eds.), 963–1006

Nagayo T, Ito A, Yamada S (1970) Accelerated induction of hepatoma in rats fed N,N`-2,7-fluorenylenebisacetamide by X-irradiation to the target area. Gann 61, 81–84

Neta R, Oppenheim JJ (1991) Radioprotection with cytokines: learning from nature to cope with radiation damage. Cancer Cells 3, 391–396

Neugut AI, Murray T, Santos J, Amols H, Hayes MK, Flannery JT, Robinson E (1994) Increased risk of lung cancer after breast cancer radiation therapy in cigarette smokers. Cancer 73, 1615–1620

Nomura T, Nakajima H, Hatanaka T, Kinuta M, Hongyo T (1990) Embryonic mutation as a possible cause of in utero carcinogenesis in mice revealed by postnatal treatment with 12-O-Tetradecanoylphorbol-13-acetate. Cancer Research 50, 2135–2138

Omenn GS, Goodman GE, Thornquist MD et al. (1996) Effects of a combination of beta carotene and vitamin A on lung cancer and cardiovascular disease. N. Engl. J. Med. 334 (18), 1150-1155

Painter RB (1980) Effect of caffeine on DNA synthesis in irradiated and unirradiated mammalian cells. J. Mol. Biol. 143, 289–301

Pershagen G, Akerblom G, Axelson O, Clavensjö B, Damber L, Desai G et al. (1994) Residential radon exposure and lung cancer in Sweden. N. Engl. J. Med. 330, 159–164

Pierce DA, Shimizu Y, Preston DL, Vaeth M, Mabuchi K (1996) Studies of the mortality of atomic bomb survivors. Report 12, Part I. Cancer: 1950-1990. Radiat. Res. 146, 1–27

Preston DL, Kusumi S,Tomonaga M. et al. (1994) Cancer incidence in atomic bomb survivors. Part III. Leukemia, Lymphoma and Multiple Myeloma, 1950-1987. Radiat. Res. 137, 68–97

Ron E, Kleinermann JD, Boice Jr, LiVolsi VA, Flannery JT, Fraumeni JF Jr. (1987) A population-based case-control study of thyroid cancer. J. Nat. Canc. Inst. 79, 1–12

Rossman TG (1981) Enhancement of UV-mutagenesis by low concentrations of arsenite in E.coli. Mutat. Res. 91, 207–211

Rothman KJ (1976) The estimation of synergy or antagonism. Am. J. Epidemiol. 103, 506–511

Sanders CL, Cannon WC, Powers GJ (1978) Lung carcinogenesis induced by inhaled high-fired oxides of beryllium and plutonium. Health Phys. 35, 193–199

Schmahl W, Kriegel H (1978) Oncogenic properties of transplacentally acting ethylnitrosourea in NMRI-mice after antecedent X-irradiation. Zeitschrift für Krebsforschung und Klinische Onkologie 91, 69–79

Selikoff I, Hammond EC, Chug J (1968) Asbestos exposure, smoking and neoplasia. J. Am. Med. Assoc. 204, 106–112

Shellabarger CJ, Stone JP, Holtzman S (1983) Effect of interval between neutron radiation and diethylstilbestrol on mammary carcinogenesis in female ACI rats. Environ. Health Perspect. 50, 227–232

Shubik P, Goldfarb AR, Ritchie AC, Lisco H (1953) Latent carcinogenic action of beta radiation on mouse epidermis. Nature 171, 934–935

Silverman J, Shellabarger CJ, Holtzman S et al. (1980) Effect of dietary fat on X-ray-induced mammary cancer in Sprague-Dawley rats. J. Nat. Canc. Inst. 64, 631–634

Stammberger I, Schmahl W, Nice L (1990) The effects of X-irradiation, N-ethyl-N-nitrosourea or combined treatment on O^6-alkylguanine-DNA alkyltransferase activity in fetal rat brain and liver and the induction of CNS tumours. Carcinog. 11, 219–222

Steel G, Peckham J (1979) Exploitable mechanisms in combined radiotherapy-chemotherapy: The concept of additivity. Int. J. Radiat. Oncol. Biol. Phys. 5, 85–91

Steenland K, Thun M (1986) Interaction between tobacco smoking and occupational exposures in the causation of lung cancer. J. Occup. Med. 28 (2), 110–118

Steenland KA (1994) Age specific interactions between smoking and radon among United States uranium miners. Occup. Environ. Med. 51, 192–194

Streffer C (1982) Some fundamental aspects of combined effects by ionizing radiation and chemical substances during prenatal development. in: Kriegel H, Schmahl W, Kistner G and Stieve FE (eds.) Developmental Effects of Prenatal Irradiation. Fischer, Stuttgart, pp. 267–285

Streffer C, Müller W-U (1984) Radiation risk from combined exposures to ionizing radiations and chemicals. Adv. Radiat. Biol. 11, 173–210

Streffer C (1991) Stochastische und nichtstochastische Strahlenwirkungen. Nucl. Med. 30, 198–205

Streffer C (1997a) Threshold dose for carcinogenesis: What is the evidence? In: Goodhead DT, O'Neill P, Menzel HG (eds.) Microdosimetry. An Interdisciplinary Approach. The Proceedings of the Twelfth Symposium on Microdosimetry, September 1996 in Oxford, UK; pp. 217–224

Streffer C (1997b) Strahlenexpositionen in der Umwelt und am Arbeitsplatz. In: Konietzko J, Dupius H (eds) Handbuch der Arbeitsmedizin, II-3.3.1, 19. suppl. del.11, 1–10

Streffer C (1997c) Wirkungen ionisierender Strahlen unter besonderer Berücksichtigung von Expositionen am Arbeitsplatz und in der Umwelt. Konietzko J, Dupuis H (eds.) Handbuch der Arbeitsmedizin (eds.), IV-3.8.1, 19. suppl. del.11, 1–13

Streffer C (1997d) Genetische Prädisposition und Strahlenempfindlichkeit bei normalen Geweben. Strahlentherapie und Onkologie 9, 462–468

Swift M, Morrell D, Massey RB, Chase CL (1991) Incidence of cancer in 161 families affected by ataxia-telangiectasia. New Engl. J. Med. 325, 1831–1836

Talbot RJ, Morgan A, Moores SR et al. (1987) Preliminary studies of the interaction between 239-Pu-oxide and cigarette smoke in the mouse lung. Int. J. Radiat. Biol. Rel. Stud. Phys. Chem. Med. 51, 1101–1110

Tanooka H and Ootsuyama A (1993) Threshold-like dose response of mouse skin cancer induction by repeated beta irradiation and its relevance to radiation-induced human skin cancer. Recent Results Cancer Res. 128, 231–241

Taylor PR, Qiao YL, Schatzkin A, Yao S-X, Lubin J, Mao B-L, Rao J-Y, McAdams M, Xuan Z and Li J-Y (1989) Relation of arsenic exposure to lung cancer among tin miners in Yunnan Province, China. Br. J. Ind. Med. 46, 881–886

Thomas DC, McNeill KG, Dougherty C (1985) Estimates of lifetime lung cancer risks resulting from Rn progeny exposures. Health Phys. 49, 825–846

Thompson D, Mabuchi K, Ron E et al. (1994) Cancer induction in atomic bomb survivors. Part II: Solid tumors. 1958-87. Radiat. Res. 137, 17–67

Tucker MA (1993) Secondary cancers. In: DeVita Jr. VT, Hellman S, Rosenberg SA et al. (eds.) Cancer: Principles and Practice of Oncology, 4. edition, J.B. Lippincott Co. Philadelphia, pp. 2407–2416

United States Department of Health and Human Services (1988) The surgeon general's report on nutrition and health. Public Health Service, DHHS, Publication (PHS) 88–50210, Washington, D.C.

UNSCEAR (1986) United Nations Scientific Committee on the Effects of Atomic Radiation. Sources and Effects of Ionizing Radiation. United Nations, New York

UNSCEAR (1988) United Nations Scientific Committee on the Effects of Atomic Radiation. Sources and Effects of Ionizing Radiation. United Nations, New York.

UNSCEAR (1993) United Nations Scientific Committee on the Effects of Atomic Radiation. Sources and Effects of Ionizing Radiation. United Nations, New York

UNSCEAR (1994) United Nations Scientific Committee on the Effects of Atomic Radiation. Sources and Effects of Ionizing Radiation. United Nations, New York

Vallee BL, Ulmer DD (1972) Biochemical effects of mercury, cadmium and lead. Annual Review of Biochemistry 41, 91–128

van Bekkum DW (1969) Bone marrow transplantation and partial body shielding for estimating cell survival and repopulation. in: Bond VP and Sugahara T (eds.), Comparative Cellular and Species Radiosensitivity. Igaku Shoin, Tokyo, pp. 175–192

van Leeuwen F, Klokmann W, Stovall M (1995) Roles of radiotherapy and smoking in lung cancer following Hodgkin`s disease. J. Nat. Canc. Inst. 87, 1530–1537

Vogel HH, Zaldivar R (1971) Co-carcinogenesis: The interaction of chemical and physical agents. Radiat. Res. 47, 644–659

Vorobtsova IE, Aliyakparova LM, Anisimov VN (1993) Promotion of skin tumors by 12-O-tetradecanoylphorbol-13-acetate in two generations of descendants of male mice exposed to X-ray induction. Mutat. Res. 287, 207–216

Wei LX, Zha YR, Tao ZF, He WH, Chen DQ, Yuan YL (1990) Epidemiological investigation of radiological effects in high background radiation areas of Yangjiang, China. J. Radiat. Res. Tokyo, 31 (1), 119–136

Weisburger JH (1991) Nutritional approach to cancer prevention with emphasis on vitamins, antioxidants and carotenoids. Am. J. Clin. Nutr. 53, 226S–237S

Welsch CW (1987) Enhancement of mammary tumorigenesis by dietary fat: review of potential mechanisms. Am. J. Clin. Nutr. 45, 192–202

Wingren G, Hatschek T, Axelson O (1993) Determinants of papillary cancer of the thyroid. Am. J. Epidemiol. 138, 482–491

Xuan X-Z, Lubin JH, Jun-Yao L, Li-Fen Y, Sheng LQ, Lan Y, Jian-Zhang W, Blot WJ (1993) A cohort study in southern China of workers exposed to radon and radon decay products. Health Phys. 64, 120–131

Yao SX, Lubin JH, Qiao YL (1994) Exposure to radon progeny, tobacco use and lung cancer in a case-control study in southern China. Radiat. Res. 138, 3260–336

Yokoro K, Ito T, Imamura N et al. (1969) Synergistic action of radiation and virus induction of leukemia in rats. Canc. Res. 29, 1973–1976

Yokoro K, Niwa O, Hamada K (1987) Carcinogenic and co-carcinogenic effects of radiation in rat mammary carcinogenesis and mouse T-cell lymphomagenesis: a review. Int. J. Radiat. Biol. Rel. Stud. Phys. Chem. Med. 51, 1069–1080

2.4
Combined Exposure to Chemical Substances

2.4.1
Introduction

Nearly all limit values established so far apply to single substances only, mainly for three reasons: The first limit values aimed at preventing health damage at the workplace. The impetus came from cases where chemical workers suffered poisoning. Although several substances were involved quite frequently, the rule was that such incidents could mostly be attributed to a single chemical, on which the efforts of injury prevention then concentrated. Occupational exposure limits, which were the only limits in existence for seven decades (SRU 1996), were behind the types of limit values introduced later, e.g. the limits on general air pollution, food contaminants, etc. The dominating toxic component was regarded as representative for the effect of multiple combinations.

The second, equally important reason derives from side effects of the exponential expansion of chemical innovation: Ever more substances were discovered, developed into useable products and introduced into various spheres of life. In the beginning, observations on human beings were called upon for finding limit values, but later in the course of the development of a toxicology guided by the "polluter pays" principle, the animal experiment came to the fore, more and more used as a predictive tool. New substances had to be tested for possible harmful effects *before* they were used and marketed. Traditionally, however, procedures based on animal experiments were designed to test individual, single substances. Standard models for examining combinations did not exist. Another problem was that for any new substance it was all but impossible to say which other substances it could be accompanied by, and with which substances it could act in combination. Therefore, German acts of parliaments, notably the *Chemikaliengesetz* ("chemicals act"), merely prescribe single-substance tests, with pharmaceuticals still being the sole exception from this rule. The *Arzneimittelgesetz* ("Pharmaceuticals Act") of 1976 regulates that any new drug to be licensed must be tested for possible or foreseeable interactions – but only interactions with other pharmaceuticals and with alcohol.

As a third reason we must note the complexity of real or conceivable combined exposures. The components in mixtures of chemical substances, which humans are exposed to at work and in everyday life, present an almost unlimited variety with regard to their number and – far more important – to their effect types and contributions. Another aggravating factor is that not only substances foreign to a living organism (xenobiotics) can interact with each other; the components of common food, physical exertion, varying hormone levels, medical disorders and many other factors can have a crucial impact on harmful effects of chemicals. Moreover, the scenario is not complete without physical harm factors: High-energy radiation, heat, noise, weather conditions and even purely mechanical strains on body functions can supersede and modulate toxic effects to a significant degree. This variety and, above all, variability has thwarted every attempt so far to develop a simple, formalized, and hence practical test strategy for establishing limit values on combined exposures to harmful substances. Simple mathematical models assuming additive effects, as they were suggested in the context of workers' protection (e.g. the Threshold Limit Values List, TLV, for the USA), do not stand up to scientific scrutiny; they presume the pure additivity of substances, hence making an arithmetic link between substance [dose] increase and effect increase, neglecting a fundamental principal of the dependence between dose and effect, and failing to cover the especially important case of super-additive increase in effect.

2.4.1.1
The Categorical Heterogeneity of Limit Values

On the other hand, it has to be stated that there is not a single valid report in existence citing health damage caused by a combination of several substances, when the limits for the single substances involved were adhered to. Even if this does not prove the absence of such effects, it rather supports than contradicts the suggestion that the existing system of limit values is efficient. Pharmaceuticals, again, are the exception, where, although the "standard doses" for the single substances were observed, side effects occurred that have to be attributed to the combined exposure. One reason for the lack of findings on harmful combination effects below the single-substance limits may be the non-existence of limit values for scores of environmental pollutants. More plausible, however, is the explanation that most limit values were set under application of "safety margins", pushing the limit values out of the dose range where harmful effects can be expected. The notion of safety margins goes back to the American suggestion to call on animal experiments for deriving limits on residual contents of supplements and environmental pollutants in food (A. Lehman, 1953): In a chronic experiment, a threshold value is established (NOAEL = No Observed Adverse Effect Level), and the limit value is set at a hundredth of this threshold value. The safety margin is justified by the possibility that humans could be more sensitive than test animals, by measuring uncertainties and lack of reproducibility of toxicological experiments, by interindividual variations of the sensitivity in human populations and by other experimental conditions not known (yet) or not open to standardization – and by the principle that it is better to be on the "safe side", when there is any doubt. The choice of the factor 100 is completely arbitrary; it is scientifically justified in principle, but not in its numerical value.

2.4.1.2
Problems concerning "Safety Margins"

Presently existing categories of limit values are based on a great variety of safety margins (also called "uncertainty margins"). Figure 2.4-1 shows some relevant examples of this fact. The range is extremely wide, spanning more than five orders of magnitude. In workers' protection (MAK values, BAT values), nearly full use is made of the range ending at the effect threshold. For standards on ambient air, one keeps to the same principle while moving away from the threshold established in experiments by a factor of up to ten, depending on the data available for individual substances and taking into account that people of different ages and health conditions will be affected. For food, the American suggestion of a factor 100 has been accepted worldwide; in some countries, a lower (10) or higher (1000) factor is applied in regulations for certain cases. The concept of the recently introduced soil conservation value aggregates the entire span in four multiplicative extrapolation steps of 10, so up to 10,000. For pesticide residues in drinking water, the EU has set a limit value of 0.1 µg/litre. This value only takes into account the chemical-analytical detection limit. For the most important pesticides, the factors between the limit value and the effect threshold determined in animal experiments range from 1,000 to 1,000,000. The extreme case, for which the EPA (US) has applied the factor 1,000,000, is dioxin (TCDD = 2,3,7,8-Tetrachloro-dibenzo-dioxin). If only the principle of precaution is applied – i.e. if one moves away from any effect relationship – there is, in theory, no upper limit; approaching zero-tolerance means factors rising towards infinity.

The other extreme is the case of "negative" safety factors as applied to pharmaceuticals. In that field, one always expects unwelcome side effects, insofar as a drug has been judged to be of a reproducible therapeutic effect. This is a matter of a conscious decision to tolerate the situation, as the (therapeutic) benefit outweighs any harm.

The scale of extreme safety margins is the product of a historic development, not of rational argument. Non-specialists will find such margins impossible to understand, and from the scientific as well as from the political perspective, they appear questionable. Concerning the question if and how limit values for combined exposures can be established, we can draw two conclusions: Risk comparisons based on existing limits belonging to different systems are not feasible. Still, there are too many efforts of selling simple, so-called "pragmatic" solutions, which appear tempting because they can be arrived at with such ease: Take a simple arithmetic model, like using existing limit values in contributions weighted according to the respective mass contributions in a substance mixture, and calculate a "mixture limit". Such approaches however lack any legitimacy. Limit values for combined exposures can only be guided by quantitative effect parameters gained through identical or at least comparable methods (see section 1.2.1).

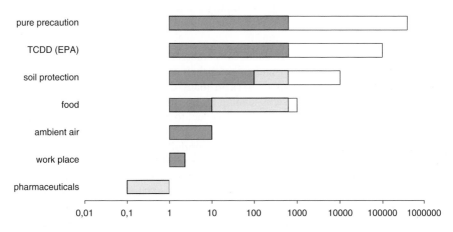

Fig. 2.4-1 Examples of "safety margins" for the definition of limit values. The value spans are displayed in light grey.

2.4.1.3
Toxicological Systematics of Combined Exposures

When an organism is exposed to two (or more) substances, there are, in principle, two ways in which a toxic effect can evolve: (1) The substances do not influence each other and develop their effects independently, as if they were acting alone; (2) in the course of substance movement and substance transformation in the organism, the substances enter interactions that change the strength of their effects, causing either amplification or attenuation (for the terminology, see sections 1.2.3 and 2.2.2).

The interactions referred to can be extremely complex with regard to where they take place and of which type they are, but they must be understood for any rational approach to assessing limit values for combined exposures. To this end, we are going to line out the cascade of single steps involved in the passage of a harmful substance through the organism (figure 2.4-2). The possibilities are as follows:

1. Substances can undergo purely chemical reactions even before they enter the organism (prior to absorption): Two components of a mixture can deactivate each other (hydrocarbons, for instance, remove ozone from the atmosphere), or form a new product with perhaps stronger, novel effects (e.g. the formation of peroxyacyl nitrates from NO_x, ozone and olefins). Such primary processes can occur within the organism, too, e.g. the nitrosation of amines in food through nitrite with the creation of carcinogenic nitrosamines.

2. During absorption itself, there is the theoretical possibility of interactions. At the very low concentrations to be considered in the context of limit values, such interactions are negligible, with one important exception: An airborne irritant can cause an increased production of phlegm in the respiratory tract and thus reduce the intake of a combination partner.

3. During their transport through blood and lymph channels, many substances bond to blood proteins and other carriers (e.g. fat droplets). The bond is reversible and follows the law of mass action. Hence, two substances may compete at the same bonding site, which can lead to an amplification of the effect of one of the partners. For practical purposes, this phenomenon of "displacement" is of low relevance in the region of limit values.

4. Many harmful substances are chemically transformed by enzymes in the organism, which can have two consequences for their effect: Amplification (bioactivation) or attenuation (biodegradation). In many cases, the transformation takes place in several sequential steps leading to the emergence of the "ultimal" toxic form, on one hand, and to a non-toxic product to be excreted by the kidneys, on the other. The important point is that activation and deactivation take place one after the other, meaning that bioactivation products once formed can be detoxicated by enzymes again. The extent of the harmful effect is then determined by the balance of all activating and deactivating processes. This balance can be established through experiments, if an "ultimal" metabolite firmly bonds to a biomolecule and this reaction product is accessible, like a carcinogenic substance bonded to DNA as an "adduct". Effect amplification can also arise from the strengthening of the activating enzymes by one of the mixture components (enzyme induction). In turn, attenuation can result, when one component inhibits

the enzyme responsible for activating another component. Of even more practical importance is the case of both substances competing for the bond to the enzyme, which precedes the chemical transformation (competitive inhibition), because most of the transformations relevant for the toxic effect take place at the oxidizing enzymes (mixed-function oxygenases), with the main activity in the liver.

5. The interactions described under point 3 can also occur during the transport of the toxic effect form from the site of formation (mostly in the liver) to the target (effect) site.

6. At the effect site (organ, tissue, cell, sub-cellular structures), substances can be either amplified or attenuated in their effects. The direction of this change is determined by the molecular mechanism of the interaction. If a receptor mediates the effect, as is often the case for reversible effects, the following rule applies: If the antagonistic effects result in attenuation.

7. Damage, once inflicted, can be repaired, especially where the damage is due to irreversible (e.g. genotoxic) effects. It is often the case that one substance inhibits the repair of damage caused by another substance, which results in amplification. Less often, the opposite happens: One substance stimulates the repair (induction), which leads to an attenuation of the total effect.

From this mechanistic basis, *various scenarios* can be derived, which allow us to *make predictions about interactions and their consequences for combined effects* as a basis for establishing limit values:

- Substances in combination attack as such – i.e. independent of each other, in their unmodified form – at different cell structures and produce different harm-

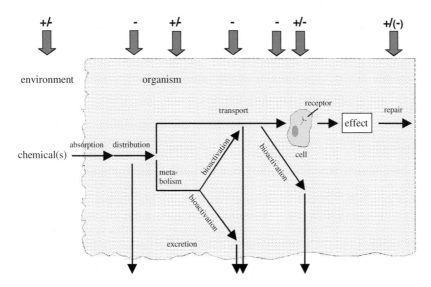

Fig. 2.4-2 Steps of the substance movement and transformation leading to a harmful effect of chemicals in an organism. – Upper arrows: Sites and types of interactions between several substances; +, amplification; –, attenuation of the effect. Explanations in the text.

ful effects. Regarding the relative specificity of the effects of harmful substances, this should be the most common case, in which effect-related limits on the single substances are also valid for combinations of the substances.

- Substances, in their unmodified form, independently produce the same effect at the same target site. The effects are assumed to behave additively. Limit values for the single substances may not offer sufficient protection, if they are guided by effect thresholds. Where necessary, the limits have to be lowered.
- Substances cause their effects only after an enzymatic biotransformation. If one component inhibits the formation of the effective form of the other, effect attenuation is to be expected. Existing limit values offer sufficient protection.
- If one substance inhibits the otherwise highly effective enzymatic deactivation of another substance, strong effect amplification can occur. Even if it is rare, this type of combination is the most important for purposes of practical regulation, since the amplification can be overproportionate.
- In theory, the possibility exists that one effective substance increases the effect of another substance by inducing the bioactivation enzymes of the latter. Since, as a rule, limit values for such substances (e.g. tumour promoters, see section 2.3.3.2), are set lower than the threshold values for the induction, this case is only of minor importance in the real world.

2.4.1.4
Present Knowledge on Combined Effects

Epidemiology offers few insights into physically harmful combined effects of environmental pollutants. There are no validated findings at all for exposures close to threshold values. An exception is met with carcinogenic chemicals. In this domain, other fundamentals apply than those valid for conventional, reversible damage (also see chapter. 2.3), which is why we will discuss them separately.

Toxicology has not covered the field of combined effects comprehensively, firstly because of its complexity, but also due to its intellectual unattractiveness: Studying the very low dose range close to threshold values, one easily ends up with a "data graveyard" full of missing effects ("non-toxicology") – not a good result for a young researcher to build a career on. Success is more likely to come from the discovery and mechanistic explanation of massive toxic effects. Existing studies only cover small sectors of the field and were carried out for a variety of reasons: practical relevance of common mixtures, political requirements, the need to explain an unexpected effect and, above all, the scientific interest in understanding effect mechanisms. Consequently, the experimental approaches chosen are very varied too, which makes comparisons extremely difficult or, in some cases, impossible. Studies on mechanisms mostly require very high doses in order to make certain that an effect does occur, which can be analyzed further. Any extrapolation from there to the dose range close to the threshold is fraught with difficulties. The a priori aim of defensively minded investigations is to exclude harmful combined effects; the doses applied can be too low, the effects too small and too doubtful to draw conclusions from them. For a quantitative risk appraisal, one needs to determine the dose-response relationship for every component of a mixture down to the lower dose range. Due to the many dose points to be established, experiments of this kind involve a lot of effort and expense.

In the following, we will present an exemplary evaluation of the literature with regard to the question if the theoretical predictions described above are correct or otherwise.

2.4.2
Conventionally Toxic Effects

Our assumption is that "effect thresholds" can be found, for the majority of harmful chemical substances, through established scientific methods. Based on these thresholds, safe limit values can be set and justified, on condition that the toxic effects relevant in the dose or concentration range near the limit values are reversible. Such effects are referred to as conventionally toxic effects. Substances acting as "genotoxics" present us with a fundamentally different situation. They react and form stable chemical bonds with the building blocks of the genetic material; in this way they produce premutagenic transformations that may become manifest as genetic leaps (mutations). In principle, these mutations must be considered as irreversible, even if some of them are repairable. The most important final manifestation of a genotoxic effect is the development of cancer. Cancer is defined as the unfettered, tissue-destroying and replacing growth of certain cell and tissue complexes, mostly accompanied by the spread of daughter growths (metastases) ending in death. For this effect category, safe limit values cannot be formulated, since the effects are irreversible in principle and therefore lead to a cumulation of individual effects over long periods. If one sets "single reasonably acceptable" limit values for carcinogenic substances despite of this situation, the adherence to these limits still involves the acceptance of a certain if minimal, but in most cases hardly quantifiable "residual risk".

These principal differences between effect characteristics imply different approaches to a risk assessment, including different criteria and procedures for establishing limit values. Hence, our literature survey is subdivided into two parts: conventionally toxic effects and carcinogenic effects.

The literature on combined effects of chemical substances includes numerous review articles. Only a few of those strive to clarify if and how limit values for the combined exposure to mixtures of chemical substances can be established and scientifically justified. Instead, the main interest was directed towards the elucidation of effect modes and mechanisms of combined effects. Mostly high doses were applied to produce toxic effects reliably, enabling the researchers to perform comparative evaluations with statistical methods. Because of their simplicity and clarity, binary mixtures were preferred for this kind of investigation; multi-component mixtures, which would be prevalent under realistic conditions, have hardly ever been studied in this way.

Another class of experiments is far less common in its approach. In general, such experiments were triggered by pressing questions, for instance which health effects, if any, a plethora of contaminants in drinking water could cause, or if the population in the vicinity of certain chemical-waste disposal sites faces health hazards. Such experimental approaches may mimic "natural" conditions of exposure, but they neglect mechanistic considerations e.g. structure-effect relationships. Thus, we are faced with major difficulties in the interpretation of combined effects in cases where such effects are actually found.

Therefore, our evaluation of the literature concentrates on quoting and analyzing papers containing useable information for dealing with the following question: How can we find and justify limit values for complex mixtures of or exposures to chemical substances?

An extensive analysis including studies preformed under the *National Toxicology Program* (NTP) arrives at the following conclusion: 95% of 122 studies under the NTP and of 151 projects outside the NTP deal with single substances. Of the only 5% concerned with combinations, the overwhelming majority of studies aimed at finding and analyzing strong, acutely toxic effects (Yang 1994). Our selection of investigations discussed in the following is guided by the principal pattern given by Yang.

2.4.2.1
A Complex, mostly Undefined Mixture: Love Canal

A risk analysis was to be performed at the "Love Canal", a multisubstance disposal site close to the Niagara Falls. Originally, the canal had formed part of a civil engineering project started in 1894 with the aim to develop an industrial town near Niagara Falls. The canal, through which part of the Niagara River was to be rerouted, would have solved the problem of energy supply and at the same time would have provided a transport line. In the end, economic depression as well as technological developments enabling the low-loss, long-distance transport of electric energy led to the project to be stopped, since it was not necessary any longer to erect industrial installations close to energy sources of sufficient capacity. From 1942 to 1954, the unfinished canal was used for the disposal of chemical wastes. After that period, it was covered with a layer of soil 0.5–1.5 m thick and sold to the Niagara School Board. Despite of some reports, according to which minor explosions and smoke development had been observed at the site, a school was erected there and a residential settlement came into being.

Not before 1976, the canal attracted attention through the discovery that the Niagara River was polluted with the pesticide, mirex. The assumption was that the particularly high levels of rain and snowfall in that year could have caused substances from the old disposal site to be washed into the river. Although checks in the area did not produce any evidence for the presence of mirex, about 80 other organic substances were detected at the site, some with known toxic effects. Later investigations produced evidence of more than 200 different chemical substances, predominantly chlorinated benzene and toluol compounds and lindane. As a consequence of this discovery, extensive studies were commissioned in the following years in order to determine, among other things, the extent of the contamination and the potential hazard for residents and ecosystems. In the end, the canal was sealed with a ca. 6 meters thick clay cap, and a new canal system was constructed for draining off and collecting the contaminated water.

Initial medical examinations among residents produced conflicting results, not least because many of the components of the chemical mixture could not be identified with the analyzing techniques of the day, and because different routes of exposure, like the inhalation of fumes, absorption through the skin etc. were conceivable. These circumstances made difficult both the experimental investigations and the derivation of the theoretical foundations from treating combined chemical exposures.

In 1984, a study was performed with the aim to appraise the total toxicity of the mixture and its volatile components, respectively, to identify the main target organs and, if possible, to attribute the toxic effect to one or several components of the mixture. For a period of 90 days, mice were kept in cages with contaminated soil. Some of the mice were in direct contact with the soil, enabling an oral as well as a dermal uptake. Another group of mice was kept on a grid above the soil in order to examine the effects of the volatile components of the mixture. Some of the cages were enclosed in large polypropylene containers increasing the concentration of volatile substances by factors between 5 and 10. To establish a dose-response curve, the animals were exposed to the volatile substances in the enclosed cages for their whole life and checked at monthly intervals. (For the open cages, the concentration of the substances was too low to produce detectable effects). In this way, the effects of substances slowly accumulating in the body could be investigated, too; any change of the mixture through artificial, possibly selective accumulation processes for certain components could be excluded as well.

During the experiments, the changes detected in the animals only exposed to the volatile components consisted of an increase of the relative weight of the liver, the spleen and the thymus and a decrease of the residual nitrogen content in the blood serum. After the end of the experiments, however, these quantities went back to normal without leaving any changes detectable by histopathological means.

Animals that had been in direct contact with the contaminated soil showed a slight increase in their body weight at the end of the experiment, compared to control animals, which was however accompanied by a markedly increased weight of the liver and by histopathologically detectable liver damage (hypertrophy of centrilobular liver cells and focal necrosis). On this basis, the liver was identified as the main target organ. In all exposed groups, the effects were more distinct than in the animals kept from direct contact. Death due to the treatment was not observed.

Further experiments concerning the dose-response curve showed a dose-dependent increase in the body weight or a relative decrease of the weight of the kidneys, respectively. The suspicion was that hexachlorobenzene was responsible, since at its measured concentration in the soil, 27,000 ppm (mg/kg), an oral intake of just 0.05 g of the mixture per day and per animal would be enough to explain the observed effect quantitatively as well as qualitatively.

An epidemiological study performed on the residents over the period 1940-1978 (Vianna and Polan 1984) showed a small, but statistically significant decrease in the birth weight (most distinct in 1950). This result instigated experiments aiming to determine the effects of the mixture on reproduction. The investigations involved female, pregnant Sprague-Dawley rats treated with various doses (25, 75 and 150 mg per kg of body weight and per day) of a laboratory-prepared solvent extract from contaminated soil and of a sample of the organic phase of the contents of the collection vessel of the drainage system (10, 100 and 250 mg/kg per day) on days 6–15 of pregnancy. The laboratory extract contained hexachlorobenzene as its main component, while the canal sample mainly contained tetrachloroethane, but also ca. 3 ppm 2,3,7,8-tetrachlorodibenzo-p-dioxin (TCDD, commonly known as dioxin) and other traces.

Tests with the laboratory extract lead to the confirmation of the epidemiological finding of a lowered birth weight of the offspring; this could, however, only be achieved with the highest dose, which also killed 67% of the mother animals. The residents near the Love Canal had never suffered such a high, acute exposure. Apart from the effects already described in connection with the previous experiment, no further changes – neither in the mother animals nor in the 836 animals of the F1 generation examined – was found. An analysis of the xenobiotics deposited in the liver showed 90% hexachlorobenzene. This finding supports the earlier conclusion from the experiments with mice that this substance is mainly responsible for the effects observed.

In the experiments with the canal sample, on the other hand, changes known from TCDD were observed (hydronephrosis, cleft palates etc. among the F_1 animals), while some expected hexachlorobenzene effects did not materialize. Quantitative evaluations led to the result that the effect of the canal sample could be attributed, qualitatively as well as quantitatively, to its TCDD content, whereas the toxicity of the laboratory extract can be explained through the hexachlorobenzene content in it.

After further experiments had shown that TCDD bonds to the aromatic hydrocarbon receptor (AHR), which in turn changes its conformation and bonds to the DNA, triggering the TCDD effect as such, reproduction experiments were performed with two inbred mouse strains that differed genetically in the affinity of the AHR to TCDD only. On days 6–15 of pregnancy, the mother animals were given either pure TCDD (0.5–4 µg/kg per day) or the canal sample (0.1–2 g/kg per day). In this way, it was hoped to gain information on which effect is attributable to TCDD alone and which effect is caused by other components of the mixture as well as by possible interactions between TCDD and other components. Chemical analyses had shown that the mixture did not contain – according to the knowledge of the day – any other substances in concentrations strong enough to open the AHR route.

A comparison of the ED_{50} in both strains with regard to the hydronephrosis induced by TCDD and by the canal sample, respectively, showed increasing differences in the sensitivity of both strains when using the canal sample. However, if there was no interaction between TCDD and other components, the differences in the sensitivity should remain constant. This observation led to the conclusion that the other components of the canal sample caused either an attenuation of the TCDD effects on the less sensitive strain or amplification in the more sensitive strain. The increase in the relative thymus weight as well as some immunotoxic effects were attributed to TCDD. Finally, the calculations of the respective ED_{50} and the comparison between the sensitivities led to the conclusion that an attenuation of the TCDD effect by other components of the mixture is more likely in the less sensitive strain.

This hypothesis had to be modified, however, as soon as a new analysis of the mixture with superior chemical detection methods showed that it contained only about 0.74 ppm TCDD instead of 3 ppm as initially assumed. This finding presented the researchers with the problem that the canal sample contained less TCDD than required – according to data on the single effect – to cause the observed effects. Further experiments gave rise to the hypothesis that other components of the canal sample amplify the TCDD toxicity through an AHR-mediated mechanism. One can still not say if the interactions are dose-additive or super-additive, since we do not know well enough the components of the mixture and the exact mechanism.

These examples demonstrate, on the one hand, how important mechanistic information on the effect of certain substances is for a risk appraisal, and highlight, on the other, some problems faced when investigating undefined mixtures. They also point to possible solutions like using extracts for identifying the component(s) responsible for the toxic effect and the kind and strength of the interactions. Another interesting finding is that one of the least dominant components regarding its concentration, in this case TCDD, is mainly responsible for some of the most prominent effects e.g. the cleft palates. The fact that these effects were found for rats, but not for mice, underlines the importance of selecting the "right" test species, a selection that must be based on mechanistic information on the effects of the respective substances.

2.4.2.2
A Complex, Definite Mixture: Contaminations in Drinking Water

Since the majority (ca. 75%) of the urban population in the US obtain their drinking water from ground water, the NTP (National Toxicology Program) commissioned a study to identify the chemical substances prevalent in the ground water, regarding quantity and risk potential, and to reveal possible combination effects in a simultaneous exposure. Of particular interest were substances that can infiltrate the ground water through agricultural activity – especially the use of fertilizers and pesticides – and by being washed out from landfill sites.

In 1989, there were about 30,000 waste disposal sites in the USA. Samples could not be taken from all of them. Therefore, an effort was made to find criteria by which a representative mixture could be collated for experimental studies. The substance selection took into account published chemical analyses from ground water samples from the vicinities of about 1,000 waste disposal sites all over the USA, firstly in order to find the "important" substances, secondly to be better equipped for defining and reproducing "environmentally relevant" concentrations.

The advantages of using a complex but well-defined mixture are obvious. For instance, it is far easier to establish dose-response relationships, if all components are known, as diluted or concentrated solutions of the base mixture can be prepared without any difficulty. There is no need for complex and perhaps selective enrichment processes. In addition, the dose incorporated can be determined more precisely e.g. by measuring the amounts of various metabolites in blood or urine samples or by measuring concentrations at the target site of the effect. Furthermore, one can fall back on substances for which experimental data on the effect of the single agent are readily available. Relying on such information, substances can be selected which are more likely to enter interactions. If interactions do occur in an experiment, the mixture can be subdivided as required for calculating the influence of a substance on the total effect. This is important for the establishment of limit values, since it allows determining the type of interaction. Because of the particular route of exposure, through drinking water, the experiments should also take into account the possibility of a long-term or even lifelong exposure, as pollutions already present can be only be reduced very slowly or not at all, even if no new contaminations are added.

The final mixture consisted of 19 organic substances and 6 inorganic ones (see appendix A), which were examined by different laboratories using different experimental set-ups. The experiments included diluted or concentrated solutions of the base mixture (concentration 1), in most cases the 0.1-, 10- and 100-fold concentration with the relative contributions of the components unchanged.

For most of the end points of the toxic effects investigated, no harmful effects could be detected – perhaps a reassuring result in the first instance, considering that a 100-fold concentration was included in the experiments. However, humans and laboratory animals differ in their sensitivities; in every population, there are particularly sensitive groups like children, the elderly or infirm individuals etc. Therefore, some of the results relevant in this context will be presented here.

In an early phase of experimenting, the inhibition of the division of bone marrow stem cells in mice was studied. Over 90 days, the base mixture of 25 chemicals and 5- and 10-fold concentrated solutions were administered to the animals, leading to the detection of a dose-dependent inhibition through the concentrated solutions. The base mixture did not show any detectable effect. As the bone marrow stem cells form part of the immune system, the resistance of the treated mice against various bacterial infections was then investigated. Applying *Plasmodium yoelii* (the malaria pathogen) in combination with a 10-fold concentrated mixture did indeed lead to statistically significant increase in the sensitivity measured by counting the number of germs in the blood 10, 12 and 14 days after the injection of 10^6 infected erythrocytes.

Giving a mixture of 6 of the chemicals in the base mixture at a concentration in the ppb(μg/kg)-range and nitrate (10,000 ppb) in drinking water produced another surprising result: In rats, the concentrations 1, 10 and 100 of the mixture, in mice only the concentration 100 led to a slight, still significant increase in the rate of chromosomal aberrations (sister chromatid exchange) in cells of the spleen, although all other parameters measured remained unchanged in comparison with the control animals. It is still unclear what this finding could mean, but the observation alone that miniscule doses of chemical substances can be sufficient to cause such changes calls for further investigations.

Other findings came from an experiment in which mice were treated with the complete mixture over 15 weeks, followed either by a recovery period of 10 weeks or by full-body irradiation (200 rad) in weeks 2 and 9. Radiation doses like this are used e.g. in tumour therapy. In the animals not treated with radiation, the cell counts and various blood parameters went back to normal after the 10 weeks. Animals that had been irradiated after pre-treatment with a 5- or 10-fold concentrated mixture showed, compared to animals treated with radiation only, a statistically significant increase in the bone marrow toxicity and a longer regeneration period. In this case, too, the base mixture did not show any detectable effect. According to literature, none of the single substances in the experiments described above could have triggered one of the said reactions at the concentrations applied in each case.

From this and other experiments, Yang (1994) developed the concept of the "generic promoter". The basic idea had been discussed earlier in the context of "pathobiosis": The homeostasis i.e. the dynamic balance in an organism, which must be maintained through continuous regulating processes in the body, is thought to be disturbed, even if one cannot detect discrete changes with the experimental methods available at present. Only when the body is further subjected to chemical, physical or biological agents, the "subliminal" changes manifests itself in the emergence of stronger effects than in an organism not treated with further agents.

2.4.2.3
Conclusions

The two large-scale studies on complex mixtures described above yielded the following results: In experiments with environmentally relevant concentrations, no combination effects or, occasionally, very slight combination effects were detected. The combined effects were judged to be dose-additive at most. Only when excessive doses of the mixtures were applied, super-additive effects were found in some cases. Nevertheless, the question remains open if super-additive effects can still occur in the environmentally relevant lower dose range. The answer might be found through formulating hypotheses derived both from the findings with the complex mixtures and from the theoretical considerations discussed earlier, followed by further experiments testing these hypotheses in systematically prepared set-ups. The experiments and findings, respectively, should be presented in the following form: (1) Support for the hypotheses, (2) (apparent) disagreements with the hypotheses, and (3) open questions.

2.4.3
Hypotheses on Combined Effects in the Range of "Effect Thresholds"

The following hypotheses have been formulated:

- If the concentrations of all components in a mixture are below the respective effect thresholds, the complete mixture will not have any detectable effect either.
- Interactions between substances acting through the same mechanisms on the same target structures can be predicted on the basis of dose additivity.
- Super-additive effects are to be expected only if the concentration of at least one component exceeds the respective effect threshold.

In the assessment of experimental data, one should bear in mind that measurements of effect thresholds are fraught with uncertainties, since any changes arising in the relevant concentration range are very slight in most cases. Toxicological end points of insufficient sensitivity, a lack of accuracy in the measuring method, the insufficient length of the measuring period or interindividual variations in the sensitivity of the test animals may lead to cases where a minimal effect is not recognized as such and the respective concentration is classed as ineffective. This would be particularly important if the principle of the "generic promoter" mentioned above was correct, meaning that the effect thresholds for single substances could be lowered or the dose-response curve could be modified so that a given dose, which was regarded as ineffective on the basis of data from single-agent experiments, still causes an effect.

2.4.3.1
Difficulties of Finding Limit Values for Combined Exposures

The choice of how to express findings should be guided partly by practical aspects of solving the problem of finding limit values for complex mixtures. Such considerations will have to be taken into account at a later point. The principal modes of interaction inside and outside an organism, as listed earlier, should be considered too, even if they need to be seen in a new context for discussing limit values. The following list was suggested by Bolt (1993):

- Inert substances i.e. substances that, on their own, do not have any toxic effect and thus are not assigned any limit values may modify the toxicity of other components in a mixture e.g. through physicochemical interactions changing the solubility properties and hence the absorption of a toxic substance. This may lead to a strong increase in the concentration at the target structure and thus to a much stronger effect. For instance, from animal experiments it was extrapolated that an orally taken dose of 20–30 g of the insecticide dichlorodiphenyltrichloroethane (DDT) is probably lethal for humans. However, if the substance is incorporated orally in an oily solution, the lethal dose is expected to fall to 3–6 g. Similar effects were observed in connection with detergents (often in pesticide preparations) and dermal exposures to harmful substances.
- When setting emission limits (e.g. for industrial exhaust gases), not only substances or mixtures emitted into the environment directly must be considered, but also agents evolved from these substances through physical, chemical or biological transformation (especially by microbes). Due to the multitude of possible transformations, this is a very difficult task.
- Interactions between different substances do not necessarily require a simultaneous exposure. Sequential exposures, too, can involve interactions especially for substances that remain in the organism for long periods and/or accumulate in certain tissues. This complicates the task of establishing limit values. Thus the lethal dose of the fertilizer calcium cyanamide, for instance, falls from ca.

Table 2.4-1 Substances and doses in the mixture experiments with 8 randomly chosen substances

Substance	Single-substance concentration in mixture (ppm)				Main target organ at MOAEL
	$^1/_{10}$ NOAEL	$^1/_3$ NOAEL	NOAEL	MOAEL	
KNO2	10	33	100	300	Adrenal glands
SnCl2Σ2H2O	100	330	1000	3000	Haemoglobin, body weight
Na2S2O5	500	1670	5000	20000	Red blood cells, stomach
Metaldehyde	20	70	200	1000	Liver
Loperamide hydrochloride	0.5	1.7	5	25	Body weight
Mirex	0.5	1.7	5	80	Liver, body weight
Lysino-alanineΣ2HCl	3	10	30	100	Kidneys
Di-n-octyl-tin dichloride (DOTC)	0.6	2	6	30	Thymus

40–50 g to ca. 0.3–0.4 g, if the exposure to the fertilizer is followed by a small alcohol intake, like two glasses of beer (Hald et al. 1949).

- Substances that – due to their chemical stability and their solubility properties – accumulate in the entire body or in individual organs can come to exceed the effect threshold within an organism in the course of a long-term exposure, even if their concentration in the environment is low. Limit values for this area must be set with prolonged exposures in mind.
- Certain substances cause irreversible changes in the organism. A long-term exposure to such substances can lead to an accumulation of damage that would be insignificant for a single intake. This is the case for e.g. carcinogenic substances. Due to the lack of hard data, the advice is to minimize such exposures.

For the purpose of monitoring, the substance or metabolite concentration in the organism, not in the environment, should be considered for the last two points (biomonitoring) as far as possible.

2.4.3.2
Findings Supporting the Hypotheses

Here we are mostly dealing with studies carried out with relatively simple, definite mixtures (of few components). The most fruitful work for our purposes was done by the group around V. J. Feron in the Netherlands, who proposed working hypotheses similar to the ones presented in this volume and who examined their hypotheses in a series of experiments – covering various toxicological end points – carried out between 1990 and 1996.

2.4.3.2.1 Eight Randomly Chosen Chemicals with Different Organ Specificities of their Effects: In an initial model investigation with rats, a mixture of 8 chemicals was used, randomly chosen from substances for which there were sufficient data on their toxic effects as single agents and which, when applied separately at MOAEL (Minimum Observed Adverse Effect Level), did not influence the food and water intake of the test animals. Having determined the MOAEL and the NOAEL (No Observed Adverse Effect Level) doses in preliminary experiments, mixtures were prepared and given to the respective test groups (10 male and 10 female animals per group) with their food – or with their drinking water, in the case of KNO_2 – over a period of 4 weeks. The substances and doses applied are listed in table 2.4-1.

By including the MOAEL, it was made certain that an effect should be detectable in every case. Thus, interactions in this group can be found and examined in detail. The 1/3 and 1/10 NOAEL groups were included in order to approximate concentration relevant in the environment.

More than 40 end points altogether were evaluated, including body and organ weights, food and water intake during the experiment, various blood parameters (e.g. the number of red and white blood cells, the haemoglobin, glucose and urea level, and a variety of enzyme activities) as well as complete morphological and histopathological examinations of all organs.

Tables 2.4-2 (a and b) show the most important results of the mixture experiments for male and female animals respectively. The statistically significant differ-

Table 2.4-2 Results of the mixture experiments with 8 randomly chosen substances

a) Male animals		Body weight (g)		Organ weights (g/kg)		Blood levels			Histopathology	
		Day 14	Day 28	Kidney	Liver	Hb*	Glc*	ALP*	SCH**	SCN**
Control		127 ± 2.4	197 ± 4.2	7.31± 0.16	46.2 ± 1.0	8.3 ± 0.1	3.1 ± 0.0	334.2 ± 18.9	0	0
Mixtures	$^1/_{10}$ NOAEL	129 ± 3.2	199 ± 4.7	7.51 ± 0.13	46.4 ± 0.8	8.3 ± 0.1	3.3 ± 0.1	317.9 ± 16.0	0	0
	$^1/_3$ NOAEL	132 ± 2.2	203 ± 3.0	7.38 ± 0.10	48.6 ± 0.6	8.2 ± 0.1	3.6 ± 0.1^a	376.6 ± 26.4	0	0
	NOAEL	129 ± 2.1	196 ± 3.9	7.86 ± 0.12^a	50.8 ± 1.1	8.0 ± 0.1^a	3.5 ± 0.1^a	328.9 ± 13.6	0	0
	MOAEL	98 ± 3.1^b	136 ± 7.1^b	8.39 ± 0.20^b	84.8 ± 2.3^b	7.9 ± 0.1^b	3.8 ± 0.1^b	266.8 ± 14.5^a	9^b	3
b) Female animals		Body weight (g)		Organ weights (g/kg)		Blood levels			Histopathology	
		Day 14	Day 28	Kidney	Liver	Hb*	Glc*	ALP*	SCH**	SCN**
Control		119 ± 1.8	157 ± 2.7	7.81 ± 0.13	43.9 ± 1.0	8.6 ± 0.1	4.0 ± 0.1	244.3 ± 10.7	0	0
Mixtures	$^1/_{10}$ NOAEL	119 ± 1.6	154 ± 2.3	7.90 ± 0.20	45.2 ± 0.9	8.6 ± 0.0	3.9 ± 0.1	206.6 ± 9.1^a	0	0
	$^1/_3$ NOAEL	117 ± 1.9	151 ± 2.2	8.42 ± 0.18^a	43.8 ± 0.4	8.7 ± 0.1	3.9 ± 0.1	247.2 ± 14.5	0	0
	NOAEL	118 ± 2.7	153 ± 3.3	7.76 ± 0.15	46.3 ± 0.8	8.5 ± 0.1	4.0 ± 0.1	206.3 ± 8.8^a	0	0
	MOAEL	96 ± 1.8^b	119 ± 4.6^b	8.39 ± 0.15	81.6 ± 2.1^b	8.1 ± 0.1^b	3.8 ± 0.1	188.3 ± 7.4^b	6^a	7^b

Mean values ± standard deviation. n = 10 animals per group. Statistically significant deviations from the control experiments:
$^a = p < 0.05$; $^b = p < 0.01$; $^c = p < 0.001$.
*   Hb = haemoglobin concentration (mmol/l); Glc = glucose concentration (mmol/l); ALP = alkaline phosphatase aktivity (u/l).
**   SCH = swollen centrilobular hepatocytes; SCN = single cell necroses. The numbers in the table are the numbers of animals in which the respective transformation of the liver cells were detected.

ences are highlighted by a grey background. Administering the MOAEL mixture led to the detection, as expected, of discrepancies with the control groups for almost every measured parameter; in some cases, gender-specific differences in the sensitivity of the animals were found, too. Moreover, the authors reported stronger as well as entirely new effects of the mixture compared to the single components. Some expected effects, on the other hand, did not occur or only occurred in an attenuated form. Apparently, both amplifying and attenuating interactions take place, though the attenuating interactions seem to be less frequent. Overall, the interactions lead to a (dose sub-additive) amplification of the mixture toxicity and to a broader effect spectrum.

In the NOAEL group, only 2–4 end points showed minor differences to the control groups. Hence, one can derive a common MOAEL for the mixture. In the two other groups (1/3 and 1/10 NOAEL) no changes due to the treatment were found in comparison to the control groups. The findings in the NOAEL group may be interpreted in two ways:

- The mixture is slightly more toxic than the single substances. The higher toxicity could be attributable to an amplifying interaction, although the combined effect is below dose additivity.
- The NOAEL's were not determined with sufficient accuracy, so that at least one component was applied at a dose above its effect threshold.

Since all the observed effects were minimal – and non-existing in the 1/3 NOAEL group – the second interpretation appears more likely.

2.4.3.2.2 Nine Randomly Chosen Substances Relevant for Humans: As the next step, the 4-weeks experiment was modified in the way that only male rats were used (8 animals per group) and that the selection of the substances was guided by their relevance to humans. As a further modification, "satellite groups" – 15 groups of 5 animals each, treated with a mixture of 4 and 5 chemicals respectively – were included now. By choosing the chemicals appropriately, all possible interactions can be investigated without having to include every conceivable combination ($2^9 = 512$

Table 2.4-3 Substances and doses used in the mixture experiments with 9 randomly chosen substances relevant to humans

Substance	Single-substance concentration in mixture*			Main target organ at MOAEL
	MOAEL	NOAEL	1/3 NOAEL	
Aspirin	5000	1000	330	Liver, stomache
Cadmium chloride ($CdCl_2$)	50	10	3	Red blood cells, liver
Tin chloride ($SnCl_2$)	3000	800	260	Red blood cells
Loperamide	30	6	2	Liver
Spermine	2000	400	130	Heart, liver
Butyl hydroxyanisole (BHA)	3000	1000	330	Stomach
Di(2-ethylhexyl)-phthalate (DEHP)	1000	200	65	Liver
Dichloro methan	500	100	30	Blood
Formaldehyde	3	1	0.3	Nose

* Quantities in mg/kg food or ppm (ml/m^3) in air for formaldehyde and dichloromethane

groups) in the experiment ("fractionated factorial design"). The main groups were given the complete mixture at MOAEL, NOAEL and 1/3 NOAEL doses, respectively, of each of the single substances; the substances (except for formaldehyde and dichloromethane, which were inhaled with air) were administered with food.

The analysis covered more than 70 end points. Apart from general observations on the appearance and the behaviour of the animals, the analysis included pathological findings, like body and organ weight, histopathological examinations of tissues, and a variety of substance concentrations and enzyme activities in blood and urine samples.

Table 2.4-4 shows, again, the most important results of the experiments. For reasons of clarity, the data for the satellite groups, which were needed for a more detailed analysis of the interactions, are not shown here; they can be found in the original paper.

As expected, changes were observed in the MOAEL group for almost every end point. The analysis of the interactions mostly resulted in sub-additive or dose-additive interactions partly of 4 or more substances, where some of the interactions led to an amplified total effect while others showed an attenuating influence. In one case, however, a super-additive interaction was found: $CdCl_2$ and Loperamide led to an aspartate-aminotransferase activity 4.74 units/litre higher than expected for dose additivity.

The NOAEL group showed, as in the previous experiment, only slight changes in few parameters. Again, there are two possible explanations, the "inaccurate determination of the NOAEL's" or the "occurrence of amplifying, dose-sub-additive interactions below the single-substance effect threshold". In any case, the observed total effect of the mixture is far weaker than any effect that could be calculated

Table 2.4-4 Reslts from the experiments with 9 substances relevant for humans

Test group	Body weight (g)	(relative) Organ weights			Blood levels			Histopathology	
		Kidneys (g/kg)	Liver (g)	Spleen (g)	CO-Hb*(%)	GLC*	ALP*	LH**	HRE**
Control	304.3 ± 21.1	7.29 ± 0.26	10.9 ± 0.97	0.534 ± 0.034	0.8 ± 1.8	9.52 ± 0.83	306 ± 50	0	0
1/3 NOAEL	302.5 ± 14.3	7.73 ± 0.31^a	10.95 ± 0.81	0.526 ± 0.033	1.1 ± 1.0	8.96 ± 0.70	278 ± 16	0	1
NOAEL	297.6 ± 20.7	7.91 ± 0.36^b	11.17 ± 1.05	0.509 ± 0.033	3.7 ± 1.5^b	9.47 ± 0.39	254 ± 45	4	1
MOAEL	262.8 ± 20.4^b	8.52 ± 0.21^b	11.51 ± 1.03^b	0.464 ±0.033^b	11.0 ± 1.3^b	8.28 ±0.32 b	227 ± 21^b	8^a	8^a

Mean values ± standard deviation. n = 8 animals per group. Statistically significant deviations from the control results: $a = p < 0.05$; $b = p < 0.01$.

* CO-Hb = percentage concentration of carboxy haemoglobin; GLC = glucose concentration (mmol/l); ALP = alkaline phosphatase activity (u/l);

** LH = liver cell hypertrophy; HRE = hyperplasia of the respiratory epithelia of the nose. The numbers represent the numbers of animals with detected changes.

under the assumption of dose additivity. This could be interpreted as an indication that the super-additive interactions observed in the MOAEL group do not occur here. In the 1/3 NOAEL group, only the relative kidney weight showed a slight change of low statistical significance.

2.4.3.2.3 Four Chemicals Affecting the Same Organ through Different Mechanisms: The chemicals used in the mixtures for the experiments described so far were chosen randomly. The aim of the following experiment, again with male rats, was to test if the hypothesis that effect amplification will not be found for a mixture of sub-NOAEL single components is also valid if all substances attack at the same target organ – but at different target substructures and thus through different mechanisms, or if dose additivity can be detected in this case. To this end, 4 substances with the kidneys as their main target organ were selected. The mixtures contained the single substances at the LONEL (Lowest Observed Nephrotoxic Effect Level) and at the NONEL (No Observed Nephrotoxic Effect Level) respectively. The substances and doses used are listed in table 2.4-5.

Table 2.4-5 Substances and doses used in the mixture experiments with 4 substances affecting the same organ through different mechanisms

Nephrotoxic substances used	Doses in food (ppm)		
	1/4 NONEL	NONEL	LONEL
Hexachlor-1,3-butadiene (HCBD)			
Mercuric chloride (HgCl$_2$)	3.75	15	120
LysinoalanineΣ2HCl	7.5	30	240
d-Limonene	125	500	4000

Three groups of test animals were given one of the mixtures each with their food over a period of 4 weeks; other groups were given a single substance each at their respective LONEL and NONEL doses. 40 parameters capturing the kidney function and morphology were measured, including blood and urine analyses, histopathological examinations of kidney tissue, food and water intake, and relative and absolute organ weights.

Table 2.4-6 shows some selected results for male test animals, which are generally known to react more sensitively to nephrotoxic substances than females. Mixture doses of 1/4 LONEL did not lead to any statistically significant deviations from the control results and are therefore not included here. Statistically significant deviations are highlighted by a grey background.

In agreement with expectations, single-substance doses at LONEL led, for some end points, to small deviations from the control group results. LONEL doses of the mixture, on the other hand, caused significant changes in 20 of the 40 measurement parameters. As in previous experiments, both a broadening of the effect spectrum and an effect amplification – however less than dose additivity – were detected.

The NONEL mixture led, as was suspected following earlier results, to slight changes, which were found in similar form, even if not always statistically signifi-

Table 2.4-6 Results from the experiments with 4 components with the same target organ and effects through different mechanisms

Male animals Mixture given with food over 4 weeks		Final body weight (g)	Final kidney weight (g/kg)	Urine analysis 1. week		Urine analysis * 4. week epithelial cells	Histopathology*		
				Volume (ml)	Density (kg / l)		Basophile tubuli	Regeneration	Protein deposits
	Control	372 ± 10	6.05 ± 0.14	3.0 ± 0.3	1.056 ± 0.003	1	3/10	0/10	0/10
Combi-nations in	NONEL	355 ± 8 b	6.82 ± 0.12^a	3.0 ± 0.2	1.051 ± 0.003	2^a	7/10	0/10	4/10
	LONEL	313 ± 6 b	8.07 ± 0.28^b	5.3 ± 0.5 ^{b}	1.041 ± 0.002 ^{b}	4^c	10/10 ^{b}	10/10 ^{b}	8/10 ^{b}
Single subst. in NONEL	HCBD	365 ± 7	6.49 ± 0.09	2.7 ± 0.2	1.060 ± 0.004	1	0/5	0/5	0/5
	HgCl$_2$	365 ± 7	6.53 ± 0.16	2.8 ± 0.4	1.061 ± 0.005	1	0/5	0/5	0/5
	d-li-monene	369 ± 9	6.25 ± 0.22	3.6 ± 0.8	1.049 ± 0.003	3^b	3/5	0/5	5/5^b
	Lysino-alanine	378 ± 10	5.74 ± 0.08	2.8 ± 0.3	1.059 ± 0.003	1	0/5	0/5	0/5
Einzel-stoffe in LONEL	HCBD	355 ± 9^a	6.67 ± 0.11^a	2.5 ± 0.3	1.063 ± 0.003	2	4/5	0/5	1/5
	HgCl^2	352 ± 9^b	7.07 ± 0.19^b	3.2 ± 0.4	1.055 ± 0.005	1	5/5^b	0/5	0/5
	d-Li-monene	353 ± 8	6.59 ± 0.36	3.2 ± 0.2	1.051 ± 0.001	4^c	5/5^b	0/5	5/5^b
	Lysino-alanine	348 ± 11^b	6.41 ± 0.22	3.4 ± 0.1	1.053 ± 0.003	1	0/5	5/5^b	0/5

Mean values ± standard deviation. Control/Combination group n = 10, single substances n = 5 animals. Statistically significant deviations from the control results: a = $p < 0.05$; b = $p < 0.01$; c = $p < 0.001$.
*   Number of animals with above-average numbers of epithelial cells in the urine or with the given histopathological changes in the kidneys, respectively.

cant, for with the single substances, too, and thus could not be taken as an indication of amplifying interactions.

For the interpretation of these findings, one must recall that all four substances develop their effects in the same organ, but via different mechanisms (attacking different target cells or sub-cellular structures). Hence, it does not surprise that dose additivity was not found at or below NONEL doses. Instead, the initial hypothesis of independent effects was confirmed.

2.4.3.2.4 Four Chemicals Affecting the Same Organ through the Same Mechanism:
The last experiment of this series was carried out with a mixture of substances that develop their effects at the same target structures through the same mechanism. All the substances attack the cells of the proximal kidney tubules. The mechanism involves a complex, enzymatic bioactivation through coupling to glutathione, formation of a cysteine adduct therefrom, fission of that adduct by b-lyase in the tubulus cells accompanied by the release of a very reactive, halogenated, vinylic thiol, from which a halothioketene is formed that damages essential tubulus cells (for an overview, see Vamvakas et al. 1993). In this case, dose additivity should be expected, even if each component of the mixture is applied below its respective NONEL.

In this experiment – in contrast to the previous one – only female Wistar rats were used as test animals, for two reasons: First, male animals excrete larger amounts of protein with their urine even under normal conditions, making the detection of effects caused by the treatment more difficult; second, the induction of the so-called α_{2u}-globulin route, which can only occur in male animals, had to be prevented. This experiment, too, lasted for 4 weeks and used a mixture of 4 nephrotoxicants at the doses listed in table 2.4-7, orally administered in an oily solution.

Table 2.4-7 Substances and doses used in the mixture experiments with 4 substances affecting the same organ through the same mechanism

Substance	Abbrev.	Dose (mg / kg)	
		NONEL	LONEL
Tetrachloroethylene	TETRA	600	2400
Trichloroethylen	TRI	500	2000
Hexachlor-1,3-butadiene	HCBD	1.5	6
1,1,2-trichlor-3,3,3-trifluorpropene	TCTFP	1	4

Each single-substance group was given the substances in LONEL and NONEL doses. The mixture groups were treated with a mixture of all substances at 1/2 LONEL and NONEL (= 1/4 LONEL) respectively, or a mixture of 3 substances, each at 1/3 LONEL. Thus, all mixture groups were given one "toxic unit" each, except for the 1/2-LONEL group, which received 2 units.

The same parameters were measured as in the previous experiment. In addition to that, however, the urine of the test animals was collected and examined with regard to volume, density, total contents of protein and glucose, and to various enzyme activities over 24 h each at the end of weeks 1 and 4.

The most important results are listed in table 2.4-8. The single-substance tests at NONEL did not show any significant deviation from the control groups, which is why those results are not shown here. The fields highlighted in grey contain deviations of statistical significance.

The 1/3-LONEL mixtures caused significant deviations from the controls for, depending on the group, 3–7 parameters out of the total of 40 parameters investigated. The NONEL mixture only caused a lower creatine level in the urine and an increase in the relative weights of liver and kidneys. The increase in the relative kidney weight was similar for all mixture groups treated with one toxic unit. For this

Table 2.4-8 Results from the experiments with 4 components with the same target organ attacked through the same effect mechanism

Female animals only 4 weeks oral intake dissolved in oil		Final kidney weight (g)	Final liver weight (g)	Water intake (g/anim. /day)	Urine analysis (24-h sample; 4. week)				Histopathology	
					Volume (ml)	Glucose (mmol/ 24h)	NAG* (mU)	LDH* (mU)	Multi-focal** vacuolation	Karyo-megaly**
	Control	7.46 ± 0.08	34.4 ± 0.7	17.2	5.3 ± 0.4	11.37 ± 0.68	67.9 ± 4.2	305 ± 26	0/10	0/10
Complete mixture	NONEL	8.41 ± 0.08 b	43.8 ± 0.6 b	23.8	5.6 ± 0.8	8.88 ± 1.18	71.2 ± 10.0	297 ± 74	0/5	0/5
	LONEL	8.36 ± 0.17 b	54.5 ± 0.9 b	32.6 b	15.6 ± 2.3 b	20.01 ± 0.78 b	140.2 ± 7.3b	549 ± 29 b	5/5 b	2/5
Mixtures of 3 substances at 1/3 LONEL	without HCBD	8.19 ± 0.14 b	48.4 ± 0.9 b	22.4	7.5 ± 0.5	17.28 ± 2.23 a	100.7 ± 5.0a	432 ± 26	2/5	3/5 a
	without TCTFP	8.21 ± 0.08 b	48.0 ± 1.1 b	20.0	6.6 ± 0.9	13.13 ± 1.33	82.6 ± 7.9	310 ± 35	4/5 b	5/5 b
	without TRI	8.06 ± 0.18	40.4 ± 0.4 b	21.1	7.2 ± 0.3	14.31 ± 0.09	94.5 ± 4.9	386 ± 40	1/5	0/5
	without TETRA	8.39 ± 0.13 b	41.8 ± 0.9 b	19.8	7.3 ± 0.4	14.96 ± 1.60	99.8 ± 4.7a	381 ± 91	1/5	1/5
Single sub-stances in LONEL	TETRA	8.36 ± 0.14 b	50.4 ± 0.9 b	29.9 b	13.8 ± 2.9 b	16.39 ± 1.9	136.1 ± 9.4 b	683 ± 50 b	4/5 b	4/5 b
	TRI	8.26 ± 0.32 b	51.6 ± 1.8 b	22.1	10.8 ± 1.0 b	17.53 ± 2.34a	114.0 ± 6.7 b	362 ± 70	3/5 a	5/5 b
	TCTFP	8.03 ± 0.20	36.5 ± 1.2	19.1	5.3 ± 0.4	11.43 ± 2.05	77.4 ± 12.2	302 ± 55	1/5	4/5 b
	HCBD	8.40 ± 0.23 b	32.7 ± 0.7	19.8	7.4 ± 1.4	13.43 ± 2.21	88.7 ± 4.8	307 ± 35	0/5	1/5

Mean values ± standard deviation. Control group n = 10, treatment groups n = 5 animals. Statistically significant deviaitons from the control results: a = p < 0.05; b = p < 0.01.
*   LDH = lactate dehydrogenase, NAG = N-Acetyl-b-glucosaminidase
** Number of animals showing the given histopathological changes

end point, dose additivity was found, in contrast to all other end points, which only showed sub-additive interactions.

In the 1/2-LONEL group (2 toxic units), almost every parameter investigated showed significant deviations from the control group. However, the interpretation of the results from this group is made difficult by the fact that no other group in the experiment received a dose like this, so that direct comparisons cannot be drawn. Still, the total effects observed seem to be weaker than expected under the assumption of dose additivity.

2.4.3.2.5 Conclusions: The results from these experiments can be summarized as follows: Where the components of a mixture applied in doses according to their respective NOAEL develop their effect in the same target organ and through the same mechanism, the mixture causes a stronger effect than the single substances, because the effects of the latter act additively. Although each component was applied in sub-toxic amounts in the last experiment, the mixtures produced a toxic effect, even if the effect was not very distinct. In the lower dose range, the interactions resulting from the components acting in the same direction can be predicted under the assumption of dose additivity. We must note, however, that even in this case the interactions were found to be dose-sub-additive for most toxicological end points. The reason behind this finding is still unclear. Future investigations have to show if experimental problems were responsible (e.g. insufficient sensitivity of the end points), or if other factors were at work.

Dose-super-additive effects were found only where at least one component in a mixture was applied at a dose above its effect threshold. This finding is in agreement with the notion of a "generic promoter".

When components of different effect characteristics are present in a mixture, the model of independent effects appears to produce better predictions for appraising the health risk than the model of dose additivity (see section 2.2.3.6), if there are any differences at all in the effect strength predicted by the two models. For toxic effects, the dose additivity model typically produces stronger effects in such cases than the independence model, leading to an overestimation of the risk. (In the lower dose range, independent combined effects equate to the sum of the single effects.) Hence, the hypotheses formulated at the beginning are supported by these experiments.

2.4.3.3
Further Experiments Supporting the Hypotheses

The hypotheses stated above were also supported by an investigation performed by Gaido et al. at the Chemical Industry Institute of Toxicology (CIIT) in 1997. The occasion for that study was an experiment with four chloro-organic pesticides (endosulfan, dieldrin, toxaphene and chlordane) carried out by Arnold et al. in 1996. In that experiment, the estrogen-like effect of these substances in yeast cells examined by, among other methods, comparing the affinities of the pesticides and the natural ligand, estradiol, to the estrogen receptor. Synergistic effects of binary mixtures were reported, with various effects known from single-substance experiments caused by doses 5 to 1,600 (!) times smaller when applied in a mixture. The reactions to these findings ranged from uncertainty to disbelief among scientist and from worries to scares among the general public. Since such amplifications are not covered by far even the most stringent existing limit values on mixtures, the results by Arnold et al. were to be checked in a series of experiments under the auspices of the CIIT. The substances named above as well as two hydroxylated, polychlorated biphenyls (2',4',6'-Trichloro-4-biphenylol and 2',3',4',5'-Tetrachloro-4-biphenylol) were investigated at various laboratories and compared, in analogue experiments, with the effect of the natural receptor ligand, estradiol. In the following, three experiments performed at four different laboratories – out of the total of nine experiments presented in the study – will be discussed in more detail.

2.4.3.3.1 Investigation of the receptor bonding of pesticides in yeast cells: A test was carried out similar to the one performed by Arnold et al. (1996). As yeast cells do not possess steroid receptors of their own, mammal receptors can be expressed in them in a controlled manner, through transformation by plasmids. In this case, receptors from mice and from humans, respectively, were used. Estrogenic activity was detected by means of another plasmid carrying an easily detectable reporter gene, which was combined with the regulatory sequences, to which the activated estrogen receptor normally bonds. The reporter gene chosen was β-galactosidase. The enzyme converts the colourless original product o-nitrophenyl-β-o-galactopyranoside (ONPG) into the yellow end product o-nitrophenol. The progress of this reaction– and thus the strength of the induction of the β-Galactosidase by the activated estrogen receptor – can be measured with ease by photometric methods. The effect of the chloro-organic pesticides was studied with single substances as well as with equimolar, binary mixtures in the (total) concentration steps 10^{-4} M, 10^{-5} M and 10^{-6} M and then compared with the effect of estradiol. The photometric measurement followed a 24-hour incubation period. The concentration of the free estrogen receptor was further varied by changing the concentration of copper ions in the medium. These ions represent a co-factor stabilizing the receptor conformation required for the DNA bonding.

As expected, the strength of the induction depended on the receptor concentration. However, in contrast to the findings by Arnold et al., the cells transformed with mouse receptors only showed minimal activity now, and neither did the binary mixtures show any indication of synergistic interactions. The cells transformed with human receptors showed hardly any reaction to the pesticide treatment either. The activity of the β-galactosidase could only be detected when using endosulfan alone, in a 10^{-4} M concentration, or an equimolar mixture of endosulfan and dieldrin. The effect of the mixture, which contained only half of the amount of endosulfan used in the single-substance measurement, turned out to be roughly half as strong as the effect of endosulfan alone. Since all the substances act through the same receptor, a dose additive effect amplification should be expected, but the results rather lead to the conclusion that the endosulfan effect in the other binary mixtures was inhibited by the respective combination partner (antagonism). Such inhibition could arise e.g. if the other components, while having a higher affinity to the estrogen receptor than endosulfan itself, are not able to induce the change in the conformation of receptor, which is needed to continue the reaction cascade. In this way, most of the receptors for endosulfan would be blocked. Hence the synergism reported by Arnold et al. could not be confirmed.

2.4.3.3.2 Investigation of the Influence of Receptor Concentrations in Human Cells: In this experiment, human hepatoma cells, HepG2, were transformed by various concentrations of the estrogen receptor. The concentrations used were 270, 27, 2.7, and 0.27 ng / preparation. The transformation effectivity was controlled through constitutively expressed β-Galactosidase. The reporter gene was luciferase. Again, the hydroxylated, polychlorinated biphenyls mentioned above were studied, again applied on their own and in equimolar mixtures in the (total) concentration steps of 10^{-6}, 10^{-7}, 10^{-8} and 10^{-9} M; the colour changes were measured photometrically after an incubation period of 24h. The results were related to receptor concentration, which was determined through the β-galactosidase activity.

In agreement with expectations, the luciferase activity fell with decreasing receptor concentrations. It was further shown that the activity induced by the mixture ranged either between or below the activities for the single substances. The use of equimolar mixtures results in maximal dose additivity, as one would expect on the basis of the above hypotheses and for effects acting through the same mechanism. This means that the results by Arnold et al. were not confirmed by this experiment either.

2.4.3.3.3 Full-animal studies examining the strength of the estrogen-like effect: The female B3C6F1 mice used in these studies had not reached sexual maturity yet and therefore featured a relatively low estrogen level. The strength of the estrogen-like effect of toxaphene and dieldrin or of an equimolar mixture, respectively, was compared with the effect of estradiol. The animals were given daily doses of 60 μmol/kg of body weight and were killed and examined after three days of treatment.

In this model, estradiol led to a rise in the uterus weight and to an increased concentration of the progesterone receptor as well as a strengthening of the peroxidase activity in the uterus. The experiment yielded only minimal evidence for the estrogen-like activity for the single substances as well as for the mixture, meaning that all three of the parameters measured for this activity were only in minimal disagreement with the results obtained with the control group. The data may be in broad agreement with dose additivity, although they must probably be interpreted as being caused by dose-sub-additive interactions. Due to the weakness of the effects, this cannot be judged with sufficient certainty, leaving us with the conclusion that the synergistic interactions reported by Arnold et al. were not confirmed in this full-animal experiment. We will not discuss the results from the remaining 6 investigation models in any detail here, because they essentially reflect the findings described above.

2.4.3.3.4 Conclusions: Not one of the experiments could prove the existence of a super-additive interaction; even the most distinct effect could be explained in agreement with the dose additivity model. The most remarkable discrepancy with the results by Arnold et al. (1996) was definitely noticed in specialists' circles and by environmentalist organizations. In the media, however, which had produced such sensationalist reports about a new principle of potentiation of environmental pollutants in the context of male infertility, these new results found hardly any interest. The reason behind the flagrant contradictions between the findings by Arnold et al. (1996), on the one hand, and by Gaido et al. (1997), on the other, could still not be explained – or the explanation has not been published. However, since 4 different laboratories took part in the investigations by Gaido et al. (1997), all of which arrived at comparable results, the obvious conclusion would be that the experiments by Arnold et al. must be at fault. The fact, that the original authors have officially withdrawn the paper in question seems to support this suggestion (McLachlan 1997).

Since all of the substances investigated here develop their effect through the same mechanism, the results are in agreement with the predictions based on the above hypotheses. The finding that combined effects are often not more than sub-additive coincides with the results of the Dutch research group and could be explained by some of the substances being removed (detoxicated or excreted) too soon after incorporation, so that only part of the amount of substances taken in have a chance to enter interactions at the receptor. Because the calculation was based on the full substance amount administered, the results seem to indicate not completely dose-additive interactions.

2.4.3.4
Apparent Conflicts with the Hypotheses

Some other findings do not agree with the hypotheses formulated at the beginning. For the question we are trying to answer here, cases where dose-super-additive effects were reported are of particular importance. In the following, we will discuss some of those reports, paying special attention to the question if any general rules applicable for the establishment of limit values for substance mixtures.

2.4.3.4.1 Carbon Tetrachloride + Chlordecone: Already in 1979, Curtis et al. performed an experiment with carbon tetrachloride (CCl_4) and chlordecone (CD), looking at combined effects of the two substances in the lower dose range. It had already been known that a number of substances are able to modify the strength of the toxic effect of CCl_4 on the liver. Phenobarbital, DDT and some polychlorinated biphenyls (PCBs), for instance, amplified the effect, while 3-methylcholanthrene, cystamine and cysteine attenuated it.

In the experiment, male Sprague-Dawley rats were treated with 10 ppm (mg/kg) of chlordecone in food over 2 weeks. The appearance and behaviour of the animals, their daily intake of food and their body weight were measured on days 1, 8 and 15. After that, the animals were given ip injections of 0, 25, 50, 100 or 200 µl CCl_4, dissolved in oil, per kg of body weight. 24 h later, the animals were anesthetized and given 3 g phenolphthalein-glucuronide (PG) intravenously, the concentration of which was measured in arterial blood samples 2, 5, 10, 15, 30 and 60 min. after the PG infusion. The same blood samples were used to determine the enzyme activities of serumglutamate-pyruvate transaminase (SGPT) and serumglutamate-oxalacetate transaminase (SGOT). The excretion of PG through the liver was monitored by analyzing gall fluid taken every 15 min. After these tests lasting for 60 minutes, the animals were killed and examined for histopathological liver damage.

For animals that had only had the CD pre-treatment, the analysis of the PG content in the gall fluid showed a significant inhibition of the excretory liver function (59% of the control group) and a slight depletion of the glycogen in the liver cells. The other parameters did not show any difference to the control group.

Giving CCl_4 alone did not lead to any sign of acute damage. The PG content in the gall fluid showed a significant, dose-dependent inhibition of the excretory function, setting in at a dose of 100 μl CCl_4 / kg (76% of the control group). The elimination half-life was not prolonged significantly. The pre-treatment with CD plus the CCl_4 injection led to a further significant inhibition of the PG elimination compared to treatment with CCl_4 alone. In this case, 50 μl CCl_4 / kg caused a reduction to 45% of the control group, 100 μl CCl_4 / kg a reduction to 2%. The enzyme tests showed, for only CCl_4 being given, significant increases in the activity only for the highest dose. CD plus CCl_4 resulted in significant, dose-dependent rises in the activity even at the lowest dose.

The histopathological examination of animals that had only received CCl_4 showed deviations from the control group for doses of 100 and 200 μl/kg. Centrilobular necrosis and cytoplasmic fat deposits were detected. In the groups treated with CD and CCl_4, deviations from the controls were detectable for doses from as small as 50 μl/kg, with the necrosis limited to the central regions for this concentration. At 200 μl / kg, on the other hand, a majority of the cells of the central and intermediate regions were found to be necrotic. The treatment with CD and CCl_4, in doses of 100 or 200 μl/kg, also caused lethargy in the test animals. The relative liver weight rose significantly in the groups treated with 50, 100 and 200 μl/kg, where the same final value was reached in all three groups. The production of gall liquid was inhibited for 100 and 200 μl/kg (27 and. 15% of the control group respectively).

Hence, the sequential administration of both substances led to the detection of super-additive interactions, although CD was applied in a concentration of only 10 ppm in food. We know from other investigations that the CCl_4 effect can be attributed to a conversion of CCl_4 at enzymes of the P-450 family to the CCl_3 radical responsible for the toxic effect. It is further known that CD can induce the P-450 enzymes of the liver, which explains the super-additive effect amplification. This type of potentiation was predicted by the hypotheses formulated at the beginning. The surprising result of this experiment is the very low dose of CD already developing a significant interaction effect.

It should be noted that the single doses of CD and CCl_4, respectively, used in these experiments must not be regarded as NOAEL doses, despite of the low concentrations, since they caused detectable, if not very distinct deviations from the control groups in the single-substance experiments. Thus, the hypotheses above are still not disproved, as the respective conditions are not met. Still, even if we consider the doses as MOAEL, the observed effect amplification is remarkable, because the two substances develop their effects in the same target organ, but through different mechanisms. Therefore, this is a model experiment: The use of substances affecting each other's biotransformation through enzymatic induction can lead to super-additive interactions.

2.4.3.4.2 Damage to the Haematopoiesis by Benzene – Interactions of Metabolites: Another example emerged from investigations on the benzene metabolism. Various experiments had produced indications that the toxic effect of benzene is not initiated by benzene itself, but must be attributed to one or several metabolites. In single-substance experiments with known metabolites (including phenol, hydroquinone and catechol), however, the effects of benzene could not be reproduced even by using higher concentrations. Therefore, Eastmond et al. suggested in 1987 that the effect might be attributable to interactions between different metabolites.

Their argument was that phenol is known to be a substrate of myeloperoxidases and that hydroquinone and catechol are suspected to be converted by those peroxidases too, because studies appraising the carcinogenic potential of benzene had produced indications of a conversion by the peroxidases as well. Hence, the benzene metabolites were investigated in a number of experiments looking at the single substances phenol, hydroquinone and catechol as well as at the three possible combinations. As the publication summarizes several experiments, there is no precise information on concentrations or doses. For some experiments, they range from 25 to 150 mg/kg of body weight. B6C3F1 mice, 4–6 weeks of age, were given twice-daily, with an interval of 6 h, ip-injections of the mixtures or the single substances in a buffered sodium chloride solution over a period of 12 or 36 days respectively.

Suspending and counting the cells of the bone marrow, where benzene develops its main effect, led to the result that none of the concentrations used had caused a decrease in the cell count, when only phenol had been injected. Catechol, too, did not show any effect. Hydroquinone caused a slight decrease, which became noticeable – but still not significant in a statistical comparison with the control group – from day 3 after the injection. The mixtures phenol + catechol and hydroquinone + catechol, respectively, did not produce any significant deviations either. Phenol + hydroquinone at different proportions, on the other hand, lead to a significant fall in the cell count, which was detectable as a dose-dependent effect from day 12.

In a model investigation with horseradish peroxidase (HRP) it was shown that hydroquinone is converted into 1,4-Benzoquinone in a H_2O_2-dependent reaction. Therefore, the assumption is in order that such a conversion can be achieved by the myeloperoxidases as well. The presence of phenol ensured a faster elimination of hydroquinone from the reaction. A dose-dependent stimulation was detectable from a phenol concentration of 100 µM Phenol. At low HRP concentrations, phenol induced an increase in the reaction rate by 200%.

Studies on protein bonding with radioactively marked hydroquinone showed that, in the presence of phenol, a larger proportion of the radioactivity remained bonded to proteins. This result indicates that the hydroquinone metabolism in the presence of phenol leads to the production of larger amounts of reactive metabolites, which then bond to proteins. The reverse approach (radioactively marked phenol) showed that hydroquinone induces the competitive inhibition of the phenol metabolism through HRP.

Hence, the outcome of these experiments was that the specific toxicity of benzene is triggered not by benzene itself, but by the interaction between two metabolites (hydroquinone and phenol), with the mixture causing markedly stronger effects as the single components. This could be a case of potentiating, since hydroquinone alone only develops a low myelotoxicity, and phenol does not cause any effect as a single substance. The interaction between the two substances, on the other hand, leads to the same effect that was observed in experiments using phenol only. When the experiments were performed with isolated, human myeloperoxidase instead of HRP, low enzyme concentrations in the presence of phenol even lead to the detection of an increase in the hydroquinone metabolism by 400%. It is assumed that this increase is initiated by phenoxyl radicals (from the phenol metabolism), which have the capacity to oxidize hydroquinone directly and thereby to increase the progress of a rate-setting reaction within the hydroquinone metabolism. Other experiments had already indicated that the actual toxic metabolite of hydroquinone as well as of benzene is in fact 1,4-benzoquinon, the production of which is accelerated in the presence of phenol.

We draw the following conclusion: Since there is no precise information on the concentrations used, this experiment cannot be interpreted as a disproval of the above hypotheses. However, if we consider the low concentrations, at which benzene is still effective, and the fact that even then the effect must have arisen from interactions between hydroquinone and phenol, we see ourselves faced with a case of immediate consequence for the lower dose range: There is a super-additive effect. Another important conclusion is that even for characterizing the effect of a single substance, we have to take into account chemical interactions between several metabolites.

2.4.3.4.3 Benzene and Bone Marrow Damage: Synergisms and Antagonisms: A series of experiments performed by Gad-El-Karim et al. (1984) also had benzene as its subject. In these experiments, the modification of the so-called myeloclastogenic effect – including the induction of sister chromatid exchange (SCE) – on cells of the bone marrow by the pre-treatment with five other substances influencing the benzene metabolism was to be investigated. The five substances were phenobarbital (PB), 3-methylcholanthrene (3-MCA), SKF-525A, aroclor 1254 and toluolene. Apart from that, the route of exposure was to be checked for any influence on the effect. The test animals were CD-1 mice, 7 - 10 weeks old, a subgroup of which received one of the following pre-treatments:

- PB: either 0.1 g/100 ml drinking water for 7 days or ip-injections of 80 mg per kg of body weight and per day, for 3 days.
- SKF-525A: 80 mg ip per kg of body weight, 2 h prior to each benzene treatment.
- Aroclor 1254: 5 days prior to the start of the treatment, single ip-injection of 100 mg per kg of body weight, in olive oil.
- 3-MCA: 1 day before starting the benzene treatment, followed by an ip-injection of 30 mg/kg body weight, in olive oil.

After that, in the main experiment, the animals were given two oral or ip doses of a solution of benzene (8.8, 44, 88, 220, 440 or. 880 mg per kg) and/or toluolene (1,720 mg / kg) in olive oil, with an interval of 24 h between the dose pairs. The control animals received cytoxan (= cyclophosphamide) as a positive control. 30 and 54 h later, respectively, the animals were killed, the bone marrow was removed from the thigh, and the cells were subjected to a micronucleus test and a metaphase analysis. In the micronucleus test, 1,000 polychromatic erythrocytes were examined with regard to the number of polychromatic erythrocytes with micronuclei present (MPCE) in the sample. In the metaphase analysis, the number of chromosomal aberrations (CA) was counted.

The results were as follows: *Benzene alone:* Compared to the control group, a dose-dependent, statistically significant increase in the CA- and MPCE-number was observed. The number of MPCE was higher for oral application than for ip-injection. For the CA-number, no dependence on the way of incorporation was found. The pre-treatments alone did not lead to any statistically significant differences to the results from the control group. The pre-treatments with PB, SKF-525A or Aroclor 1254 followed by doses of benzene did not lead to any statistically significant deviation from the results achieved with benzene alone.

Benzene + toluolene: The numbers of CA and MPCE were higher than the numbers measured in the control group, but lower than those observed for benzene alone – toluolene inhibits the effect of benzene. Toluolene alone did not lead to deviations from the control results.

3-MCA + benzene: The increase of the MPCE-number significantly surpassed the increase initiated by benzene alone, with the oral administration causing stronger effects than the ip-injections. The CA-number increased stronger than for benzene alone, only if the mixture was given orally. The injections did not produce any detectable difference to the case of benzene alone. Dose-response studies showed the differences between benzene alone and 3-MCA + benzene increased with increasing benzene doses.

3-MCA + toluolene + benzene: No difference to benzene alone, meaning that the inhibition through toluolene compensates for the effect amplification initiated by 3-MCA. 3-MCA + toluolene does not produce any deviations from the results for toluolene alone or from the control results.

In all groups, the female animals were less sensitive than the male specimens were. The variation of the treatment periods did not result in any significant discrepancies concerning the MPCE-number between comparable groups. With regard to the CA-number, on the other hand, the 54-h groups registered significantly lower values than the corresponding 30-h groups. Commencing the 3-MCA pre-treatment 48 h prior to giving benzene led to significantly stronger effects than the "normal" pre-treatment with 3-MCA 24 h before administering benzene.

This work confirms again that the effect is caused not by benzene itself, but by metabolites. In all likelihood, the toluolene, which has a very similar structure, is metabolized initially by the same enzyme systems. Thus, the benzene conversion is inhibited by competition from toluolene, which reduces the production of benzene metabolites. 3-MCA on its own did not initiate any effect, although it is known as an inductor of the P-448 mono-oxygenase, which is involved in the benzene conversion too. By making available sufficient numbers of bonding points for the parallel conversion of toluolene and benzene, the enzyme induction triggered by 3-MCA cancels out the competitive inhibition caused by toluol.

These findings lead to the following conclusions: Since the aim of the experiment was to illuminate the mechanism of the benzene effect, the concentrations applied were far above the environmentally relevant range. The antagonistic effect of toluolene with regard to benzene would probably not occur in the lower dose range, where the capacity of the metabolizing enzymes should suffice to convert both substances independently. The inclusion of antagonistic effects when establishing limit values would be questionable in any case, because at the workplace, too, the composition of chemical mixtures will be too variable. Hence, there is no specific need for action in this respect.

The enzyme induction through 3-MCA presents us with a different situation, since the pre-treatment alone did not cause any detectable changes. Therefore, this dose cannot be regarded as NOAEL with regard to the end points studied, although the animals were given a total of 60 mg 3-MCA per kg of their body weight. Nevertheless, the homeostasis of the organism was disturbed by a change in the enzymatic inventory of the target cells. It is therefore conceivable that giving benzene at NOAEL, if that had ever been done, could have caused a toxic effect. It was shown, on the other hand, that the effect amplification became more significant with increasing benzene concentrations. The explanation could be that the capacity of the metabolizing enzymes runs into saturation only at relatively high concentrations, so that only then the enzyme induction triggered by 3-MCA becomes noticeable. If this interpretation were correct, a mixture of benzene at NOAEL and 3-MCA would probably not cause an effect.

A NOAEL can always only be measured in relation to certain endpoints. Thus, in the context of limit values, we have to be extremely careful to choose end points as sensitive as possible, which in turn can only be found on the basis of a mechanistic understanding of the effect. Hence, this investigation does not disprove the initial hypotheses either.

2.4.3.4.4 Drinking Water Contaminants and their Effects on Reproductive Processes:

In 1995, Narotsky et al. investigated 3 substances that were often found in the ground water close to waste disposal sites: Trichloroethylene (TCE), which is used as a solvent and degreaser, di(2-ethylhexyl)phthalate (DEHP), a softener in the manufacture of plastic products, and the insecticide, heptachlor (1,4,5,6,7,8,8-heptachlor-3a,4,7,7a-tetrahydro-4,7-methanoindene; HEPT). The substances act towards different end points and through different mechanisms. As test animals, 90 days old, pregnant F344-rats were selected for examining the effects of the mixture on the feti. Dose-response relationships for the single substances were determined in preliminary experiments. To this end, the animals were given various concentrations of the single substances in oral doses, dissolved in oil. Then, in order to determine the maximum concentration to be used in the main experiment, the dose required to cause the mother animal to loose 2 g of body weight on day 6 of the pregnancy was calculated for each of the substances. This dose, referred to as "level 4" was then reduced in steps by the factor 1.33 defining the levels 3, 2 and 1. "Level 0" means that the substance referred to was not present in the mixture. The doses arrived at in this way are listed in table 2.4-9.

In the main experiment, every possible combination was applied ("full factorial design"), making possible the examination of all combination effect theoretically conceivable. The mother animals were given the mixture on days 6–15 of pregnancy; they were killed and examined 6 days after delivery of the offspring (corresponding to day 22 of pregnancy). Among the end points and parameters analyzed

Table 2.4-9 Substances and doses in the mixture experiments with 3 drinking-water contaminants (mg per kg and day).

Substance	Level 1	Level 2	Level 3	Level 4
Trichlorethylene (TCE)	475	633	844	1125
Di-(2-ethylhexyl)-phthalate (DEHP)	333	500	750	1125
Heptachlor (HEPT)	5.1	6.8	9.0	12.0

and evaluated statistically were the body weights – or any change therein – of mother animals and newborns, the number of deaths, the number of newborns delivered and of pre- or postnatal deaths respectively, as well as various clinical and morphological parameters.

From the number of mother animals having died (77% of the animals for the highest dose), proof of a statistically significant, synergistic interaction between DEHP and HEPT could be derived. With regard to the increase in the mother animals' body weight on days 6–8 of pregnancy, such an interaction was detected for TCE and DEHP, with the highest dose causing a weight rise to 9.5 g compared to 2 g observed in the control group. DEHP and HEPT showed antagonistic effects for this parameter. For the number of prenatal deaths of the offspring and for the body weight of the offspring on day 6 after delivery, too, TCE and DEHP showed statistically significant, synergistic interactions. For DEHP-HEPT and TCE-HEPT, antagonistic effects were found with regard to the complete resorption off all implanted embryos. This antagonism for HEPT and another component is, however, doubtful, since the data contained indications of saturation effects at least for some of the end points. In a numerical evaluation, such effects can cause a synergistic interaction to look like an antagonism. The original data seem to indicate, too, that HEPT has in fact a potentiating effect on the other two components.

The interpretation of these findings is problematic in some points. The publication fails to give any information concerning the effect thresholds of the single substances. Hence, it cannot be assessed if – or which, if any – concentration steps may be regarded as NOAEL. These effect thresholds, however, play a crucial role in our hypothesis tests, even if it can be said that lowest concentrations are indeed relatively low.

Another problem is that almost all of the numbers given in the publication are the products of a computer-aided, statistical evaluation. Only a few examples of raw data are cited mostly without any reference to data from the control groups and those are not in most cases. For these reasons, we cannot give any perspicuous examples to demonstrate the respective interactions in this discussion.

It should further be noted that in *this* experiment (only) the terms "synergism" and "antagonism" relate to deviations from the model of effect additivity. Under this definition, dose additivity, as compared to effect additivity, represents a synergistic interaction (see section 2.2.2). The fact that, for instance, cases of complete resorption were also observed in a group that was given a mixture of all substances at their lowest concentration level, indicates that at least some of these "synergistic" interactions can be regarded as synergistic in the framework of dose additivity, too, which is the framework used by the authors of this text.

On the whole, the findings of the last paper are not in contradiction to our initial hypotheses, because of the difficulties described above. However, they present further indications of dose super-additive interactions in the lower dose range, which have yet to be studied in more detail especially with regard to the effect thresholds of the component substances.

2.4.3.5
Summary and Assessment of Combined Exposures to Substances with Conventionally Toxic Effects

First, it ought to be pointed out once more that most studies on combined effects of chemical substances were carried out with high doses far above those to be expected in the region of the "effect thresholds" i.e. of the concentrations relevant for establishing limit values. Consequently, most of the data need to be extrapolated from the higher dose range to lower doses, before they can be of help in establishing limit values. Further auxiliary constructions are required if one wants to draw conclusions from binary mixtures, as they have mostly been applied in mechanistic studies, to more complex blends. The data to base such conclusions on is desperately thin; a systematic model experiment combining several binary systems so that the validity of the data for complex mixtures can be assessed has not been performed yet.

We can still conclude that the initial hypotheses based on mechanistic considerations were essentially confirmed:

- If the individual components in mixtures of chemical substances attack different structures, and if the affects are based on different mechanisms, one must expect independent effects of the single components; neither amplification nor attenuation of the effects will take place.
- This rule applies only if the substances themselves cause toxic effects. In cases where the components are chemically transformed through the xenobiotic metabolism in the organism, amplification (in case of the metabolism leading to bioactivation) or attenuation (in case of biodegeneration) of the single-component effects can occur; both inductions and inhibitions can play a role in these processes. In cases of complex cascades of (sequential or simultaneous) enzymatic conversion steps, all steps must be known for a sound prediction.
- If two or more components of a mixture attack at the same target structure (organ, tissue, cell type and sub-cellular structures), and if the effects are driven by the same mechanism, dose or effect additivity has to be expected.
- The extent of the additive combination of effects of the single components appears to be dose-dependent: In the high dose range (preferred by experimentalists), it is usually larger than in the (practically relevant) lower dose range.

From these fundamental rules, the following recommendations can be derived: Where limit values (with some "safety margins" attached to them) have already been set, these limits should be kept unchanged for mixtures as well. Cases where a super-additive effect has been observed or suspected require special attention – and special regulation for practical purposes. Existing knowledge about effect mechanisms of toxic substances in general, and about the situation for the individual components of a mixture in particular, still allow a practical prediction for the overwhelming majority of these special cases, provided the most important, basic information is available regarding the type of incorporation, the distribution, the metabolization and excretion, the attack site and the attack type. Since these exceptions from the general rules formulated above are the most relevant cases in terms of hazard prevention, the elucidation of effect mechanisms is of extraordinary importance for the research of combined exposures.

On examination of the work discussed above, the general conclusion must be that the hypotheses formulated in the beginning were confirmed in their essence; they were definitely not disproved. This would not justify the reverse conclusion, however: We are not asserting the absolute truth of the hypotheses. As was already shown in the discussions of the individual experiments and studies, the main difficulty is that investigations were usually performed with higher concentrations, which cannot be regarded as environmentally relevant, or that concentrations were not given with reference to the effect thresholds of the single substances. The problems in connection with the extrapolation to lower dose ranges remain too. Nevertheless, it can be concluded that the results gained in the higher dose range rather overestimate the risk. In this sense, we are still on the "safe side".

Since the hypotheses refer exclusively to mixtures in which – in terms of the definition of "safe limits" – all components are present in concentrations below their respective effect thresholds, high-dose experiments are only suited to support hypotheses, but not to disprove them. They can only point to tendencies towards interactions. Hence, it appears reasonable for future research to set up experiments in a way that they establish a better foundation on which to base an assessment of interactions in the lower dose range.

Indications of the existence of dose-super-additive effect amplifications will have to be further investigated. Until further results are available in this field, the indications referred to should be taken account of by applying safety margins.

2.4.3.6
Appendix

Composition of the mixture of 25 ground water contaminants according to Yang, R.S.H., Toxicology of Chemical Mixtures, 1994:

Chemical	Average EPA (ppm)	Dose range (ppm)
Acetone	6.9	1.59–53
Arochlor 1260	0.21	0.0003–0.01
Arsenic (trioxide)	30.6	0.27–9
Benzene	5.0	0.375–12.5
Cadmium (acetatehydrate)	0.85	1.53–51
Carbon tetrachloride	0.54	0.012–0.4
Chlorobenzene	0.1	0.01–0.3
Chloroform	1.46	0.21–7
Chromium (chloridehexa-hydrate)	0.69	1.08–36
Di-(2-ethylhexyl)-phthalate (DEHP)	0.13	0.0005–0.015
1,1-dichloroethane	0.31	0.042–1.4
1,2-dichloroethane	6.33	1.2–40
1,1-dichloroethylene	0.24	0.015–0.5
1,2-trans-dichloroethylen	0.73	0.075–2.5
Ethyl benzene	0.65	0.009–0.3
Lead (acetate-trihydrate)	37.0	2.1–70
Mercury (chloride)	0.34	0.017–0.5
Methylene chloride	11.2	1.125–37.5
Nickel (acetate-tetrahydrate)	0.5	0.204–6.8
Phenol	3.27	0.87–29
Tetrachloroethylene	9.68	0.102–3.4
Toluol	5.18	0.21–7
1,1,1-trichloroethane	1.25	0.06–2
Trichoroethylene	3.82	0.195–6.5
Xylene (and derivatives)	4.07	0.048–1.6
Total concentration	131.05	11.3428–378.025

The evaluation for the US Environmental Protection Agency (EPA) was done by Lockheed Engineering and Management Service Company in July 1985. The numbers are concentrations averaged from 14 to 3,011 positive analyses for each chemical taken from ground water samples collected close to waste disposal sites all over the USA (Yang and Rauckmann, 1987). The metal concentrations refer to the ionized form; the respective salt is specified in brackets (from Yang, 1994).

The base mixture for the experiments was prepared from the average EPA doses.

2.4.3.7
Literature

Introduction and Problem Description

Vianna MJ, Polan AK (1984) Incidence of low birth weight among Love Canal residents. Science 226, pp. 1217–1219

Yang RSH (ed.) (1994) Toxicology of Chemical Mixtures: Case Studies, Mechanisms and Novel Approaches. Academic Press, San Diego, New York, Boston, London, Sydney, Tokyo, Toronto

Hypotheses and Effect Thresholds

Bolt HM, Westphal G, Riemer F (1993) Kenntnisstand und Bewertungskriterien für Kombinationswirkungen von Chemikalien, study report for the committee of inquiry (Enquête-Kommission) "Schutz des Menschen und der Umwelt" *(The Protection of Humankind and the Environment)* of the German Bundestag

Hald J, Jacobson E, Larsen V (1949): Mechanisme de l'hypersensibilité à alcool dans les intoxications par la cyanamide (mal rouge). Arch. Mal. profess. 10, 232–42

Supporting the Hypotheses

Arnold SF, Klotz DM, Collins BM, Vonier PM, Guillette LJ Jr., McLachlan JA, (1996) Synergistic Activation of Estrogen Receptor with Combinations of Environmental Chemicals. Science, Vol. 272, pp. 1489-1492 (triggered the experiments of Gaido et al.)

Feron VJ, Groten JP, van Zorge JA, Cassee FR, Jonker D, van Bladeren PJ, (1995) Toxicity Studies in Rats of Simple Mixtures of Chemicals with the Same or Different Target Organs. Toxicol. Letters, 82/83, pp. 505–512

Feron VJ, Jonker D, Groten JP, Horbach GJMJ, Cassee FR, Schoen ED, Opdam JJG (1993) Combination Toxicology: From Challenge to Reality. Toxicol. Tribune 14 (supplements to the previous articles)

Gaido KW, McDonnell DP, Korach KS, Safe SH (1997) Estrogenic Activity of Chemical Mixtures: Is there Synergism? CIIT Chemical Industry Institute of Toxicology, Vol. 17, 2, pp. 1–6

Groten JP, Schoen ED, van Bladeren PJ, Kuper CF, van Zorge JA, Feron VJ (1997) Subacute Toxicity of a Mixture of Nine Chemicals in Rats: Detecting Interactive Effects with a Fractionated Two-Level Factorial Design. Fundam. And Appl. Toxicol., Vol. 36 pp. 15–29

Jonker D, Jones MA, van Bladeren PJ, Woutersen RA, Til HP, Feron VJ (1993) Acute (24 hr) Toxicity of a Combination of Four Nephrotoxicants in Rats Compared with the Toxicity of the Individual Compounds. Food Chem. Toxicol., Vol. 31, 1, pp. 45–52

Jonker D, Woutersen RA, Feron VJ (1996) Toxicity of Mixtures of Nephrotoxicants with Similar or Dissimilar Mode of Action. Food Chem. Toxicol., Vol. 34, pp. 1075–1082

Jonker D, Woutersen RA, van Bladeren PJ, Til HP, Feron VJ (1993) Subacute (4-wk) Oral Toxicity of a Combination of Four Nephrotoxins in Rats: Comparison with the Toxicity of the Individual Compounds. Food Chem. Toxicol., Vol. 31, 2, pp. 125–136

Jonker D, Woutersen RA, van Bladeren PJ, Til HP, Feron VJ (1990) 4-week Oral Toxicity Study of a Combination of Eight Chemicals in Rats: Comparison with the Toxicity of the Individual Compounds. Food Chem. Toxicol., Vol. 28, 9, pp. 623–631

McLachlan JA (1997) Synergistic Effect of Environmental Estrogens: Report Withdrawn. Science Vol. 227, 462–463

Apparent Contradictions with the Hypotheses

Curtis LR, Williams WL, Mehendale HM (1979) Potentiation of the Hepatotoxicity of Carbon Tetrachloride following Preexposure to Chlordecone (Kepone) in the Male Rat. Toxicol. Appl. Pharmacol., Vol. 51, pp. 283–293

Eastmond DA, Smith MT, Irons RD (1987) An Interaction of Benzene Metabolites Reproduces the Myelotoxicity Observed with Benzene Exposure. Toxicol. Appl. Pharmacol., Vol. 91, pp. 85–95

Gad-El-Karim MM, Harper BL, Legator MS (1984) Modifications in the Myeloclastogenic Effect od Benzene in Mice with Toluene, Phenobarbital, 3-Methylcholanthrene, Aroclor 1254 and AKF-525A. Mutat. Research, Vol. 135, pp. 225–243

Narotsky MG, Weller EA, Chinchilli VM, Kavlock RJ (1995) Nonadditive Developmental Toxicity in Mixtures of Trichloroethylene, Di(2-ethylhexyl) Phthalate and Heptachlor in a 5 x 5 x 5 Design. Fundam. And Appl. Toxicol., Vol. 27, pp. 203–216

2.4.4
Combined Exposure to Carcinogenic Substances

In principal, typical carcinogenic substances are essentially different from conventionally toxic substances insofar as their effects are irreversible. Irreversible means here that out of the changes effected in an organism, a residual effect persists notwithstanding the possibility of repair of the transformations. Through cumulation of the effect instances in an exposure over prolonged periods, total doses can become high enough to cause the development of cancer in a measurable fraction of a collective of exposed individuals. Even if for a certain total dose no cases of cancer are observed yet in a collective of a given size, this does not imply that the previous doses had no effect. It rather means that other influences can trigger the manifestation of a tumour, once a disposition to develop cancer, which otherwise would remain "silent", has been installed.

Cancerous diseases stand out against other important endemic diseases by a number of biological and medical criteria:

- The trauma suffered by the affected individuals and families is incomparably worse. It is fed by the initial uncertainty about the outcome and, in the terminal stages, by the extremely painful decline of life processes.
- The relative number of cured cancer cases is still disappointingly low. The most common cancer types, in particular, are lethal in large proportion of cases. The intensive efforts of recent decades to find new therapeutic approaches – and to enhance proven ones – may have brought progress concerning some tumour types, but these cancers are mostly rare, which is why the advances have little effect on the overall balance including all tumour sites.
- Some carcinogenic substances are transferred from the mother to the fetal organism during pregnancy (transplacental carcinogenesis). Numerous animal experiments have left no doubt that an organism in prenatal development is many times more sensitive than an adult one. Carcinogenic substances incorporated by a pregnant woman will be a carcinogenic burden for the child.
- Cancer develops from damage to the genetic material in the somatic cells of an organism. If germ cells are affected as well and if mutations occur in these cells, such mutations can be transferred, as genetic leaps, to future generations. So far, this phenomenon may have been demonstrated only in animal experiments with extremely high doses, but experimental experience as a whole and the theories developed on the basis of experiments suggest that genotoxic transformations in germ cells present a risk for humans too.

These criteria, as well as the other peculiarity of characteristics of carcinogenic substances already discussed – that ineffective and therefore safe threshold doses cannot be determined and justified scientifically – impart a special importance to considerations on how to limit combined effects of carcinogenic substances.

Cancers are stochastic events, meaning that an individual might develop cancer or not. Still, the development of cancer takes place in a sequential process through several steps, each irreversible, and can extend over long periods (the "latency period of the cancer"). In turn, the emergence of the respective next step is conditional on the conclusion of the respective previous step. When considering the combined exposure to chemical carcinogens, one must bear in mind that the genotoxic attack of these substances can occur not only at the initial stage ("initiation"), but also at every one of the successive stages i.e. "promotion" and "progression" leading to the final manifestation of the malignant tumour. Both initiation and progression are regarded as irreversible processes. At the promotion stage, on the other hand, reversible phenomena have been observed too, showing strict dose dependence: The smaller the dose and the shorter the period of exposure, the higher is the probability for the restoration of the original condition (reversibility) once the exposure is no longer present. This fact is of special importance for the discussion if for notoriously carcinogenic substances, which however only attack at the promotion stage, "threshold values" can be postulated or determined experimentally, and if for this special category of carcinogens limit values can be established that can be regarded as safe with regard to human health. Still, for numerous carcinogenic substances, the analysis of their modes of action has shown that most of them have both initiating and promoting properties – they are "complete carcinogens".

2.4.4.1
Attack Sites and Factors of Influence

Factors controlling the attack sites and the type of the transformations effected in the course of the sequential process of chemical carcinogenesis are referred to as causal factors. Apart from that, there are a number of other factors of influence unconnected to the exposure to the carcinogenic substance. Those factors are listed in the table below *(Landesverband Rheinland-Westfalen der Gewerblichen Berufsgenossenschaften* 1996). They can have a promoting (age, hormone balance) or an inhibiting (protective) influence.

    I. Endogenous factors (individual habitus, "pre-disposition"):

- Age
- Hormone balance (gender)
- Endogenous enzyme equipment
- Endogenous immune defences

   II. Exogenous factors of influence and causal factors:

     1. Individual freedoms and the environment:
- Smoking habits
- Eating and drinking habits
- General environmental pollution
- Natural and other radiation
- Pharmaceuticals consumption
- Viral and bacterial infections

2. Workplace:
- Unusual chemical exposures (both carcinogenic and non-carcinogenic hazardous substances)
- Unusual physical exposures
- (carcinogenic noxae)
- Unusual biological exposures (e.g. viruses)

As in the case of conventionally toxic substances, the enzyme equipment of an organism plays a special role in the bioactivation of chemical carcinogens. This can be demonstrated with the following example:

Some genotypically designed enzymatic polymorphisms strongly modulate the risk of falling ill with certain cancers. In epidemiological studies with humans, for instance the influence of the genotype regarding the enzymes CYP1A1 (cytochrome P450 isoenzyme 1A1), GSTM1 (gluthathione-S-transferase isoenzyme M1) and NAT1/NAT2 (N-acetyl-transferase isoenzyme 1 or 2) was investigated. CYP1A1 oxidizes various xenobiotic substances, in this way often activating them to become ultimal metabolites, whereas GSTM1 deactivates xenobiotic compounds by bonding to glutathione. NAT, too, deactivates xenobiotic substances, in this case by bonding to an acetyl residual. The isoenzymes 1 and 2 differ in their reaction kinetics, which is why one also refers to slow (NAT1) and fast (NAT2) acetylaters respectively.

For individuals of a CYP1A1-positive and GSTM1-negative genotype, the exposure to polycyclic aromatic hydrocarbons (PAH) was shown to cause an up to 40-fold increase in the risk of lung cancer (Alexandrie et al. 1994, Raunio et al. 1995). In the same way, smokers of the genotype NAT2-positive and GSTM1-negative demonstrably face a 10-fold increased risk of developing colon tumours (Probst-Hensch et al. 1995).

In the following, we will briefly comment on some other modulating factors:

DNA repair (also see section 2.3.3): Since the manifestation of mutation precedes or accompanies the tumour formation, DNA repair processes can influence, at least in theory, the carcinogenesis. First, an "uncatalyzed repair" is conceivable, where DNA adducts, for instance, decay automatically due to their chemical instability before they become manifest through cell division; on the other hand, cells have a variety of enzyme systems at their disposal, which have the capacity to repair faults in the DNA structure. These repair systems act in a relatively specific manner and probably evolved mainly for repairing defects caused by endogenous factors. A multitude of model experiments with animals have shown that changes of the genetic material due to exogenous factors can be repaired as well. In humans, the repair of e.g. a chemical transformation at the base guanine caused by aflatoxin, one of the strongest carcinogenic substances, can be tracked by measuring the excision product in the urine. According to estimations, about 10,000 oxidative transformations – caused by the endogenous metabolism of oxygen or by natural radiation – have to be repaired in each cell of the human body, even in the absence of exogenous influences. Considering this high repair rate, it is still under discussion which role the repair systems could play in a carcinogenesis following a combined exposure to substances.

Intercellular communication: Individual cells communicate with each other by exchanging certain messenger substances, which activate or inhibit transduction cascades in the target cells. Within a tissue, the messenger substances are mainly transferred through gap junctions i.e. specialized membrane areas enabling the direct exchange of substances dissolved in the cytosol. In this way, a signal received in a cell can "jump" to the neighbouring cells through the transfer of transduction molecules. For the gap junctions to function correctly, cell-adhesion molecules, like N-CAM or E-cadhedrin, have to be present. The communication between tissues is controlled by hormones.

We know that in tumor tissue the communication of cells between each other and the reaction to endogenous signals from other tissues is vastly reduced. At the initiation stage, this should not be of great importance. At the promotion stage, on the other hand, this communication plays a central role, because one of its functions is to regulate proliferation and differentiation processes. Hence, when the functioning of the gap junctions is inhibited, it becomes more likely that an initiated cell evades the control of the neighbouring cells and embarks on an uncontrolled division. There is, for instance, a proven connection between the function of E-cadhedrin and the malignancy of tumours in humans (Consensus Report 1992, Holder et al. 1993). Another important event in this context is the genetically programmed cell death (apoptosis). As soon as a cell within a tissue formation behaves abnormally, signals from the surrounding cells can cause it to self-destruct, which can also prevent the development of a tumour. This self-protecting system of the organism ceases functioning, when the affected cell does not react on the signals or when these signals are superseded by exogenous factors like the inhibition of the related signal transduction cascade.

Activation/inhibition of immune defences: There are indications that defects of the immune system promote the emergence of certain cancers. Giving drugs to suppress reactions of the immune system (immunosuppressants), for instance rejection processes following organ transplantations, can lead to an increased incidence of tumours in some tissues e.g. of carcinomas on exposed areas of the skin of kidney transplant patients (zur Hausen 1991). The immune system has to fulfil the task of killing both resident and migrating tumour cells early in their development. A weakened immune system will totally or partly fail in this task and thus promote the development of cancer indirectly. It was further found that the incidence of certain cancers rises in connection with virus infections. This has been demonstrated e.g. for certain liver tumours coinciding with hepatitis B infection (Koch et al. 1995) or for bladder tumours in connection with the papilloma virus (Anwar et al. 1992). According to a paper by zur Hausen (1994), in 15% of all cancer cases worldwide, a correlation with viral infections should be detectable, the most likely reason being a weakening of the immune system caused by the viral infection.

Cytotoxic effect: In a similar way as the immune system, substances with (nonspecific) cytotoxic effects can inhibit the process of carcinogenesis by killing cells at the initiation or promotion stage, before they can develop into tumours. For typical carcinogenic substances, initiating effects are mostly accompanied by cytotoxic ones. Again, we must note that effect thresholds can be established for cytotoxic effects, but not for initiators.

Low concentrations of harmful substances in the target tissue damage the cells just slightly (initiation), whereas high concentrations lead to cell death. These facts are particularly important for experimental practice, since there is always the risk to drown out the initiating effects of a substance by using excessively high doses or concentrations. Hence, researchers experimenting with mixtures find themselves in the paradox situation that two or more components may amplify each other's effects, thereby surpassing the threshold of cytotoxicity and thus causing an antagonistic effect overall with regard to the tumour incidence.

2.4.4.2
Establishing the Carcinogenic Properties of Chemical Substances: Epidemiology vs. Experiment

For chemical substances, too, limit values are a useful and necessary tool for dealing with cancer risks. They can reduce and sometimes minimize such risks, but they will never completely exclude residual risks. Therefore, quantitative risk data are indispensable as a basis on which to establish such limit values. Hence, in the following discussion our emphasis will be on the question if the current methods of detecting carcinogenic properties of chemical substances are apt to provide quantitative data in the form of dose-response curves.

Investigations into the carcinogenic potentials of chemical substances make use of either of three basic methods: The epidemiological study, the animal experiment and the in vitro model experiment. All three approaches have their advantages and drawbacks.

- Short-term in vitro tests put emphasis on detecting the malignant transformation (in certain cell culture systems) or on finding mutations and chromosomal damage in microbes and body cells. Their advantage is the low expense in materials and time, which is why they are particularly suitable for screening large series of chemical compounds. The drawbacks are the following: Only 50% to 90% of such tests, depending on the test method, produce the correct result regarding carcinogenic potentials. Especially the "false negative" results – i.e. negative results for substances positively known as carcinogens – give reason for concern, as even a strong carcinogenic potential might be overlooked. Short-term in vitro tests do produce valuable indications to be followed-up, however, by full-animal experiments.
- Animal experiments as tests for carcinogenic potential are performed with large collectives of inbred rat or mouse strains, and they extend over their full lifetime. The current strategies for such studies generally use two doses, the maximum tolerated dose (MTD) and half of it (MTD/2). In the vast majority of experiments, both doses are several orders of magnitude higher than those found in practical circumstances. The high doses are chosen in order to maximize the measuring sensitivity. At the same time, they constitute a crucial drawback: Extrapolations from the high dose range to the lower dose range – which is the range relevant for realistic risk analyses – are fraught with difficulties. An additional problem is that the sensitivity of a test animal to carcinogenic stimuli can be considerably, in some cases vastly higher or lower than the corresponding sensitivity of humans. The great advantage of animal experiments is their feasibility for advance tests of newly developed compounds (e.g. pharmaceuticals, industrial chemicals, pesticides, household cleaners etc.), meaning that products can be checked for carcinogenic properties before they are introduced and brought into contact with humans. This makes animal experiments the predominant predictive tool for dealing with cancer risks potentially posed by chemical substances.

- In epidemiology, groups of humans exposed to chemicals are compared with (as far as possible) identical groups free of the same exposure. The tumour incidence is measured for both groups and risk numbers are computed. The main advantage of this approach is that uncertainties like those introduced by extrapolations from animal experiments are excluded from the start. Another advantage is the possibility of producing dose-response curves by including several groups with different exposure cases – provided the tumour incidence is sufficiently high in both groups. These few advantages are offset by a number of substantial drawbacks: From first principles, epidemiological studies will never be able to prove causal relationships between exposure and tumour development; they can only point out correlations, from which causal relationships can be derived, again, only indirectly and only supported by other criteria of biological plausibility, which in turn emerge from animal experiments or in-vitro tests. This leads to the most important conclusion that epidemiology is not only unable to prove cancer risks; it is also unable to exclude such risks. Further disadvantages are the following: In general, it is very difficult to determine the kind and intensity of an exposure to chemical substances, mainly because, due to the long latency period, the conditions of several past decades have to be reconstructed to this end. Extensive, chemoanalytical determinations are required to obtain reliable data on the strength of an exposure and on the amounts of harmful substances incorporated. This precondition for quantifying the risk is rarely fulfilled. One of the few occasions, where this was the case, was the counting of asbestos fibres, which accumulate in lung tissue virtually without any losses over several decades. It is also difficult, in most cases impossible to determine the everyday and occupational factors of influence explained above, and to standardize them in a way so that they are evenly distributed through test and control groups. Many positive findings from epidemiological studies are largely or exclusively attributable to those factors of influence, not to the chemical compounds that had actually been studied. The most crucial drawback is, however, that epidemiology has to rely on data from damage that has already occurred. For risks from carcinogens, this means counting cancer cases, which mostly are cases of death. Due to the long latency periods of cancers caused by chemicals – often spanning several decades – the epidemiological approach lacks the preventive function fulfilled by animal experiments.

As preventive measures based on results from predictive toxicology (mainly animal experiments) are steadily improved, it will become more and more difficult to perform successful epidemiological studies. Both at the workplace and in everyday life, exposures to carcinogenic substances are continuously reduced. The risk reduction resulting from this development limits the prospects of detecting the remaining risks by – generally quite insensitive – epidemiological methods. In fact, epidemiology still mainly relies on massive "legacy" exposures of decades ago.

2.4.4.3
Results from Epidemiological Studies

Only a small number of studies have produced validated results with regard to combined effects of chemical carcinogenics. Such combined effects can only be found if there are strong carcinogenic potentials causing high tumour incidences with, at the same time, large differences between the tumour incidence following the combined exposure and the single exposure to the individual components of a mixture. So far, epidemiology has produced positive results only for a handful of binary combinations.

2.4.4.3.1 Cigarette Smoke and Alcohol Consumption – Increased Risk of Esophageal Cancer: The extreme example of a risk increase caused by the combined exposure to two carcinogenic noxae was reported by Tuyns et al. (1977). The comparison between four groups of differently strong smokers related to four groups with different (rising) daily alcohol intakes led to the results displayed in fig. 2.4-3. There are clear dose-response dependencies for the development of esophageal carcinomas in relationship to the number of cigarettes smoked per day, on the one hand, and to the extent of alcohol consumption, on the other. With both agents in combination, the relative risks due to higher alcohol consumption increase up to 70-fold, while the risk increase correlated to increased tobacco use amounts can be up to 8-fold. The reasons for this multiplicative risk increase for a rare cancer are unknown yet. One suspects that a local formation of reactive metabolites – acetaldehyde, for alcohol, and oxidative activation products from the many carcinogenic components of tobacco smoke – might be responsible, but that has not been proven yet.

2.4.4.3.2 Tobacco Smoke and Asbestos Fibre Dust: The perhaps best-known example of a syncarcinogenic effect is the almost multiplicative increase in the tumour risk faced by smokers exposed to asbestos. The first publication on this issue was submitted by Selikoff et al. (1968). Their work has been confirmed repeatedly, e.g.

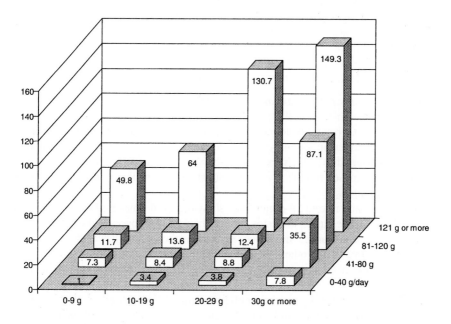

Fig. 2.4-3 Increase in the risk of esophageal cancer through the combined exposure to tobacco smoke and alcohol intake, plotted over the respective doses; from Tuyns et al. (1977).

recently by Vainio and Boffetta (1994). We quote from the most extensive study (Hammond et al. 1979) in this discussion, because the length of the observation period and the very large number of individuals looked at ensure the highest statistical reliability of the results.

The observation period covered the years from 1959 to 1972. The analysis included 17,800 men who had been exposed, by working in occupations connected to building insulation, to asbestos fibre dust for at least 20 years, and a control group of 468,688 men and 610,206 women from 25 states in the US, who had not suffered such an exposure. All groups were subdivided into

- non-smokers,
- former smokers who had stopped smoking at least 10 years before,
- smokers who smoked less than 20 cigarettes a day, and
- smokers who smoked 20 or more cigarettes per day.

The quantity derived was the risk of death by lung cancer, where the relative risk for the non-smoking population that had not been exposed to asbestos either was defined as 1. Without asbestos exposure, the relative risk of death by lung cancer amounted to 3.6 for former smokers, 9.2 for smokers of less than 20 cigarettes per day, and 10.4 for smokers of more than 20 cigarettes per day. With asbestos exposure, the relative risk for non-smokers was found to be 5.2, while it was 36.6 for former smokers, 50.8 for smokers of less than 20 cigarettes per day and 87.4 for smokers of more than 20 cigarettes per day.

Thus, there is a proven super-additive, almost multiplicative effect, when the two causal factors, asbestos fibre dust and tobacco smoke, coincide. All the results cited above are of a high statistical significance, and they prompted intensive protective and preventive measures worldwide. However, it has not been investigated yet if such super-additive combination effects also occur in the range of the low asbestos doses prevalent in non-occupational situations. The current ideas about the mechanism of this effect amplification will be discussed later (see 2.4.4.4).

2.4.4.3.3 Asbestos Fibre Dust and Other Hazardous Substances: In a study published by Bittersohl et al. (1972), 30,000 workers in the large-scale chemical industry in the then German Democratic Republic (East Germany) were examined. 746 cases of carcinomas in various organ systems were analyzed. The tumours induced by asbestos fibre dust, which are marked by their organ specificity, were compared with groups exposed to other hazardous substances as well (isobutyl oil, ammonia, coal gas and tar).

The carcinoma rate in all organs was shown to be 6 times higher on average for workers in factories under asbestos hazard than for employees in factories without such hazard. The cancer incidence among workers simultaneously exposed to asbestos and one of the other substances named above, on the other hand, was ca. 9–14 times higher than for unexposed workers. However, that study lacks a deeper statistical analysis.

Nevertheless, the existence of an effect amplification caused by mixtures could be proven, even if it could not be quantified due to the lack of comparable numbers for single exposures to one of the other substances (isobutyl oil, ammonia, coal gas, tar) without an exposure to asbestos at the same time. If the combination of asbestos and ammonia really leads to such a, in mechanistic terms, largely unexplained risk increase is yet unclear.

2.4.4.4.4 Further Results from Epidemiological Studies: In a case-control study with 300 individuals fallen ill with a histologically confirmed ureothelic carcinoma, super-additive effects through the exposure to cigarette smoke combined with the occupational exposure to aromatic amines were detected (Bolm-Audorff et al. 1993). Individuals were classified under the risk groups "smoker" and/or "occupationally exposed" according to their own statements about smoking for at least 6 months and working for at least 6 months, respectively, in one of the risk occupations, chemical worker, rubber manufacturing, vulcanizing worker, chemist, chemical technician, chemical laboratory assistant, painter and decorator, and hairdresser. According to that investigation, smoking led to a 3.5-fold increase (95% confidence interval: 2.16 - 5.81) of the relative risk of falling ill with ureothelic carcinomas; the occupational exposure led to a 3.8-fold increase (95% confidence interval: 1.06–13.95). For occupational exposed smokers, the relative risk of falling ill rose to 16.17. Hence, when both causal factors coincided, a statistically significant, multiplicative increase was observed.

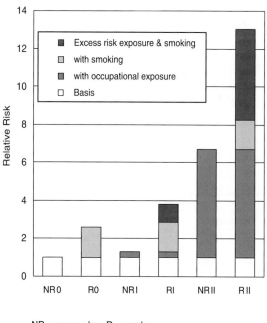

Fig. 2.4-4 Super-additive, syncarcinogenic risk of falling ill with carcinomas of the respiratory tract in connection with a coincidence of partly carcinogenic substances in coke distillery emissions in dependence of the smoking habits of employees of the *Hamburger Kokereibetriebe* (Manz 1986).

A cohort study performed with employees of the *Hamburger Kokereibetriebe* (a coke distillery) showed super-additive, syncarcinogenic effects for carcinomas of the respiratory tract in connection with a synchronous combined exposure to cigarette smoke and coke emissions (Manz 1986, see fig. 2.4-4). Especially for smokers subjected to a strong occupational exposure, we note that the influences of smoking (slanted hatching) and of the occupational exposure (black) in relation to the number of carcinomas of the respiratory tract not only add up, but lead to super-additive

Table 2.4-10 Overview of epidemiologically confirmed interaction effects, from Popp and Norpoth (1996).

Epidemiologically confirmed, super-additve interaction effects				
Risiko related to agent 1	**Risiko related to agent 2**	**Total risk**	**Organ with tumour**	**Author**
Asbestos	**Smoking**			
<1[a]	8.1	92	Lung	Selikoff et al. (1968)
15	4.3	78.1	Lung	Selikoff et al. (1980)
2.7**-	6.7*	17.4**	Lung	Rösler et al. (1993)
Arylamine	**Smoking**			
3.8***	3.5	16.2	Bladder	Bolm-Audorff et al. (1993)
<1[a]	11.1	31.5	Bladder	Bi et al. (1992)
1.9	1.8	4.6	Bladder	Cartwright (1982)
2.0****	4.4	7.8	Bladder	Gonzales et al. (1989)
4.6	2.8	11.3	Bladder	Thäriault et al. (1984)
1.7	4.4	6.8	Bladder	Vineis et al. (1982)
2.5	2.8	3.7	Bladder	D'Avanzo et al. (1990)
7.2	3.6	13.6	Bladder	Tremblay et al. (1995)
63.4	6.2	152.3	Bladder	Wu (1988)
Smoking	**Alcohol**			
3.7	8.0	44.2	Mouth	Maier and Sennewald (1994)
3.4	2.3	15.6	Mouth	Tuyns (1990)
Alcohol	**Loe carotene intake**			
1.9	2.4	7.3	Esophagus	Tavani et al. (1994)

*   modest, ** high exposure to asbestos fibre dust; *** chemistry occupations, **** textiles industry

[a]   Risk numbers < 1 are due to statistics. Selikoff et al., for instance, extrapolated the number of deaths to be expected in a group of 87 non-smokers exposed to asbestos on the basis of the "US life table for white males" of 1964. The result, 0.18, is due to the fact that there were no deaths in the observation periode: It is a consequence of the insufficient size of the group considered.

if not multiplicative effects. In table 2.4-10, epidemiologically confirmed interaction effects from two coinciding carcinogenic agents are summarized (Popp and Norpoth 1996).

Other connections e.g. the combined exposure to cigarette smoke and Diesel-engine emissions, investigated in a case-control study with lung cancer patients (Emmelin et al. 1993), are still in question. In the study, smoking resulted in a relative increase to 3.7 in the risk of lung cancer; Diesel-engine emissions correlate to a relative risk number of 2.9; the combined exposure to both agents appeared to increase the relative risk to 28.9.

In the same way, the risk of lung cancer due to smoking cigarettes was found to rise to 4.8, while the risk due to an occupational exposure to pyrolysis products in foundries and to general air pollution amounted to 1.8 (Jedrychowski et al. 1990). The combination of the two agents resulted in a risk increase to 8.5 in this case. Since the carcinogenic potential of cigarette smoke is attributable mainly to the polycyclic aromatic hydrocarbons (PAHs) present in the smoke, it appears to make sense to transfer the results found in this context to other mixtures with similar PAH contents produced by burning organic matter.

2.4.4.4
Animal Experiments

Many animal experiments have been performed with the aim of discovering and analyzing combined effects of carcinogenic substances. A recent review article (Berger 1996) classifies the experiments according to the target organs investigated; these can be the same or different ones for the individual components of a mixture. Another criterion for the selection of substances – which had not been applied in studies on the effect mechanisms of individual components in mixtures until recently – is the chemical substance class, which is of great importance in experimental cancer research. Mixtures containing more than two components were investigated less often. A small number of experimental set-ups were guided by practical situations (mixtures of pesticides and pyrolysis products of proteins).

2.4.4.4.1 Binary Mixtures: For reasons of brevity and clarity, the results published up to the year 1996 are summarized in table 2.4-11; the sources can be found in review articles (Arcos et al. 1998, Berger 1996).

2.4.4.4.2 Mixtures with Three or More Components: The analysis of experiments with 3 or more components (Berger 1996) led to the conclusion that for complete carcinogens, the way or sequence of administering the mixture determines which combination effects occur.

For instance, a single dose of benzo[a]pyrene (BaP) combined with a coal extract containing a complex mixture of organic carcinogens applied to the skin of mice, followed by 24 weeks of promotion by 12-O-tetradecanoyl phorbol-13-acetate (TPA), led to a lower tumour rate than giving BaP without the complex organic mixture, again using promotion by TPA. The analysis of the mixture showed its antagonistic, effective components to be PAHs and nitrogenous, polycyclic aromatic compounds. In an analogue experiment with a lifelong application of BaP and a PAH mixture on the skin of NMRI mice, dose additivity was found.

Assuming that the antagonism observed after the single application can be explained in terms of competition for bonding sites at metabolizing enzyme systems, the lifelong treatment appears to lead to changes in the enzyme equipment of the affected cells, caused by interference in the protein synthesis. With an increased number of metabolizing enzymes, enough bonding sites are available, so that competition and hence antagonism do not occur anymore.

With regard to the experimental research into new substances and combinations of substances, the subdivision into substance classes is of help, since it becomes possible to arrive at sound prediction of combined effects to be expected and to set up the experiments accordingly. However, as the author himself regards the finding as non-representative – due to the sometimes relatively small number of experiments concerning individual substance classes – Berger's review offers not more than limited information on which substance classes might have to be treated with special caution when setting limit values. Any given substance or mixture will still have to be examined in experiments, before reliable statements can be made concerning their carcinogenic potential and their combined effects.

The results of the analysis of N-nitroso, hydrazo and azoxy compounds, in particular, are worth mentioning. Although the significance of the reported findings is limited, the investigation produced clear indications of an increased likelihood of super-additive effects for binary mixtures containing a component belonging to one of the two substance classes. When regulating on limit values for substances from these classes, super-additive combined effects must be taken into account.

An increased occurrence of super-additive effects was found with mixtures of N-nitroso and halogen compounds and, to a lesser degree, with aromatic amines and N-nitroso compounds. Provided these findings are valid, it appears reasonable to set mixture limits covering the combinations of substances from these classes.

The frequent finding of sub-additive effects in binary mixtures of various PAHs or of one PAH and one azo dye, surprises at first, since one would rather expect additivity for substances so similar chemically. The explanation might lie in the competition for bonding sites at metabolizing enzymes, but only if the substances considered show major differences in their carcinogenic potential. In such cases, the less carcinogenic substance can reduce the effect of its combination partner, if its affinity to the enzyme is stronger than that of its partner. More likely, the experiments included in the analysis used relatively high doses leading to saturation effects, which would not occur in the lower dose range relevant for humans. Hence, in that dose range, dose additivity should be predominant. No analysis looking at substance-class correlations and exclusively based on experiments in the lower dose range has been published so far.

Table 2.4-11 Results from animal experiments with binary mixtures, published up to 1996

Results of analyses of binary mixtures		
(Substance class)	**Organs affected most often**	**Distribution of combined effects**
Polycyclic aromatic hydrocarbons (PAH)	Skin (most frequently used route of exposure), mammary glands, liver, respiratory tract	Sub-additive effects were found 1.6 times more often than super-additive effects. Sub-additive effects were found more frequently with PAHs or azo dyes as combination partners.
Aromatic amines	Liver, kidneys/ ureter/ bladder, mammary glands, gastrointestinal tract	Occurrence of sub-additive and super-additive effects at similar frequencies. Sub-additive effects were found (22%) more often with N-nitroso compounds as combination partners.
Azo dyes	Liver (high organ specificity of substances)	Occurrence of sub-additive and super-additive effects at similar frequencies. Sub-additive effects show a tendency to occur more frequently with PAHs as combination partners.
N-nitrosamines/ N-nitrosamines	Liver, kidneys/ ureter/ bladder, respiratory tract etc. (widely spread organ specificity)	Super-additive effects were found 3.6 times more often than sub-additive effects. Super-additive effects were found more frequently with halogen compounds and natural components.
Hydrazo and azoxy compounds	Gastrointestinal tract, liver (high organ specificity of substances)	Super-additive effects were found 5.5 times more often than sub-additive effects. No substance-class dependence of incidence rates was found.
Carbamates	Skin, blood and lymph channels, respiratory tract	Super-additive effects were observed 1.3 times more often than sub-additive effects. No increased incidence was found for certain combinations.
Halogen compounds	Liver, skin	On the whole, super-additive effects occurred 1.7 times more often than sub-additive ones. For cases where the liver is affected, the rate rises to 3.8. Concerning super-additive effects, a rate increase of 37% was found for N-nitroso compounds as combination partners.
Natural substances	Liver, skin, mammary glands	Sub-additive effects and super-additive ones occur with similar frequencies. Super-additive effects show a slight tendency to occur with N-nitroso compounds and natural substances, respectively, as combination partners.
Inorganic carcinogens	Respiratory tract, kidneys/ ureter/ bladder	Super-additive effects were observed 4.5 times more often than sub-additive effects. Super-additive effects occur more frequently with PAHs and N-nitroso compounds as combination partners.
Others	Liver, skin, kidneys/ ureter/ bladder	Super-additive effects were observed 1.7 times more often than sub-additive effects. For sub-additive effects, an increased incidence was found with azo dyes as combination partners. For super-additive effects, an increased incidence was observed with N-nitroso compounds.

2.4.4.4.3 Apparent Contradictions in the Results from Different Studies, and Possible Explanations: For some of the substance combinations considered, the type of combination effect observed varied with the applied dose in various experiments (for an overview, see Berger 1996). In a mixture of 7,12-dimethylbenz(a)anthracene (DMBA) and methylnitrosourea (MNU) for instance, with the mammary glands as the main target organ of both substances, super-additive effects were detected when low MNU doses were applied, while dose-sub-additive effect were observed for higher MNU doses. The reason for this could be the cytotoxic effect of MNU in higher doses, which can lead to the death of the affected cells before they can develop into tumour cells. The carcinogenic effect is superseded in this way. A similar combination effect was observed for diethylnitrosamine (DEN)-induced liver tumours and polychlorated biphenyls (PCB's). After giving the PCB's before or together with DEN, sub-additive effects were detected, whereas administering the PCB's after giving DEN led to super-additivity.

For an explanation, one could refer to the enzyme-inducing effect of various PCB's, which affects both activating and degrading enzyme systems. From the review article, it is not clear if the PCB treatment prior to the DEN initiation activates degrading enzyme systems or if it inhibits detoxicating ones. The modulation of the effect of one combination partner through enzyme inhibition/activation was already discussed in the context of conventionally toxic substances.

2.4.4.4.4 The Combination of 3 Nitrosamines at Very Low Doses: After several qualitative studies had indicated that the simultaneous exposure to several carcinogens mainly attacking the same target organ leads to dose-additive effect amplification, the suggestion was tested in a quantitative study involving Sprague-Dawley rats of about 100 days of age (Berger et al. 1987). This relatively old study will be described here for the following reasons:

- The animal groups experimented on were relatively large (at least 80 animals per group and 500 animals in the control group, with a total of 1,800 animals).
- The three substances used were administered to the animals through their *full lifetime.*
- The doses applied were very low.

Hence, the experimental set-up almost matched the real, human exposure situation and, at the same time, provided a basis for statistically reliable results.

The substances, N-nitrosodiethylamine (NDEA), N-nitrosodiethanolamine (NDElA) and nitrosopyrolidine (NPYR), were given with drinking water over the life-span of the test animals (max. 170 weeks). In addition, the animals received special food low in nitrosamine (< 2 ppb) in order to prevent distortions of the results. While the main target organ of all the substances is the liver, various secondary target organs, e.g. the gastrointestinal tract, were determined in experiments with the individual substances.

The results are listed in table 2.4-12. In each case, the groups designated HD (high dose) were given the lowest doses still having shown a carcinogenic effect in earlier experiments. MD (medium dose) and LD (low dose) amounted to 1/3 or 1/10 of HD respectively. HCD (highest combination dose) contained each of the single substances at MD; MCD (medium combination dose) was composed of all the substances at LD, and LCD (lowest combination dose) contained them at 1/3 LD each. The size of the respective groups was determined according to the tumour incidence estimated on the basis of previous dose-response experiments with the single substances.

Table 2.4-12 Results of the mixture experiments with three nitrosamines at very low doses

Treatment groups and doses applied (mg/kg)								
Group number	Number of test animals	Single dose per day				Average total dose		
		NDEA	NPYR	NDElA	Designation	NDEA	NPYR	NDElA
1	500	-	-	-	Control	-	-	-
2a	80	0.1-	-	-	HD	61.1	-	-
2b	80	0.032	-	-	MD	20.3	-	-
2c	80	0.01	-	-	LD	6.5	-	-
3a	80	-	0.4	-	HD	-	272	-
3b	80	-	0.133	-	MD	-	85.2	-
3c	80	-	0.04	-	LD	-	26.4	-
4a	80	-	-	2	HD	-	-	1327
4b	80	-	-	0.63	MD	-	-	420.8
4c	80	-	-	0.2	LD	-	-	130.9
5a	100	0.032	0.13	0.63	HCD	21	85	409
5b	240	0.01	0.04	0.04	MCD	6.7	27	135
5c	240	0.0032	0.013	0.013	LCD	2.1	8.7	42.2

The appearance and behaviour of the animals were examined twice daily; the body weight was checked once every month. After the death of an animal, a complete, macroscopic, pathological analysis was carried out, and histological examinations of liver and brain sections as well as of every macroscopic abnormality detected.

Because the animals were kept in good conditions, they achieved the very high life expectancy of 1,000 to over 1,200 days. This explains the relatively high incidence of animals with at least one malignant tumour in the control group, totalling 29%. The treatment did not reduce the average life expectancy of the animals, with the exception of groups 2a (reduction by 77 days) and 2b (reduced by 52 days), meaning that the treatment only changed the *cause* of death. The results are shown in table 2.4-13.

For the liver as the main target organ, a comparison between the mixture effect and the effects of the single substances led to the conclusion that the combination causes dose-additive effect amplification. Summed up over all single-substance MD experiments, the proportion of animals with liver tumours was 10.1%. For the HCD group (having received all the single substances at the same time, each at the

Table 2.4-13 Number and proportion of animals with tumours in selcted organ systems when treated with mixtures of three nitrosamines at very low doses

Treatment	# of animals	Liver		Gastro-intestinal tract		Nerve tissue		Blood and lymph system		Urinary channels	
		%	#	%	#	%	#	%	#	%	#
Control	500	0.6	3	5.2	26	10.8	54	4.6	23	0.2	1
NDEA HD	80	45	36	31.3	25	5	4	5	4	1.3	1
NDEA MD	80	3.8	3	8.8	7	13.8	11	5	4	1.3	1
NDEA LD	80	2.5	2	11.3	9	12.5	10	2.5	2	2.5	2
NPYR HD	80	21.3	17	7.5	6	12.5	10	5	4	0	0
NPYR MD	80	5	4	8.8	7	11.3	9	6.3	5	1.3	1
NPYR LD	80	1.3	1	7.5	6	6.3	5	7.5	6	2.5	2
NDElA HD	80	7.5	6	7.5	6	13.8	11	8.8	7	1.3	1
NDElA MD	80	1.3	1	3.8	3	20	16	6.3	5	1.3	1
NDElA LD	80	2.5	2	8.8	7	10	8	10	8	0	0
HCD	100	16	16	9	9	17	17	3	3	3	3
MCD	240	4.2	10	7.9	19	10	24	4.2	10	0.8	2
LCD	240	1.7	4	5.8	14	12.1	29	7.9	19	0.4	1

MD dose) 16% was the detected proportion (grey background). The difference of 6% is not, the deviation from the control is statistically significant (p < 0.001).

In the single-substance LD experiments, the overall proportion came to 6.3% compared to 4.2% in the corresponding MCD mixture group (dark grey background). In this case, too, the difference is of no statistical significance. Again, the deviation from the control group was significant, with p < 0.001.

Even in the LCD group (combination of the single substances at 1/3 LD), a slight increase in the tumour rate was detected, when results were compared with the control group, however with a lower statistical significance (p < 0.1). This important finding shows that combined effects of carcinogens can be detected carcinogens even at relatively low doses. Although single-substance experiments have not been carried out at 1/3 LD, the results from the other groups as well as the order of magnitude of the observed increase suggest the existence of dose-additive effects in this case too.

For the secondary target organs, neither clear dose-response relationships could be established, nor was there any detectable additive effect amplification.

The conclusion has to be the following: Even when relatively low concentrations were used in the mixture experiments, dose-additive effect amplification was detected for the main target organ of the three nitrosamines. Hence, dose additivity will also occur in the lower dose range relevant for humans. For the secondary target organs, dose additivity was not detected. This finding supports the assumption that carcinogens attacking several different organs do not influence each other. An analogue finding was already stated for conventionally toxic substances.

2.4.4.4.5 The Combination of 5 Heterocyclic Amines with Carcinogenic Effects: In the heat treatment of proteins in meat and fish, heterocyclic amines emerge as pyrolysis products from certain amino acids. Heterocyclic amines have a high mutagenic potential and always proved to be carcinogenic in long-term animal experiments. For a study performed in Japan (Ito et al. 1991), five of these chemically defined amino compounds were combined and examined in a short-term carcinogenesis experiment. The observed parameter was the occurrence of liver cell foci, a precursor stage of liver cancer, in rats after an 8-week treatment following an initial ip-injection of 200 mg/kg diethylnitrosamine (DEN). In initiation-promotion experiments like this, a notoriously liver-carcinogenic dose is given (DEN) before administering the amines to be tested for their promoting (in this case amplifying) effect over a longer period starting 14 days after initiation. The following 5 heterocyclic amines (at the concentrations given in brackets, in ppm), were given as single substances and in combination:

- 3-amino-1,4-dimethyl-5H-pyrido[4,3-b]indole (Trp-P-1, as Trp-P-1-acetate; 150 ppm),
- 2-aminodipyrido[1,2-a:3',2'-d]imidazole (Glu-P-2, as Glu-P-2-HCl; 500 ppm),
- 2-amino-3-methylimidazo[4,5-f]quinoline (IQ; 300ppm),
- 2-amino-3,4-dimethylimidazo[4,5-f]quinoline (MeIQ; 300ppm),
- 2-amino-3,8-dimethylimidazo[4,5-f]quinoxaline (MeIQx; 400ppm).

Apart from the unit doses given in brackets, dilutions of 1/5 and 1/25 of each preparation were tested too (see table 2.4-14).

The animals used in the experiments were Fisher-344 rats, which were given – after the initial DEN dose and an interval of 2 weeks – the single substances and the mixtures for a period of 8 weeks each, administered with food. At the end of the third week of the experiment, 2/3 of the livers were resected in order to add another carcinogenic stimulus to the increased, compensatory cell division triggered in this way.

The observed quantities during the lifetime of the animals were their body weight and food intake. After killing the animals at the end of week 8 of the experiment, the absolute and relative liver weights were measured, and an immuno-histochemical examination of liver sections was performed. In the course of these tests, the GST-P (glutathione S-transferase, placental form)-positive foci were evaluated with regard to their number and size.

The single-substance experiments with Trp-P-1, IQ and MeIQx (all at dose 1/1) resulted in a decreased body weight, while MeIQ (dose 1/1) caused a rise in the body weight. Both the 1/5-mixture and the 1/25-mixture caused a weight gain, too, which was however of no statistical significance in comparison to the control group. The food intake in the DEN-initiated group was generally somewhat less than in the corresponding non-initiated group. Concerning the liver weight, no deviations from the control results were detected. The absolute liver weight ranges between 7 and 8 grams in all groups (i.e. 2.7–3.3 % of the body weight).

Table 2.4-14 The number and total area of GST-P-positive foci in rat livers following treatment with mixtures of 5 heterocyclic amines and initiation with DEN.

Treat-ment group	Dose	No. of ani-mals	Number* of foci / cm^2	Number of foci minus control	Total area of the foci **	Total area of the foci** minus control
Control		15	20.38 (4.58) [a]	-	0.8 (0.45)	-
Trp-P-1	1/1	14	69.55 (14.79) [a]	49.17	3.43 (0.94) [a]	2.63
Glu-P-2	1/1	14	29.78 (5.90) [a]	9.40	0.95 (0.19)	0.15
IQ	1/1	13	85.70 (14.39) [a]	65.32	4.08 (1.30) [a]	3.28
MeIO	1/1	15	37.82 (7.15) [a]	17.44	1.66 (0.83) [b]	0.86
MeIQx	1/1	14	48.74 (11.19) [a]	28.36	1.86 (0.42) [a]	1.06
Average			54.3 [a]	33.9	2.38 [a]	1.59
Trp-P-1	1/5	14	28.38 (5.65) [a]	8.00	1.02 (0.32)	0.22
Glu-P-2	1/5	14	22.24 (3.80)	1.86	0.85 (0.28)	0.05
IQ	1/5	13	28.27 (5.00) [a]	7.89	1.10 (0.30)	0.30
MeIO	1/5	14	34.68 (6.77) [a]	14.30	1.25 (0.30) [b]	0.45
MeIQx	1/5	14	21.55 (3.88)	1.17	0.79 (0.37)	-0.01
Average			28.9 [a]	6.64	1.00	0.2
Mixture	1/5 each	18	56.85 (9.69) [a]	36.47	2.02 (0.45) [a]	1.22
Trp-P-1	1/25	13	21.20 (3.88)	0.82	0.72 (0.16)	-0.08
Glu-P-2	1/25	14	20.18 (4.48)	-0.20	0.66 (0.25)	-0.14
IQ	1/25	15	21.07 (3.94)	0.69	0.67 (0.25)	-0.13
MeIO	1/25	14	19.49 (4.05)	-0.89	0.78 (0.35)	-0.02
MeIQx	1/25	15	20.02 (2.45)	-0.36	0.78 (0.24)	-0.02
Mixture	1/25 each	14	26.80 (4.86) [b]	6.42	1.39 (0.92) [c]	0.59

*   Mean value ± standard deviation (in brackets)   ** Numbers in mm^2 / cm^2
a: Significant deviation from the control group; $p < 0.001$
b: Significant deviation from the control group; $p < 0.01$
c: Significant deviation from the control group; $p < 0.05$

The results of the foci analysis are summarized in table 2.4-14. Only the results for the DEN-initiated groups are listed here, since the effects were stronger in these groups than in the corresponding non-initiated groups. Furthermore, the average values for the 1/1- and 1/5-single-substance experiments, respectively, were computed in order to obtain a better comparison between the effect of the single substances and of the respective mixtures.

The comparison between the average results for the single-substance groups and the corresponding results from the mixture groups (see highlighted fields in the table) supports the hypothesis that carcinogenics attacking the same target organ behave dose additively. The single substances at the 1/1-dose, for instance, cause the same increase in the number of positive foci ($33.9/cm^2$) as the 1/5-mixture group ($36.47/cm^2$). In the same way, the total area of the foci was increased by the 1/1-single-substance groups by the same extent on average ($1.59 \ mm^2/cm^2$) as by the 1/5-mixture group ($1.22 \ mm^2/cm^2$). Corresponding results were obtained for the 1/5-dose single substances compared to the 1/25-mixture group, with regard to the number as well as to the total area of the foci. The total increase in the number of foci ($33.22/cm^2$) triggered by the 1/5-diluted single substances is in agreement, too, with the increase caused by the 1/5-mixture group ($36.47/cm^2$).

It is also remarkable that there are indications of dose additivity even in the lowest-dose mixture group. In the 1/25-single-substance experiments, for instance, no significant difference to the control group was found, whereas the 1/25-mixture group showed a marked increase in the number of positive foci. This demonstrates that, when several carcinogenic substances are given simultaneously, dose additivity can occur in the lower dose range too, concerning both the number and the area covered by the foci. The low statistical significance of some of the results for the 1/25-group can be explained by the observation that the effects caused by doses that low are very low as well and hence difficult to detect.

The experimental protocol of looking at transformed liver cell foci in order to detect the early stages of carcinogenesis has thus proven its value. Its advantages are the lower experiment costs and the early availability of results, enabling the investigation of a larger number of combinations in a short time.

A dose-super-additive effect of the mixture, as seen in some epidemiological studies, was not found in these experiments, though some of the substances used are chemically similar and develop their effect in the same target organ.

2.4.4.4.6 The Combination of 20 or 40 Pesticides, Some with Carcinogenic Effects:
The aim of another experiment performed in Japan was to examine the pesticides most commonly used in that country at present for possible carcinogenic potentials of combinations between them. The doses were guided by the so-called ADI values (Acceptable Daily Intake, established by the Food and Agriculture Organization (FAO) of the United Nations). The mixtures tested were composed of components in doses below their respective ADI values (Ito et al. 1995).

From its basic design, the experiment followed the initiation-promotion model using Fisher 344 rats with a 2/3-liver resection. The experimental protocol was subdivided into three different set-ups:

- 20 pesticides with the liver as their main target organ; initiation with diethylnitrosamine (DEN, 200 mg/kg)

- 20 pesticides with different target organs; initiation with 5 compounds (DMBDD: diethylnitrosamine/DEN, N-methyl-N-nitrosourea/MNU, 1,2-dimethyl-hydrazine/ DMH, N-butyl-N-4-(hydroxybutyl)nitrosamine/BBN, 2,2'-dihydroxy-di-n-propyl-nitrosamine/DHPN); initiators partly ip-applied (DEN and MNU), partly in drinking water (DMH, BBN, DHPN)
- 40 pesticides with different target organs; initiation as in the second set-up.

The details of the substances and their doses are listed in the appendix. The results can be summarized as follows: In the experiment with 20 pesticides and the liver as their main target organ, only doses 100 times higher than the ADI-values led to a statistically significant increase in the number and size of liver foci. The area covered by foci rose from $3.5/cm^2$ in the control group to about $4.5/cm^2$ in the treatment group; the average size of the foci increased from ca. $0.28 \ mm^2/cm^2$ to ca. $0.44 \ mm^2/cm^2$. In the two experimental set-ups with 20 and 40 pesticides, respectively, and with different target organs, no differences between control groups and treatment groups were found.

From these findings, we draw the following conclusions: As long as the components of the mixtures were applied at their respective ADI doses, none of the groups showed effects attributable to the treatment. Since substances with proven carcinogenic effects were used in the first part of the experiment, this finding might appear surprising at first, especially as some of the substances must be assumed to attack the same target organs. One would expect, therefore, at least partially dose-additive effects of the mixture.

The explanation could lie in antagonistic effects occurring in the mixture experiments. Another possibility is that the ADI doses – which were set with a safety margin of generally 1:100 – pushed the single-substance concentrations to values so low that the combination even of 20 or 40 pesticides did not lead to a sum effect causing a detectable cancer development within the time frame of the experiments.

This explanation is also supported by the finding that the 100xADI-group in the first part of the experiment caused detectable effects due to the treatment. In the field of conventionally toxic substances, too, mixtures of components at their respective "no observed adverse effect levels" (NOAEL) already showed minimal effects in some cases. Then, the reason was that at least one component dose was just above the effect threshold, making possible some interactions..

As a rule, the 100xADI-doses roughly equate to the doses established as NOAEL in single-substance experiments. Since carcinogens cannot have an effect threshold and therefore all substances attacking the same target organ are available for interaction effects, the results from the 100xADI-group indicate amplifying combination effects of mixtures of carcinogens in the lower dose range. However, since the experimental protocol does not refer to the effects of the components as such, we cannot identify the type of the combination effects (sub-additive, dose-additive or super-additive). The obvious suspicion would be that some components have an amplifying influence on the total effect and others act as attenuators.

On the whole, this study appears to support the approach of regulating carcinogens by setting ADI's. The component concentrations are kept so low by the ADI's that even if amplifying combination effects occurred, these would not be strong enough to produce an effective mixture.

2.4.4.4.7 The Combination of Two Carcinogens with Different Main Target Organs: Various studies involving animal experiments, including the work by Schmähl (1976), had produced indications that dose-additive effects in mixtures of carcinogens occur only if the component substances attack the same main target organ. Combined doses of, for instance, 4-dimethylaminoazobenzene and diethylnitrosamine (DEN) – both compounds with the liver as their main target organ – lead to dose-additive effects. A study by Odashima (1962) showed, on the other hand, that additive effects also take place, when 4-dimethylaminoazobenzene and trans-4-dimethylaminostilbene, which causes tumours at the zymbal gland of the auditory canal, are given simultaneously.

In these early experiments, no difference was made between the initiating and the promoting effects of carcinogens. Hence, the conflicting results could be explained if trans-4-dimethylaminostilbene acted as an initiator and a promoter in the ear while also having an initiating effect in the liver, and if the latter was not detected in the early experiments just because of the absence of a promoting activity.

Therefore, the question was asked in the following experiment (Hammerl et al. 1994) if additive effects are detectable for a similar mixture of 2-acetylaminofluorene (AAF, main target organ liver, initiating and promoting activity) and trans-4-acetylaminostilbene (AAS, main target organ ear, initiating and promoting activity, plus an initiating effect in the liver). It was further examined if, for sequential doses of the two compounds, the sequence of the doses has any influence on the total effect.

In the first part of the experiment, newborn Wistar rats were used ("newborn model"). The substances were given orally, dissolved in evaporated milk, on days 6, 8, 12 and 14. Group 1 received AAS (50 μmol/kg day) on days 6 and 8, and AAF (125 μmol/kg day) on days 12 and 14. Group 2 was given a mixture of both compounds every day (AAS 25 μmol/kg day, AAF 62.5 μmol/kg day), so that the total doses were equal for groups 1 and 2. With these two groups, the research was going to look for differences between the effects of sequential and simultaneous doses. Group 3 was given AAS only (50 μmol/kg day) and group 4 only AAF (125 μmol/kg day, see table). From the end of the suckling period, the animals received 500 ppm of phenobarbital (PB) with their drinking water, until they were sacrificed after 14 or 24 weeks respectively. A comparison group only was given PB in the drinking water for a period of 15 months. These animals were sacrificed two years after the start of the experiment and examined for liver tumours in order to measure the impact of the PB treatment. After sacrifying the animals of groups 1-4, their body and liver weights were measured. Liver sections were prepared and tested, by immuno-histochemical means, for γ-glutamyl-transpeptidase (GGT)-positive liver cell foci or for the placental form of glutathione-S-transferase (GST-P)-positive foci respectively. For every animal, a total section surface of at least 3 cm^2 from different liver lobes was surveyed microscopically; areas containing at least 8 positive cells were counted as positive foci, by definition.

In the second part of the experiment (2/3-partial liver resection model), 9–10 weeks old female Wistar rats were used as test animals. The substances were administered orally, dissolved in 1,3-propanediol, in 2 phases on days 1, 4, 7, 10 and 15, 18, 21, 24. AAS was applied in doses of 20 μmol per kg and day, AAF in doses of 100 μmol/kg day. Group 1 was treated with AAS in the first phase and with AAF in the second; group 2 received AAF in the first phase and AAS in the second; groups 3 and 4 given only AAS or AAF, respectively, in both phases. On day 27, the 2/3-partial liver resection was performed for promotion. From day 37 until the end of

Table 2.4-15 Number, size and total area of pre-neoplastic liver cell foci in male and female rats following treatment with two carcinogens attacking different target organs (newborn rats)

Male rats							
Duration of experiment	Treatment	GGT			GST-P		
		Number (n/cm^2)	Size (mm^2)	Total area*	Number (n/cm^2)	Size (mm^2)	Total area*
14 weeks	Control	0.4 ± 0.2	0.01 ± 0.01	0.04 ± 0.02	0.3 ± 0.2	0.01 ± 0.01	0.06 ± 0.04
	AAS-AAF	2.8 ± 0.9	0.02 ± 0	0.5 ± 0.2	3 ± 1	0.02 ± 0	0.7 ± 0.3
	AAF/AAS	3.6 ± 0.6	0.02 ± 0	0.83 ± 0.09	7 ± 2	0.02 ± 0	1.3 ± 0.3
	AAS-AAS	**9 ± 1**	0.05 ± 0.01	**4.18 ± 0.07**	**15 ± 4**	0.04 ± 0.01	**6 ± 1**
	AAF-AAF	2.4 ± 0.5	0.02 ± 0	0.4 ± 0.1	4 ± 1	0.02 ± 0	0.7 ± 0.2
24 weeks	Control	0.59 ± 0.06	0.01 ± 0	0.06 ± 0.01	1.8 ± 0.9	0.01 ± 0	0.3 ± 0.1
	AAS-AAF	6.0 ± 1	0.04 ± 0.01	2.3 ± 0.9	12 ± 2	0.03 ± 0.01	3.4 ± 0.4
	AAF/AAS	11.0 ± 2	0.03 ± 0.02	3.0 ± 0.6	13 ± 3	0.03 ± 0	3.4 ± 0.8
	AAS-AAS	**8.0 ± 3**	0.05 ± 0.01	**5 ± 3**	**13 ± 1**	0.04 ± 0.02	**6 ± 3**
	AAF-AAF	5.0 ± 0.3	0.03 _ 0	1.3 ± 0.2	6 ± 0.2	0.03 ± 0	1.6 ± 0.2
Female rats							
Duration of experiment	Treatment	GGT			GST-P		
		Number (n/cm^2)	Size (mm^2)	Total area*	Number (n/cm^2)	Size (mm^2)	Total area*
14 weeks	Control	0	0	0	0	0	0
	AAS-AAF	2.2 ± 0.9	0.03 ± 0.01	0.6 ± 0.2	2.8 ± 1.2	0.03 ± 0.01	0.6 ± 0.2
	AAF/AAS	3.7 ± 0.1	0.02 ± 0.01	0.7 ± 0.2	7 ± 2.5	0.02 ± 0.01	1.0 ± 0.3
	AAS-AAS	**16.7 ± 1.4**	**0.10 ± 0.02**	**16.9 ± 2.1**	**20.6 ± 0**	**0.09 ± 0.01**	**18.9 ± 2**
	AAF-AAF	1.6 ± 1.3	0.02 ± 0	0.4 ± 0.3	2.4 ± 2.1	0.02 ± 0	0.5 ± 0.4
24 weeks	Control	0.2 ± 0.2	0.01 ± 0.01	0.1 ± 0.1	0.4 ± 0.4	0.01 ± 0.01	0.1 ± 0.1
	AAS-AAF	6.0 ± 0.5	0.04 ± 0.02	2.2 ± 0.6	8.0 ± 1.0	0.03 ± 0.01	2.4 ± 0.6
	AAF/AAS	6.3 ± 2.5	0.05 ± 0.01	2.5 ± 1.0	6.8 ± 3.1	0.04 ± 0.01	2.1 ± 0.7
	AAS-AAS	**17.8 ± 2.4**	**0.13 ± 0.1**	**25 ± 18**	**24.4 ± 3.2**	**0.13 ± 0.1**	**27.4 ± 19**
	AAF-AAF	6.3 ± 0.4	0.03 ± 0.01	1.5 ± 0.1	6.2 ± 0.3	0.03 ± 0.01	1.6 ±0.2

Mean value ± standard deviation, n = 3 (male animals), n = 2 (female animals)
* Total area of foci, in ‰ of the area surveyed

the experiment after 18 or 31 weeks, respectively, the animals received 250 ppm PB with their drinking water. In this part of the experiment, too, another group, which was given only PB for 15 month, was sacrified 2 years after the start of the experiment and surveyed for the development of liver tumours. The same cytological end points were examined as in the first part of the experiment; in addition, haematoxylin-eosin dyeing was performed.

The results of the first part of the experiment ("newborn model") are listed in table 2.4-15. The body and liver weights did not show any significant deviations from the control group. In the number, the average size and the total area surface of the foci for male and female animals, respectively, some differences were found (bold-faced numbers in the table).

Table 2.4-16 Number, size and total area of pre-neoplastic liver cell foci following treatment with two carcinogens attacking different target organs (adult rats)

Duration of experiment	Treatment	GGT			GST-P			Hematoxylin-eosin		
		Number (1/cm²)	Size (mm²)	Area* (‰)	Number (1/cm²)	Size (mm²)	Area* (‰)	Number (1/cm²)	Size (mm²)	Area* (‰)
18 weeks	Contr.	0.02 ± 0.02	0.01 ± 0.01	0.02 ± 0.02	0.7 ± 0.4	0.02 ± 0.02	0.1 ± 0.1	0	0	0
	AAS-AAF	**3.7 ± 0**	0.13 ± 0.1	4.8 ± 0.2	**9.4 ± 1.1**	0.08 ± 0.02	6..5 ± 0.6	1.8 ± 0.03	0.18 ± 0.02	3.0 ± 0.1
	AAF-AAS	**10.8 ± 1.5**	**0.05 ± 0**	5.3 ± 0.8	**18.3 ± 0.7**	**0.04 ± 0.01**	6.8 ± 0.3	3.6 ± 1.3	0.08 ± 0.03	3.3 ± 2.1
	AAS-AAS	**9.1 ± 1.8**	0.10 ± 0.03	7.9 ± 0.6	**12.5 ± 1.8**	0.08 ± 0.02	9.1 ± 0.8	3.6 ± 0.9	0.15 ± 0.05	5.5 ± 3.0
	AAF-AAF	**3.5 ± 0.3**	0.15 ± 0	5.1 ± 0.4	**9.9 ± 1.7**	0.07 ± 0..01	6.9 ± 0.1	1.7 ± 0.8	0.13 ± 0.07	2.8 ± 2.2
31 weeks	Contr.	0.2 ± 0.2	0.02 ± 0.02	0.08 ± 0.06	1 ± 1	0.01 ± 0.01	0.2 ± 0.2	0	0	0
	AAS-AAF	**5.9 ± 0.5**	0.32 ± 0.04	19 ± 2	**13 ± 2**	0.19 ± 0.05	21 ± 1	5.6 ± 0.5	0.3 ± 0.08	17 ± 4
	AAF-AAS	**11 ± 2**	**0.09 ± 0.02**	11 ± 4	**21 ± 8**	**0.06 ± 0**	13 ± 5	4 ± 1	0.08 ± 0.02	2.6 ± 0.6
	AAS-AAS	**8 ± 2**	0.22 ± 0.06	16 ± 2	**21 ± 5**	0.13 ± 0.05	22 ± 2	5 ± 3	0.4 ± 0.3	12 ± 3
	AAF-AAF	**4 ± 0.6**	0.13 ± 0.05	5 ± 2	**12 ± 3**	0.08 ± 0.03	10 ± 4	2.3 ± 0.4	0.3 ± 0.1	6 ± 1

Mean value ± standard deviation, n = 2 animals (18 weeks), n = 3 animals (31 weeks)
*   Total area of foci, in ‰ of the total area surveyed.

In the second part of the experiment (2/3-partial liver resection model), as in part 1, no significant differences from the control group were found concerning the body and liver weights of the treated animals. The deviations observed for the liver cell foci are highlighted by a grey background in table 2.4-16: A comparison between the two parts of the experiment shows that the 2/3-partial liver resection model generally led to stronger effects concerning the number as well as the size of the foci. Hence, this experimental set-up is more sensitive than the newborn model. In both experiments, the substances caused a significant rise in the *number* of positive foci, with AAS causing the stronger effect in every case. The GST-P dyeing constituted the most sensitive end point, followed by dyeing with GGT and with haematoxylin-eosin respectively. Concerning the *total area* of the foci, too, both substances caused a significant increase compared to the control groups. AAS showed the stronger effect in the second part of the experiment. In the first part, no significant differences were found between different treatment groups. GGT dyeing was more sensitive with regard to the area than dyeing with GST-P.

The sequential administration of two substances led to a total effect on the number of foci largely determined by the substance given last. In each of the experiments spanning 31 weeks, GST-P dyeing led to the detection of ca. 20 foci/cm^2 (GGT ca. 10/cm^2), when AAS was given as the second substance; when AAF was administered as the second substance, only about 12-13 foci/cm^2 (GGT ca. 4–5/cm^2) were counted. Hence, with regard to the number of foci, dose-additive effects were shown to be at work.

Comparisons of the foci numbers and sizes for simultaneous and sequential (AAS-AAF) doses of both substances showed a tendency, but not a statistically significant trend towards higher values for simultaneous administration. Still, in the second part of the experiment, the sequence AAF-AAS was shown to have a considerably stronger effect concerning the size of the foci. The finding from the first part of the experiment can be interpreted in two ways: Either the said tendency reflects dose additivity and lacks statistical significance only because of the insufficient group size, or, when giving the substances simultaneously, antagonistic effects occurred. Unfortunately, the group AAF-AAS was not examined in the first part of the experiment, so that a direct comparison – and thereby a decision between the two possible explanations – cannot be made.

The AAF-AAS groups developed foci of a significantly smaller average size than the AAS-AAF groups. Interestingly, the values for the groups AAS-AAS and AAF-AAF, respectively, did not show any significant differences from each other; both were of a similar order of magnitude as in the AAS-AAF groups. AAS given as the second substance seems to have an antagonistic effect on AAF with regard to the size of the foci. However, since the substances are metabolized via different routes, according to the authors, competition for bonding sites cannot be the reason. A mechanistic explanation of this finding has still not been found. The lack of a promoting activity of AAS in the liver cannot be the reason either, since, firstly, promotion was initiated by a partial liver resection and treatment with PB and, secondly, because the average sizes of the foci in the AAS-AAS groups were similar to the sizes in the AAF-AAF groups.

For the purposes of this monograph, the data from both experiments were subjected to a deeper statistical analysis, the results of which are shown in figures 2.4-5

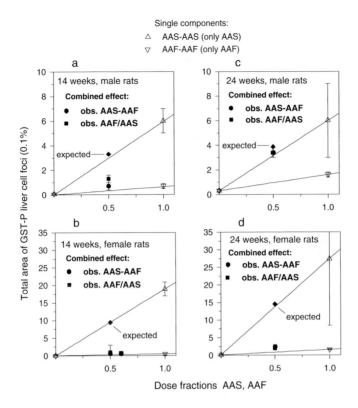

Fig. 2.4-5 AAS and AAF in the "newborn model" (see text and the corresponding tables). Comparison between the observed combined effects and the values expected for a dose additive combined effect assuming a linear dose-response relationship. Under this assumption, the expected combination effect equals the sum of the effects of the single components. The observed combination effect appears to be clearly sub-additive in a, b and d, while it looks additive in graph c.

and 2.4-6. In the "newborn model" (fig. 2.4-5), only for male rats examined after 24 weeks, dose and effect additivity are in (approximate) agreement (graph c); in the other cases, the contribution the "zymbal gland" carcinogenic AAS is dominant by far (a,b,d). In the "liver resection model", agreement between dose and effect additivity was found for both examination dates (fig. 6 a,b).

2.4.4.4.8 Mineral Fibre Dust + Cigarette Smoke: Epidemiological studies on the two causal factors, smoking and exposure to asbestos, in coincidence have pointed to super-additive effects (see 2.4.4.3.2). The mechanistic background was researched in animal experiments by Morimoto et al. (1993).

The macrophages in the lung alveoli are known to play an important role in the development of lung fibroses ("asbestosis") caused by mineral fibres. The macrophages phagocytize the fibres, destroying some of the cells, which then release substances causing damage to the lung tissue; this damage (e.g. lysosomal enzymes)

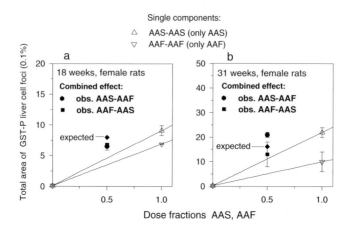

Fig. 2.4-6 AAS and AAF in the "partial liver resection model" (see text and the corre-spond-ing table). Comparison between the observed combination effects and the values expected for a dose additive combined effect assuming a linear dose-response relationship. Under this assumption, the expected combination effect equals the sum of the effects of the single com-ponents. The observed combination effect appears to be approximately additive in graphs a and b.

may then initiate the carcinogenesis. Under the combined exposure to mineral fibres and cigarette smoke, the additional effect can occur that carcinogenic smoke components (condensate) are deposited on the surface of the fibres and infiltrate the macrophages from there. Apart from that, there are indications that further stim-ulation of the immune system takes place, and the homeostasis mediated by cell-cell interactions becomes disrupted. The cell-cell interactions are mediated by cytokines and other substances. Hence, for the study described in the following, the influence of an exposure to cigarette smoke and/or mineral fibres on the release of the tumour-necrosis factor (TNF), an important cytokine involved in the develop-ment of lung fibroses, was investigated in cell-biological experiments.

The test animals were 9 weeks old Wistar rats (5 animals per group). One group was exposed to the smoke of altogether 40 cigarettes (condensate concentration 10 mg/m^3) over a period of 8 hours in a smoke-flooded inhalation chamber; another group was used for clean-air control measurements. After the smoke exposure, the animals were sacrificed and their lungs were irrigated with 50 ml of a physiological saline solution in order to obtain free macrophages. After centrifugation and re-sus-pension into a solution containing 10^6 cells/ml, mineral fibres were added, at con-centrations of 25, 50 and 100 µg/ml, to half of the cell cultures produced in this way. The fibres used were "Canadian chrysotile fibres" and "aluminum silicate ceramic fibres", both pulverized and prepared as respirable aerosols. Following an incuba-tion period of 24 h, the supernatant of the cell culture was extracted for the TNF experiment. The cell cultures (3×10^5 cells/ml) of the L-929 mouse fibroblasts were used for this purpose because of their particularly high sensitivity to the cytotoxic effect of TNF. Various dilutions of the TNF supernatant were introduced into the supernatant of these cultures and incubated for 18 h. Then, through dyeing and photometric evaluation, the lysated L-929 cells were counted. The amount of TNF present in the preparation was measured in activity units, where one unit was defined as the TNF activity required for killing 50% of the L-929 cells. In addition to that, electron-microscopic examinations of the two fibre types were carried out, because there had been indications of the carcinogenic potential of mineral fibres being co-determined by their physicochemical properties including their surface structure and shape.

The electron-microscopic examinations showed that the ceramic fibres are stick-shaped with a relatively smooth surface, whereas the chrysotile fibres are of very irregular shapes – e.g. axially split, turned or twirled – and have a relatively rough surface. The lung irrigation showed significantly more macrophages (ca. 4.5×10^7/ml) for the groups that had been exposed to cigarette smoke than for those groups which had not been exposed to smoke (ca. 2.5×10^7/ml). Without the mineral fibre treatment, no significant differences in the TNF activities in the two test groups ("smokers" and "non-smokers") were detected.

In the groups treated with chrysotile fibres, a dose-dependent increase of the TNF production was noticed. In the "non-smokers" group, a significant increase of the TNF activity occurred, from about 100 units/ml in the control group to ca. 400 units/ml for 50 µg of fibres/ml and to ca. 700 units/ml for 100 µg of fibres/ml. In the smokers group, the activities rose to ca. 900 units/ml and 1,300 units/ml respectively. The rise in the smokers group is statistically significant in comparison both with the controls and with the non-smokers group. The use of 25 µg of fibres/ml also led to increased activities, which were, however, of no statistical significance in either group (at $p < 0.05$).

The application of ceramic fibres led to the TNF production increasing with the fibre concentration, and the TNF production was higher in the smokers group than in the non-smokers group, but the differences to the control group did not reach statistical significance in this case.

The conclusion was that cigarette smoke, which on its own did not trigger an increase in the TNF production, amplifies the TNF activity gains initiated by mineral fibres. An investigation by McFadden et al. (1986) showed that cigarette smoke inhibits the elimination of mineral fibres, thereby increasing the effective dose of the fibres in the lung. However, fibrosis was also found in smokers exposed to relatively small concentrations of asbestos, whereas it did not develop in other smokers, who were exposed to higher concentrations. Hence, the amplifying effect of smoking must have other, possibly endogenous reasons too.

According to Kelley (1990), the fibre-induced increase in the TNF production also attracts neutrophils, which, by releasing superoxides and proteases, can damage the lung tissue to such an extent that they cause asbestosis. In the study described here, too, it was shown that cigarette smoke increases the number of macrophages in the lung. The TNF production could be increased indirectly through the rising number of macrophages producing TNF, and the increased number of macrophages leads to more macrophages killed off by the mineral fibres and releasing larger amounts of substances harmful for the lung tissue cells (e.g. lysosomal enzymes). Other studies (e.g. Baughman et al. 1986, Davis et al. 1988) have further shown that cigarette smoke induces morphological transformations in the macrophages, leading to the release of lactate dehydrogenase and lysosomal enzymes as well as to an increased production of superoxides. Carcinogenic substances from cigarette smoke can also get deposited on the mineral fibres and in this way infiltrate the macrophages in larger numbers. This rise in the concentration [of carcinogenic agents] should further increase both the release of lysosomal enzymes mentioned before and the number of dead macrophages. In the presence of the fibres, this multitude of effects of cigarette smoke finally leads to increased damage to the lung tissue and to a higher risk of asbestosis.

The stronger effect of the chrysotile fibres can probably be attributed to their rough surface and irregular shape causing more damage to the macrophages, on the one hand, and to the chemical components of the fibres (high levels of magnesium and silicate crystals), on the other, which cause additional strains on the lung tissue. There are two reasons for a stronger damage due to the rough surface and irregular shape. First, mechanical damage to the cells is more likely in this case than for smooth fibres; second, the more structured and larger surface area offers more space for other substances e.g. from the cigarette smoke to attach themselves to. The interaction can only be called synergistic, since cigarette smoke alone does not cause any increase in the TNF activity, though it amplifies the fibre-induced increase.

2.4.4.5
Summary Assessment and Recommendations for Action

The assessment models valid for conventionally toxic substances cannot be applied to carcinogenic substances, since there are threshold values for conventional toxic effects, below which toxic effects do not occur. Such threshold values have not been found for typical carcinogenic effects – neither in epidemiological studies nor in animal experiments – and would not have any foundation in theory, which regards the effects as irreversible. Hence, alternative approaches must be found for carcinogenic substances, as even for very low doses a residual, if not directly quantifiable risk is to be expected.

Neither the data from epidemiology nor those from research involving animal experiments reveal any simple rules, according to which effects would evolve following an exposure to combinations of carcinogenic compounds. There are three possible types of combined effects:

- Sub-additive (to the extreme of antagonism),
- additive, and
- super-additive (up to multiplicative).

Super-additive effects are the type most relevant both for assessments and for recommendations for actions. Epidemiology has provided drastic examples of super-additivity (combined effects of asbestos + tobacco smoke causing lung cancer or of alcohol + tobacco smoke with regard to esophageal carcinomas). Experimental research has provided only partial explanations for these phenomena, and there are no signs of any decisive progress of mechanistic research in the near future.

Considering this situation, it would appear inappropriate to formulate simple solutions covering every scenario of a combined exposure to carcinogenic substances. Any approach to regulating for combinations must be guided by the criteria established for single substances. The aim must be to define a socio-economically acceptable residual risk, which for each case has to be quantified on the basis of biological data, even if this cannot be done without calling on complex auxiliary constructions. The ALARA approach ("as low as reasonably achievable") is unsatisfactory, since it does not provide any risk quantification and can lead to unacceptable strains in terms of socioeconomic implications.

Super-additive effects in mixtures of carcinogens have been observed in dose ranges leading to high incidences of tumours. The limit of detectability for an increase in tumour incidences is in the region of 0.1–2 %, both for epidemiological studies and for investigations involving animal experiments. For practical regulations, on the other hand, residual risks several orders of magnitude below the threshold of detectability have been discussed worldwide for some decades. The EPA (US Environmental Protection Agency) model, for instance, considers a tumour incidence of 10^{-5}–10^{-6} as acceptable, which is 2–4 orders of magnitudes below the limit of detectability. There are several biomedical findings justifying the assumption that super-additive effects do not occur in this lowest dose range to any significant degree:

- The degree of super-additivity – as measured in the high dose range – rises with increasing doses (examples: asbestos + tobacco smoke, alcohol + tobacco smoke). There is an exponential dependence leading to the conclusion: The smaller the dose, the smaller the probability of super-additive effects.
- According to experience so far, super-additivity is mostly found in tumour-promoting effects. For such effects, however, effect thresholds can be found and theoretically understood (see section 2.3.3.2).
- In realistic mixtures, not only super-additive combinations between some components but also sub-additive (antagonistic) components have to be expected. The latter will at least partially compensate for super-additive effects, which still may occur and hence cannot be excluded for the lower dose range.

We therefore propose the following approach to regulating for combined exposures: For each single component, the carcinogenic risk is determined or extrapolated. One approach to this end, which is often used, is calculating the so-called "unit risk". The individual unit risks are fragmented according to the mass contributions of the components in the mixture and summed up to a total, which then is discussed or defined as the acceptable "residual risk". Following this model, the German state commission for protection from exposures (LAI, *Länderausschuss für Immissionsschutz*) has proposed an acceptable carcinogenic risk posed by pollutants in urban air. It makes use of the assumption of linear dose-response relationships in the lower dose range and of the additive effect of the single substances, meaning: It reflects a worst-case scenario. For the majority of combinations relevant in real life, it will overestimate the actual risk. Nevertheless, considering the gaps in our knowledge on the mechanisms of carcinogenic effects, such overestimation can be tolerated.

2.4.4.6
Appendix

List of the 20 pesticides and their doses in the first part of the experiment described in the text:

Chemical	ADI (mg / kg and day)	Concentration in food (ppm)	Carcinogenicity and target organ (species)
Acephate*	0.03	0.3	Liver (mouse)
Chloropyrifos*	0.01	0.1	–
Chlorofenvinphos*	0.0015	0.015	?
Dichlorvos*	0.0033	0.033	Pancreas (rat), stomach (mouse)
Dimethoate*	0.01	0.1	Gastrointestinal tract, lung, ovaries (rat and mouse)
Endosulfan*	0.006	0.06	–
Etrimfos*	0.003	0.03	?
Fenitrothion*	0.005	0.05	–
Isoxathion*	0.003	0.03	?
Malathion*	0.02	0.2	–
Methidathion*	0.001	0.01	Liver (mouse)
Pirimiphosmethyl*	0.01	0.1	?
Prothiophos*	0.0015	0.015	?
Pyraclofos*	0.001	0.01	?
Trichlorofon*	0.01	0.1	–
Vamidothion*	0.008	0.08	?
Edifenphos**	0.0025	0.025	?
Iprobenphos**	0.003	0.03	?
Tolclofos-methyl**	0.064	0.64	?
Butamifos***	0.0016	0.016	?

List of the 20 pesticides and their doses in the second part of the experiment described in the text:

Chemical	ADI (mg / kg and day)	Concentration in food (ppm)	Carcinogenicity and target organ (species)
Acephate*	0.03	0.3	Liver (mouse)
Dichlorvos*	0.0033	0.033	Pancreas, leukemia (rat), stomach (mouse)
Dicofol*	0.025	0.25	Liver (mouse)
Cypermethrin*	0.05	0.5	liver (mouse)
Permethrin*	0.048	0.48	Liver, lung (mouse)
Phosmet*	0.02	0.2	Liver (mouse)
Amitraz*	0.0012	0.012	Suspected carcinogenic effect
Clofentezin*	0.0086	0.086	Suspected carcinogenic effect
Propoxur*	0.063	0.63	Suspected carcinogenic effect
2,4-D**	0.3	3	Brain (rat)
Glyphosat**	0.15	1.5	Kidneys (mouse)
Trifluralin**	0.0075	0.075	Various organs (rat, mouse)
Metalochlor**	0.097	0.97	Positive (rat)
Dichlobenil**	0.004	0.04	Suspected carcinogenic effect
Captafol***	0.05	0.5	Various organs (rat, mouse)
Propiconazol***	0.018	0.18	Liver (mouse)
Fosetyl***	0.88	8.8	Bladder (rat)
Triadimefon***	0.012	0.12	Suspected carcinogenic effect
Mancozeb***	0.05	0.5	Suspected carcinogenic effect
Maneb***	0.005	0.05	Suspected carcinogenic effect

* insecticide, ** fungicide, *** herbicide,
? not reported, to the authors' knowledge

List of the 40 pesticides and their doses in the second part of the experiment described in the text:

Chemical	ADI (mg / kg and day)	Concentration in food (ppm)	Carcinogenicity and target organ (species)
Acephate	0.03	0.3	Liver (mouse)
Chlorobenzilat	0.02	0.2	Liver (mouse)
Cypermethrin	0.05	0.5	Lung (mouse)
Permethrin	0.048	0.48	Liver, lung (mouse)
Clofentezin	0.0086	0.086	Suspected carcinogenic effect
Oxamyl	0.02	0.2	–
Chlorpyrifos	0.01	0.1	–
Cyhalothrin	0.0085	0.085	–
Diflubenzuron	0.012	.012	–
Pirimifos methyl	0.01	0.1	–
Bendicarb	0.004	0.04	–
Malathion	0.02	0.2	–
Fenbutantinoxid	0.03	0.3	?
Cyfluthrin	0.02	0.2	?
Trichlorfon	0.01	0.1	?
Fenvalerat	0.02	0.2	?
Flucythrinat	0.0125	0.125	?
Mepiquat chloride*	0.075	0.75	?
Glyphosat**	0.15	1.5	Kidney (mouse)
Metalochlor**	0.097	0.97	Positive (rat)
Sethoxydim**	0.14	1.4	–
Thiobencarb**	0.009	0.09	–
Pendimethalin**	0.043	0.43	–
Metribuzin**	0.0125	0.125	–
Quinclorac**	0.029	0.29	?
Chlorpropham**	0.1	1	?
Bensulide**	0.04	0.4	?
Bentazon**	0.09	0.9	?
Propiconazol***	0.018	0.18	Liver (mouse)
Maneb***	0.005	0.05	Suspected carcinogenic effect
Triadimefon***	0.012	0.12	Suspected carcinogenic effect
Imazalil***	0.025	0.25	–
Zineb***	0.005	0.05	–
Vinclozolin***	0.1215	1.215	–
Fenarimol***	0.01	0.1	–
Flutolanil***	0.08	0.8	–
Metalaxyl***	0.019	0.19	–
Chinomethionat***	0.006	0.06	?
Pyrifenox***	0.1	1	?
Myclobutanil***	0.012	0.12	?

* plant growth regulator, ** herbicide, *** fungicide
? not reported, to the authors' knowledge

2.4.4.7
Literature

Foundations

Alexandrie AK, Ingelman-Sundberg M, Seidegard J, Tornling G, Rannung A (1994) Genetic susceptibility to lung cancer with special emphasis on CYP1A1 and GSTM1: A study on host factors in relation to age at onset, gender and histological cancer types. Carcinogenesis. Vol. 15, pp. 1785–1790

Anwar K, Naiki H, Nakakuki K, Inuzuka M (1992) High frequency of human papillomavirus infection in carcinoma of the urinatry bladder. Cancer. Vol.10 pp. 1967–1973.

Consensus Report: Multistage and multifactorial nature of carcinogenesis. In: Vaino H, Magee PN, McGregor DB, McMichael AJ (eds.) (1992) Mechanisms of carcinogenesis in risk identification. IARC Scientific Publications 116. IARC. Lyon, pp. 9–54

12. Duisburger Gutachtenkolloquium (1996) arbeitsmedizinischer Teil, Synkarzinogenese, LVBG (Landesverband Rheinland-Westfalen der gewerblichen Berufsgenossenschaften)

zur Hausen H (1994a) Krebsentstehung durch Infektionen – ein wichtiger, noch wenig beachteter Sektor der Krebsforschung, Dt. Ärztebl. 91, pp. 738–740

Holder JW, Eimore E, Barrett JC (1993) Gap junction function and cancer. Cancer Res. Vol. 53., pp. 3475–3485

Koch K (1995) Herpes-Viren jetzt nachgewiesen. Dt. Ärzteblatt 92, C–208

Probst-Hensch NM, Halle RW, Ingles SA, Longnecker MP, Han CY, Lin BK, Lee DB, Sakamoto GT, Frankl HD, Lee ER, Lin HJ (1995) Acetylation polymorphism and prevalence of colorectal adenomas. Cancer Res. Vol 55, pp. 2017–2020

Raunio H, Husgafvel-Pursiainen K, Anttila S, Hietanen E, Hirvonen A, Pelkonen O (1995) Diagnosis of polymorphisms in carcinogen-activating and inactivation enzymes and cancer susceptibility, a review. Gene Vol. 159, pp. 113–121

Woitowitz HJ (1988) Die Problematik der konkurrierenden Kausalfaktoren, Kolloquium Krebserkrankungen und berufliche Tätigkeit. Südd. Eisen- und Stahl-BG, Mainz, 13.07.1988, pp. 37–61

Epidemiological Studies

Bi W, Hayes RB, Feng P, Oi Y, You X, Zhen J, Zhang M, Qu B, Fu Z, Chen M, Chien HTC, Blot WJ (1991) Mortality and incidence of bladder cancer in benzidine-exposed workers in China. Am. J. Ind. Med. 21, pp. 481–489

Bittersohl G (1972) Epidemiologische Untersuchung über Spätkomplikationen nach Asbestexposition, lecture note, 1972, no bibliographic details

Bolm-Audorff U, Jöckel KH, Kilguss B, Pohlabeln H, Siepenkothen T (1993) Bösartige Tumoren der ableitenden Harnwege und Risiko am Arbeitsplatz. Report of the Bundesanstalt für Arbeitsschutz, Germany. Wirtschaftsverlag NW, Bremerhaven

Cartwright R (1982) Occupational bladder cancer and cigarette smoking in West Yorkshire Scand. J. Work Environ. Health. Vol. 8, suppl. 1, pp. 79–82

Emmelin A, Nyström L, Wall S (1993) Diesel exhaust exposures and smoking: a case-referent study of lung cancer among swedish dock workers. Epidemiology Vol. 4, pp. 237–244

Gonzáles CM, López-Aberite G, Errezola M, Escolar A, Riboli E, Izarzugaza I, Nebot M (1989) Occupational and bladder cancer in Spain: a multi-centre case-control study. Int. J. Epidemiol. Vol. 18, pp. 569–577

Hammond EC, Selikoff IJ, Seidmann H (1979) Asbestos exposure, cigarette smoking and death rates. Ann. N. Y. Acad. Sci. 330 pp. 472–490

Jedrychowski W, Becher H, Wahrendorf J, Basa-Cierpialek Z (1990) A case-control study of lung cancer with special reference to the effect of air pollution in Holland. J. Epidemiol. Comm. Health Vol. 44, pp. 114–120

Maier HE, Sennewald E (1994) Plattenepithelkarzinome. HVBG, St. Augustin

Manz A (1986) Berufsbedingte Erkrankungen im HNO-Gebiet, paper presented at the 20. Fortbildungsveranstaltung des Berufsverbandes der HNO-Ärzte, Essen, 30.10.–1.11.1986

Rösler JA, Lange HJ, Woitowitz RH (1993) Asbesteinwirkung am Arbeitsplatz und Sterblichkeit an bösartigen Tumoren in der Bundesrepublik Deutschland. Research report Asbest IV, HVBG, St. Augustin

Selikoff IJ, Hammond EC, Churg J (1968) Asbestos exposure, smoking and neoplasia. JAMA Vol. 204, pp. 104–110

Selikoff IJ, Seidmann H, Hammond EC (1980) Mortality effects of cigarette smoking among amosite asbestos factory workers. JNCI Vol. 65, pp. 507–513

Tavani A, Negri E, Franceschi S, La Vecchia C (1994) Risk factors for esophageal cancer in life-long nonsmokers. Cancer Epidemiol. Biomark. Prevent. Vol. 3, pp. 387–392

Thäriault G, Tremblay C, Cordier S, Gingras S (1984) Bladder cander in the aluminium industry. Lancet, April 28, pp. 947–950

Tremblay C, Armstrong B, Thäriault G, Brodeur J (1995) Estimation of risk of developing bladder cancer among workers exposed to coal tar pith volatiles in the primary aluminium industry. Am. J. Ind. Med. Vol. 27, pp. 335–348

Tuyns AJ, Esteve J, Raymond L, Berhno F, Benhamou E, Blanchet F, Boffetta P, Crosignani P, del Moral A, Lehmann W, Merletti F, Pequignot G, Riboli E, Sancho-Garnier H, Terracini B, Zubiri A, Zubih L (1988) Cancer of the larynx/ hypopharynx, tobacco and alcohol. Int. J. Cancer Vol. 41, pp. 483–491

Vainio H, Boffetta P (1994) Mechanisms of the combined effect of asbestos and smoking in the etiology of lung cancer. Scand. J. Work Environ. Health Vol. 20 pp. 235–242

Vineis P, Terracini B, Costa G, Merletti F, Segnan N (1982) Interaction between occupational risks and cigarette smoking in bladder cancer: a case-control study. In: ILO: Prevention of occupational cancer – International Symposium. Occupational Safety and Health Series 46. Geneva, 327–331.

Wu W (1988) Occupational cancer epidemiology in the Peoples's Republic of China. China. J. Occup. Med. Vol. 30, pp. 968–974

Studies Involving Animal Experiments

Arcos JC, Woo YT, Lai DY (1988) Databas on Binary Combination Effects of chemical Carcinogens. Environ. Carcino Revs. C6 1. Special Issue

Baughman RP, Corser BC, Strohofer S, Hendricks D (1986) Spontaneous hydrogen peroxide release from alveolar macrophages of some cigarette smokers. J. Lab. Clin. Med. Vol 107, pp. 233-237

Berger MR, Schmähl D, Zerban H (1987) Combination experiments with very low doses of three genotoxic N-nitrosamines with similar organotropic carcinogenicity in rats. Carcinogenesis. Vol. 8, 11, pp. 1635–1643

Berger MR (1996) Synergism and Antagonism between chemical Carcinogens. In: Arcos et al.: Chemical Carcinogenesis

Davis WB, Pacht ER, Spatafora M, Martin WJ (1988) Enhanced cytotoxic potential of alveolar macrophages from cigarette smokers. J. Lab. Clin. Med. Vol 111, pp. 293–298

Hammerl R, Kirchner T, Neumann HG (1994) Synergistic effects of trans-4-acetylaminostilbene and 2-acetylaminofluorene at the level of tumor initiation. Chem.-Biol. Int. Vol 93, pp. 11–28

Ito N, Hagiwara A, Tamano S, Hasegawa R, Imaida K, Hirose M, Shirai T (1995) Lack of carcinogenicity of pesticide mixtures administered in the diet at acceptable daily intake (ADI) dose levels in rats. Toxicology Letters Vol. 82/83 pp. 513–520

Ito N, Hasegawa R, Shirai T, Fukushima S, Hakoi K, Takabe K, Iwasaki S, Wakabayashi K, Nagao M, Sugimura T (1991) Enhancement of GST-P positive liver cell foci development by combined treatment of rats with five heterocyclic amines at low doses. Carcinogenesis Vol. 12, 5, pp. 767–772

Kelley J (1990) Cytokines of the lung. Am. Rev. Respir. Dis. Vol 141, pp. 471–501

McFadden D, Wright JL, Wiggs B, Churg A (1986) Smoking inhibits asbestos clearance. Am. Re. Respir. Dis. Vol. 133, pp. 372–374

Morimoto Y, Kido M, Tanaka I, Fujino A, Higashi T, Yokosaki Y (1993) Synergistic effects of mineral fibres and cigarette smoke on the production of Tumor Necrosis Factor by alveolar macrophages of rats. Brit. J. Industr. Med. Vol. 50, pp. 955–960

Odashima S (1962) Combined effect of carcinogens with different actions. Development of liver cancer in the rat by the feeding of 4-dimethylaminostilbene following initial feeding of 4-dimethylaminoazobenzene. Gann Vol 53 pp. 247–257

Schmähl D (1976) Combination effects in chemical carcinogenesis (experimental results). Oncology, Vol. 33, pp 73–76

Vamvakas S, Dekant W, Henschler D (1993) Nephrocarcinogenicity of Haloalkenes and Alkynes. In: Anders M.W. et al.: Renal Disposition and Nephrotoxicity of Xenobiotics. Acad. Press, San Diego

2.5
Quantitative Relationships between Mixed Exposures and Effects on Plants

2.5.1
Introduction

Finding quantitative relationships between air pollutants and their effects on plants counts among the most important tasks of air pollution ecology, a field that emerged from research into smoke damage ("Rauchschadenskunde"). Research into effects in this branch of ecotoxicology have to

- identify and assess the factual consequences of anthropogenic exposures for soil, water, air, plants and animals,
- prognosticate possible consequences of environmentally relevant actions according to their type, intensity, special distribution and probability,
- find and predict present and future origins and originators of air pollution effects and
- help developing strategies, methods and tools of remedy.

Knowing dose-response relationships is of principle importance for preventive protection measures against exposures. Expert committees distil findings drawn from scientifically motivated experimental and epidemiological studies into "value-free" guidelines ("air quality criteria") e.g. in form of "maximum immission values" (VDI 1983), "air quality guidelines" (WHO 1987) or "critical levels" (UNECE 1988a) and "critical loads" (1988b). The guideline values thus arrived at serve the decision-makers responsible for immission protection as markers for setting statutory limits on allowable immissions. Studies on plants are of particular importance in this context, because they are more sensitive to pollution components as diverse and widespread as ozone (O_3), sulphur dioxide (SO_2), the oxygenic nitrogen compounds (NO_Y) and ammonia (NH_3) as well as ammonium (NH_4^+ ; $NH_3 + NH_4^+ = NH_Y$) than heterotrophic organisms. The adherence to guideline values for these components for the protection of vegetation would also protect humans and animals.

However, establishing dose-response relationships turns out to be extremely difficult, since a large number of individual factors are involved in this process, especially when dealing with combined effects. By evaluating the current literature on air pollution exposures, on the one hand, and on the reactions of plants, on the other, we will examine if the existing guideline values can still be regarded as valid or if modifications are necessary. Proceeding from there, we will further examine if and to which degree the hazard potential of mixed exposures can be estimated. These tasks require

- an effect-representative characterization of the air pollution load expressed through exposure indices,
- the selection and description of the objects to be protected,
- the selection of criteria for detecting and assessing effects, and
- the assessment of results from effect research with regard to their relevance in practical circumstances.

These steps to finding relationships between air pollution and effect must be followed both for single components and for mixed exposures. The following discussion of the procedure for single components and the evaluation of the guideline values established for them also provide foundations for the analysis and assessment of combined effects. Hence, knowing the quantitative relationship between the single components and their effects is an essential precondition for the appraisal of combined effects.

2.5.2
Establishing Dose-Response Relationships and Evaluating Guideline Values for Single Components

2.5.2.1
The Relationship between Exposure and Effect

The reaction of a plant to a given exposure depends on its autonomous and environmental resistance. This degree of resistance is determined primarily by the genetic inventory and the stage of development the plant is in at the time of exposure and, secondarily, by the modifying influence of various outside factors on the effect (Guderian et al. 1985). Depending on the intensity and constellation of the exposure, the development stage of the plant and the environment constituted by the soil, the

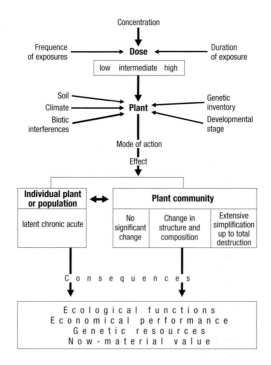

Fig. 2.5-1 Reactions of plants – individually, in populations and as communities – to air pollution in relation to various factors determining the effects (Guderian et al. 1983, modified)

climate and by biotic interferences, individual plants suffer acute, chronic or latent damage, while plant communities undergo changes of varying significance in their composition and structure (see fig. 2.5-1). The consequences can be disruptions in the ecological functions and the economic performance of the affected plants as well as damage to their non material value or the loss of genetic information.

2.5.2.2
The Present Situation Concerning Air Pollution and Effects

During the second half of the twentieth century, the situation concerning emissions underwent important changes due to the fact that, apart from the number and density of stationary emittants, road and air traffic have led to a massively growing number of line sources, while the intensification of agricultural activity has brought a high growth rate in area sources (Friedrich and Obermeier 2000). This entails an increased total amount of emitted substances as well as a greater number of emission components, which in some cases show new effect characteristics.

Therefore, the present air pollution climate is characterized by:

- complex types of pollution from gases, aerosols and dusts with different effects on the atmosphere and the biosphere; substances influencing the climate, degrading ozone and having ecological, ecotoxicological and trophic effects occur simultaneously;
- as a consequence of high emission and emittant densities, today we face not only regional and supraregional exposures, but also global ones manifesting themselves as changes in the UV-radiation and higher CO_2-concentrations in the atmosphere.

Table 2.5-1 Environmentally relevant properties of important trace gases in the atmosphere (JPCC 1990, Houghton et al. 1996, supplemented)

Component	CO_2	CH_4	N_2O	FCKW	O_3	SO_2	NO NO_2
Concentration (units)	ppmv	ppbv	ppbv	ppbv	ppbv	ppbv	ppbv
Concentration							
1750-1800	280	700	280	0	5-15	up to ~0.3	up to ~1
1990	358	1720	310	0.5 F12	30-50	up to 500	up to ~50
Annual increase							
absolute	1.5	13	0.75	0.02 F12	0.25		
percentage	0.4	0.8	0.25	4	0.5		
Average retention period in the atmosphere	100 a	15 a	120 a	100 F12	0.1 a	days	days
Relative greenhouse potential	1	24.5	320	8500 F12	1800		
Contribution to anthrop.	61	15	4	11	7		

Depending on type and range, we can distinguish between two groups of air pollutants, one having continental or even global range, the other spreading locally, regionally or supraregionally. Naturally, the pollutants present worldwide in increased or still increasing concentrations demand special attention, although effect research has mainly concerned itself with, apart from ozone, the oxygenic sulphur and nitrogen compounds occurring regionally and supraregionally. These pollution components have a decisive influence on the air pollution climate in the industrial belt of Europe, stretching from southeast England through northern France, Belgium, the Netherlands, the industrial regions of western Germany down to southeast Germany and the neighbouring regions in the Czech Republic and Poland (Last 1989). Since the beginning of 1990s, apart from SO_2, which had already been in decline in the old *Länder* of the Federal Republic of Germany since about ten years earlier, the emissions of NO_X and of volatile hydrocarbons (NMVOC = non methane volatile organic compounds) also showed falling values (UBA 1997).

Figure 2.5-2 gives a basic picture of the reactions and interactions of the sulphur and nitrogen compounds, the reactive hydrocarbons and the photooxidants with their exposure routes within the terrestrial ecosystem. The entire inventory of organisms is affected directly or indirectly, including producers, consumers and destruents, as well as the soil and the climate including several substance cycles. Also shown is the accumulation of components relevant to the climate and responsible for the breakdown of ozone in the stratosphere.

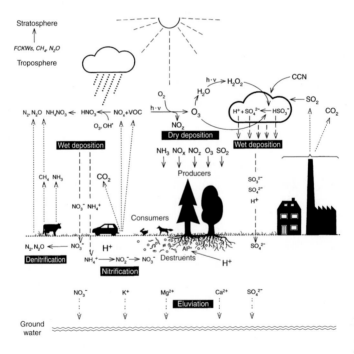

Fig. 2.5-2 Exposure sites of air pollutants containing sulphur and nitrogen and of photooxidants in the terrestrial ecosystem (Guderian and Bücker 1995, modified)

According to Guderian and Ballach (1989), the effect situation under the current exposures is characterized by the following features:

- Direct and indirect effects on the inventory of organisms
- Accumulation of pollutants and of substances with trophic effects in plants and soil
- Chronic and latent effects with the potential to reduce performance and to cause functional disruptions
- Combined effects of two or more components having different effect properties
- Combined strain due to air pollution, unfavourable environmental factors and harmful organisms.

The relationship pattern of cause and effect becomes even more complex as long-term mixed exposures interacting with other abiotic and biotic stress factors trigger effect chains or complex damage, which are difficult to detect.

Presently, the most important hazards for the vegetation in Central Europe and in North America are ozone, oxygenic sulphur compounds and nitrogen immissions (US-EPA 1993, 1996), the effects of which have been studied in extensive experimental and epidemiological investigations (Guderian et al. 1985, US-EPA 1986, Winner et al. 1985, US-EPA 1993, Fangmeier et al. 1994, Segschneider 1995). All of these pollutants are of different effect types: O_3 and SO_2 are phytotoxic; SO_2 (with sulphur deficiency) and the nitrogen compounds are trophic up to a certain exposure level.

For the pollutants mentioned so far, guideline values for the protection of vegetation have been established by various expert groups (see section 2.5.2.6).

2.5.2.3
Protected Objects

From the ethics point of view, the protection of living things from air pollution should include every species. Plants, however, show individual differences in their resistance apart from differences depending on species. According to the current thinking of practical players in the field of immission protection, the individual protection of plants is not feasible (VDI 1988). On the other hand, not only individuals and homotypical collectives are affected, but also communities of living organisms. Due to exposure-related shifts in the conditions of interspecific competition, they often react more distinctively than any single species in a community. Given the vast number of plant communities, it is extremely difficult to identify the biocenoses which are particularly sensitive and hence at risk. Still, the increasing number of synecological investigations undertaken in recent years already provides a broader basis for assessments (Fangmeier 1996).

The long-term exposures of terrestrial ecosystems predominant at present – the introduction of nutrients and acids through dry and wet deposits – mainly threaten long-lived natural and close to natural communities of plants. Apart from important plants cultivated by agriculture and horticulture, woodlands including their ground vegetation, the oligotrophic (poor in nutrients) heathlands, semidry meadows (van der Eerden 1990, Webb 1990) and the ombrotrophic (nourished by deposits from the air) moorland vegetation are particularly well suited for detecting and assessing the effect potential of the selected immission components.

Compared to human toxicology, ecotoxicological investigations have the advantage that epidemiological surveys as well as experimental studies can be performed with the protected species itself under more or less well defined conditions. As a result, the process of establishing guideline values is supported by a large number of experiments with individual plants or homotypical plant collectives. For plant communities and ecosystems, on the other hand, we have to rely largely on outdoor surveys in exposed areas.

Another particularity of ecotoxicology is that, as a rule, no safety margins are applied when guideline values are derived from it. The lowest doses still causing an effect are used as reference points for setting numerical limit values.

2.5.2.4
Criteria for the Detection and Assessment of Effects

For the most important air pollutants harmful to plants, detailed findings about their effect modes on different organizational levels, from sub-cellular to ecosystemic, are already available. The effects following the uptake of pollutants range from biochemical and molecular-biological processes to effects on integrated performances like growth and yield (table 2.5-2).

Effects on the food chain can be left aside here, since ozone decays on absorption and SO_2 and the gaseous nitrogen compounds at current immission concentrations do not accumulate in toxic form. Applying the classification presented in table 2.5-2, we will examine the validity of the guideline values listed in section 2.5.2.6 under present conditions. Special emphasis, in view of the assessment of plant reactions under combined exposures, will be put on possible differences between the effect mechanisms of single immission components.

The assessment of immission effects on plants must proceed from the various functions the plants fulfil in the ecosystem. Pollution above a certain degree influences or damages the performance of the vegetation as a primary producer, as an economic asset, as part of healthy economic landscapes and residential areas and as a gene reserve. Hence, if one aims to record these influences as a basis for establishing remedies, one has to select criteria characterizing the consequences for the use value, measured by ecological and economic standards, of individual plant species and of plant communities. The effects outlined above on each level of organization in the ecosystem are of varying significance for the assessment of their consequences for the use value of the plants considered (fig. 2.5-3).

In order to prevent effects, investigations on air pollution impacts on plants must serve the purpose of protecting the vegetation with its various benefits as an economic asset and as part of functioning ecosystems. This implies that of the effects on each hierarchical level of the ecosystem outlined in table 2.5-2, the integrative criteria are of special importance. For the assessment of effects both autecological and demecological, meaning effects on the individual organism and the homotypical collective respectively, the following criteria are crucial:

- Growth
- Yield, its amount and quality
- The capacity for reproduction
- Vitality

In addition, on the synecological and ecosystemic levels, the following, community-shaping criteria must be considered:

- Intra- and interspecific competitiveness
- Bio-geochemical substance cycles (e.g. through a slow-down of litter decomposition)
- Rejuvenation and succession
- Reproduction strategy of producers and consumers (r- and K-strategists)
- Biocenotic connexes in form of symbiosis (tuber bacteria, mycorrhiza) or parasitism (e.g. fungi, animal pests)
- Diversity of species, biotopes and ecosystems

Injury is defined as all plant reactions to harmful substances. For instance, they include reversible metabolic changes irrelevant for growth and yield as well as impairments of photosynthesis, leaf necroses, leaf shedding, quality losses or growth reduction. The immediate effects of air pollutions (primary damage) can diminish the use value of vegetation, characterized by its economic performance, its ecological functions, its non-material value and its importance as a gene pool. With the exposures predominant today, secondary damage due to a lowered resistance of plants against biotic and abiotic stresses caused by immissions is becoming more and more important (Guderian and Fangmeier 2001). The detection of such damage is however fraught with particular difficulties concerning both diagnosis and quantification.

2.5.2.5
Assessment of the Results regarding their Relevance for Determining Dose-Response Relationships

As a basis for determining quantitative relationships between immission and effect, the results from effect research must meet three conditions:

1. They must have been confirmed by causal analysis, i.e. the cause of any effect detected must be known definitely.
2. The results drawn from experimental investigations must be applicable to practical situations.
3. Both the exposure to and the effects of immissions must be expressed in numbers and units.

Investigations on the effects of immissions on plants are carried out in various systems of exposure, ranging from strictly reproducible but unnatural experiments in climatic chambers to geobiological surveys at the natural site of the plants in an exposure area. While results from climatic chamber experiments can be applied to outdoor situations only under certain conditions, finding the cause of an effect in field investigations often suffers from insurmountable difficulties (Jäger and Guderian 2001).

When determining the dose-response relationships for ecotoxicologically effective components and selecting the exposure systems, the following conditions must be considered:

- The plants reaction to immissions can be represented by a sigmoid curve only above a certain effect threshold; ecologically relevant stochastic relationships

Table 2.5-2 Classification of the effects of immissions on plants (Guderian et al. 1983, modified)

	Level of organization		
Cell	Tissue and organ	Organism	Plant community/ecosystem
Changes in the membrane integrity	Damage to epidermal tissue	Changes in the growth of the plant	Changes in the production of biomass
Changes in enzyme activities	Changes in photosynthesis, respiration, and transpiration	Changes in yield and quality	Degradation of the original biocenoses
Changes in the metabolic regulation	Changes in the translocation and allocation of plant substances	Loss of vitality	Secondary successions
Physicochemical changes	Changes in growth and development of individual organs	Changes in the predisposition towards biotic and abiotic stress factors	Changes in the incidence (abundance) of individual species and in the diversity of species
Sub-structural transformations of cell organelles	Effects on root symbioses	Changes in the fructification	Changes in host-parasite relationships and in symbioses
Disruption of the cell function	Discolouring and chloroses	Changes in competitive capacities	Changes in the structure of populations
Cell death	Necroses	Plant death	Changes in food networks with consequences for consumers and destruents
	Early and accelerated senescence		Changes in substance cycles
	Death or abscision of organs		Effects on the performance of ecosystems including their stability and capacity for self-regulation

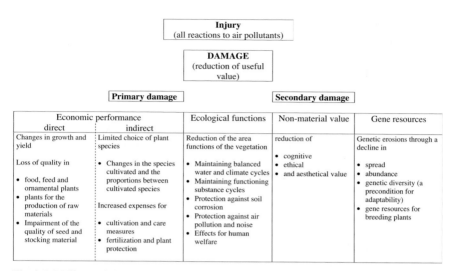

Fig. 2.5-3 Effects of air pollution on the performance of plants in the ecosystem (Guderian et al. 1983).

with effects on integrative performance markers like growth and quality have never been observed.

- For gaseous immission components like NO_X, ozone and SO_2, the relationship between concentration and exposure time does not follow the Bunsen-Roscoe law (Mohr and Schopfer 1992) stating that a certain product of concentration (c) and time (t) always leads to the same effect. For the said components, the effect of any product $c \times t$ rises exponentially with the concentration c (van Haut 1961). Due to the different effect mechanisms in the higher and lower concentration ranges, one cannot extrapolate from effects of high concentrations to effects of lower concentrations at longer exposure times.
- Plants are not only resistant to varying degrees depending on species and variety; their resistance also depends on the age and the stage of development of the individual plant and its organs. For technical reasons, only plants up to a certain size, but not at the later development stages of trees and shrubs, can be investigated.
- Neither can certain exogenous abiotic and biotic factors of influence, which have a crucial impact on the amounts of pollutants absorbed and thereby on the dose-response relationships, be reproduced in controlled experiments.
- In most cases, experiments cannot simulate outdoor immission situations with regard to their actual composition and constellation.
- Findings from epidemiological investigations performed in exposure situations, on the other hand, on the quantitative relationship between exposure and effect of mixed immissions especially at low concentrations, often do not provide definitive information about which component dominates the effect.

Only by considering the results from both categories of exposures in a sufficiently differentiated manner, the above postulates of representativity and causal-analytical definitiveness can be met.

In deriving guideline values for the protection of humans, on the one hand, and of material assets like vegetation, on the other, standards of different rigorousness are applied due to the different degrees of protection afforded to humans and plants, respectively. Moreover, in the case of vegetation, investigations are carried out with the protected species itself, whereas in research for the protection of humans, animal experiments are widely used.

2.5.2.6
Evaluation of the Guideline Values for Single Components

Taking into account the requirements on investigative methods formulated in section 2.5.2.5, current reports of experimental and epidemiological work are selected and examined with regard to their relevance for existing guideline values. Hence, we are dealing with the question if these guidelines still reflect the present state of knowledge and if they could serve as a basis for assessing combined effects.

For the guideline values for the protection of vegetation, one distinguishes, depending on the type of deposition, between critical [concentration] levels of pollutants in the air and critical loads of pollutants getting from the air into the soil. Both are threshold values for the effect of air pollutants. A critical level is defined as the "lowest concentration of a pollutant in the atmosphere that, according to current knowledge, would cause direct, negative effects on receptors like plants, ecosystems or matters" (UNECE 1988a). A critical load is, by definition, "the deposited amount of one or several pollutants, below which no significant injurious effects on specifically sensitive constituents of ecosystems will occur according to present knowledge" (UNECE 1988b).

2.5.2.6.1 Nitrogenous Air Pollutants: Under the present conditions of exposure, terrestrial ecosystems are influenced by a variety of nitrogenous compounds. Many effects are initiated by the oxygenic nitrogen compounds, NO_Y (US-EPA 1994). Due to their supraregional spread in relatively high concentrations, their degree of phytotoxicity, their contribution to acid precipitation and their function as a plant nutrient as well as a precursor for the formation of photooxidants, nitrogen oxide (NO) and nitrogen dioxide (NO_2; $NO + NO_2 = NO_X$) have to be regarded as the most important among these compounds in terms of their pollution hazard to vegetation. At the same time, they influence the nitrogen balance in plants and soil following their deposition. Further direct and indirect effects are caused by ammonia (NH_3) and ammonium (NH_4^+). Both for NO_X and for NH_3, critical levels to limit the concentration in the air and recommendations for critical loads to limit the N-deposition in soil have been established (tables 2.5-3 and 2.5-4).

The critical levels derived by UNECE (1988a) and WHO (1987) from experimental and epidemiological investigations, respectively, are both given as a tolerance value of 30 µg m^{-3} (annual average) for NO_X exposures, with growth as the decisive effect criterion. The UN committee set an additional value <30 µg m^{-3} for biochemical and physiological reactions, without giving any concrete information on the relevance of these reactions for the use value and function of the plants affected. The committees of both institutions also took into account possible combination effects of simultaneous SO_2 and/or O_3 exposures when setting their guideline values for NO_X ($NO_X = NO + NO_2$, added in ppb and expressed as NO_2, in μg m^{-3}. 30 μg m^{-3}, on average over a year, are considered as tolerable for all those areas where the guideline values for SO_2 and O_3 are not exceeded (see table 2.5-3 and tables 2.5-5/2.5-6).

The body of research published since the establishment of the guideline values quoted above, summarized by the US Environmental Protection Agency (US-EPA 1993), gives no reason to change the critical levels for NO_X in areas with tolerable SO_2 and O_3 exposures. For instance in the open-top chamber (OTC) experiments by Wenzel (1992), where cereal species were exposed to filtered and unfiltered air, no negative effects occurred below ca. 30 μg m^{-3} NO_X in the air, with SO_2 and O_3 both below their guideline values. Any effects observed below the NO_X threshold rather were of a more or less growth-stimulating nature.

Nitrogenous compounds influence the vegetation not only directly through the contact between air and above-surface plant organs, but also indirectly, following their deposition into the soil. In our humid ecosystems, where nitrogen widely acts as a limiting growth factor, the following reactions are to be expected as conse-

Table 2.5-3 Guideline values for the protection of vegetation against NO_X ($NO + NO_2$, added in ppb and expressed as NO_2, in μg m^{-3}) and NH_3 (in μg m^{-3})

Component	Protected object	Effects and criteria	Guideline value	Source
NO_2	Plants in general	Growth reduction	30 μg m^{-3} on annual average and 95 μg m^{-3} on average over 4 hours with simultaneous exposure to O_3 in concentrations under 60 μg m^{-3} and to SO_2 in concentrations not above 30 μg m^{-3}	WHO (1987)
NO_2	Plants in general	Growth reduction	30 μg m^{-3} on annual average with simultaneous exposure to O_3 and/or SO_2 in concentrations below their critical levels	(UNECE 1988a)
			< 30 μg m^{-3} on annual average and 95 percentile	
		Biochemical and physiological changes	< 95 μg m^{-3} with simultaneous exposure to O_3 and/or SO_2 in concentrations below their critical levels	
NH_3	Plant species of the (semi)natural vegetation	negative effects on the level of individual plants or plant communities	3,300 μg m^{-3} on hourly average 270 μg m^{-3} on daily average 23 μg m^{-3} on monthly average 8 μg m^{-3} on annual average	(UNECE 1992)

quences of anthropogenic N-depositions (Augustin 1997, van der Eerden et al. 1999):

- Growth enhancement
- Changes in the composition of vegetation
- Physiological reactions of the plants
- Changes in both quantity and quality of the humus
- Nutrient imbalances
- Soil depletion and acidification
- Increased N_2O emission
- Ground water exposures
- Eutrophication of surface waters

While critical levels help regulating the actual exposure of the air itself, critical loads are a tool for controlling the long-term take-up of accumulating substances e.g. nitrogenous compounds by the soil (table 2.5-4). Hence, these two guideline values not only refer to different receptors, but also to different time scales. Consequently, the critical-load concept represents the far more stringent standard of limitation under the exposure regime dominant at present (i.e. long-term exposure to relatively low concentrations with hardly any direct, injurious effect).

Table 2.5-4 General characteristics of critical levels (CLE) and critical loads (CLO, from van der Eerden et al. 2001, modified)

Feature	CLE	CLO
Duration of exposure	Short-term (1 year or less)	Long-term (10 - 100 years)
Unit for the exposure	Concentration ($\mu g\ m^{-3}$)	Deposition ($kg\ ha^{-1}\ a^{-1}$)
Protected object	Sensitive species	Ecosystem / function of ecosystem
Combination of pollutants	Components are evaluated separately; interactions (e.g. synergisms) are considered too	Nitrogenous compounds are summed up (assuming an additive effect); the same applies to acidic deposits
Methods of assessment	Laboratory experiments and field research; lowest relevant and effective exposure	Empirical data and steady-state soil models

In contrast to the critical levels, which are valid on a national level without any further spatial differentiation, critical loads refer to specific types of ecosystems of varying spatial extent. Three groups of factors decide if a given deposit, or the accumulation of nitrogen resulting from it, produces positive or negative effects. Depositions in agricultural ecosystems contribute to the growth and yield enhancements aimed at through fertilization, which is why they are generally regarded as positive.

For natural and close to natural ecosystems, the tolerable level of deposition depends on the way the ecosystems are used, the soil condition (Agren and Bosatta 1988) and the type of vegetation (table 2.5-5).

More recently, the Nordic Council of Ministers arrived at a more differentiated proposal for the establishment of critical N-loads (Bobbink et al. 1992). Van der Eerden et al. (2000) have produced a summary table of the recommendations given there (table 2.5-6).

Table 2.5-5 Critical N-loads for ecosystems of varying productivities, in kg N ha^{-1}a^{-1} (UNECE 1988b)

Ecosystem	Critical load
Deciduous woodlands	5 - 20 kg N ha^{-1} a^{-1}
Coniferous woodlands	3 - 15 kg N ha^{-1} a^{-1}
Neglected grasslands (e.g. mesobrometum)[2]	3 - 10 kg N ha^{-1} a^{-1}
Moorlands[2]	3 - 5 kg N ha^{-1} a^{-1}

[1] In the terminal or decomposition phase, the load should be close to zero.

[2] For negligible N-removal through utilization of the ecosystem. Assuming the ECE standard of 50 mg nitrate l^{-1} in the ground water and a precipitation surplus of between 100 and 400 mm a^{-1}, one arrives at a critical deposition of 10 to 40 kg N ha^{-1} a^{-1}

The new proposals identify a differentiated spectrum of protected objects with special consideration of various soil characteristics, albeit with major uncertainties with regard to reliability in some cases. Moreover, they cite indicators of any disruption when the limit values are exceeded. For the foreseeable future, nitrogen depositions of an ecologically and ecotoxicologically relevant degree have to be expected. Ombrotrophic ecosystems like moorlands and oligotrophic ecosystems like plant communities on heathlands and neglected grasslands are threatened in their existence in large areas (van der Eerden et al. 2001). Hence, it is even more important for establishing reliable standards to limit the spread of tolerable deposition numbers.

In order to avoid exceeding nitrogen depositions identified as tolerable, anthropogenic emissions from industry, transport and agriculture must be reduced accordingly. To this end, precise information on the quantitative relationship between the amount deposited and the concentration of nitrogen compound in the air, depending on climatic conditions, especially on the type, amount and distribution of precipitations, must be available (Gravenhorst 2000). These important tasks for conserving diversity in nature and landscapes can only be fulfilled if differentiated experimental and epidemiological investigations in various types of ecosystems and for different forms of utilization are undertaken.

Table 2.5-6 Summary of the recommendations on critical loads for nitrogen depositions (kg N ha^{-1} a^{-1}) in (semi)natural ecosystems (from Bobbink et al., 1992, and van der Eerden et al. 2001)

Ecosystem	Critical load (kg N ha^{-1} a^{-1})	Indicators of disruptions
Ombrotrophic moorlands	5-10#	Loss of sphagnum and secondary species; increase of gramineous species
Alkaline, multispecies grassland	14-25##	Increase in tall grasses, decrease in diversity
Neutral to acid multispecies grasslands	20-30#	Increase in tall grasses, decrease in diversity
Alpine/sub-alpine grasslands	10-15(#)	Increase in tall, gramineous species, decrease in diversity
Low-lying, dry heathlands	15-20##	Transformation of heathlands into grass communities
Low-lying, wet heathlands	17-22##	Transformation of heathlands into grass communities
Multispecies heathlands/ acid grasslands	7-15/20#	Loss of sensitive species
Arctic and alpine heathlands	5-15(#)	Loss of lichen, mosses and perennial dwarf shrubs, increase in grasses and herbs
Coniferous woodlands on acid soils, cultivated	0-15#	Tree health; nutrient imbalances (low rate of nitrification)
Coniferous woodlands on acid soils, cultivated	15-20##	Changes in the ground vegetation
Coniferous woodlands on acid soils, cultivated	20-50#	Tree health; nutrient imbalances (medium to high rate of nitrification)
deciduous woodlands on acid soils, cultivated	0-15#	Tree health; nutrient imbalances; shoot/root ratio
deciduous woodlands on acid soils, cultivated	15-20#	Changes in the ground vegetation
Woodlands on alkoline soils, non-cultivated	unknown	unknown
Woodlands on alkoline soils	15-20(#)	Changes in the ground vegetation

##: reliable, #: reasonably reliable, (#): best numbers available

2.5.2.6.2 Ozone: The tropospheric ozone concentration above the northern hemisphere has doubled to tripled during the past 200 years and is now rising by ca. 0.5%, or according the WMO (1992), even by 1% annually (table 2.5-1). Today, ozone concentrations have reached phytotoxic levels worldwide (Fuhrer 1996) and are regarded as the most important components of pollutions harmful to plants (US-EPA 1996). According to investigations under the umbrella of the National Crop Loss Assessment Network (NCLAN), a reduction of the ozone exposure in the USA by 40% would result in a national profit of 3 billion US dollars for the 8 most important cultivated plants alone (Heck et al. 1983, 1984a, 1984b; Adams et al. 1989). For the Netherlands, van der Eerden et al. (1988) calculated that ozone diminishes the agricultural yield by 3.4%, leading to an additional cost of about 220 million US dollars for the consumer.

One example of the grave consequences of ozone exposures for woodland ecosystems is the widespread damage done to multispecies coniferous forests in the San Bernadino Mountains in Southern California. The dominant ponderosa pine (*Pinus ponderosa* Laws.) is particularly affected. The high ozone exposures with daily averages up to 580 ppb (Miller 1973) reached during prolonged inversions are attributed to massive emissions of precursors by road traffic and industry in the Los Angeles basin to the west of the mountains. The changes of components, structures and processes in the San Bernadino National Park, which have been confirmed by extensive, interdisciplinary studies, are in broad agreement with the trends predicted by Odum (1985) for ecosystems under stress (US-EPA 1986, 1996):
Low efficiency of the transformation of energy into organic matter

- Reduced nutrient exchange between different trophic stages
- Lower availability of nutrients due to a slow-down of litter decomposition
- Increased fraction of r-strategists
- Reduced size and shortened life expectancy of organisms and their organs
- Reversal of the autogenous succession trends
- Fewer symbioses and increased parasitism

This brief list already demonstrates how complex immission-related changes on the ecosystem level can be. From the results only outlined here, we can conclude that a sustained exposure to harmful ozone immissions transforms highly diverse woodlands into less diverse forests of trees more resistant to ozone.

The type and degree of plant damage caused by ozone was the subject of extensive research leading to a vast number of original publications both on experimental and epidemiological studies now being available. Hence, we are well informed about issues ranging from the effect mechanism on the level of molecular biology (e.g. Ernst et al 1993, 1996; Galliano et al. 1993) to the effects on individual plants and populations (Guderian et al. 1985, Sandermann et al. 1997, Mehlhorn 2000). On the quantitative relationship between ozone exposure and plant reactions, too, a large body of literature can be called upon (US-EPA 1996). Nevertheless, the appraisal of the hazard potential of ozone still suffers from considerable uncertainties.

The first threshold values for the protection of plants against acutely damaging ozone exposures were derived as early as the mid-seventies, by Linzon et al. (1975), Heck and Brandt (1977) and by Jacobson (1977). The progress of research in this field, which is continuing and has concentrated on the assessment of long-term exposures in recent years, has led to improved conditions for an appraisal of the hazard potential of ozone. In table 2.5-7, we have collated the current guideline values for the protection of plants against ozone, derived for the region of Europe by various institutions.

The table does not only list various numbers, it also demonstrates the process of finding guideline values. The ozone concentration was limited to exposure times of single hours in order to prevent acute, harmful effects manifesting themselves as leaf necroses due to high concentrations during short periods of exposure. In this way, the overall effects as the integral of acute, chronic and latent effects were not taken into account. Therefore, it was tried at first (see WHO 1987, UNECE 1988a) to include this hazard potential, too, by quoting arithmetic mean values for the duration of a vegetation period. Averaging over periods that long, however, does not even come close to characterizing the, in the case of ozone, extremely variable exposure dynamics in an effect-adequate manner. In their quest for more suitable indices for long-term ozone exposures, American experts found that the summation of ozone doses above a defined concentration (AOT = accumulated over threshold) best describes the quantitative relationship between long-term exposure and effect (Lee et al. 1988, Lefohn et al. 1988). Starting from this result, the effort was made at the UNECE Convention on Long Range Transboundary Air Pollution during the Second UNECE-Critical Level Workshop (1992) in Egham, UK, to use this procedure for deriving numerical values for the protection of agricultural and silvicultural plant species.

Values justified in more detail were presented at the later workshops in Berne (UNECE 1994) and, in a modified form, in Kuopio, Finland (UNECE 1996). The examination of various threshold values, above which the dose accumulated over the entire exposure period showed the best correlation to the effect, resulted in an hourly average of 40 ppb. Proceeding from there, accumulated doses tolerable for agricultural plant species, on the one hand, and for forestry coppices, on the other, were defined as AOT-40 values for a number of different boundary conditions. While for the forestry plants an exposure period of 6 months was chosen, the quarterly period from May to July was used as a reference frame for the agricultural plant species. In contrast to the approach for forestry plants, only the daylight hours with a global irradiation of more than 50 W m^{-2} were taken into account for the agricultural species. In order to prevent a loss of yield of more than 5% caused by long-term effects on agriculture, an AOT-40 value of 3,000 ppb h was set. Additional limit values were established for the prevention of acute plant damage to agricultural plant species (see table 2.5-7).

The guideline values published as critical levels over the years reflect the gradual effort to characterize the exposure by indices so as to reduce the degree of uncertainty about the relationship between the air pollution exposure and the effect on the plant. In this respect, applying the AOT 40 derived from the OTC experiments, which came relatively close to natural conditions, to agricultural plant species represents an improvement compared to the earlier short- and long-term averages. Significant losses in yield – more than 5% or 10% respectively – should not occur if the critical level of 3,000 or 6,000 ppb h, respectively, are not exceeded. However, the actual ozone dose in the months May through July, calculated according to AOT 40, is not about 6,000 ppb h, but above 25,000 ppb h (Fuhrer 1996). Based on the open-top chamber experiments, this would mean that the current ozone exposures in Central Europe cause a yield reduction of about 50% for summer wheat, as illus-

Table 2.5-7 Guideline values for ozone, for the protection of vegetation

Value($\mu g\ m^{-3}$)	Time frame	Protected asset	Institution
400	0.5 h average	Very sensitive plants	VDI (1987)
600	0.5 h average	Sensitive plants	
1,000	0.5 h average	Less sensitive plants	
200	1.0 h average	Very sensitive plants	
400	1.0 h average	Sensitive plants	
600	1.0 h average	Less sensitive plants	
140	2.0 h average	Very sensitive plants	
300	2.0 h average	Sensitive plants	
500	2.0 h average	Less sensitive plants	
100	4.0 h average	Very sensitive plants	
240	4.0 h average	Sensitive plants	
460	4.0 h average	Less sensitive plants	
60	8.0 h average	Very sensitive plants	
200	8.0 h average	Sensitive plants	
400	8.0 h average	Less sensitive plants	
200	1 h average	Terrestrial vegetation	WHO (1987)
65	24 h average	Terrestrial vegetation	
60	average over vegetation period	Terrestrial vegetation	
300	0.5 h average	Sensitive plants and plant communities	UNECE (1988a)
150	1 h average		
110	2 h average		
80	4 h average		
60	8 h average		
50	Average over the vegetation period (averaged over 7 h daily from 9:00–16:00 hrs)		
AOT 40 = 700 ppb h	Sum of the hourly ozone averages above 40 ppb (AOT 40) during daylight on three days, thereby avoiding acute plant damage	Agricultural plants	UNECE 1994
AOT 40 = 5,300 ppb h	Sum of the hourly ozone averages above 40 ppb (AOT 40) during daylight in the months May to June, for preventing yield losses < 10%	Agricultural plants	
AOT 40 = 10,000 ppb h	Sum of the hourly ozone averages above 40 ppb (AOT 40) during the vegetation period (April to September)	Forestry plants	
AOT 40 = 3,000 ppb h	Sum of the hourly ozone averages above 40 ppb (AOT 40) during the daylight phase in the months May to July	Annual plants in near-natural ecosystems	Supplemented by UNECE (1999)

trated in fig. 2.5-4 (Fuhrer et al. 1997). Hence, these AOT 40 values represent a distinct overestimation of the hazard potential of ozone especially in Central and Northern Europe, since neither climatic nor plant-specific factors, which influence the pollutant-receptor relationship, have been considered.

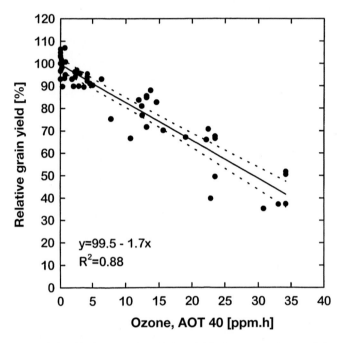

Fig. 2.5-4 The relationship between the relative yield of summer wheat and the ozone exposure expressed as AOT 40 (Fuhrer et al. 1997)

Thus, there is a vast chasm between prognosis and reality; the actual yield losses are of a range where [statistically] significant results are difficult to obtain (van der Eerden 1988).

At the workshop in Gerzensee (Switzerland) in April 1999, the basic suitability of the AOT-40 concept was confirmed (UNECE 1999), though corrections to the numerical values were considered necessary. These corrections were to be carried out in two ways, by applying the flux model, on the one hand, and by using the level-II approach, on the other.

Previously, the quantitative relationship between immissions and effects on plants was determined by relating a plant reaction to the concentration [of a pollutant] in the ambient air. The concentration in air is however an inadequate measure of the hazard potential of a given immission exposure. The effect is determined by the amount of pollutant absorbed per unit of plant matter and per unit of time. The deposition rate i.e. the flux from the ambient atmosphere into the plant varies widely with the internal and external growth factors. As a result, the levels derived from OTC experiments often overestimate the phytotoxic potential of the ozone exposures present in open air, as already mentioned. Important factors determining the intake and the effect of pollutants, like air humidity, water and nutrient supply, wind speed and radiation conditions, can differ extremely between the two exposure systems (see chapter 1.1). Grünhage and Jäger (1994) measured the ozone concentration at a standard height over the earth surface as well as the turbulent exchange properties of the atmosphere and the accompanying sinking properties of the individual plant and of the plant population. In this way, they found that ozone concentrations between 50 and 90 ppb hold a particularly high phytotoxic potential. The immense practical importance of this concentration range is explained by the flux density of ozone from the atmosphere into the leaf, which often is quite high under open-air conditions, and by the fact that ozone concentrations in this range are very common. Concentrations above 90 ppb, on the other hand, are associated with lower flux densities and therefore with a lower phytotoxicity than to be expected for their dose, because of the lower air humidity predominant under such conditions. Hence, one of the main reasons for the overestimation [of the effects] of high concentrations is that even the highest concentrations coincide with maximum flux densities in OTC experiments and therefore cause harmful effects of a corresponding strength.

For the reasons explained above, the flux model will produce factors shifting the conventional AOT 40 to higher values i.e. towards their true hazard potential. However, only if all the factors determining the flux of pollutants from the ambient atmosphere into the plant are applied, one can expect satisfactory results for the required correction factors in the long term. Therefore, in the first instance the level-II approach will be pursued in parallel, taking into account factors of particular importance for absorption and effect of the pollutant, like the height over ground, the air and soil humidity and the wind speed. While in the level-I approach just one critical level is supposed to protect the vegetation in a political region, the level-II approach produces a range of critical levels dependent on the type of vegetation as well as soil and climate conditions (UNECE 1992; Fuhrer et al. 1997). It appears more than doubtful if such a procedure can be practical. The data required for this approach are neither available nor can they be analyzed. As a consequence, the level-II approach will probably be discarded as an unsuitable method for predicting hazards for vegetation.

So far, only the economically most important plant species from agriculture and forestry were considered when setting guideline values, but not the near-natural vegetation. At the workshop in Gerzensee, guideline values for these plants, too, were established for the first time. As few dose-response relationships are available for this purpose, the guideline values from agricultural plants were adopted (AOT 40 of 3,000 ppb h and 6,000 ppb h for annual plant species and perennials respectively).

As already explained, the large number of factors influencing the flux of pollutants to the effect site and determining the actual degree of tolerance of a plant leads to a situation where the quantitative relationship between immission and effect can only be approximated. If one wants to examine the question if a guideline value derived from dose-response relationships actually ensures the intended protection, the procedure referred to as bioindication offers a useful set of tools for deferred immission protection (Guderian et al. 1985, Arndt et al. 1987). The reaction of homogeneous plant matter of a high and component-specific sensitivity subjected to standard exposures represents the integral effect of all immission components and factors of influence at the site of exposure. Hence, in a procedure like this, referred to as active bio-monitoring, the exposed plants act as sensitive and reliable early indicators, when the threshold value for the substance they are particularly sensitive to is exceeded. If the indicator plant shows harmful effects, a comparison with the measured immission values has to decide whether the existing guideline value must be corrected downwards. It is therefore imperative to strengthen the efforts to develop indicators representing the vegetation in Central Europe. Cloned varieties of the Lombardy poplar *(Populus nigra* L; Bücker et. al. 1993) and the birch (*Betula pendula* Roth; Pääkönen et al. 1997*)* appear to be particularly suitable for these purposes.

2.5.2.6.3 Sulphur Dioxide: Already in the first half of the nineteenth century, the transition from crafts and agriculture to industrial production made sulphur dioxide the most important component of air pollution harmful to plants. Up to the 1960s, typical examples of extreme SO_2 exposures included highly visible damage to vegetation in the vicinity of single emittants, so-called *Rauchblößen* or "smoke wastelands" (Guderian and Stratmann 1968). Through measures to maintain clean air and following the closure of many metal-roasting installations, these areas of extremely high, punctual exposure have now disappeared. However, with the increasing emission accompanied by an increasing density of emittants, the second half of the twentieth century saw the development of supraregional exposure areas. One example of extremely strong, "classic" air pollution effects is the large-area forest damage with, in some cases, total loss of all tree and shrub vegetation in the upper reaches of Central European mountain ranges like the Oar Mountains, the Jizera Mountains, the Bohemian Massive and the Beskids, which Kluge (1993, p. 463) called "the greatest Central European forest catastrophe in human memory". Particularly strong effects are visible in the Oar Mountains. The fir began to die out already at the end of the nineteenth century, soon followed by the first, localized damage to the spruce. The large-area dying of spruce set in during the 1950s. About 60,000 ha had to be cleared – 45,000 on the Czech side, 15,000 in Germany. In the 1980s, 200,000 ha around the damage zones were severely affected (Wentzel 1984). The decisive factor in this large-area forest destruction is the high SO_2 exposure originating mainly from lignite burning in thermic power plants in the Bohemian basin of the Czech Republic (Liebold and Drechsler 1991). Decades of immission of acid-forming substances also led to serious changes in the soil chemistry for instance an increased acidity in the humus layer and the A-horizon, an extreme reduction in the supply of nutrient cations available to plants (Mg^{2+}, Ca^{2+} and K^+) and the rise in easily soluble aluminium compounds (Materna 1987).

Even moderate exposures, though, in form of sulphurous and nitrogenous immissions and depositions can severely affect woodland ecosystems directly or indirectly (Guderian and Tingey 1987, Fangmeier et al. 1994, Mohr 1994, Augustin 1997), mainly by

- soil acidification introduced through protons and anions of strong mineral acids,
- changes in the nutrient balance (accumulation of nitrogen and sulphate, increased concentrations of Ma-cations, washing out of Mb-cations, nutrient imbalances),

- changed succession: Increased transformation into grasslands by *Calamagrostis villosa* on the higher reaches of the Oar and Harz Mountains, by *Avenella flexuosa* at montane and sub-montane heights and by *calamagrostis epigajos* in the pine forests of Saxony-Anhalt, Brandenburg and Mecklenburg-Vorpommern;
- increased incidence of nitrophilic shrubs such as raspberry, blackberry, elder and the late flowering bird cherry,
- deterioration of rejuvenation conditions,
- decreased water seepage and ground water formation,
- disruptions of the water balance of soil and plant, and
- increased susceptibility of trees to frost and secondary pests.

The high phytotoxic potential of SO_2, on the one hand, and its widespread occurrence in concentrations harmful to plants, on the other, gave rise to intensive research into harmful effects as early as the mid eighteen hundreds. Thus, SO_2 counts among the best-researched air pollution components internationally. In parallel to experimental and epidemiological studies, efforts were made to elucidate the quantitative relationships between the SO_2 concentration in the air and its effect on plants (Wislicenus 1898). However, sound numerical values for the protection of vegetation could only be derived – as guideline values – by national and international institutions during the second half of the twentieth century (table 2.5-8).

The guideline values for SO_2 are mainly based on averages over long time intervals like a vegetation period or a year. In this way, the immission type with long-term, relatively uniform exposures in the range of concentrations with chronic effects is taken into account. The blue alga lichen proved to be extremely sensitive.

They are already at danger from only 10 µg m^{-3} of air per year, whereas forestry ecosystems and natural vegetation can tolerate SO_2 doses twice that high, according to the UNECE (1992).

While the "level-II approach" is still under discussion in the case of ozone, it has already produced a guideline value for SO_2. This number takes into account recent findings on the influence of temperature on the degree of phytotoxicity of SO_2. High, acutely damaging concentrations in the short term develop the strongest effect at higher temperatures i.e. in the summer half-year (Guderian 1977). Long-term exposures at concentrations with chronic effects, on the other hand, are most effective in the winter half-year (Baker et al. 1982). The strength of effects of gaseous air pollutants on plants are determined by two parameters, the amount of pollutant absorbed per unit of time and the physiological activity of the plant during exposure. The mostly good water supply in the winter half-year promotes the intake of pollutants. At the same time, the physiological activity is largely reduced due to the lower temperatures, resulting in a slow-down of the metabolization and break-up of toxic substances. Under SO_2 exposure, large amounts of sulphite accumulate in leaf tissue in such conditions (Kropff et al. 1990), whereas sulphite is detoxified relatively effectively by oxidation to sulphate or by reduction to hydrogen sulphide during the vegetation period (Winner et al. 1985). Hence, under the SO_2 exposure currently predominant in Europe, the vegetation is in greater danger during the winter months than in summer. The guideline value of 15 µg m^{-3} of air during the win-

Table 2.5-8 Summary of the guideline values for the protection of vegetation against SO_2

Value ($\mu g\ m^{-3}$)	Time frame	Protected plants	Institution
50	Average over the vegetation period (7 months)	Very sensitive plant species	VDI (1978)
250	97,5-percentile		
80	Average over the vegetation period (7 months)	Sensitive plant species	
400	97,5-percentile		
120	Average over the vegetation period (7 months)	Less sensitive plant species	
600	97,5-percentile		
50	Annual average	Forests	IUFRO (1978)
150	97,5-percentile		
25	Annual average	Forest at extreme sites	
75	97,5-percentile		
30	Annual average	Terrestrial vegetation	WHO (1987)
100	Daily average	Terrestrial vegetation	
20	Annual average	Sensitive plants	UNECE (1988)
30	Annual average	Sensitive cultivated plants / forests	
70	Daily average	Sensitive cultivated plants / forests	
10	Annual average	Blue alga lichen	UNECE (1992)
15	Annual and half-year average (October - March)	Forestry ecosystems* (ETS < 1000 C° days)	
15	Annual and half-year average (October - March)	Natural vegetation* (ETS < 1000 C° days)	
20	Annual and half-year average (October - March)	Forestry ecosystems	
20	Annual and half-year average (October - March)	Natural vegetation	
30	Annual and half-year average (October - March)	Agricultural plants	

• ETS = Effective temperature sum above 5 C° on daily average.

ter half-year (October to March), as laid down by the UNECE (1992), takes this finding into account. This value applies whenever the effective temperature sum (ETS) above a daily average of 5 °C does not exceed 1,000 °C over the winter half-year.

2.5.3
Prospects for the Appraisal of the Hazard Potential of Mixed Exposures

The discussion of guideline values for the single components nitrogen oxides, ozone and sulphur dioxide has made clear the difficulties in characterizing, by practical exposure indices, existing immission exposures with regard to their effects. The main problems result from the necessity to derive representative numerical values for prolonged exposure times – a vegetation period or a year – from experimental and epidemiological studies on dose-response relationships.

Under the current exposure conditions, further difficulties arise in the differentiation between athropogenic influences and the natural variation and fluctuation in time of the structure and function of the individual plant, the population and the biocenosis. This differentiation becomes even more difficult for combined exposures, especially since substances with ecotoxicological and trophic effects can be present at the same time. Small changes in concentrations already decide if the effects will be positive or negative.

In contrast to the situation under high, acutely damaging concentrations, air pollution under conditions of latent exposure with chronic effects is only one among many stressors. One typical disease is replaced by essentially multicausal processes, in which a more or less large number of components affect the plant simultaneously or in succession. For this phenomenon, referred to as chain or complex diseases in the case of trees, Manion (1981) has introduced the term

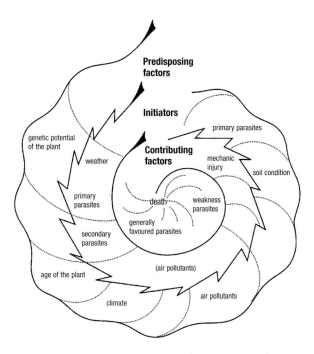

Fig. 2.5-5 The decline spiral (from Manion 1981)

"decline", for which he also provided a model explanation, the "decline spiral" (fig. 2.5-5). Manion distinguishes between three classes of factors, predisposing, initiating and contributing factors. Tesche (1991) has extended this model by introducing the eustress region with its potential for an adaptive increase of the resistance of plants (fig. 2.5-6). When and under which conditions air pollutants are to be classed under one or the other of these categories, can be determined representatively for practical situations only by calling on major input from epidemiological studies.

2.5.3.1
Examples of Combined Effects

Following the overview of the status concerning guideline values for the protection of vegetation for the single components selected (see section 2.5.2), in this section we will examine to what extent these values take into account combined exposures too and whether the information available is sufficient to derive further limit values.

Due to their ecotoxicological importance as single components, there exist numerous results from experimental and epidemiological studies on the effects of combinations of O_3, SO_2 and NO_2. By means of a selection of such investigations, we will point out the different effect modes possible under combined exposures.

2.5.3.1.1 Combined Effects of O_3 and SO_2: Investigations into combined effects on plants were initiated by a publication by Menser and Heggestad (1966), on the

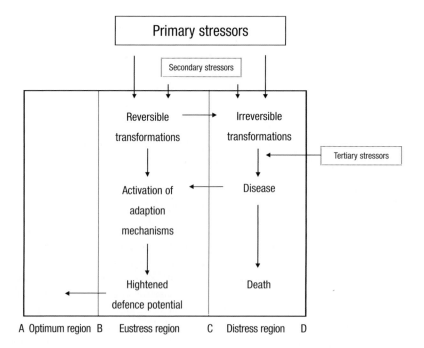

Fig. 2.5-6 Stress and decline reactions of plant systems (Tesche 1991, modified)

impact of sulphur dioxide (240 ppb) and ozone (30 ppb) on tobacco plants. While the single components had not caused any external signs of damage after 2 and 4 hours, respectively, all three tobacco varieties investigated showed leaf damage under combined exposure. The authors described this reactive behaviour as a synergistic (super-additive) effect. In the period following this discovery, many experiments were performed with different concentration/time patterns and various plant species, in which the other effect modes perceivable i.e. additive and sub-additive effects occurred as well. In ecotoxicology, the term synergism has only been applied to super-additive effects since then. However, for reasons of uniformity, another definition of synergism is used in all contributions to this volume: An effect is a case of synergism if the combined effect of two components exceeds the effect of the component with the stronger single effect (also see the glossary). In the "short-term investigations" with exposure times of ≤ 1 day, the external leaf damage (table 2.5-9) were used as effect criterion; in the "long-term investigations" ≥ 1 day, the effects on growth or yield were used for this purpose (table 2.5-10).

Both in the short-term and the long-term experiments, super-additive responses were observed under combined exposure, dependent on the plant species and the constellation of the exposure. Figure 2.5-7 illustrates that super-additive effects are dominant where the single components are present in concentrations that only cause minor acute harm. For leaf damage $> 40\%$ caused by the single component, virtually no super-additive effects could be observed, as shown by the experiments with soybeans by Heagle and Johnston (1979). The yield data, too, showed a response into the same direction (see table 2.5-10).

Unambiguous biochemical explanations of the phenomenon of sub-additive, additive and super-additive effects are still not available. Due to the phytotoxic properties of ozone and sulphur dioxide, various approaches to this question are conceivable. For instance, one can compare the experimentally observed combined effects with effects computed under the assumption of an independent combination of the components. For agents attacking in different ways, a stronger than independent combination effect is only expected, if no interaction occurs between the single components (see section 2.2.2.6 for methods of calculating such effects). This means that whenever stronger effects occur than calculated for an independent combination of the single components, an amplifying interaction must have taken place.

According to present knowledge, ozone develops its phytotoxic effect mainly through the formation of highly reactive radicals ($OH\cdot$, $O_2^-\cdot$), which in turn can attack a multitude of cell components. The plasma membrane is considered as particularly susceptible to radical attacks. Increased repair activity of such damage puts an additional strain on the energy balance. Insufficient repair of the plasma membrane should lead to effects on the osmotic potential. SO_2 dissolves in water, causes acid anions to accumulate, and hence leads – after exhausting the buffer capacity – to the acidification of the cell sap. Moreover, SO_2 promotes the forming of radicals, too, especially during the photooxidation of sulphite into sulphate in the chloroplasts ($O_2^-\cdot$; Okpodu et al. 1996).

Thus, both pollutants, SO_2 and O_3, attack the same targets, the osmotic potential and the energy balance of the cell (Heath 1980, Olszyk and Tingey 1985). Both SO_2

Table 2.5-9 Leaf damage (percentage of leaf surface) caused by O_3 and SO_2 as single agents and in combination, compared to the computed, independently combined effect (indep.); combined effects stronger than the single effects are printed in italics, combined effects stronger than for independent combination in bold type

Plant species	O_3 (ppm)	SO_2 (ppm)	Exposure time	O_3	SO_2	O_3 + SO_2	indep.	Source
Malus domestica cv. Vance Delicious	0.4	0.4	O_3: 4h/d (1x) SO_2: 4h/d (1x)	24	8	*26*	30	Shertz et al. (1980a)
cv. Imperial McIntosch				30	9	22	36	
cv. Golden Delicious				27	19	19	41	
Vitis vinifera cv. Ives	0.4	0.4	O_3: 4 h/d (1x) SO_2: 4 h/d (1x)	27	18	***47***	40	Shertz et al. (1980b)
cv. Delaware				1	1	***4***	2	
Raphanus sativus	0.15	0.15	O_3: 6 h/d (5x) SO_2: 4 h/d (5x)	13	1	***30***	14	Beckerson and Hofstra (1979)
Cucumis sativus				27	9	***54***	34	
Vicia faba				19	0	0	19	
Elatior begonia cv. Schwabenland	0.25	0.5	O_3: 4 h/d every 6 days (4x) SO_2 : 4 h/d every 6 days (4x)	54	2	***67***	55	Reinert and Nelson (1980)
cv. Wisper '0'				25	1	***58***	26	
cv. Fantasy				2	0	***13***	2	
cv. Renaissance				15	0	***18***	15	
cv. Turo				8	0	***12***	8	
Pisum sativum	0.13	0.40	O_3: 4h (1x) SO_2: 4h (1x)	0	0	***32***	0	Olszyk and Tibbits (1981)

and O_3 exposure have been shown to reduce the net rate of photosynthesis, meaning that a [dependent] response stronger than predicted by the independent model can occur when the two components are present in combination. Table 2.5-9 shows that such responses really take place. However, the statistical average of the experimental combined effect (29% leaf damage) is approximately equal to the calculated effect for an independent combination of O_3 and SO_2 (23%).

The effect of SO_2 and O_3 on growth and yield has been demonstrated by a large number (40) of individual experiments listed in table 2.5-10. On average, a combined effect of 23% was measured. As in table 2.5-9, this only slightly exceeds the effect of the component with the stronger single effect (21%). Figure 2.5-8a shows

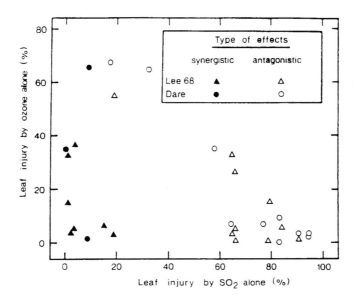

Fig. 2.5-7 The relationship between the degree of leaf damage (%) under single exposure to O_3 and SO_2 and under combined exposure of two varieties of soybeans, in relation to the sum of the single effects (Heagle u. Johnston 1979)

an analysis of the individual results in form of a comparison between the effect of the stronger component and the respective combined effect. According to the terminology used in this volume, stronger combined effects correspond to a synergism, while weaker effects point to an antagonism. On average, no synergism was found over the entire effect range.

2.5.3.1.2 Combined Effects of NO_2 and SO_2: Initial studies by Tingey et al. (1971) showed that a combined exposure to nitrogen dioxide and sulphur dioxide could cause harmful effects already at lower doses than a single exposure to one of the components. Since then, this original finding has been largely confirmed by experiments with various grass and tree species, mostly in the UK (table 2.5-11).

In biochemical terms, the super-additive response is, according to Wellburn et al. (1981), attributable to SO_2 inhibiting the nitrite reductase (NiR) activity, which normally increases under NO_2 exposure. Thus, excessive amounts of nitrite, which under NO_2 exposure alone is fed into the nitrogen metabolism through NiR and the enzyme system GS/GOGAT (glutaminesynthetase/ glutamate oxoglutarate aminotransferase) and shows trophic effects to a certain degree, accumulates in the chloroplast under the additional exposure to SO_2. There, nitrite interferes, as a so-called decoupler, with the proton gradient at the thylakoid membrane, which is essential for the formation of ATP, and affects, via the energy balance of the cell, the growth of the plant, which is already impaired by the impact of SO_2 on the capacity

Table 2.5-10 The effects of O_3 and SO_2 as single agents and in combination: The percentage reduction marks an increase in growth and yield. The effect of O_3+SO_2 is compared with the computed effect of an independent combination (indep.). DM = dry matter; FW = fresh weight

Plant species	O_3 (ppm)	SO_2 (ppm)	Exposure time	Criterion	O_3	SO_2	$O_3 + SO_2$	indep.	Source
Raphanus sativus	0.05	0.05	8 h/d 5d/ week (5x)	Shoot DM	10	0	10	10	Tingey et al. (1971a)
				Root DM	50	17	*55*	58	
Medicargo sativa	0.05	0.05	8 h/d 5d/week (12x)	Shoot DM	12	26	18	35	Tingey a. Reinhardt (1975)
				Root DM	22	29	24	45	
Vicia faba	0.05	0.05	7 h/d 5d/ week (3x)	Shoot FW	2	-5	*12*	-3	Tingey et al. (1973)
				Root FW	3	0	*24*	3	
Vicia faba	0.1	0.1	7 h/d 5d/ week (up to harvest)	Shoot FW	65	-3	52	64	Heagle et al. (1974)
				Seed	54	4	*63*	56	
Nicotiana tabacum	0.05	0.05	7 h/d 5d/week (4x)	Leaf DM	1	14	*30*	15	Tingey a. Reinert (1975)
Picea abies	8:00 16:00: 0.09 16:00– 8:00: 0.045	0.035	2 vegeta-tion periods	Needle DM	34	24	*43*	50	Guderian et al. (1985)
				Shoot DM	33	41	*54*	60	
				Root DM	17	29	*53*	41	
Fesuca pratensis	0.1	0.1	O_3 and SO_2 6h/d 1d/w (12x)	Stipule ears (number)	-1	6	4	5	Flagler a. Youngner (1982a)
				Shoot DM	-3	5	*18*	2	
	0.2	0.1		Stipule ears (number)	6	6	-12	12	
				Shoot DM	19	5	19	23	
	0.3	0.1		Stipule cars (number)	-5	6	*19*	1	
				Shoot DM	18	5	*53*	22	

Medicargo sativa	0.05	0.05	O$_3$: 6 h/d (68x) SO$_2$: 68d	Leaf DM	49	0	46	49	Neely et al. (1977)
Pinus taeda	0.035	0.022	6 h/d 35d	Number of short roots	0	0	*23*	0	Mahoney et al. (1985)
				Shoot DM	9	0	9	9	
				Root DM	22	22	*35*	39	
				Shoot length	0	8	0	8	
Begonia	0.15	0.60	5 d	Dry weight	5	10	4	15	Gardener a. Ormrod (1976)
Glycine max	0.25	0.25	4 h/d 3d/w 11x	Shoot (DM)	45	20	*63*	56	Reinert a. Weber (1980)
				Root (DM)	40	27	*70*	56	
				Plant (DM)	45	11	*65*	51	
Populus deltoides x trichocarpa	0.062	0.062	27 d	Leafs (DM)	11	0	*24*	11	Reich et al. (1984)
				Shoot spindles (DM)	10	-1	*51*	9	
Populus deltoides x trichocarpa	0.25	0.5	12 h/d 24x	Plant (DM)	78	13	60	81	Noble a. Jensen (1980)
Populus deltoides x trichocarpa	0.15	0.25	12 h/d 7 w	Growth rate	62	28	*98*	73	Jensen (1981)
Glycine max	0.25	1	1.5 h	Shoot (DM)	9	-9	-24	1	Heagle a. Johnston 1979
	0.50	1			-12	-9	-35	-59	
	0.25	0.5			0	-7	*7*	-7	
	0.25	1			0	-4	-9	-4	
	0.25	1.5			0	-51	-43	-51	
	1	0.5			-41	5	-38	-34	
	1	1			-41	2	-49	-38	
	0.25	1			2	-9	-18	-7	
	0.25	1.5			2	-77	-48	-73	

Stronger combined effects are printed in italics, stronger effects than calculated for an independent combination are in bold type

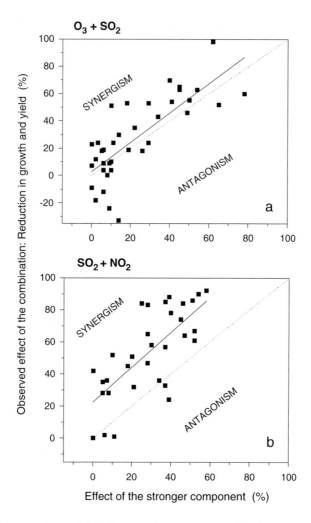

Fig. 2.5-8 Comparison of the effects of the component with the stronger effect and the observed combined effect. The solid line represents the computed regression curve; the dotted line separates synergism from antagonism effects. a: Combination of O_3 and SO_2 (see table 2.5-10); b: combination of NO_2 and SO_2 (see table 2.5-11).

for photosynthesis. Hence, it is no surprise that this combination, in particular, led to distinct amplification (table 2.5-11). The experimentally observed combined effect was 53% stronger, on average, than the effect of the stronger component (25%). However, it also exceeded significantly the computed effect of an independent combination (39%); the 95%-confidence intervals do not overlap. Accordingly, the graphical evaluation of the individual results (fig. 2.5-8b) shows a completely different picture from fig. 2.5-8a: The regression line clearly lies in the synergism region.

Remarkably, the distinct combination effect of SO_2 and NO_2 was only observed for plants exposed in winter in many cases. The reason for the increased occurrence of synergistic effects of these two pollutants must be the overproportionate reduction in the detoxification activity of the plants due to the low temperatures (Kropff et al. 1990). For perennial cereal species, for instance, the increased combination effects can disappear again during the spring and summer months, despite a continuing exposure (e.g. Pande and Mansfield 1985, Wenzel 1992).

Regarding the practical relevance of these results, we must state that the findings described above were achieved with concentrations rarely found in today's Central and Western Europe.

2.5.3.1.3 The Combined Effect of O_3 and NO_2: Compared to the combined effects of O_3 and SO_2 as well as of NO_2 and SO_2, the combined effect of O_3 and NO_2 has been hardly investigated. The study by Kress and Skelly (1982), for which they examined a number of different tree species for combined effects of these two pollutants, may serve as an instructive example for such attempts (table 2.5-12). As the table shows, stronger combination effects than to be expected for a combination of independent effects were found only occasionally. On average, the growth decreased by 23% under the combined exposure to the two pollutants; the calculation for the independent response resulted in a decrease by 28%. Thus, the effects were mostly sub-additive. Ito et al. (1985), on the other hand, reported super-additive effects in experiments with green beans (*Phaseolus vulgaris)*, however at extremely high NO_2 concentrations of 2 and 4 ppm, respectively, over 7 days. With both ozone (50-60 ppb) and NO_2 (30 - 40 ppb) in realistic dose ranges, neither the single exposure nor the combined exposure resulted in a negative effect; increased growth was observed instead (Bender et al. 1991). The biochemical mechanisms behind the combined effect of O_3 and NO_2 are still unknown.

More recent studies on the combined effect of O_3 and NO_2 at realistic concentrations rightly point out that, with increasing ozone concentrations, the concentrations of NO_2 as a precursor of the ozone formation fall so sharply that both agents being present at high concentrations would be a rare coincidence. Sequential exposures are therefore more important than simultaneous ones. A series of experiments on this matter (see table 2.5-12a) produced sub-additive effects in five cases and a super-additive response in only one case.

2.5.3.1.4 Combined Effects of O_3, SO_2 and NO_2: The combination of three air pollutants caused very varied effects including sub-additive, additive and super-additive effects (table 2.5-13). On average, the observed effect, at 27%, was in agreement with the calculated independently combined effect of 29%. Hence, we may assume that this combination broadly behaves like an independent combination – despite the fact that SO_2 and NO_2 in binary combination had caused distinct amplification effects (see table 2.5-11 and fig. 2.5-8b).

2.5.3.2
Models for Dose-Response Relationships in Combined Exposures

Sporadically in the 1980s, models were suggested to derive the quantitative relationship between the exposure of a plant to more than one pollutant, on the one

Table 2.5-11 Growth-inhibiting effects of NO_2 and SO_2 as single agents and in combination (percentage deviation from the control plants). Negative numbers mean increased growth. The effects of $NO_2 + SO_2$ are compared with the computed effect of an independent combination (indep.)

Plant species	NO_2 (ppm)	SO_2 (ppm)	Expo- sure time	Criterion	NO_2	SO_2	NO_2+ SO_2	indep.	Source
Dactylis glomerata	0.07	0.07	104 h/week (20x)	Leaves	7	28	*83*	33	Ashenden 1979
				Stalk	46	52	*67*	74	
				Roots	11	37	*85*	44	
Poa praten- sis	0.07	0.07	104 h/week (20x)	Leaves	29	39	*88*	57	
				Stalk	27	37	*57*	54	
				Roots	47	54	*90*	76	
Lolium multiflorum	0.07	0.07	104 h/week (20x)	Leaves	20	28	*65*	42	Ashenden and Williams (1980)
				Stalk	5	-3	*28*	2	
Phleum pratense	0.07	0.07	104 h/week (20x)	Leaves	-14	25	*84*	15	
				Stalk	12	47	*64*	53	
				Roots	-1	58	*92*	58	
Poa praten- sis	0.07	0.07	104 h/week (20x)	Shoot	45	45	*74*	70	Whitmore and Mansfield (1983)
Hordeum vulgare				Leaves	1	18	*45*	19	
				Stalk	2	5	*35*	7	
				Shoot	-18	20	*51*	6	
Populus nigra	0.07	0.07	22 weeks	Plant	0	28	*47*	28	Freer-Smith (1984)
Tilia cor- data	0.062	0.062	60 weeks	Plant	0	37	33	37	
Malus do- mestica				Plant	0	34	*36*	34	
Betula pendula				Plant	-15	39	24	30	
Betula pubescens				Plant	0	52	*61*	52	
Populus nigra				Plant	0	0	*42*	0	
Alnus incana				Plant	-65	0	0	-65	

Lycopersico *escu-lentum*	0.11	0.1	4 weeks	Leaves	-6	11	1	6	Marie and Ormrod (1984)
				Stalk	-10	8	*28*	-1	
				Root	-8	7	*36*	0	
Betula *pendula*	0.05	0.05	12 h/d 4 weeks	Plant	-49	6	*22*	-40	Freer-Smith (1985)
Dactylis *glomerata*	0.068	0.068	20 weeks	Plant	22	40	*78*	53	Ashenden and Mansfield (1978)
Lolium *multi-* *florum*					10	5	*52*	15	
Phleum *pratense*					1	51	*86*	51	
Poa pra- *tensis*					38	46	*84*	67	
Hordeum *vulgare*	0.1	0.1	20 days	Shoot spindle (DM)	2	5	*35*	7	Pande and Mansfield (1985)
				Leaves (DM)	2	18	*45*	20	
Sorghum *bicolor*	0.1	0.1	24 days	Leaves (DM	7	21	*32*	27	Pande (1984)
Betula *pendula*	0.062	0.062	10 months	Plant (DM)	13	30	*58*	39	Wright (1987)

Combined effects stronger than the single effect are printed in italics, effects stronger than for an independent combination in bold type

hand, and the response of the plant, on the other, especially for the two exposure types SO_2 + NO_2 and O_3 + SO_2. The response criteria were, as for the dose-response relationships for single agents, integrative responses like visible damage or growth. The following approaches were chosen:

1. Whitmore (1985) added up the concentrations of SO_2 and NO_2 in order to determine the dose of the combined exposure of bluegrass (*Poa pratensis*). For this purpose, she selected the concentrations of the individual pollutants in a way that they were at the same level in each experimental preparation (SO_2 = NO_2). Again, growth served as the effect criterion. In the lower dose range (< 2000 nl l^{-1} × d), she found that the growth of the grass was stimulated; at higher dose, the growth became more and more inhibited (fig. 2.5-9). While the stimulation should be attributable to the trophic effect of NO_2 in the lower dose ranges, the inhibition results from the combined effect of the two gases, as described above.

Table 2.5-12a Growth-inhibiting effects of O3 and NO2 as single agents and in combination (percentage deviation from the control plants). Negative numbers mean increased growth. The effects of O3 + NO2 are compared to the computed effect of an independent combination (indep.). DM = dry matter (Kress and Skelly, 1982)

Plant species	O3 (ppm)	NO2 (ppm)	Exposure time	Criterion	O3	NO2	O3 + NO2	indep.
Pinus taeda	0.1	0.1	6 h/d (28x)	Shoot length	17	15	*39*	29
				Shoot DM	21	22	*26*	38
				Root DM	13	17	*26*	28
Pinus taeda (6-13 x 2-8)	0.1	0.1	6 h/d (28x)	Shoot length	25	11	24	33
				Shoot DM	11	10	4	20
				Root DM	31	14	17	41
Pinus rigida	0.1	0.1	6 h/d (28x)	Shoot length	14	16	*26*	28
				Shoot DM	-14	20	11	9
				Root DM	0	11	**15**	11
Pinus virginiana	0.1	0.1	6 h/d (28x)	Shoot length	11	13	*23*	23
				Shoot DM	2	1	1	3
				Root DM	19	7	*19*	25
Liquidamber styraciflua	0.1	0.1	6 h/d (28x)	Shoot length	27	32	28	50
				Shoot DM	30	25	21	48
				Root TS	45	27	*48*	60
Fraxinus americana	0.1	0.1	6 h/d (28x)	Shoot length	20	-5	16	16
				Shoot DM	37	1	37	38
				Root TS	55	37	52	72
Fraxinus pennsylvanica	0.1	0.1	6 h/d (28x)	Shoot length	19	-1	**22**	18
				Shoot DM	17	10	**29**	25
				Root DM	12	18	*19*	28
Quercus phellos	0.1	0.1	6 h/d (28x)	Shoot length	-5	10	**14**	6
				Shoot DM	-1	24	13	23
				Root DM	-11	14	12	5

Combined effects stronger than the single effects are printed in italics, effects stronger than for an independent combination in bold type

It is out of the question to transfer this dose-response model to practical conditions. Apart from the fact that the results were not obtained under realistic exposure and environmental conditions, insurmountable reservations arise from the assumption made in the model that identical concentrations of NO_2 and SO_2 cause identical effects.

Table 2.5-12b: Summary of open-top chamber experiments with various agricultural crops, concerning the effect of a sequential exposure to O_3 and NO_2 (Bender and Weigel 1994)

Species	Criterion	Type of interaction
Phaseolus vulgaris 'Rintintin'	Biomass	sub-additive
	N-metabolism	super-additive
Triticum aestivum 'Turbo'	1000-corn weight	sub-additive
	Shoot dry weight	sub-additive
	Yield	sub-additive
Triticum aestivum 'Star'	1000-corn weight	sub-additive
	Shoot dry weight	sub-additive
	Yield	sub-additive
Hordeum vulgare 'Arena'	1000-corn weight	sub-additive
	corn number per ear	sub-additive
Hordeum vulgare 'Alexis'	1000-corn weight	sub-additive
Brassica napus 'Callypso'	Yield	sub-additive
	Length of husk	sub-additive

2. Posthumus (1982) and, continuing his work, van der Eerden and Duym (1988) followed a different approach to finding a guideline value for the injurious combination of the gaseous pollutants SO_2 and NO_2. They evaluated their experiments and those by other authors with regard to the dose at which no damage occurred yet (table 2.5-14). To this end, they plotted the dose of the pollutant effect in a three-dimensional graph with one axis representing the SO_2 concentration, the second the concentration of NO_2 and the third being a time axis (fig. 2.5-10). The guideline level for the protection of vegetation against this mixture of harmful gases, as read from this plot, is not represented by a line, as it usually is in studies on single components, but by a surface of values.

3. Larsen et al. (1983) derived a mathematical model for the combined effect of O_3 and SO_2 from a series of experiments with the two soybean varieties Lee 68 and Dare. As the response parameter, they chose the percentage of necrotized leaf surface. This model led to the conclusion that for a SO_2 concentration causing about 25% leaf damage (ca. 1.5 ppm SO_2) and a concentration of O_3 causing the same proportion of leaf damage (ca. 0.7 ppm O_3), the combined effect should be 50% i.e. exactly additive, for an exposure time of 45 minutes (fig. 2.5-11, top). For concentrations of both gases below the values given above, the model predicts a super-additive effect; for concentrations above the given levels, the predicted damage is sub-additive, similar to the findings by Heagle and Johnston (1979; fig. 2.5-7). The lowest concentration of SO_2, which theoretically could cause leaf damage in a mixture with subacute O_3 concentrations under a 3-hour exposure (fig. 2.5-11, bottom) would be ca. 300 ppb, a value extrapolated from the model suggested by Larsen et al. (1983). The correspon-

Table 2.5-13 Growth-inhibiting effects of O_3, SO_2 and NO_2 as single agents and in three-component combination (percentage deviation from control plants). Negative numbers mean increased growth. The effects of $O_3 + NO_2 + SO_2$ are compared to the computed effect of an independent combination (indep.)

Plant species	O_3 (ppm)	SO_2 (ppm)	NO_2 (ppm)	Exposure time	Criterion	O_3	SO_2	NO_2	O_3 +SO_2 +NO_2	indep.	Source
Vicia faba	0.15	0.15	0.15	4 h; 3/ week (4x)	Fresh weight	27	9	-12	**27**	26	Reinert and Heck (1982)
Tagetes patula	0.3	0.3	0.3	3 h/d; 3d/ week (3x)	Flower weight	20	47	-16	20	51	Reinert and Sanders (1982
Tagetes patula	0.3	0.3	0.3	3 h/d; 3d/ week (1x)	Flower weight	41	49	23	20	77	Sanders and Reinert (1982)
Raphanus sativus	0.3	0.3	0.3	3 h/d; 3d/ week (1x)	Hypocotyl	30	-21	-10	**65**	7	Sanders and Reinert (1982a)
Raphanus sativus	0.4	0.4	0.4	3 h + 6 h (1x)	Hypocotyl	20	4	0	*36*	23	Reinert and Gray (1981)
Azalea floribunda	0.25	0.25	0.25	3 h/d; 6d/ week (4x)	Leaves	6	7	0	**27**	13	Sanders and Reinert (1982b)
Poa pratensis	0.15	0.15	0.15	10 days; O_3: h/d; SO_2: 24h/d; NO_2: 4h/d	Leaf surface	5	12	6	*16*	21	Elkiey and Ormrod (1980)
Agrostis alba	0.15	0.15	0.15		Leaf surface	14	12	12	28	33	
Agrostis palustris	0.15	0.15	0.15		Leaf surface	7	18	8	26	30	
Agrostis tenuis	0.15	0.15	0.15		Leaf surface	15	6	13	27	30	
Festuca rubra	0.15	0.15	0.15		Leaf surface	16	0	0	**22**	16	
Lolium perenne	0.15	0.15	0.15		Leaf surface	20	-7	2	13	16	

Combined effects stronger than the single effects are printed in italics, effects stronger than for an independent combination in bold type.

Table 2.5-14 Concentrations of injurious gases without a detectable effect on sensitive plants (Posthumus 1982)

NO$_2$	SO$_2$				
(ppm)	0.015	0.038	0.095	0.190	0.380
0.021	>100 d	30 d	15 h	3.5 h	1.4 h
0.053	>100 d	10 h	4 h	2 h	1.2 h
0.133	7 d	5 h	3.1 h	1.8 h	1.2 h
0.266	38 h	3.8 h	2.5 h	1.7 h	1.2 h
0.532	13 h	3 h	2.1 h	1.7 h	1.2 h

ding O$_3$ concentration would be ca. 60 ppb. In practical experiments, however, only the concentration intervals of 250 to 1,000 ppb O$_3$ and 500 to 1,500 ppb SO$_2$, respectively, were investigated.

2.5.3.3
Filtered-Unfiltered Experiments

Finally, to circumvent the experimental and mathematical difficulties in finding critical values for combinations of harmful gases, plant responses to filtered and unfiltered ambient air, respectively, were compared. In this approach, plants were

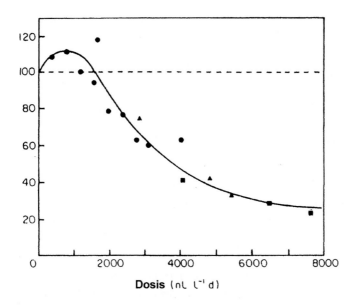

Dosis (nl l^{-1} d)

Fig. 2.5-9 The dose-response relationship for the effects of SO$_2$+NO$_2$ mixtures on the growth of *Poa pratensis*. The plants were exposed under controlled conditions (7 nl l^{-1} NO$_2$ and SO$_2$, and 40 (●) 70 (▲) and 100(■) nl l^{-1} each for both gases) for periods of 4 to 50 days (Whitmore 1985).

Table 2.5-15 Summary of the results of exposures to realistic long-term immissions leading to the occurrence or absence, respectively, of yield reductions (Wenzel 1992)

Plant species	O_3 (ppb)	SO_2 (ppb)	NO_2 (ppb)	Yield reduction
Triticum aestivum				
cv. Okapi	13	5	24	no
cv. Kraka	11	7	11	no
Hordeum vulgare				
cv. Sonate	12	17	13	no
cv. Corona	11	7	11	no
Avena sativa				
cv. Flämingsvita	39	3	13	no
cv. Flämingsnova	22	4	23	no
Hordeum vulgare				
cv. Golf	39	3	13	no
cv. Klaxon	39	3	14	no
Avena sativa				yes
cv. Flämingsvita	22	4	23	
cv. Flämingsnova	22	21	13	yes
Hordeum vulgare				
cv. Golf	22	21	13	yes
cv. Klaxon	22	21	13	yes

exposed in a standardized manner in a number of open-top chambers distributed over Central Europe. They were exposed to the ambient air at their location, which typically contained a mixture of several pollutants (O_3, SO_2, NO_X and NH_Y), as well as to additional doses of individual harmful components. The response was determined by means of control data obtained through a defined filtering of the ambient air. Again, the response parameters were integrative quantities like growth and yield. Table 2.5-15 summarizes a series of experiments performed in Essen, Germany (Wenzel 1992).

Significant effects on the yield performance of cereals only occurred under combined exposures to O_3 + SO_2 + NO_2 at concentrations of 22 + 4 + 23 and 22 + 21 + 13 ppb respectively. The advantage of this approach is that pollutant mixtures of more than two components are considered, and that the climate conditions at the respective sites contribute to the results, too. Using the ambient air at the site of the experiment also ensures a realistic exposure situation under the local, coincident conditions of air pollution load and the climate.

2.5.3.4
Possible Interactions between the Effects of O_3, SO_2 and NO_X, of Increased CO_2 Concentrations and Changes in the UV Radiation

Our knowledge about the present situation concerning exposures and effects is essentially based on the findings on effects of SO_2, O_3 and NO_X. However, these exposures, which are present regionally and supraregionally, are superseded by continental and even global changes in the atmosphere, for instance the rising CO_2 con-

Time in days

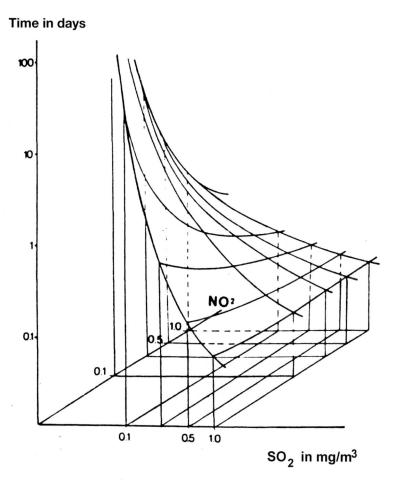

Fig. 2.5-10 The threshold surface for combined effects of SO_2 and NO_2. Exposition-time patterns below the surface are phytotoxic (van der Eerden and Duym 1988)

centrations and changes in the UV radiation caused by the ozone break-down in the stratosphere.

As shown in table 2.5-1, the CO_2 concentration has risen from 280 to 360 ppm over the past two centuries and is still rising by about 0.4% annually. By now, there have been scores of investigations on the effect of increased CO_2 concentrations, most of which, however, looked at short-term exposures, so that conclusions regarding possible long-term effects can only be drawn with reservations. Under this proviso, the available findings can be summarized by the following principal statements (Krupa and Kickert 1993; Fangmeier and Jäger 2000; Guderian and Fangmeier 2001):

- Increased CO_2 concentrations usually enhance the general growth performance of plants.

- Increased CO_2 concentrations tend to raise the water use efficiency, so that plants at arid sites find better growth conditions. Nevertheless, Mansfield (1998) points out that forestry cultures exposed to increased CO_2 concentrations can show disruptions in the control mechanism of the stomata, leading to increased transpiration and hence to a decrease in the water use efficiency.
- Exposed to increased CO_2 concentrations, the plant is able to meet its carbon requirements through uptake from the air in a shorter time period. One result of a gas exchange such reduced is a lower uptake of pollutants into the plants, which in turn reduces the danger to plants. This point is relevant for the evaluation of dose-response relationships for the three air pollutants discussed here.
- Increased CO_2 concentrations have different effects on the growth performance of individual plant species and hence affect the competition between species. C3 plants are promoted to a larger extend than C4 plants. The overall effect of these shifts in the conditions of interspecific competition is expected to be changes in the composition of the vegetation cover.
- Possible effects on biocenotic connexes.

In contrast to the effects of increased CO_2 concentrations, a higher intensity of electromagnetic radiation in the UV-B region (wavelengths of 280–315 nm) tends to bring forth negative effects, as investigations with, up to now, about 400 plant species and culture crops have shown (Tevini 2001). In most cases, a suppression of leaf and shoot growth, a reduction in the photosynthesis activity and lower biomasses were found, as was the case for the air pollutants O_3 and SO_2. Moreover, germination was impaired in many cases. On the biochemical level, UV-B particularly leads to an accumulation of secondary plant constituents like anthocyanins and etheric oils.

There has been little research yet on the question in which way changes in the UV-B radiation in combination with increased CO_2 and/or ozone or SO_2 concentrations affect plants and ecosystems respectively. Initial approaches focussed on so-called stress proteins (catalase, peroxidases, chitinase, glucanase), which are genetically induced both by UV-B irradiation and by O_3 or pathogen exposure. The findings available so far are rather sketchy and contradictory, for which reason they do not allow any conclusions with regard to the establishment of guideline values.

2.5.4
Summary Assessment and Recommendations for Action

Ozone, the oxygenic sulphur compounds and nitrogen immissions are the greatest hazards for vegetation in Central Europe and North America at present, and there are copious findings from extensive experimental and epidemiological research on these pollutants as single components. Hence, various specialist committees were able to derive guideline values for the protection of vegetation against said single agents. For those reasons, these three components were chosen as examples of the evaluation of combined effects in ecotoxicology.

Combinations of these pollutants cause sub-additive, additive and super-additive responses of plants. Super-additivity was mainly observed for weakly damaging effects, which only slightly exceed additivity in many cases. Almost every combination results in combined effects that (largely) equal independent effects and hence

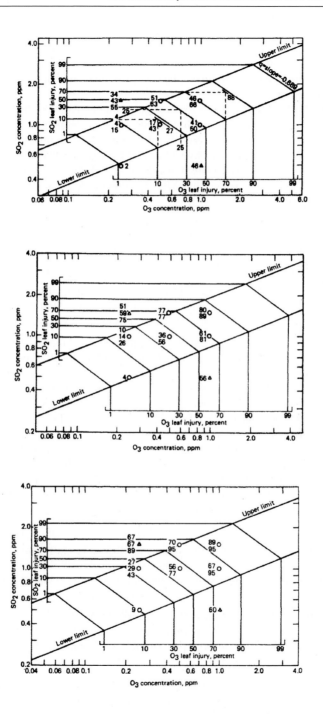

Fig. 2.5-11 Percentage leaf injury for the soybean varieties Lee 68 and Dare following 45-minute (top), 1.5-hour (centre) and 3-hour exposures (bottom) to SO_2 or O_3, or to both pollutants

point to the absence of particular interactions. One exception is the combination of SO_2 and NO_2 at concentrations so high that they are, under present conditions, unrealistic for practical purposes. However, the experiments often showed a synergism in the sense of a super-additive effect. The reason behind this amplifying interaction is still under discussion.

The various effect properties of the selected components (SO_2 and O_3 are phytotoxic, SO_2 under low-sulphur conditions and the N compounds, up to a certain exposure level, act trophically) result in small shifts of the concentration proportions already leading to qualitative changes in the plant responses.

Guideline values for the protection of vegetation against mixed exposures can be derived from results obtained through three methodical approaches:

- Comparative experimental investigations on the effect of two or three components as single agents and in combination
- Comparative epidemiological studies on the effects of filtered and unfiltered air in exposure areas
- Models allowing to derive still tolerable concentration-time patterns.

The results from these investigations provide no basis for setting lower guideline values for combined exposures to the selected components than the values for the single components already established, for the following reasons:

- When setting the guideline values for single agents, the expert committees responsible took into account, more or less explicitly, results from experiments with combined exposures.
- In combination experiments with harmful gases, the lowest concentrations still leading to sub-additive, additive or super-additive effects always ranged above the established guideline values for the protection of vegetation against the component pollutants.
- As a rule, plants are subject to higher deposition rates – meaning a higher effective dose at the same concentration of pollutants – under exposure in chamber systems than in the open air. Since guideline values are essentially derived from experimental and epidemiological investigations using chamber systems, a safety margin should be applied to any guideline value thus arrived at.

Any hazard the air pollutants discussed here pose to vegetation does not primarily result from uncertainties in the findings concerning combined effects. There rather is the general problem of characterizing an existing exposure, single or mixed, by effect-representative exposure indices. This leads to uncertainties at three stages:

- in measuring and assessing air pollution exposures,
- in measuring and assessing air pollution effects, and
- in modelling concentration-time patterns for investigations into dose-response relationships.

The efforts to characterize exposures to single or mixed immissions must be continued in international collaboration. Chamber-free systems for the exposure of plants to filtered and unfiltered air in exposure areas as well as to given immission constellations have to be developed and applied to extensively. Through the analysis of predominant immission situations, models for the simulation of exposure scenarios for fumigation experiments must be developed, especially for experiments under near-natural conditions. The concentration in the ambient air is an inadequate measure for establishing the hazard potential of air pollution exposures. Concerning the effect, the decisive parameter has to be the concentration of pollutants at the target. Therefore, detailed studies on the flux of a pollutant from the atmosphere, through the stomata, to the effect site in the mesophyll, which have begun only recently, have to be pursued at high priority. Results from these investigations should be suitable to appraise the degree of effect amplification under chamber exposure.

Especially for long-lived plant species, the long-term exposure to low concentrations of single components or mixtures leads to vitality losses and secondary damage through an increased infestation with harmful organisms or through frost injury. The further development of methods for the causal verification both of the vitality loss and its knock-on effects is an important research objective under the exposure conditions predominant at present.

The ability to adhere to the nitrogen depositions established as tolerable, i.e. to critical loads, depends on the exact knowledge of the quantitative relationship between the amounts deposited and the concentration of nitrogen compounds in the air. Investigations in this respect must be performed have to consider climate conditions like the type, the amount and the distribution of precipitations.

The development and application of specific methods of active biomonitoring is crucial for answering the question whether existing guideline values ensure the intended protection of vegetation. Special attention should be directed to procedures that allow quantifying the effect contribution of the single components in a mixture of injurious agents.

Due to the lack of results in this respect, it is still impossible to say if and to what degree global changes characterized by increased CO_2 concentrations and the changed UV-B radiation modulate the quantitative relationship between conventional immission components and their effects on plants. Investigations on this issue are urgently needed.

2.5.5
Literature

Adams RM, Glyer JD, Johnson SL, McCarl BA (1989) A reassessment of the economic effects of ozone on US agriculture. JAPCA 39: 960–968

Agren GJ, Bosatta E (1988) Nitrogen saturation of terrestrial ecosystems. Environ Pollut 54: 185–197

Arndt U, Nobel W, Schweizer B (1987) Bioindikatoren. Möglichkeiten, Grenzen und neue Erkenntnisse. Ulmer, Stuttgart

Ashenden TW (1979) The effects of long-term exposures to SO_2 and NO_2 pollution on the growth of Dactylis glomerata L and Poa pratensis L. Environ Polut 18: 249–258

Ashenden TW, Mansfield TA (1978) Extreme pollution sensitivity of grasses when SO_2 and NO_2 are present in the atmosphere together. Nature 273: 142–143

Ashenden TW, Williams IAD (1980) Growth reductions in Lolium multiflorum Lam. and Phleum pratense L as a result of SO_2 and NO_2 pollution. Environ Pollut 21: 131–139

Augustin S (1997) Forstbodenkunde. In: Umweltbundesamt (ed.) Auswertung der Waldschadensforschungsergebnisse (1982–1992) zur Aufklärung komplexer Ursache- Wirkungsbeziehungen mit Hilfe systemanalytischer Methoden. UBA, Berlin 6/97

Baker CK, Unsworth MH, Greenwood P (1982) Leaf injury on wheat plants exposed in the field in winter to SO_2. Nature 299: 149–151

Beckerson DW, Hofstra G (1979) Response of leaf diffusive resistance of radish, cucumber, and soybean to O_3 and SO_2 singly or in combination. Atmos Environ 13: 1263-1268

Bender J, Weigel HJ, Jäger HJ (1991) Response of nitrogen metabolism in beans (Phaseolus vulgaris L) after exposure to ozone and nitrogen dioxide, alone and in sequence. New Phytol 119: 261–267

Bender J, Weigel H-J (1994) The role of other pollutants in modifying plant responses to ozone. In: Fuhrer J; Achermann B (eds.) Critical levels for ozone, 240–246

Bobbink R, Boxman D, Fremstad E, Heil G, Houdijk A, Roelofs J (1992) Critical loads for nitrogen eutrophication of terrestrial and wetland ecosystems based upon changes in vegetation and fauna. In: Grennfelt P a. Thörnelöf E (eds.) Critical loads for nitrogen. Tema Nord (Miljörapport) 41: 111–159. Nordic Council of Ministers, Copenhagen

Bücker J, Guderian R, Mooi J (1993) A novel method to evaluate the phytotoxic potential of low ozone concentrations using poplar cuttings. Water, Air u. Soil Pollution 66: 193–201

Eerden Van der LJM, Duym N (1988) An evaluation method for combined effects of SO_2 and NO_2 on vegetation. Environ Pollut 53: 468–470

Eerden van der LJM, Dueck TA, Elderson J, Van Dobben HF, Berdowski JJM, Latuhihin M, Prins AH (1990) Effects of NH_3 and $(NH_4)_2SO_2$ deposition on terrestrial semi-natural vegetation on nutrient-poor soils. Project 124/125, IPO-Report R 90/06, RIN-Report 90/20, Wageningen

Eerden van der LJM, Tonneikk AEG (1988) Crop loss due to air pollution in The Netherlands. Environ Pollut 53: 365–376

Eerden van der LJM, Jäger HJ, Fangmeier A (2001) Wirkungen von Stickstoffdepositionen auf terrestrische Ökosysteme. In: Guderian R (ed.) Handbuch der Umweltveränderungen und Ökotoxikologie. Vol. 2B. Terrestrische Ökosysteme. Springer, Berlin

Elkiey T, Ormrod DP (1980) Response of turfgrass cultivars to ozone, sulfur dioxide, nitrogen dioxide, or their mixture. J Amer Soc Hort Sci 105: 664–668

EPA (United States Environmental Protection Agency 1986) Air quality criteria for ozone and other photochemical oxidants. Volume III of V; EPA 600/8-84/020cF; Center of Environmental Research Information, Cincinnati OH 45268

EPA (United States Environmental Protection Agency 1993) Air Quality criteria for oxides of nitrogen. Volume II of III, EPA 600/8-91/049bF, Center of Environmental Research Information, Cincinnati OH 45268

EPA (United States Environmental Protection Agency 1996) Air Quality criteria for ozone and related photochemical oxidants. Volume II of III. EPA/600P-93004bF, Office of Research and Development, Washington D.C. 20460

Ernst D, Bodemann A, Schmelzer E, Langebnartels C, Sandermann Jr H (1996) β1,3-glucanase mRNA is locally, but not systemically induced in Nicotiana tabacum L cv BEL W3 after ozone fumigation. J Plant Physiol 148: 215–221

Ernst D, Schraudner M, Langebartels C, Sandermann Jr H (1992) Ozone-induced changes of mRNA levels of β1,3-glucanase, chitinase and 'pathogenesis-related' protein 1b in tobacco plants. Plant Mol Biol 20: 673–682

Fangmeier A, Hadwiger-Fangmeier A, Van der Eerden L, Jäger HJ (1994) Effects of atmospheric ammonia on vegetation – A review. Environ Pollut 86: 43–82

Fangmeier A (1996) Postdoctoral thesis, Universität Gießen

Fangmeier A, Jäger HJ (2001) Wirkungen erhöhter CO_2-Konzentrationen. In: Guderian R (ed.) Handbuch der Umweltveränderungen und Ökotoxikologie Vol. 2A; Springer, Heidelberg

Flagler RB, Youngner VB (1982) Ozone and sulfur dioxide effects on tall fescue: I. Growth and yield responses. J Environ Qual 11: 386–389

Freer-Smith PH (1984) The responses of six broadleaved trees during long-term exposure to SO_2 and NO_2. New Phytol 97: 49–61

Freer-Smith PH (1985) The influence of SO_2 and NO_2 on the growth, development and gas exchange of Betula pendula Roth. New Phytol 99: 417–430

Friedrich R, Obermeier A (1999) Emissionen von Spurenstoffen. In: Guderian R (ed.) Handbuch der Umweltveränderungen und Ökotoxikologie: Springer, Berlin, Heidelberg

Fuhrer J (1996) The critical level for effects of ozone on crops, and the transfer to mapping. In: Critical levels for ozone in Europe: testing and finalising the concepts. UNECE Workshop report, University of Kuopio, Kuopio, Finland

Fuhrer J, Skärby L, Ashmore MR (1997) Critical levels for ozone effects on vegetation in Europe. Environ Pollut 97: 91–106

Galliano H, Cabané M, Eckerskorn C, Lottspeich F, Sandermann Jr H, Ernst D (1993) Molecular cloning, sequence analysis and elicitor-/ozone-induced accumulation of cinnamyl alcohol dehydrogenase from Norway spruce (Picea abies L). Plant Mol Biol 23: 145–156

Gardner JO, Ormrod DP (1976) Response of the Rieger begonia to ozone and sulfur dioxide. Scientia Hort 5: 171–181

Grafenhorst G (2001) Trockene und nasse Deposition von Spurenstoffen aus der Atmosphäre. In: Guderian R (ed.) Handbuch der Umweltveränderungen und Ökotoxikologie. Vol. 1 B Springer, Berlin

Grünhage L, Jäger H-J (1994) Influence of the atmospheric conductivity on the ozone exposure of plants under ambient conditions: Considerations for establishing ozone standards to protect vegetation. Environ Pollut 85: 125–129

Guderian R (1977) Air pollution. Phytotoxicity of acidic gases and its significance in air pollution control. Ecological Studies 22; Springer, Berlin.

Guderian R, Ballach HJ (1989) Aufgaben und Probleme der Wirkungsforschung als Grundlage für den praktischen Immissionsschutz. Verh Ges Ökol 18: 289–297

Guderian R, Fangmeier A (2001) Interaktionen zwischen Luftverunreinigungen, Wirtspflanzen und Parasiten. In: Guderian R (ed.) Handbuch der Umweltveränderungen und Ökotoxikologie, Vol. 2 A, Springer, Berlin

Guderian R, Tingey DT, Rabe R (1985) Effects of photochemical oxidants on plants. In: Guderian R (ed.) Air pollution by photochemical oxidants. Ecological Studies 52, Springer, Berlin

Guderian R, Bücker J (1995) Ökotoxikologische Risikobewertung - naturwissenschaftliche Probleme von Grenzwertvorschlägen. In: Arndt U et al (eds.) Hohenheimer Umwelttagung 27 (Grenzwerte und Grenzwertproblematik im Umweltbereich): 32–34

Guderian R, Küppers K, Six R (1985) Wirkungen von Ozon, Schwefeldioxid und Stickstoffdioxid auf Fichte und Pappel bei unterschiedlicher Versorgung mit Magnesium und Kalzium sowie auf die Blattflechte Hypogymnia physodes. VDI-Berichte 560: 657–701

Guderian R, Tingey DT, Rabe R (1983) Wirkungen von Photooxidantien auf Pflanzen. In: Umweltbundesamt (ed.) Luftqualitätskriterien für photochemische Oxidantien. Erich Schmidt, Berlin, Berichte 5/83: 205–427

Haut van H (1961) Die Analyse von Schwefeldioxidwirkungen auf Pflanzen im Laboratoriumsversuch. Staub 21: 52–56

Heagle AS, Body DE, Neely GE (1974) Injury and yield responses of soybean to chronic doses of ozone and sulfur dioxide in the field. Phytopathol 64: 132–136

Heagle AS, Johnston JW (1979) Variable responses of soybeans to mixtures of ozone and sulfur dioxide. JAPCA 29: 729–732

Heath RL (1980) Initial events in injury to plants by air pollutants. Ann Rev Plant Physiol 31: 395–431

Heck WW, Brandt CS (1977) Effects on vegetation: Native crops, forests. In: Air Pollution Vol II, Academic Press, New York pp. 157–229

Heck WW, Adams RM, Cure WW, Heagle AS, Heggestad HE, Kohut RJ, Kress LW, Rawlings JO, Taylor OC (1983) A reassessment of crop loss from ozone. Environ Sci Technol 17: 572A–581A

Heck WW, Cure WW, Rawlings JO, Zaragoza J, Heagle AS, Heggestad HE, Kohut RJ, Kress LW, Temple PJ (1984a) Assessing impacts of ozone on agricultural crops: I Overview. JAPCA 34: 729–735

Heck WW, Cure WW, Rawlings JO, Zaragoza J, Heagle AS, Heggestad HE, Kohut RJ, Kress LW, Temple PJ (1984b) Assessing impacts of ozone on agricultural crops: II. Crop yield functions and alternative exposure statistics. JAPCA 34: 810–817

Houghton JT et al. (eds.) (1996) The science of climate change (IPCC/WMO/UNEP) Univ. Press Cambridge

IPCC (Intergovernmental Panel on Climate Change) (1990) The IPCC Scientific Assessment WMO/UNEP Univ. Press Cambridge

Ito O, Okano K, Totsuka T (1985) Effects of NO_2 and O_3 exposure alone or in combination on kidney bean plants: amino acid content and composition. Soil Sci Plant Nutr 32: 351–363

IUFRO (International Union of Forestry Research Organisation 1978) Resolution über maximale Immissionswerte zum Schutz der Wälder. IUFRO-Fachgruppe S 2.09.00

Jacobson JS (1977) The effects of photochemical oxidants on vegetation. VDI-Berichte 270: 163–173

Jäger HJ, Guderian R (2001) Expositionssysteme. In: Guderian R (ed.) Handbuch der Umwelt-veränderung und Ökotoxikologie. Vol. 2 A. Terrestrische Ökosysteme. Springer, Berlin

Jensen KF (1981) Growth analysis of hybrid poplar cuttings fumigated with ozone and sulfur dioxide. Environ Pollut 26: 243–250

Kluge H (1993) Nur die Buchen überlebten. Forst und Holz 48, 462–466

Kress LW, Skelly JM (1982) Response of several eastern forest tree species to chronic doses of ozone and nitrogen dioxide. Plant Dis 66: 1149–1152

Kropff MJ, Smeets WLM, Meijer EMJ, van der Zalm AJA, Bakx EJ (1990) Effects of sulfur dioxide on leaf photosynthesis: the role of temperature and humidity. Physiol Plant 655–661

Larsen RI, Heagle AS, Heck WW (1983) An air quality data analysis system for interrelating effects, standards, and needed source reductions: Part 7. An O_3-SO_2 leaf injury mathematical model. JAPCA 33: 198–207

Last FT (1989) Experimental investigation of forest decline: the use of open-top chambers. Lecture during the 5. presentation of reports (Statuskolloquium) of the PEF, 5-9 March, at the Nuclear Research Centre Karlsruhe (KfK), PEF Reports

Lee EH, Tingey DT, Hogsett WE (1988) Evaluation of ozone exposure indices in exposure response modelling. Environ Pollut 53: 43–62

Lefohn AS, Laurence JA, Kohut RJ (1988) A comparison of indices that describe the relationship between exposure to ozone and reaction in the field of agricultural crops. Atmos Environ 22: 989–995

Liebold E, Drechsler M (1991) Schadenszustand und -entwicklung in den SO_2-geschädigten Fichtengebieten Sachsens. AFZ 10, 492–494

Linzon SN, Heck WW, MacDowall FDH (1975) Effects of photochemical oxidants on vegetation. In: Photochemical air pollution: Formation, transport, effects. National Research Council, Canada, S 89–142

Mahoney MJ, Chevone BI, Skelly JM, Moore LD (1985) Influence of mycorrhizae on the growth of Lobolly pine seedlings exposed to ozone and sulfur dioxide. Phytopathology 75: 679–682.

Manion PD (1981) Tree disease concepts in relation to forest and urban tree management practice. Prentice Hall

Mansfield TA (1998) Stomata and plant water relations: does air pollution create problems? Environ. Pollut. 101: 1–11

Marie BA, Ormrod DP (1984) Tomato plant growth with continuous exposure to sulfur dioxide and nitrogen dioxide. Environ Pollut 33: 257–265

Mehlhorn H (1998) Ozon. In: Guderian R (ed.) Handbuch der Umweltveränderung und Öko-toxikologie. Vol. II. Terrestrische Ökosysteme. Springer, Berlin 2001

Menser HA, Heggestad HE (1966) Ozone and sulfur dioxide synergism: Injury to tobacco plants. Science 153: 424–425

Miller PL (1973) Oxidant-induced community change in a mixed conifer forest. In: Nägele JA (ed.) Air pollution damage to vegetation. Washington DC: American Chemical Society: Advances in chemistry 122: S 101–117

Mohr H, Schopfer P (1992) Lehrbuch der Pflanzenphysiologie. Springer, Heidelberg

Neely GE, Tingey DT, Wilhour RG (1977) Effects of ozone and sulfur dioxide singly and in combination on yield, quality and N-fixation of alfalfa. In: Dimitriades B (ed.) International conference on photochemical oxidant pollution and its control: proceedings. EPA -Report 600%3-77-0016b; pp. 663–673

Noble RD, Jensen KF (1980) Effects of sulfur dioxide and ozone on growth of hybrid poplar leaves. Amer J Bot 67: 1005–1009

Odum EP (1985) Trends expected in stressed ecosystems. BioScience 35: 419–422

Okpodu CM, Alscher RG, Grabau EA, Cramer CL (1996) Physiological, biochemical and molecular effects of sulfur dioxide. J Plant Physiol 148: 309–316

Olszyk DM, Tibbitts TW (1981) Stomatal response and leaf injury of Pisum sativum L with SO_2 and O_3 exposures. I. Influence of pollutant level and leaf maturity. Plant Physiol 67: 539–544

Olszyk DM, Tingey DT (1985) Metabolic basis for injury to plants from combinations of O_3 and SO_2. Plant Physiol 77: 935–939

Pääkkönen E, Holopainen T, Kärenlampi L (1997) Variation in ozone sensitivity among clones of Betula pendula and Betula pubescens. Environ Pollut 95: 37–44

Pande PC (1984) Sorghum development and sensitivity to SO_2 and NO_2 singly and in mixtures. Agric Ecosys Environ 11: 197–202

Pande PC, Mansfield TA (1985) Responses of spring barley to SO_2 and NO_2 pollution. Environ Pollut 38: 87–97

Pande PC, Mansfield TA (1985) Responses of winter barley to SO_2 and NO_2 alone and in combination. Environ Pollut 39: 281–291

Posthumus AC (1982) Ecological effects associated with NO_X, especially on plants and vegetation. In: Schneider T, Grant L (ed.) Air pollution by nitrogen oxides. Elsevier Scientific Pub Comp, Amsterdam, The Netherlands, Seite 45–60

Reich PB, Lassoie JP, Amundson RG (1984) Reduction in growth of hybrid poplar following field exposure to low levels of O_3 and (or) SO_2. Can J Bot 62: 2835–2841

Reinert RA, Gray TN (1981) The response of radish to nitrogen dioxide, sulfur dioxide, and ozone, alone and in combination. J Environ Qual 10: 240–243

Reinert RA, Heck WW (1982) Effects of nitrogen dioxide in combination with sulphur dioxide and ozone on selected crops. In: Schneider T, Grant L (ed.) Air pollution by nitrogen oxides. Elsevier Sci, Amsterdam, pp. 533–546

Reinert RA, Nelson PV (1979) Sensitivity and growth of five elatior begonia cultivars to SO_2 and O_3, alone and in combination. J Am Soc Hort Sci 105: 721–723

Reinert RA; Sanders JS (1982) Growth of radish and marigold following repeated exposure to nitrogen dioxide, sulfur dioxide, and ozone. Plant Dis 66: 122–124

Reinert RA, Weber DE (1980) Ozone and sulfur dioxide-induced changes in soybean growth. Phytopathology 70: 914–916

Sanders JS, Reinert RA (1982a) Weight changes of radish and marigold exposed at three ages to NO_2, SO_2 and O_3 alone and in mixture. J Amer Soc Hort Sci 107: 726–730

Sanders JS, Reinert RA (1982b) Screening Azalea cultivars for sensitivity to nitrogen dioxide, sulfur dioxide, and ozone alone and in mixtures. J Amer Soc Hort Sci 107: 87–90

Segschneider H-J (1995) Auswirkungen atmosphärischer Stickoxide (NO_X) auf den pflanzlichen Stoffwechsel: Eine Literaturübersicht. Angew Bot 69: 60–85

Shertz RD, Kender WJ, Musselmann RC (1980a) Foliar response and growth of apple trees following exposure to ozone and sulfur dioxide. J Amer Soc Hort Sci 105: 594–598

Shertz RD, Kender WD, Musselman RD (1980b) Effects of ozone and sulfur dioxide on grapevines. Sci Horti (Amsterdam) 13: 37–45

Tesche M (1991) Streß und Decline bei Waldbäumen. Forstw Cbl 110: 56–65

Tevini M (1999) Wirkungen veränderter UV-B Strahlung. In: Guderian R (ed.) Handbuch der Umweltveränderungen und Ökotoxikologie. Springer, Berlin

Tingey DT, Heck WW, Reinert RA (1971a) Effect of low concentrations of ozone and sulfur dioxide on foliage, growth and yield of radish. J Amer Soc Hort Sci 96: 369–371

Tingey DT, Reinert RA, Dunning JA, Heck WW (1971b) Vegetation injury from the interaction of nitrogen dioxide and sulfur dioxide. Phytopathol 71: 1506–1511

Tingey DT, Reinert RA (1975) The effect of ozone and sulfur dioxide singly and in combination on plant growth. Environ Pollut 9: 117–125

Tingey DT, Reinert RA, Wickliff C, Heck WW (1973) Chronic ozone or sulfur dioxide exposures, or both, affect the early vegetative growth of soybean. Can J Plant Sci 53: 875–879

UBA (Umweltbundesamt 1997) Auswertung der Waldschadensforschungsergebnisse (1982–1992) zur Aufklärung komplexer Ursache- Wirkungsbeziehungen mit Hilfe systemanalytischer Methoden. UBA, Berlin 6/97

UNECE (United Nations Economic Commission for Europe 1988a) UNECE critical levels workshop report. Bad Harzburg, FRG, March 1988

UNECE (United Nations Economic Commission for Europe (1988b) Critical Loads Workshop, Skokloster

UNECE (United Nations Economic Commission for Europe 1992) Critical levels workshop, Egham UK

UNECE (United Nations Economic Commission for Europe 1994) Critical levels for ozone. Schriftenreihe der FAC Liebefeld 16

VDI (Verein Deutscher Ingenieure 1978) Maximale Immissionswerte zum Schutz der Vegetation - Maximale Immissionskonzentrationen für Schwefeldioxid. VDI 2310, Vol. 3

VDI (Verein Deutscher Ingenieure 1983) Ermittlung von Maximalen Immissionswerten - Grundlagen. VDI 2309, VDI-Handbuch Reinhaltung der Luft Vol. 1: 1–20

VDI (Verein Deutscher Ingenieure 1988) Zielsetzung und Bedeutung der Richtlinien: Maximale Immissionswerte. - VDI 2310, Berlin (Beuth) Vol.1: 1–8

Webb NR (1990) Changes on the heathlands of Dorset, England, between 1978 and 1987. Biol Conserv. 51:273–286

Wellburn AR, Higginsdon C, Robinson D, Walmsley C (1981) Biochemical explanation of more than additive inhibitory effects of low atmospheric levels of sulfur dioxide plus nitrogen dioxide upon plants. New Phytol 88: 223–237

Wentzel KF (1984) Das Erzgebirge im Koma. In: D. Guratsch (ed.) Baumlos in die Zukunft? 49–59, Kindler, München

Wenzel AA (1992) Expositionsversuche in open top-Kammern zur Ermittlung des immissionsbedingten Gefährdungspotentials für landwirtschaftliche Kulturen in der Randzone eines Ballungsraumes. Doctoral thesis Universität GH Essen

Whitmore ME (1985) Relationship between dose of SO_2 and NO_2 mixtures and growth of Poa pratensis. New Phytol 99: 545–553

Whitmore ME, Mansfield TA (1983) Effects of long-term exposures to SO_2 and NO_2 on Poa pratensis and other grasses. Environ Pollut 31: 217–235

WHO (World Health Organization 1987) Air Quality Guidelines for Europe. WHO Regional Publications, European Series 23, Copenhagen

Winner WE, Mooney HA, Goldstein RA (1985) Sulfur dioxide and vegetation. Physiology, Ecology and Policy Issues. Stanford University Press, Stanford CA

Wislicenus H (1898) Resistenz der Fichte gegen saure Rauchgase bei ruhender und bei tätiger Assimilation. Tharandter Forstl Jb 48: 152–172

WMO (World Meteorological Organization), United Nations Environment Program UNEP, 1992). Global ozone research and monitoring project, 25 Genf

Wright EA (1987) Effects of sulfur dioxide and nitrogen dioxide, singly and in mixture, on the macroscopic growth of three birch clones. Environ Pollut 46: 209–221

3 Perception of Technical Risks

3.1
Foreword

Health risks are front-page news. Be it BSE, surface ozone or radiation from transmitter stations ofr mobile phones, the popular press puts out a constant stream of risk warnings and all-clear reports. This chapter looks at how the general public perceives and assesses such information. Although limited empirical data is available on the subject, it will focus on perception of combined risks. The term *perception* as used in cognitive psychology applies to the mental processes through which a person takes in, deals with and assesses information from the environment (physical and communicative) via the senses[1].

In this chapter, the term *risk* means the possibility of undesired adverse effects from some action or event[2]. In both natural science and engineering, *risk* is the arithmetic product of likelihood of occurrence and severity of impact (or the probability function applied across the range of potential damage). In Chapter 1, this concept of risk is termed *rational risk*. This emphasizes that risk assessment is based on the principle of transsubjective validation, i.e. it must be both justifiable and verifiable. Apart from the traditional elements of likelihood of occurrence and severity of damage with respect to health risks from exposures and dose-response functions, rational risk also involves general application of additional risk factors like the uncertainty that remains after assessing probabilities and potential for harm, the ubiquity and persistence of harmful effects independent of their severity or their delayed effects, and the scope for institutional risk management and limitation[3].

The definition of rational risk must not be confused with how the general public defines and perceives risk. Although generally applicable risk factors play an important role in how risk is perceived by individuals, groups or social institutions, subjective factors also play a part in perception and assessment[4]. The way risk is assessed differs, for example, depending on whether or not the individual or entity assessing it has self-perceived control over the degree of risk involved (personal

---

[1] Jungermann, Slovic (1993a).
[2] Fischhoff; Watson; and Hope (1984); see also Renn (1992).
[3] The German Advisory Council on Global Change (WBGU) recommended a classification system based on seven generally applicable risk factors to define various types of risk. For each type of risk, a separate strategy was developed for assessment and management of those risks. WBGU (1998).
[4] Evers, Nowotny (1987).

control and management potential)[5]. Such subjective factors should not be deemed irrational. When we assess risk, it really does make a difference whether one can personally control the degree of risk (say, during leisure activities) or whether one must passively accept a given risk. It is perfectly justified, therefore, to describe perceived risk as subjective risk rather than irrational risk. The collective term *subjective risk* is used in this chapter to include everything that individuals, social groups or institutions perceive as a potentially undesired adverse effect from an action or event. Subjective risk is a subjective assumption that an action or event could have a potentially adverse side-effect[6]. Such assumptions can be "objectified" using scientific methods, i.e. they can reflect prevailing best available collective knowledge on anticipated consequences. But they can also be based on anecdotal knowledge or social experience. Thus, the boundaries between rational and subjective risk should not be drawn too closely: Tension between rational and subjective risk is one of the main forces driving the public debate on health risks and their management.

The following presentation of risk perception aims to improve understanding of how health risks are perceived. The goal is to highlight, from both a psychological and sociological standpoint, the structures and processes of personal risk perception and how society deals with risk. A brief outline of current risk issues is followed by an explanation of intuitive risk perception, in which the main focus is placed on the cognitive and affective aspects of perception processes. A further section looks at the specific problems of risk perception in terms of interactive effects of multiple exposures to different substances or a combination of radiation and noxious chemicals. A third area of focus in this chapter is the question of the effects of interaction between noxious substances in the environment and psychological phenomena like fear, stress and panic. Finally, the chapter takes a brief look at the role of perception research in risk regulation.

3.2
Current Risk Debate

A distinguishing feature of the way in which society manages risk is selectiveness as to what risks give rise to concern in the modern world[7]. While social conflicts have general causes and bring about generally applicable consequences, they are ignited by specific subjects and topics. In the debate on risk, nuclear power, the chemicals industry and genetic engineering have become both a breeding ground for and symbolic of the conflicts surrounding social risk management. Although human health and the environment have always been exposed to risks from natural events or technological products, it is only recently that risk has become an issue of public interest. The novelty of "risk" is four-fold[8]:

---

[5] Sjöberg (1994).
[6] Obermeier (1990), p. 245f.
[7] Berger, Berger, Kellner (1973).
[8] The four factors that follow are taken from an article in Akademie der Wissenschaften zu Berlin (1992), p. 248f.

1) Since the beginning of time, people have taken precautions against risk. Their lack of anticipatory knowledge, however, led them to regard negative results less as the fruits of their own behaviour and more as "punishment inflicted by God" or as fate. Be it the plague, crop failure or flooding caused by a dam burst, all disasters that, given today's knowledge, were caused or at least influenced by human actions were usually put down to fate or punishment for sinful behaviour.

In modern times, according to sociologist Niklas Luhmann, risks that were once perceived as external and to which one believed oneself passively exposed have been transformed into personally regulatable risks[9]. According to Luhmann, danger and risk are differentiated by the degree of perceived manageability by individuals and organizations. He sees "danger" as everything that can be perceived as external danger, from which one can only flee at best. Conversely, risks are such dangers as can be managed by the perceiver or by an organization the perceiver can influence. With the first type (danger), responsibility lies with God, fate or some sinister power, while in the second (risk) it lies with oneself or an identifiable social institution. Risk management, the modern formula for proactive handling of the undesired side-effects of human activities, is thus a fitting testimony to the transformation of risk that was formerly seen as external into processable, socially influenceable and manageable activities to limit undesired outcomes from specific actions. The degree of perceived responsibility for one's own actions grows with the confidence that adverse effects can be influenced by human actions and risks transformed into dangers. Thus, the term *risk* has become central to the actions of today's society[10]. Forecasts based on methodologically proven knowledge are a necessary component of modern society's provisions for the future. Increased knowledge on cause-effect chains provides society with the instruments and institutions to anticipate harmful events and their impacts, and to develop and implement counter-measures to cope with them. At the same time, moral pressure to take risk prevention measures increases in order to evade or limit harmful events[11].

2) Advancements in technology, medicine and hygiene have both reduced the relative proportion of natural risks (like infectious diseases) and *increased* the *proportion of civilization-related risks* (from technology, nutrition and leisure activities). Some hundred years ago, premature deaths were largely due to infectious diseases that were an accepted quirk of fate in much the same way as natural disasters[12]. In contrast, accidents or environment-related risks – if thought to be related to human activities in any way at all – received little attention. Whereas today, road traffic accidents, cancer contracted through smoking and other unhealthy habits are seen as dominant, personal risk factors of modern industrialized society.

3) Technological advancement is characterized in many sectors by a tendency to *increase the disaster potential while at the same time reducing the likelihood of occurrence.* The risk of great disasters, however unlikely they may be, is accepted in

---

[9]   Luhmann (1991), p. 31ff; see also Luhmann (1993), pp. 138–185.
[10] Bechmann (1993), pp. 237-276. see also Krücken (1997), p. 28ff.
[11] See Lowrance (1976) and Renn (1984), p. 13ff.
[12] See Hohenemser, Kates, Slovic (1983).

order to keep the likelihood of individual harm to a minimum while reaping the economic benefits in the form of economies of scale[13]. Travelling by rail instead of in a
private vehicle is both cheaper in terms of the economics of resource consumption
and safer in terms of personal accident risk[14]. Nevertheless, the number of victims in
a railway accident is far higher than those in a road traffic accident. The situation is
even more dramatic with nuclear energy and major chemicals facilities. Increased
disaster potential together with simultaneous reduction of personal risk requires collective decision-making processes (in contrast to a personal decision to accept risk)
and thus, in particular, consideration of the distributed effects of risk[15]. Risk distribution issues require more than just analysis of rational risk based on the best available knowledge; in seeking their solution, subjective risk elements and fairness
principles should also be taken into account. As both are controversial in a pluralistic society, there is greater potential for conflict in the public debate on risk[16].

4) In times of economic prosperity and product diversity, marginal utility to the individual from *material goods* has fallen relative to that from general health, a cleaner
environment and mental well-being[17]. It is all the more difficult, therefore, to justify
risks when their benefits are largely of an economic nature. While environmental
topics are no longer as popular as they were a few years ago, the vast majority of
Germans are still in favour of improved environmental protection. For example, in a
nation-wide survey conducted in 1996, the German population was asked what
importance they placed on environmental protection on a scale of 0-10. The results
were clear: With 8.3 in former West Germany and 8.5 in former East Germany,
environmental protection is still a high priority among respondents[18], even though it
has slipped from first to third place in the priority ranking over surveys conducted in
previous years. There is, however, no evidence of a long-term downward trend. In
times of economic difficulty, employment concerns take first place. Once the economy revives, environmental protection receives a higher placing[19]. Environmental
awareness, therefore, is not a temporary state of affairs: It affects each and every
layer of society. This applies to both former West and former East Germany. Environmental risks thus remain an important topic for public perception and policy.

All four factors have contributed to risk being recognized as a social problem
and to its gaining in political impact. Improved forecasting and modern society's
increased moral self-responsibility to engage in risk limitation result in people having higher expectations of social groups and of policymakers in particular as
regards active future-shaping and anticipatory responses to potential dangers from
our natural and technological environments. Protection against potential risk
together with farsighted risk management are thus key issues in and across Germany's social groupings[20].

---

[13] Perrow (1984).
[14] Akademie der Wissenschaften zu Berlin (1992), p. 249.
[15] MacLean (1987).
[16] Shubik (1991).
[17] Renn, Zwick (1997), p. 49ff. See also Klages (1984), p. 107ff.
[18] Umweltbewußtsein (1996), p. 7.
[19] Knaus, Renn (1998), p. 65ff.
[20] See Klages (1984), p. 82f.

3.3
Risk Perception Models

Perceptions have a reality of their own: Just like the characters in animated films who, suspended in mid-air, do not plunge to the ground until they realize their predicament, people construct their own reality and prioritize risk according to their subjective perceptions. This type of intuitive risk perception is based on how information on the source of a risk is communicated, the psychological mechanisms for processing uncertainty, and earlier experience of danger. This mental process results in perceived risk – a collection of notions that people form on risk sources relative to the information available to them and their basic common sense[21]. Thus this section focuses on constructed reality, i.e. the world of notions and associations which help people understand their environment and on which they base their actions.

Research on risk perception has identified a range of perception models used by society in perceiving and assessing risk. As this book deals solely with risk from anthropogenic sources, the perception models that follow are ones which characterize the importance of risk in terms of anthropogenic risk (as opposed to natural risk). By way of contrast, we have included the perception of risk as fate, in which natural disasters play an important role. Under these constraints, the following perception models can be identified:

- risk as a fatal threat
- risk as fate
- risk as a test of strength
- risk as a game of chance
- risk as an early warning indicator

How do these different perceptions of risk influence how we think about and assess risk situations and risk objects? What kind of situations and objects are involved in the various risk perception models? The semantic model of *risk as an early warning indicator* plays an important role in answering these questions because it allows the best possible study of intuitive risk acceptance in terms of environmental chemicals and small doses of radiation. The other perception models also play an important role in assessment by the general public and will be outlined in brief.

3.3.1
Semantic Risk Model

Risk as a fatal threat: In many areas of technology, particularly industrial technology, major accidents involving safety system failures can have catastrophic effects on humans and the environment. In technical safety philosophy, the main aim is to reduce the likelihood of such failure occurring to ensure that the product of probability and impact is as small as conceivably possible. But the stochastic nature of such an event makes it impossible to foresee when it will actually occur. In consequence, an event could, theoretically, occur at any time although the likelihood of its occurrence is extremely low. A look at perception of rare random events shows that

---

[21] See Renn (1989).

probability plays hardly any role at all: It is the random nature of the event that poses the actual risk. Risk sources in this category include major facilities like nuclear power plants, liquid natural gas (LNG) storage facilities, chemical production sites and other man-made sources of potential danger which could have catastrophic effects on man and the environment in the event of a serious accident.

The idea that an event could impact on the population concerned at any time fosters feelings of threat and powerlessness. Instinctively, most people are better able to cope with the idea of risk mentally (whether they can in reality is a different matter) when they are prepared for and expect it. Just as the majority of people are more afraid at night than during the day (although the objective risk of coming to harm in daytime is considerably higher than at night, it is easier to be startled by potential danger during the night), most feel more threatened by the potential for danger to happen upon them unexpectedly or when they are unprepared for it than by danger that arises either on a regular basis or where there is enough time for risk control measures to be taken. Thus, the risk impact in this perception model depends on three factors: *the random nature of the event, the expected maximum impact and the time-span for risk control measures.* Conversely, the comparative rarity of an event, i.e. the statistical expected value, is of little consequence. Rather, regularly occurring events signal more of a continued sequence of harmful events for which one can prepare oneself by a process of trial and error.

While the perception of risk as an impending disaster often governs technical risk assessment, it rarely applies in assessment of natural disasters. Earthquake, flood or tornado activity follows the same determinants as technological innovation, i.e. the occurrence of such events is rare, random and allows little time for risk control measures to be taken. They are, however, assessed using a different risk model, as described below.

Risk as fate: Natural disasters are usually seen as unavoidable events with catastrophic effects, but they are also seen as *quirks of nature* or *acts of God* (and in many cases as the mythological wrath of God for collective sinful behaviour) and thus beyond man's control[22]. In terminology used by Niklas Luhmann, they are risks that humans are simply exposed to. The possibilities for controlling natural disasters and lessening their impacts have not yet anchored themselves sufficiently in people's awareness to allow the risks from natural disasters to be assessed in the same way as those from technological accidents.

Natural burdens and risks are seen as prescribed, almost inevitable fate while technical risks are seen as the consequences of decisions and actions. Such actions are assessed and legitimized according to different standards. Fate only finds 0 in mythology or religion. If none other than God can be held accountable, no amount of human activity will improve the situation. The only alternatives are either to flee from the risk situation or to deny its existence. The rarer the event, the more likely people are to deny it or suppress it; the more frequent the event, the more likely people will flee from the danger zone. It is thus understandable, although not altogether rational, that people should settle in earthquake and flood regions and will often return to those regions in the wake of a disaster. As opposed to the circumstances of

---

[22] Watson (1987).

technical risk, the random nature of the event is not the fear-triggering factor (because randomness involves fate and not the unforeseeable consequences of inappropriate action). Rather, the relative rarity of the event provides psychological reinforcement for risk denial.

With the increasing influence man's actions can have on natural disasters, the *risk as fate* model is more a mixture of risk perception elements than to do with man-made risk. This is evident when, for example, a natural disaster occurs – the question of accountability arises and cause is found in failure to implement possible control or preventive measures[23].

Risk as a personal thrill: When, despite the considerable risk, Reinhold Messner climbs the world's highest mountains without the aid of breathing apparatus, when drivers drive far faster than the speed limit allows, when people throw themselves off a mountain or a cliff-top with nothing more than a pair of artificial wings to save them and do so in the name of sport, the meaning of risk takes on a new dimension. As is often claimed, the pursuit of such leisure activities is not about accepting risk as a ticket to the pleasurable benefits (feeling the wind in one's hair or enjoying a magnificent view); instead, the benefit lies in the risk itself: The attraction of such activities is the fact that they involve risk[24].

In all these cases, people take risks in order to test their own strength and to experience triumph over natural forces or other risk factors. To pitch oneself against nature or one's competitors and to overcome a self-contrived risk situation is the major incentive in entering into the activity. It may well be that the challenges offered by our safety-conscious society are too few, so that our – often instinct-based – desire for adventure and risk goes unsatisfied. Instead, artificial situations are created to provide a calculable and, through personal effort, surmountable risk that individuals expose themselves to voluntarily. Risk as a thrill involves a range of situation-specific attributes:

- Voluntary involvement
- Personal control of and ability to influence the respective risk
- Limited period of exposure to the risk situation
- The ability to prepare oneself for the risk activity and to practice appropriate skills
- Social recognition for overcoming the risk

Risk as a thrill is such a dominant motivator that many societies have developed symbolic risk situations in the form of sports, games, speculation, investment and the rules of power acquisition as a channel for the "kick" in overcoming danger and to replace any possible negative outcome with symbolic penalties. Symbolic channelling of the "high" in taking risks also includes symbolic anticipation of real danger by way of computer simulation and hypothetical risk assessment[25]. Conventional methods of testing or finding new uses for technological innovations by means of trial and error can no longer be morally justified in a society so fixated with protecting the individual. In place of error – which always causes harm – comes symbolic anticipation of danger: Adventure holidays aim

---

[23] See Douglas (1966) and Wiedemann (1993).
[24] Machlis, Rosa (1990).
[25] Häfele, Renn, Erdmann (1990).

only to communicate the idea of risk; spare the thought that someone might get hurt. Technological systems must be so designed that no-one comes to harm as a result of system failure (learning from actual mistakes is replaced by computer simulation of hypothetical disaster scenarios). No social reform bill is passed without scientific analysis of the consequences, complete with compensation strategies for potential victims.

Increased experience of purely symbolic risk naturally increases the demands placed on technological systems. The more the thrill of risk-taking is associated with symbolic consequences for oneself and potential competitors, the more one expects that the consequences of technical sources of risk should also be symbolic. Real risk should simply not occur.

The shock of Chernobyl and other technological disasters lies largely in the outrage that the accident did not remain a purely hypothetical game of numbers but rather had very real impacts on the environment. The mix of hypothetical risk assessment and actual harm to health played a large part in causing the general confusion that abounded following Chernobyl[26]. What had long been deemed impossible both in perceived residual risk and in computerized risk simulation in the "pretend world" of hypothetical risk assessment was suddenly reality, even though the health consequences for Western Europe could only be assessed on a hypothetical basis.

Risk as a game of chance: Risk as a thrill, in which one's own abilities to cope with the risk are instrumental in entering into the activity in the first place, is not the same as what we understand by the risk involved in playing the lottery or games of chance. These involve loss or profit that is usually independent of the player's ability. While simply playing the game can in itself create a high and become the object of the exercise, it is the anticipated or desired payout, the possibility of a big win, that produces that certain "thrill" and not the actual process of playing (in contrast to games in which reward and penalties have only symbolic value).

Psychologists have long conducted in-depth studies on risk behaviour in games of chance. Firstly, the circumstances are easily simulated in the laboratory and secondly, it is easy to identify deviations from statistical expected values[27]. However, it must be pointed out that the statistical expectations provide no standard on which to base rational gaming behaviour. The stake must be kept to an absolute minimum while the main prize must be particularly attractive. Players tend to underestimate the probability of rare events and are thus more willing to play if their stake remains below their pain threshold.

The fact that there is always a winner incites the belief that we could be next. Games of chance often involve hidden distribution ideologies (like dead-cert betting systems, lucky numbers or fair shares). Some 47 per cent of Americans believe in the existence of lucky numbers and that they increase certain players' chances of winning[28]. If the random element receives any recognition, then the perceived concept of stochastic payout distribution comes closest to the technical risk model. However, this concept is not used in perception and assessment of technical risk.

---

[26] Peters, Albrecht, Hennen, Stegelmann (1987).
[27] Dawes (1988), p. 92ff; see also Kahneman, Tversky (1979).
[28] Miller (1985), Table 8-13.

Quite the opposite: A study conducted in Sweden has shown that those questioned believe it highly immoral to apply a *game of chance mentality* to technical risk sources in cases where human health and life are at stake[29].

3.3.2
Risk as an Early Warning Indicator

In recent times, public debate has acquired yet another definition of risk. Increasing reports on environmental pollution and its long-term impacts on health, life and nature have forced scientific risk assessment to adopt a role as early warning indicators.

In this risk perception model, scientific studies help early detection of lurking danger and the discovery of causal relationships between activities or events and their latent effects. This definition of risk is used, for example, in cognitive handling of low doses of radiation, food additives, chemical crop protection products and genetic manipulation of plants and animals. Perception of such risk is closely related to the need to find causes for apparently inexplicable consequences (e.g. seal deaths, childhood cancer, or forest dieback). Unlike in the technology-medical risk model, the probability of such an event is not seen as a significant deviation (i.e. it can no longer be explained by chance) from natural variation for the event in question, but rather as the degree of certainty with which a single event can be traced to an external cause[30].

The fact that cancer can be caused by ionizing radiation at least legitimizes the suspicion that all incidents of cancer that occur in the vicinity of a nuclear power plant can be explained by the fact that the plant emits radiation. Anyone who contracts cancer or is forced to watch a family member or close friend suffer from the illness will search for an explanation. In our secularized world, metaphysically based explanation patterns have lost their importance. At the same time, the best explanation supplied by current scientific knowledge, that cancer occurs at random, does little to satisfy the need for a "meaningful" explanation. There is little consolation in knowing that one has contracted cancer by way of a random distribution mechanism. If one has an actual reason – say, environmental pollution, smoking, or bad eating habits – then the illness's occurrence at least makes some sense. And if, from a subjective standpoint, one's own actions (smoking or alcohol abuse) can be ruled out as the cause of the illness and blame can be placed on external causes it may even fulfil a social purpose: that of heightening awareness in potential victims and stirring them to fight.

The often highly emotional debate on this type of risk must be viewed from this psychological standpoint. Our propensity to empathize helps us identify with the victim. Risk assessment that shows a certain probability of lurking danger due to pollution triggers identification with people affected by it. While risk analysts characterize the relative risk of events by using stochastic theories that do not take in

---

[29] Sjøberg, Winroth (1985).

[30] The results of an empirical study on the differences between lay assessment and expert assessment of toxicological knowledge and assumptions confirms the posited theory that laypeople deem causal relationships important if a relationship is seen between individual events (like exposure and illness). Significance is based on causal thinking. See Kraus, Malmfors, Slovic (1992).

direct cause-effect relationships (thus creating distance between themselves and the object of their study), the layperson sees these theories as proof of the part played by social actors in causing life-threatening diseases.

But then again, the definition of probability is the crux of the discrepancy between intuitive and technical perceptions of risk. It is difficult to give someone a plausible explanation as to why, according to assessments conducted by the US Department of Energy, some 28,000 people in Europe will contract cancer in the next 50 years as a direct result of Chernobyl, but the individual risk of dying of cancer has risen by 0.002%[31]. For the average German, this means an increased probability from around 24% today to 24.002%. So who do these 28,000 cases involve, if each potential victim is only subject to a marginally increased risk of contracting cancer? That this example (the product of low probability and a large population) also sheds light on the limitations of interpretative ability in scientific technical risk assessment goes without saying.

3.4
Intuitive Processes of Risk Perception

The semantic investigation into everyday concepts of risk leads us to the important insight that the claims to universality that are made for the technical definition of risk cannot be applied to the relative probability of negative outcomes in common language usage. Terms used in everyday language are often given multiple meanings that can easily be derived from the context of what is being said.

At the same time, terms used in everyday language are less abstract in that they do not necessarily have to be universally applicable in different contexts. The risk involved in skiing has an entirely different meaning to that involved in operating a nuclear power plant.

Although it is scientifically and technically possible, and perhaps useful in specific circumstances, to select a risk definition that combines the elements of two different situations, it is less plausible in everyday life to abstract from the context of two situations and to bring out common features that are all but meaningless in everyday life. Thus, where an intersubjectively verifiable relationship exists between the risks under comparison, risk comparison is intuitively reasonable. For example, comparison of the carcinogenic effects of formaldehyde and benzopyrenes is perfectly acceptable in the eyes of most observers because a relationship is drawn between two chemical risks in the "lurking health risk" category. Risk comparison between skiing and living next to a nuclear power station is, in contrast, intuitively less plausible because it plays no role at all in an individual's decision on whether to go skiing or to live near a nuclear power plant. Abstraction from context is only useful where it makes communication possible or easier, or enables identification of motivators to justify or change one's behaviour[32]. Thus, in Chapter 1, risk comparison is deemed useful in risk classification.

---

[31] Hohenemser, Renn (1988).
[32] Femers, Jungermann (1991) and Covello (1990).

The policy-guiding function of risk comparison across all categories should be viewed with great scepticism. The fact that, on the one hand, we accept risk in one context, perhaps even seek it, while on the other we reject the same or an even lesser risk (within the meaning of rational risk) is not necessarily proof of irrational or inconsistent behaviour. Not only do the possible benefits vary from one situation to another, but the supporting circumstances of the respective risk can make application of different assessment standards worthwhile. Using psychometric methods, psychological research has in the last two decades tried to pinpoint the importance of such supporting circumstances in risk assessment and has brought a range of interesting facts to light.

As already mentioned on several occasions, when assessing rational risk analysts equate risk with average loss expectation per time unit. In some cases, additional factors are incorporated into the definition of rational risk, either to take in the uncertainty in assessing probability and impact or to include further risk factors[33]. Laypeople, however, perceive risk as a complex, multidimensional phenomenon in which subjective loss expectations (not to mention statistically derived loss expectations) play only a subordinate role, while the context of the risk situation, as expressed in the differing semantic meanings of risk, has a significant influence on the degree of perceived risk[34]. If we compare, say, statistically derived and intuitively perceived loss expectations, the majority of studies, surprisingly enough, show a relatively good match between expert estimates and lay perceptions where an ordinal comparison standard is used (categorizing risks according to loss expectation)[35]. This means it is not so much laypeople's ignorance of the actual impact of risk from a specific risk source that results in a discrepancy between lay and expert assessment, but rather the number and type of factors used in characterizing risk. The reference point in rational risk is the expected impact, while the reference points in subjective risk also include the supporting circumstances of the risk situation, the social impact and implications of things like perceived unfair burdens. In communicating risk, this means even when someone has been given a truthful explanation of the average loss expectation, an individual may still retain his or her intuitive assessment of risk because average loss expectation is only *one* of many deciding factors in assessing the subjectively perceived degree of risk[36].

There are no dramatic differences between perceived and statistically assessed loss expectations in the majority of risk sources; however, they do show a variety of systematic characteristics that help in explaining any discrepancies that occur. These include[37]:

- The easier and faster a risk is recognized, the more conscious we are of it and the greater the chance of its probability being overestimated. If, for example, an individual has known someone who died after being struck by lightening, that individual will perceive the risk of being struck by lightening as being particularly large (availability bias).

---

[33] Bonß (1996).
[34] Jungermann, Slovic (1993b) and Slovic (1987), pp. 280–285.
[35] Pidgeon et al. (1992).
[36] See Otway, von Winterfeldt (1982).
[37] See Tversky, Kahneman (1974) and Gould et al. (1988).

- The more a risk awakes associations with known events, the more likely its probability will be overestimated. This is why, for example, the use of the term "incinerating" in waste disposal facilities readily evokes an association with harmful chemicals, especially dioxins and furans, even if there is no way that they could be released into the environment by the facilities concerned (anchoring effect).
- The more constant and similar the losses from risk sources, the more likely the impact of average losses will be underestimated. While road traffic accidents are not deemed acceptable, they are more or less passively accepted. If the average annual number of road deaths in Germany were to occur at one time instead of being spread out over the year, then a considerably greater level of rejection could be expected. Thus, people are not indifferent as regards the distribution of risks over time: They prefer even loss distribution over individual disasters[38] (risk aversion).
- The greater the uncertainty of loss expectation, the more likely the average loss assessment will be in the region of the median of all known loss expectations. In this way, loss expectations in objectively low risks are often overestimated while objectively high risks are often underestimated[39] (assessment bias).

Overestimation or underestimation of loss expectations is not, however, a significant aspect of risk perception. The context-dependent nature of risk assessment is the deciding factor. This dependence on the supporting circumstances is not random, but rather follows certain principles that can be identified by systematic psychological investigation.

Research has provided a lengthy list of supporting circumstances or *qualitative factors*[40]. Factor analysis usually reduces these lists to a few important compound factors. Studies conducted in Austria, Germany, Great Britain, the Netherlands and the US[41] have identified the following factors as particularly relevant:

- Familiarity with the risk source
- Voluntary acceptance of the risk
- Ability to personally control the degree of risk
- Whether the risk source is capable of causing a disaster (Catastrophic potential)
- Certainty of fatal impact should the risk occur (dread)
- Undesired impact on future generations
- Sensory perception of danger
- Impression of fair distribution of benefit and risk
- Impression of reversibility of the risk impact
- Congruence between benefactors and risk bearers
- Trust in state-operated risk control and risk management
- Experience (collective and individual) with technology and nature
- Reliability of information sources
- Clarity of information on risk

---

[38] This conclusion is an integral component of prospect theory, in which the assessment of risks and gains is dependent on the respective distribution of probability and the amount of gain or loss. See Kahneman, Tversky (1979).
[39] See Renn (1984), p. 151.
[40] Slovic, Fischhoff, Lichtenstein (1981).
[41] See the anthology by Renn, Rohrmann or the summary in Renn (1989).

The importance of these qualitative factors in risk assessment offers a plausible explanation for the fact that precisely those risk sources that technical risk assessment classifies as particularly low-risk are the source of greatest concern among the general public. Risk sources that are deemed controversial, like nuclear power, are very often burdened with negative attributes while leisure activities are associated with more positive ones[42].

Psychological research on risk perception brings us a step closer towards an analysis of how society really assesses risk. The observed discrepancy between the results of technical risk assessment conducted by experts and intuitive assessment of the same risk by society is not, in the first instance, due to uncertainty about statistically derived expectations or an expression of unverifiable thought processes, but rather an indication of a multidimensional assessment matrix in which anticipated harm is only one of many factors.

Studies conducted on an international scale also show that people everywhere, regardless of their social or cultural background, use practically universal risk assessment criteria in forming their opinions[43]. However, the relative effectiveness of these criteria in opinion-forming and risk tolerance varies considerably between different social groups and cultures. While the above-mentioned qualitative characteristics are accepted (often subconsciously) as assessment values for perceived risk, their relative contribution to a person's actual opinion or motivation to take action depends on individual lifestyles, threatening environmental factors and ingrained cultural values[44]. In assessing risk, people who favour alternative lifestyles tend more than others to consider both "reversibility of the consequences of risk" and "congruence between risk bearers and benefactors", while those with strong material values assess risk more by way of personal control opportunities and trust in institutional risk control[45]. As Otway and Thomas impressively showed in their studies on attitudes to nuclear energy, there is a high correlation between different value models and the relative importance that people attach to either the benefits or the risks of a specific technology[46].

The conclusion to be drawn from this is that value expectations and cultural background are significant determinants of subjective risk that do not add to the semantic and qualitative factors already described but, in effect, presuppose those factors in that they use them as channels to communicate the resulting assessment. Internalized value expectations and external circumstances can control the relative effectiveness of intuitive perception processes, but not their existence. This is not a matter of academic hairsplitting: it has direct relevance to communication and conflict management. If we assume that intuitive mechanisms of risk perception and assessment bear practically universal characteristics that can more or less be reshaped by sociocultural influences, then they can provide a fundamental basis for communication of which one can avail oneself regardless of differences between

---

[42] Jungermann, Slovic (1993)c.
[43] This theory is proven in a new anthology by Renn, Rohrmann in a range of international comparative studies.
[44] See Sjöberg (1997).
[45] Buss, Craik (1983). See also the results interpreted from the culture theory of risk: Wildavsky, Dake (1990).
[46] Otway, Thomas (1982).

the various standpoints. In addition to the pool of common symbols and rituals (shared meaning), whose importance to social integration is in constant decline in pluralistic societies, a new pool of common mechanisms of risk perception emerges that along with common sense signals the existence of supra-individual perception models.

3.5
The Media as a Source of Information

The media are the key communicators of information in society today. They are neither entirely neutral reporting bodies nor are they wholly motivated by ideological convictions, and their selection criteria influence both the choice and relevance of topics that are reported to general public (that is, they act as "gatekeepers"). By selecting and reinforcing the event in question, the media largely determine the priorities of the political agenda and communicate, to those who are not directly involved, information that is subjectively coloured, even when journalists themselves feel obliged to report objectively[47].

Contrary to frequent accusations that journalists are either friends of critical leftist ideology (one school of thought), or defenders of the status quo who are blinded, or even bribed, by the establishment (the other), most empirical studies on this subject show that selection and processing of news is based far more on professional standards of journalism than by the ideological convictions of individual journalists. To a large extent, these standards apply to all media, while some are specific to only certain media[48].

These standards include a requirement that as a rule the media should report on current affairs and largely ignore ongoing developments. This is why, for example, dioxin emissions from incinerator facilities have received wide coverage, while the ongoing technological enhancement of such facilities to reduce dioxins down to almost undetectable levels gets hardly any coverage at all. The mere fact that the media report particularly frequently on a given subject often leads their audiences to assume that the subject is particularly controversial and deserving of special attention[49].

Equally important are the selection criteria of conflict and blame apportionment. As American research shows, the intensity of reporting on catastrophes is governed less by their physical impact (the number of deaths or loss of property) than by the degree of conflict about the type of risk management required and the opportunity for proportional apportioning blame[50]. Naturally, spatial proximity to the source of the emission plays an important role – the media reflect social events, not physical impacts. If the parties to a conflict situation cannot agree on the necessary form of risk management, or if they pass the blame for negative outcomes among themselves, then such social events constitute an important trigger and amplifier for

---

[47] Peltu (1985), and Sood, Stockdale, Rogers (1987).
[48] Peltu (1989); Mazur (1987); Schanne (1996).
[49] Mazur (1984).
[50] Rubin (1987); Adams (1986).

media reporting[51]. When it comes to news value, issues like whether the conflict is real or whether its cause poses a minimal risk in "objective" terms play hardly any role at all. In essence, the media react to a social construct of reality and not to actual reality or its natural science-based explanation[52].

Applied to the debate on technology and its uses, this selection mechanism reinforces the controversial nature of the debate and has a moralizing effect on the parties involved. In the search for conflicts to report on, journalists are well served by issues of modern technology. However, they can neither assess the validity of the various standpoints by way of scientific proof, nor measure the representativeness of deviating opinions[53]. Thus, the general public is given the impression that all statements made about certain technologies are controversial.

There are, however, other selection criteria that work in the opposite way: the search for unusual events to report on directs media attention towards spectacular activities like those carried out by anarchists and militants, whose numbers are rapidly dwindling within the realm of technology activists. The effect of placing their actions at the centre of a report on a demonstration results in many media consumers being appalled by the group's actions, distancing themselves from them and identifying the movement as a whole with the demonstration[54]. At the same time, empirical research conducted in Germany has shown that while newspapers and magazines often use journalistically well-presented press releases put out by industry and public agencies without verifying their content, they do not use those of environmental associations or if they do, they cover their backs by checking with the appropriate authorities or industry representatives prior to publication[55]. Selection criteria are unwritten laws which journalists take on board in the course of their journalistic training and which apply to all ideological situations. Of course, political loyalties play an important role, especially in Germany[56]. Such loyalties are, however, characterized more by the prescribed leaning of a particular newspaper or an electronic medium than by the values and political preferences of individual journalists[57].

Up to now, no research has been conducted on media reporting of combined effects. It is to be expected, however, that the same selection criteria are applied as to other environmental topics. Events receive a lot of attention especially when they can be presented in a way to stir public curiosity. At the same time, the media strive to obtain counter-arguments for every expert opinion and to publish them in parallel regardless of how representative they may be of the relevant scientific community. Thus there is a forced emphasis on plurality and on disagreements among experts.

---

[51] Renn (1991a).
[52] Wilkins, Patterson (1987).
[53] Sharlin (1987).
[54] Guggenberger (1987).
[55] Peters (1990).
[56] Köcher (1986).
[57] Schanne (1996).

3.6
Perception of Combined Effects

In what way do psychological and social response patterns change when combined risk is involved? Until recently there was hardly any empirical evidence available, only assumptions at best. In an attempt to make good this deficit, the Center of Technology Assessment in Baden-Württemberg (Akademie für Technikfolgen-abschätzung) conducted a representative survey on this subject among the population of the state of Baden-Württemberg in 1998.

The survey questioned a representative cross-section of some 1,500 adults in Baden-Württemberg during the summer of 1998. Its main focus was society's perception and assessment of technology[58]. Within the framework of the survey, a study was also conducted of the significance of environmental risks in general and of combined environmental risks in particular. Initially, respondents were asked to assess the importance of environmental factors in health risks. According to prevailing expert opinion, more than two-thirds of all cancers can be apportioned to smoking, alcohol consumption or poor eating habits, while only a fraction of cancers can be apportioned to the impact of noxious substances in the environment[59].

How does the general public view this? Those participating in the survey were confronted with the following statement: "The health risks from environmental pollution are considerably greater than the risks from personal lifestyles like obesity, smoking, alcohol consumption or leisure activities." Some 10.7% strongly agreed and a further 15.3% tended to agree. Practically every second respondent was undecided (48.1%) and about a quarter (25.9%) thought the risks from individual lifestyles were greater than the risks from environmental pollution. While these results clearly deviate from expert assessments, they give no cause to speak of "environmental hysteria". The answers were similar across all groups; there is, for example, no difference as to whether a respondent preferred green policies or had more conservative values. The only exception was the level of education: the higher the level of education, the greater the link between health risks and personal lifestyles. Only 20% of respondents with high school diplomas but almost 50% of university trained persons expressed the opinion that environmental risks are less of a threat to human health than the risks involved in personal lifestyles[60].

A significantly bigger gap is evident between the opinions of the majority of experts and the laypeople surveyed regarding their assessment of combined environmental risks. The respondents were again confronted with a statement: "Just as combined consumption of tablets and alcohol can cause serious health problems, relatively harmless substances in the environment can also cause serious damage to one's health when they interact." Although the super-additive relationship postulated in the statement cannot be ruled out from a toxicological standpoint, the majority of studies cited in this book show that super-additive effects are more the exception than the rule. Conversely, two-thirds of respondents were convinced that such super-additive effects could be expected from interactions between multiple

---

[58]   Zwick, Renn (1998).
[59]   Henschler (1993).
[60]   Ibid., p. 44.

pollutants. Just under a quarter (23.8%) were indifferent and only 9.6% rejected the statement. There were hardly any differences between the various social groups. Age, gender, party preference, even formal education, were all variables that had no significant impact on how the respondents reacted to the statement.

On the one hand, this clear response can stem from everyday experience with medium or high doses of drugs and stimulants. On the other hand, typical aversion to the risk of lurking danger, especially intolerance of continuous uncertainty, plays a significant role. We can interpret the response to the following statement in a similar way: "Even if we don't know for sure what effect environmental pollution can have on human health, we should for economic reasons allow a certain amount of pollutants to be released into the environment." Only 8.9% of respondents agreed with this statement, a third were undecided and a large majority of 58.7% rejected the statement altogether. There was no evidence of difference according to gender, age, class or income; in response to this particular statement, only supporters of the German Green Party (Die Grünen) appeared less willing to accept risk than did the average member of the population.

Given the results of this survey, it is apparent that the vast majority of the population anticipate super-additive effects from exposures in the low-dose range. Uncertainty in assessing the risk from a combination of noxious substances in the environment is seen as an additional threat. Such expectations show that, to a large extent, the majority of people place their trust in reports on health risks from a combination of environmental pollutants, and that scientists and regulatory authorities will have a hard time convincing a sceptical public otherwise. Approaches to solving these problems by way of state-operated risk control will be revisited in the final section of this chapter.

Apart from this survey, we know of no sociological or psychological study that attempts to explain in empirical form how synergetic risk is perceived. However, there are a variety of related studies that throw additional light on the subject. The empirical works of Kraus, Malmfors and Slovic[61] on intuitive assessment of toxicological risk and the works of Jungermann, Schütz and Thüring on the question of risk perception involving the data sheets enclosed with drugs[62] all point, for example, to the fact that respondents closely scrutinize the information on a range of trigger substances and on the range of possible side-effects from individual substances. What cannot be derived from this survey is whether the increased concern is greater than the sum of possible concerns involved with each individual substance.

The results of the few available studies combined with the semantic risk models described above give rise to two interpretations. Firstly, the information on large, unimaginable quantities of chemical substances in the environment leads us to conclude that not all effects are known and that the possibilities for combination have not been fully tested. There is thus an element of uncertainty in individual perceptions, and this cannot be erased by any amount of care or knowledge about individual noxious substances in the environment. If we look at the studies on qualitative risk characteristics, perceived uncertainty as to possible impacts plays a significant part in engendering a more negative assessment of the risks, as was evident in the

---

[61] Kraus, Malmfors, Slovic (1992).
[62] Jungermann, Schütz, Thüring (1988).

results of the survey. Most importantly, perceived uncertainty leads to a desire for more stringent regulation[63]. Thus it is hardly surprising that the uncertainty surrounding combined effects tends to promote overestimation of the related risks.

Secondly, classification of combined effects in the *lurking danger* risk category gives rise to the hypothesis that causal evidence, which plays an important role in intuitive risk assessment, is more readily brought to mind when assessing combined effects than when assessing individual substances. We all see ourselves affected by omnipresent chemical substances in some way or other while at the same time we can recall at least one example of an illness that is (apparently) caused by environmental impacts but are unable to name the actual cause-effect chain. Combined effects thus appear to support easily recalled and practically irrefutable everyday hypotheses on which we base health risks from unspecific environmental impacts.

These two heurisms – rapid recall of causal experience and concern based on great uncertainty – are reinforced by perceptions of disagreement among experts[64]. As it is not possible to identify all combined effects and many theoretical possibilities for extrapolation of values are limited to the low-dose range due to a lack of significant cause-effect relationships, experts must also rely on plausible models, theories and assumptions[65]. In reality, the need for interpretation, especially where combined exposures are concerned, has contributed to a pluralization of expert opinions and assessments within the science system. There are consequences for the treatment of risk assessment both inside and outside the scientific community. Firstly, the case of scientific treatment, often referred to as an expert's dilemma of the first degree: experts conduct risk assessments in a variety of ways. Four categories can be identified in the plural sphere of expert opinion:

- Expert assessments that focus on the experimental results of risk studies and conventions drawn up by experts, and which deal with remaining uncertainties by simply ignoring rather than assessing them.
- Expert assessments that focus on the empirical results of risk studies but which hover on the border of the range of conventions drawn up by experts and, in doing so, interpret uncertainties within the meaning of those agreed conventions.
- Expert assessments that focus on the empirical results of risk studies, but which reject conventions agreed to by experts or replace them with their own interpretation models.
- Expert assessments that question both conventions and empirical results, and see their purpose in fundamental critique of the methods and interpretations of the relevant scientific community.

Without a doubt, the vast majority of experts are to be found in the first two categories, so that the conflicts that arise in reality are less strongly manifest than one would expect from all four. At the same time, the various scientific communities use methods such as consensus conferences, meta-analyses or Delphi surveys in

---

[63] Sjöberg (1994).
[64] Nennen, Garbe (1996).
[65] Peters (1996), p. 63f.

their attempts to resolve conflicts between the four types of expert opinions[66]. The multi-layered nature of risk analysis for combined pollutants makes it difficult to find a clear-cut solution for conflicting expert assessments. In addition, we have a situation where statements on risk from combined effects are difficult or impossible to falsify. The lack of opportunity to falsify statements using empirical evidence (at least in the short-term) limits the effectiveness of knowledge as a tool to evaluate risk assessment studies. Different bodies of knowledge compete with each other and the competing demands for truth cannot be met to the exclusion of all possible doubt.

Secondly, we need to consider the non-scientific consequences: the existence of discretionary freedom in the assessment of combined effects, the need to implement conventions that cannot be justified by science alone, the constant uncertainty in effects analyses and the fact that numerous scientific controversies are debated in public all have a sustained effect on people's perceptions and experience. In many ways, this involves the importance that outsiders ascribe to conflicts between experts: firstly, most people believe that, in principle, science can come up with clear and precise definitions of environmental pollution. Confronted with a large number of conflicting assessments by experts, people are thus forced to conclude that in the course of such conflict at least one or other of the parties involved is not revealing the truth, be it intentionally or unintentionally. It is not without reason that in the public eye, what experts put out is seen as a reflection of what their financial backers, their ideological preconceptions or their blinkered specialized world put in[67].

As most people are unable either to confirm or to reject the statements put out by experts by applying their own knowledge or experience, they must depend on external criteria if they want to assess the trustworthiness of those experts. Such external criteria range from assumed vested interest in the subject (an area in which industrial experts have particular difficulty) to the perceived likeableness or expertise of those involved (how they handle themselves on TV, how they dress, their debating style)[68]. Laypeople often follow an intuitive *better safe than sorry* principle. The assumption that the most pessimistic expert embodies the epitomy of trustworthiness is reinforced by the fact that in the past environmental pollution has often been reported as less harmful than it actually turned out to be. Conversely, some experts often feel pressured by public expectations to place greater emphasis on negative outcomes in order to improve their public standing. That risk assessment conducted on the basis of such external trustworthiness criteria is inappropriate and politically unsatisfactory needs no further discussion.

---

[66] See the anthology by Nennen, Garbe (1996).
[67] Covello (1992).
[68] Peters (1991).

3.7
Psychosomatic Links

There is another aspect to all this: Potential impacts on health are caused not only by pollutants, but can also be the result of psychological and psychosomatic processes[69]. Psychosomatic reactions that in the relevant literature have come to be known as *multiple chemical sensitivity syndrome* are of particular importance in this context[70]. This syndrome encompasses most symptoms of illness that can be ascribed to environmental influences. The symptoms can be measured empirically and cause somatic harm to those affected. It is unclear, however, as to what extent they can really be ascribed to chemical reactions triggered by exposures or whether they are psychosomatic reactions that are triggered by exposure perception. A series of "pseudo-exposures" in which people were not objectively exposed to the respective substances, but mistakenly assumed that they were, point to psychosomatic triggers of illness symptoms[71]. These illnesses are real and must be seen as such. It is not a matter of taking such illnesses less seriously than pathologically or physiologically induced ones. There are, however, a number of political consequences involved. In the case of toxic impacts on the environment it is exposure that must be regulated, while in the case of psychosomatic illness, focus must be placed on the associated emotional and psychological issues.

Psychosomatic response chains introduce new ambiguities into environmental policy. Are symptoms a consequence of actual or perceived exposure? Public perception of symptoms helps popularize a cause-effect relationship: a vicious circle originates . Perceived exposures, especially to combined pollutants, spark fear and unrest in those affected, which in turn leads to psychosomatic reactions in individual cases. These reactions are observed and are then seen as proof of an assumed relationship between emissions and health impairment[72]. How then, under such circumstances, can the general public be convinced that any damage to health that occurs following exposure could perhaps be the result of psychological reactions and not be attributed to physical processes? Similarly, the question of liability is often raised when psychosomatic symptoms occur. Is there any justification for someone who presents verifiable symptoms of an illness that could in principle be caused by noxious substances in the environment to submit a compensation claim against the emitting party, even if – allowing for differing individual susceptibilities – prevailing scientific opinion does not deem the level of exposure significant? The issue of liability where psychosomatic symptoms are involved is touched upon in Chapter 5 of this study. In sum, existing practice sees that someone only has a right to compensation if at a minimum they can provide plausible scientific proof that their exposure had a significant effect. Therefore, although psychological processes can be included as an amplifier there may be no serious doubt as to the actual causal relationship.

---

[69] Schultz-Venrath (1998).
[70] See the anthology by Aurand, Hazard, Tretter (1993).
[71] Küchenhoff (1994), p. 0.3-20 and I-20.
[72] Renn (1997).

This theory makes plausible the assumption that perception of risk from a combination of noxious substances in the environment triggers a greater degree of subjectively perceived threat than would the simple addition of the perceived threats of each individual substance. Firstly, people react to heightened uncertainty with heightened caution and an increased desire to avoid the issue at hand. Secondly, combined effects offer a perfect playground for popular hypotheses on the health risks caused by environmental pollution without running the danger of expert refutation. This process is further supported by psychosomatic reactions to such information.

Additionally, combined effects are excellent *subjects for political mobilization*. Firstly, they provide news content in that health and the environment are among the most popular subjects in the media[73]. Section 3.3 dealt with selective amplification by the media. The psychological and social volatility of risk perception from combined effects is increased by the debate process as a result of the selection criteria used by the media. These social amplification processes have become important components of social research. Studies on *social amplification of risk* take both a theoretical and empirical look at those elements of risk that are known to either amplify or attenuate risk[74]. Social amplification of these risks increases pressure on policymakers to take preventive action and implement regulatory legislation. Interest groups are mobilized to lobby policymakers, who in turn see themselves left with no option but to react. But such reactions can have the wrong effect: they lead to high follow-up costs, detract from the problem at hand and, in the long-term, can destroy people's trust in those at the policymaking helm.

Perception of actual illness, loss of trust in experts and discrepancies in risk models between risk experts and laypeople all given rise to public pressure being exercised on policymakers to implement more stringent regulation. But of course, such pressure leads to counter-pressure from groups who would be adversely affected by more stringent regulation. Such debate has a polarizing effect on society. Given that there are no objective standards in assessment or because they are not properly recognized, the subject becomes moralized and often trivialized. Either the good or the bad is brought to the fore depending on the respective standpoint. Harmonizing policy lacks instruments to aid negotiation between the various fronts. The result is political paralysis: there is no flexibility. The peculiarities of health risks from combined environmental impacts provide the perfect breeding ground for political paralysis. Development of strategies to combat such paralysis is thus imperative.

---

[73] Kepplinger (1993). See also Peltu (1985).
[74] Kasperson et al. (1988) and Renn (1991).

3.8
Implications for Risk Policy

Any attempt to combat political paralysis by rejecting risk perception as irrational and relying solely on expert assessment would be misguided. There is little sense in rejecting intuitive perception and assessment of risk per se just because it differs from the concept of rational risk usually applied and used in the fields of science and technology. For example, the dimensions of risk outlined in the qualitative risk characteristics are also important if risk policy is to be appropriate both to the purpose (rational risk) and to the values of those involved (subjective risk). In the following, we will refer to policy that is appropriate in both these senses as *pragmatic policy*. The term pragmatic risk policy describes a process of risk decision-making in which decision options are systematically weighed according to knowledge of the consequences and the social desirability of those consequences. Any risk policy that is founded on pragmatism would be well advised to differentiate between risks that members of society are prepared to accept and risks that involve passive third-party exposure, as highlighted by one of the qualitative characteristics. Likewise, the socio-economic costs involved in a catastrophic release of energy or material may well be higher than the costs of continually diminishing harm, even in cases where the statistical expectations may be identical. Finally, both prevailing knowledge and the scope of remaining uncertainty should be incorporated as tools for risk assessment.

Particular political weight is attached to the *distributive effects* of risk. In his book *Risikogesellschaft* (Risk Society), sociologist Ulrich Beck outlines a paradigm shift from the distribution of social wealth to the distribution of social risk[75]. This includes new distribution conflicts between the benefactors and the bearers of risk. Regional disparities, social differences and risk hypotheses for future generations are a source of conflict for future social debate in terms of the distribution of social resources and the extent to which risk caused by a third party is acceptable. According to cultural anthropologist Steve Rayner, the policy issue at hand is not "how safe is safe enough?", but rather "how fair is safe enough?"[76]. The science and technology-based risk model loses normative force in policymaking by switching the focus of the risk debate from the question of what is deemed safe or unsafe to the question of what is an acceptable way of determining the desired safety and utility levels[77]. Distribution debates are characterized less by the amount of utility or risk to be distributed than by the perception of fair or unfair distribution of that benefit among the benefactors and risk bearers. Because intuitive risk assessment also uses distribution aspects in its opinion-forming, the currently observed political emphasis on distributive effects of risk reinforces these aspects in general assessment of technical risk. Risk sources for which an imbalanced risk-benefit distribution is assumed are thus twice as difficult in terms of winning public tolerance.

What benefits can scientists and policymakers gain from the study of risk perception? What guidance can be derived from studies on intuitive risk perception for

---

[75] Beck (1986).
[76] Rayner, Cantor (1987).
[77] Krücken (1990).

risk and technology policymaking? Even if there are no targets to be derived from the current situation, analysis of risk perception still provides a number of lessons for policymakers, especially if one accepts pragmatic technology and risk policy as normative objectives.

- Science-based risk assessment is a beneficial and necessary instrument of pragmatic technology and risk policy. It is the only means by which relative risks can be compared and options with the lowest statistical expectations selected. However, it cannot and should not be used as a general guide for public action. The price for its universality is abstraction from context and the overshadowing of other rational and meaningful perception characteristics. Without taking context and situation-specific supporting circumstances into account, decisions will not, in a given situation, meet the requirement of achieving collective objectives in a rational, purposeful and value-optimising manner.
- Context and supporting circumstances are significant characteristics of risk perception. These perception patterns are not just individual perceptions cobbled together: they stem from cultural evolution, are tried and trusted concepts in everyday life and, in many cases, control our actions in much the same way as a universal reaction to the perception of danger. Their universal nature across all cultures allows collective focus on risk and provides a basis for communication[78]. While the effectiveness of these intuitive perception processes depends on ingrained values and external circumstances, they remain ever-present and measurable despite cultural reshaping[79]. Intuitive mechanisms of risk perception and assessment have practically universal characteristics that can be shaped by sociocultural influences. However, the fact that a perception model has its roots in evolution is not sufficient to lend it normative force. Such practices are at best reference points in the search for explanations that must follow the rules of normative debate, independently of the factual existence of such practices.
- From a rational standpoint, it would appear useful to systematically identify the various dimensions of intuitive risk perception and to measure those dimensions against prevailing, empirically derived characteristics. In principle, the extent to which different technical options distribute risk across the various groups of society, the extent to which institutional control options exist and to what extent risk can be accepted by way of voluntary agreement can all be measured using appropriate research tools. Risk perception supplies lessons in the need to incorporate these factors into policymaking. This is based on the view that the dimensions (concerns) of intuitive risk perception are legitimate elements of rational policy, but assessment of the various risk sources must follow rational, scientific procedures in every dimension. This applies in particular to combined effects: It makes little sense to accept the synergetic effects of drugs and stimulants as heurisms for the combined effects of low doses in the environment. When it comes to interactions between noxious substances in the environment, it is reasonable to demand

---

[78] Rohrmann (1995).
[79] Brehmer (1987).

that risk policy should measure the possible additive and super-additive effects and should regulate accordingly.

- Risk perception is no substitute for rational policy. Just as technical risk assessment should not be made the sole basis for decision-making, factual assessment of risk should not be made the political measure of its acceptability. If we know that certain risks, like passive smoking, can lead to serious illness, then policies to reduce risk are appropriate even if there is a lack of awareness of the problem among the general public. Many risks are ignored because no-one wants to deal with them. This applies especially to risks that are triggered by natural forces. To allow oneself to be guided by ignorance or obviously false perceptions hardly meets the prescription of pragmatic risk and technology policy. Knowledge of these perception patterns can, however, be used to structure and implement informational and educational measures in a beneficial manner. The inability of many people to understand probabilistic statements or to recognize the long-term risk from familiar risk sources is surely one problem area in which targeted education and information can be of benefit[80]. What is really needed is mutual enhancement between technical risk assessment and intuitive risk perception.

- Little is known about how society perceives combined effects. Our first empirical study indicates that the majority of respondents perceive the combined effects of noxious substances in the environment as being super-additive. The explanations and interpretations contained in this chapter assume that exposure to combined noxious substances triggers a particularly high degree of fear and that scientists' ability to allay that fear is limited. Such fear sends out an important signal for the need to protect ourselves against nasty "surprises". It also acts as a vent for expression of the anxiety caused by the automation and chemicalization of everyday life. But while such anxiety can be alleviated to an extent, no amount of information or education can dispel it altogether. Once relationships have been recognized, it is up to policymakers to combine scientific expertise on the possible outcomes and remaining uncertainties with the assessments and desires for change of those affected by the risks in question, and to integrate them into an holistic policy that is both knowledge and values based. Risk policy should neither be purely science-based nor purely values-based.

- Option trade-offs require policy weighting between the various target dimensions. Such trade-offs are dependent both on context and on the choice of dimension. Perception research offers important pointers concerning the selection of dimensions for focus. The fairness factor plays a significant role in such trade-offs and in weighting the various dimensions. In their roles as scientists, experts have no authority to take such things into account. This is where risk comparison reaches its limits. Even if we remain within the semantic context accepted by most people – a pool of comparable risks – multidimensionality in the intuitive risk model and variable targets in risk management prevent risk policy from focusing one-sidedly on minimization of expected impacts. A

---

[80]   Renn, Levine (1988).

breach of the minimization requirement also means acceptance of greater damage than is absolutely necessary (although this can be justified in individual cases depending on the risk situation).

But who has the authority to make these decisions and how can the decision-making process be justified? There are no general and intersubjectively binding answers. Greater involvement of those affected, greater transparency in decision-making, rational and non-hierarchical discourse, two-way risk communication – these are all potential solutions that appear in public debate on the issue.

3.9
Conclusions

Actual acceptance, as the analysis so far has shown, relies on numerous factors of which many can hardly be described as normative principles of political action. Perceptions rely partly on erroneous judgements and simple lack of knowledge, opinions on risk are often tied to symbolic attributes that are only indirectly related to the advantages and disadvantages of a specific risk source, preferences among the population are often inconsistent and, finally, the question of how to aggregate individual preferences "for the good of all" remains unsolved[81]. Should the majority decide, even if only a minority is affected? Who has the right to make collectively binding decisions? The simple solution – leaving the conflict surrounding risk to the powers that be – may well increase the acceptability of political decisions, but hardly their suitability.

Acceptability cannot be entirely removed from acceptance. In a democratic society, it is the people who determine the circumstances under which they wish to shape their future. The political task of health policy will lie in explaining the anticipated advantages and disadvantages to those affected, i.e. the risks and opportunities of the available options, and, on this basis, communicating to them the possibility of rational judgement.

The further development of pragmatically oriented risk policy will depend on how much more is learned about the causes and effects of risk perception. Available knowledge on the intuitive processes of risk perception, including the perception of combined risks, can help decision-makers and risk regulators to better anticipate conflict regarding the tolerability of risk sources and to take preventive action. Identifying the elements that lend themselves to generalization in intuitive risk perception will help society establish improved normative theories for risk source selection. Conflict resolution and risk communication programmes are likely to be rejected by the general public as long as the teaching and communicating processes are not conducted in parallel. While public perception and common sense cannot replace science and policy, they can certainly provide impetus for the decision-making process. At the same time, if decision-makers take into account the factors and needs of public perception, then public willingness to accept rational models for

---

[81] Meyer-Abich (1989), p. 39ff.

decision-making is likely to increase. Risk policy that is both rational and based on acceptability would thus appear to be founded in three principle needs of society[82]:

1. Society must renounce the postmodern belief that knowledge of any type is a social construct and that there are no universal standards of quality or universal criteria for standards of truth. In reality, people suffer and die as a result of false knowledge. Because in environmental and risk decision-making, knowledge of the consequences embraces a whole spectrum of legitimate claims on the truth, the boundaries of methodologically verifiable knowledge must be identified as clearly as possible. This is especially the case with combined risk. In contrast to assessment of individual risk, there is a higher degree of uncertainty and the scope of possible interpretations is wide open. For this reason, existing knowledge must be integrated as tightly as possible and blatantly unfounded hypotheses about human health impacts rejected – because when the scope of knowledge threatens to become boundless, pseudo-scientific explanations will be found for even the most absurd of fears about risk. The social sciences today can draw on methods and techniques like meta-analysis and Delphi surveying that, in principle, offer a more or less valid way of determining the bounds of legitimate knowledge (including verifiable dissent) without the need for a referee with overall jurisdiction[83]. Determination of the scope of methodologically verifiable knowledge should be performed by the science system itself because that is the only area with the methodological expertise and critical ability both to resolve and answer competing needs for truth. Only when it is able to identify the scope of verifiable knowledge and communicate this to the outside world can the institution of science make up the credibility it has lost over the past two decades.

2. Both expert risk assessment and lay perception are indispensable elements of risk policy that should not be played off against each other. Evaluation of discourse on the acceptability of risk clearly shows that lay participants do not try to insist that their preconceived ideas of the anticipated degree of risk be used as a measure for collective risk policy. On the contrary, the first questions are always: What is the degree of risk for myself and for others? What can the experts tell us? Once these questions are answered, the remaining policy problem is that of determining how the residual risks should be handled. This question, and especially that of prioritization, cannot be determined on the basis of expert opinions alone. Instead, expert opinions must be incorporated into risk assessment together with the criteria of fairness, institutional expertise in risk management, trustworthiness of the participating institutions and individuals, and the precautionary principle (where there is a high degree of remaining uncertainty). Credibility in risk policy presupposes that expertise is regarded as an important influencing factor, but not the sole deciding factor. Thus, pragmatic risk policy requires that for each criterion the public deems relevant, the available knowledge on the consequences and the ethical norms to provide orientation should be collected and used to make informed and unbiased decisions. These decisions

---

[82] The three points that follow have been included in an article in the German weekly paper "Die Zeit". See Renn (1996).
[83] Webler et al. (1991).

must then be communicated in an unbiased and consistent form to the members of the public who are affected by them.

3. In the end, decisions on the degree of acceptability of risk are based on a subjective trade-off between knowledge of the consequences and normative guidance. The more uncertain the interpretation of the data and the more the prevailing knowledge involves assumptions, the more important it becomes to find suitable methods for rational weighting. The less that is known about the consequences of risk and the more politicized the groups affected by the risk, the more suitable discursive forms of decision-making appear. Discursive debate that incorporates the best available knowledge of the consequences and the preferences of the people affected enables competent and fair decision-making. Discourse devoid of systematic knowledge principles has no content; discourse that overshadows the moral quality of the various options for action is reduced to mere expertocracy. The requirements of discourse are transparency of the outcome, a clear mandate and mandatory verification of knowledge elements and ethical standards. Thus, where expertise is taken into account, when the boundaries of economic efficiency are acknowledged and the legal room for manoeuvre is not overstepped, people's perceptions can take on an important action-driving part in how subjects are dealt with and decisions made.

In society today, the acceptability of risk arouses heated debate. To achieve consensus, discursive forms of joint shaping of risk are necessary and appropriate[84]. This applies especially to the question of regulating combined risks, as this is the area where uncertainties that cannot be resolved by science have the highest mobilization potential among the general public. To manage such discourse in a proactive manner, two orientation processes are helpful: firstly, productive fear of the unknown and associated recognition of the limits of organizational possibilities, and secondly, the guiding force of positive visions for the future and the potential to mobilize the necessary technological and organizational resources. Only by nailing discourse to both of these fence-posts can the requirements be met to run the delicate balancing act between "letting it happen" and "making it happen", and this includes decisions on risk. While discourse of this type cannot not do away with conflict altogether, because individual and collective rationality are not necessarily congruent, it can at least be reduced and, with good will, be turned into constructive strategies for conflict resolution. The goal should not be a conflict-free society, but rather rational debate of legitimate conflicts. This should also be possible in regulating combined effects from noxious substances that are harmful to the environment and potentially harmful to human health.

---

[84] See also Zilleßen (1993).

3.10
Literature

Akademie der Wissenschaften zu Berlin (1992), Umweltstandards, De Gruyter, Berlin

Adams WC (1986, Spring) "Whose Lives Count?: TV Coverage of Natural Disasters", Communication, 36, 2, 113–122

Aurand K, Hazard B, Tretter F (1993) Umweltbelastungen und Ängste. Westdeutscher Verlag, Opladen

Bechmann G (1993) Risiko als Schlüsselkategorie der Gesellschaftstheorie. In: G. Bechmann (ed.) Risiko und Gesellschaft. Grundlagen und Ergebnisse interdisziplinärer Risikoforschung. Westdeutscher Verlag, Opladen, pp. 237–276

Beck U (1986) Die Risikogesellschaft. Auf dem Weg in eine andere Moderne. Suhrkamp, Frankfurt am Main (English Translation: Beck U (1992): Risk Society: Towards a New Modernity. Sage, London

Berger P, Berger B, Kellner H (1973) Das Unbehagen in der Modernität. Campus: Frankfurt am Main, New York

Bonß W (1996) Die Rückkehr der Unsicherheit. Zur gesellschaftstheoretischen Bedeutung des Risikobegriffes. In: Banse G (ed.) Risikoforschung zwischen Disziplinarität und Interdisziplinarität. Edition Sigma, Berlin, pp. 166–185

Brehmer B (1987) The Psychology of Risk. In: Singleton WT and Howden J (eds.), Risk and Decisions Wiley, New York, pp. 25–39

Buss D, Craik K (1983) Contemporary Worldviews: Personal and Policy Implications, Journal of Applied Psychology, 13, 259–280

Covello V (1990) Risk Comparisons and Risk Communication: Issues and Problems in Comparing Health and Environmental Risks. In: Kasperson RE and Stallen PJ (eds.), Communicating Risk to the Public. International Perspectives Kluwer, Dordrecht, pp. 79–124

Covello VT (1992) Trust and Credibility in Risk Communcation, Health and Environmental Digest, 6, 1, 1–3

Dawes RM (1988) Rational Choice in an Uncertain World. Harcourt, Brace & Jovanovich, San Diego et al

Douglas M (1966) Purity and Danger: Concepts of Pollution of Taboo. Routledge and Kegan Paul, London

Evers A, Nowotny H (1987) Über den Umgang mit Unsicherheit. Die Entdeckung der Gestaltbarkeit von Gesellschaft. Suhrkamp, Frankfurt am Main, pp. 34 and 210

Femers S, Jungermann H (1991) Risikoindikatoren. Eine Systematisierung und Diskussion von Risikomaßnahmen und Risikovergleichen. Issue 21 (Forschungszentrum Jülich: May)

Fischhoff B, Watson, SR, Hope C (1984) Defining Risk, Policy Sciences, 17 pp. 123–129

Gould LC, Gardner GT, DeLuca DR, Tiemann AR, Doob LW, Stolwijk JAJ (1988) Perceptions of Technological Risks and Benefits. Russell Sage, New York, pp. 45–59

Guggenberger B (1987) Die Grenzen des Gehorsams. Widerstandsrecht und atomares Zäsurbewußtsein. In: Roth R, Rucht D (eds.) Neue soziale Bewegungen in der Bundesrepublik Deutschland. Campus, Frankfurt am Main, p. 330

Häfele W, Renn O, Erdmann G (1990) Risiko und Undeutlichkeiten. In: W. Häfele (ed.), Energiesysteme im Übergang unter den Bedingungen der Zukunft. Poller, Jülich and Landsberg, pp. 31–48

Henschler D (1993) Krebsrisiken im Vergleich, Folgerungen für Forschung und politisches Handeln. In: Forschungszentrum für Umwelt und Gesundheit (GSF): Mensch und Umwelt: Risiko, issue 8, p. 70

Hohenemser C, Kates RW, Slovic P (1983) The Nature of Technological Hazard, Science, 220, 378–384

Hohenemser C, Renn O (1988) Shifting Public Perception of Nuclear Risk: Chernobyl's other Legacy, Environment, 30, 3 (April), 5–11 and 40–45

Jungermann H, Schütz H, Thüring M (1988) Mental Models in Risk Assessment: Informing People about Drugs. Risk Analysis, 8, 147–155

Jungermann H, Slovic P (1993a) Charakteristika individueller Risikowahrnehmung. In: Bayerische Rückversicherung (ed.) Risiko ist ein Konstrukt. Wahrnehmungen zur Risikowahrnehmung Knesebeck, München, pp. 89–107

Jungermann H, Slovic P (1993b) Die Psychologie der Kognition und Evaluation von Risiko. In: Bechmann G (ed.) Risiko und Gesellschaft. Grundlagen und Ergebnisse interdisziplinärer Risikoforschung. Westdeutscher Verlag, Opladen

Jungermann H, Slovic P (1993c) Charakteristika individueller Risikowahrnehmung. In: Krohn W, Krücken G (ed.) Riskante Technologien. Reflexion und Regulation. Einführung in die sozialwissenschaftliche Risikoforschung. Suhrkamp, Frankfurt am Main, pp. 79–100

Kahneman D and Tversky A (1979) Prospect Theory: An Analysis of Decision Under Risk, Econometrica, 47, 263–291

Kasperson RE, Renn O, Slovic P, Brown HS, Emel J, Goble R, Kasperson, JX, Ratick, S (1988, August) The Social Amplification of Risk. A Conceptual Framework, Risk Analysis, 8, 2, 177–187

Kepplinger HM (1993) Paradigmenwechsel durch Ökologie: Umweltbotschaften in den Medien und Publikumsreaktionen. Schweizerische Gesellschaft für Kommunikations- und Medienwissenschaft, 1, 1–10

Klages H (1984) Wertorientierungen im Wandel. Campus, Frankfurt am Main and New York

Knaus A, Renn O (1998) Den Gipfel vor Augen. Auf dem Weg in eine nachhaltige Zukunft. Metropolis, Marburg, p. 65ff.

Köcher R (1986) Bloodhounds or Missionaries: Role Definitions of German and British Journalists. European Journal of Communication, 1, 43–64

Kraus N, Malmfors T, Slovic P (1992) Intuitive Toxicology: Expert and Lay Judgments of Chemical Risks. Risk Analysis, 12, 215–232

Krücken G (1990) Gesellschaft/Technik/Risiko. Analytische Strategien und rationale Strategien unter Ungewißheit. Wissenschaftsforschung Report 36. Kleiner Verlag, Bielefeld, p. 15ff

Krücken G (1997) Risikotransformation. Die politische Regulierung technisch-ökologischer Gefahren in der Risikogesellschaft. Westdeutscher Verlag, Opladen

Küchenhoff C (1994) Umwelt-Psychosomatik. In: Beyer A, Eis D (eds.), Praktische Umweltmedizin. Springer, Berlin, Heidelberg

Lowrance WW (1976) Of Acceptable Risk: Science and the Determination of Safety. William Kaufman, Los Altos

Luhmann N (1991) Soziologie des Risikos. De Gruyter, Berlin

Luhmann N (1993) Risiko und Gefahr. In: Krohn W, Krücken G (eds.) Riskante Technologien: Reflexion und Regulation. Suhrkamp, Frankfurt am Main, pp. 138–185 (English version: Luhmann, N. 1990: Technology, Environment, and Social Risk: A Systems Perspective. Industrial Crisis Quarterly, 4: 223–231

Machlis E, Rosa E (1990) Desired Risk: Broadening the Social Amplification of Risk Framework, Risk Analysis, 10, 161–168

MacLean D (1987) Understanding the Nuclear Power Controversy. in: Engelhardt HT Jr., Caplan AL (eds.) Scientific Controversies: Case Studies in the Resolution and Closure of Disputes in Science and Technology. Cambridge University Press, Cambridge, UK, pp. 567–582

Mazur A (1984) The Journalist and Technology: Reporting about Love Canal and Three Mile Island, Minerva, 22, 45–66

Mazur A (1987, Summer) Putting Radon on the Public's Risk Agenda. Science, Technology, and Human Values, 12, 3 and 4, 86–93

Meyer-Abich KM (1989) Von der Wohlstandsgesellschaft zur Risikogesellschaft. Die gesellschaftliche Bewertung industriewirtschaftlicher Risiken. Aus Politik und Zeitgeschichte, Vol. 36 (1. September 1989), 31–42

Miller S (1985) Perception of Science and Technology in the United States. Manuscript. Academy of Sciences, Washington, D.C

Nennen HU, Garbe D (eds.) (1996) Das Expertendilemma. Springer, Berlin

Obermeier OP (1990) Das Wagnis neuen Denkens – ein Risiko? In: Schüz M (ed.) Risiko und Wagnis: Die Herausforderung der industriellen Welt, Vol. 2. Gerling Akademie, Neske, Pfullingen, pp. 243–263

Otway H, Thomas K (1982) Reflections on Risk Perception and Policy, Risk Analysis, 2, 69–82

Otway H, von Winterfeldt D (1982) Beyond Acceptable Risk: On the Social Acceptability of Technologies. Policy Sciences, 14, 247–256

Peltu M (1985) The Role of Communications Media. In: Otway H, Peltu M (eds.) Regulating Industrial Risks. Butterworth, London, pp. 128–148

Perrow C (1984) Normal Accidents. Basic Books, New York

Peltu M (1989) Media Reporting of Risk Information: Uncertainties and the Future. In: Jungermann H, Kasperson RE, Wiedemann PM (eds.) Risk Communication. Forschungszentrum KFA, Jülich, pp.11–32

Peters HP, Albrecht G, Hennen L, Stegelmann HU (1987) Die Reaktionen der Bevölkerung auf die Ereignisse in Tschernobyl: Ergebnisse einer Befragung Bericht der Kernforschungsanlage Jülich. Jül-Spez-400. KFA, Jülich

Peters HP (1991) Durch Risikokommunikation zur Technikakzeptanz? Die Konstruktion von Risiko"wirklichkeiten" durch Experten, Gegenexperten und Öffentlichkeit. In: Krüger J, Ruß-Mohl St (eds.) Risikokommunikation. Edition Sigma, Berlin, pp. 11–67

Peters HP (1996) Kommentar zu Hans Mohrs Studie über das "Expertendilemma". In: Nennen HU, Garbe D (eds.) Das Expertendilemma. Zur Rolle wissenschaftlicher Gutachter in der öffentlichen Meinungsbildung. Springer, Berlin, pp. 61–74

Pidgeon NF, Hood CC, Jones DKC, Turner BA, Gibson R (1992) Risk Perception. In: Royal Society Study Group (ed.) Risk Analysis, Perception and Management. The Royal Society, London, pp. 89–134

Peters HP (1990) Technik-Kommunikation: Kernenergie. In: Jungermann H, Rohrmann B, Wiedemann PM (eds.) Technik-Konzepte, Technik-Konflikte, Technik-Kommunikation. Monographien des Forschungszentrums Jülich, Vol. 3 Forschungszentrum Jülich, Jülich, pp. 59–148

Rayner S and Cantor R (1987) How Fair is Safe Enough? The Cultural Approach to Societal Technology Choice. Risk Analysis, 7, 3–13

Renn O (1984) Risikowahrnehmung der Kernenergie. Campus, Frankfurt am Main, New York

Renn O, Levine D (1988) Trust and Credibility in Risk Communication. in: Jungermann H, Kasperson RE, Wiedemann PM (eds.) Risk Communication. Forschungszentrum, Jülich, p. 51

Renn O (1989) Risikowahrnehmung. Psychologische Determinanten bei der intuitiven Erfassung und Bewertung von technischen Risiken. In: Hosemann G (ed.) Risiko in der Industriegesellschaft. Universitätsbibliotheksverlag Erlangen-Nürnberg, Erlangen, pp. 167–191

Renn O (1991a) Risk Communication and the Social Amplification of Risk. in: Kasperson R, Stallen PJ (ed) Communicating Risk to the Public. Kluwer Academic Publishers, Dordrecht, pp. 287–324

Renn O (1992) Concepts of Risk: A Classification. In: Krimsky S, Golding D (eds) Social Theories of Risk. Praeger, Westport, pp. 53–82

Renn O (1996) Riskante Risikopolitik. Die Zeit, 39 (20.9.1996), 48

Renn O (1997) Mental Health, Stress and Risk Perception: Insights from Psychological Research. In: Lake JV, Bock GR, Cardew G (eds.) Health Impacts of Large Releases of Radionuclides Ciba Foundation Symposium 203.Wiley London, pp. 205–231

Renn O, Zwick MM (1997) Technik- und Risikoakzeptanz. Springer, Berlin

Renn O, Rohrmann B (2000) Cross-Cultural Risk Perception. A Survey of Research Results. Kluwer, Dordrecht

Rohrmann B (1995) Technological Risks: Perception, Evaluation, Communication. In: Mechlers RE, Stewart MG (eds.) Integrated Risk Assessment. Current Practice and New Directions. Balkema, Rotterdam, pp. 7–12

Rubin DM (1987) How the News Media Reported on Three Mile Island and Chernobyl. Communication, 37, 3, 42–57

Schanne M (1996) Bausteine zu einer Theorie der Risiko-Kommunikation in publizistischen Medien. In: Meier WA, Schanne M (eds.) Gesellschaftliche Risiken in den Medien. Seismo-Verlag, Zürich, pp. 207

Schultz-Venrath U (1998) Die Wechselwirkungen zwischen Umweltnoxen und psychosomatischen Symptomen. Gutachten für die Europäische Akademie zur Erforschung von Folgen wissenschaftlich-technischer Entwicklungen. Bad Neuenahr-Ahrweiler. Manuskript. Köln

Sharlin HI (1987) Macro-Risks, Micro-Risks, and the Media: The EDB Case. In: VT Covello, BB Johnson (eds.) The Social and Cultural Construction of Risk. Reidel: Dordrecht, pp. 183-197

Shubik M (1991) Risk, Society, Politicians, Scientists, and People. In: M. Shubik (ed.) Risk, Organizations, and Society. Kluwer Dordrecht, pp.7–30

Sjøberg J, Winroth E (1985) Risk, Moral Value of Actions, and Mood. Unpublished manuscript University of Gothenburg. Department of Psychology, Gothenburg

Sjöberg L (1994) Perceived Risk vs. Demand for Risk Reduction. Risk Research Report No. 18. Center for Risk Research, Stockholm School of Economics, Stockholm

Sjöberg L (1997) Risk Sensitivity, Attitude and Fear as Factors in Risk Perception. Paper at the Annual Meeting of the Society for Risk Analysis-Europe. Manuskript. Stockholm

Slovic P, Fischhoff B, Lichtenstein S (1981) Perceived Risk: Psychological Factors and Social Implications. In: Proceedings of the Royal Society, A376. Royal Society, London, pp. 17–34.

Slovic P (1987) Perception of Risk. Science, 236, 280–285

Sood R, Stockdale G, Rogers EM (1987, Summer) How the News Media Operate in Natural Disasters. Communication, 37, 3, 27–41

Tversky A, Kahneman D (1974) Judgment under Uncertainty. Heuristics and Biases. Science, 85, 1124–1131

Umweltbewußtsein in Deutschland. (1996) Ergebnisse einer repräsentativen Bevölkerungsumfrage 1996. Eine Information des Bundesumweltministeriums. Förderkennzeichen 101 07 112/05. UBA, Berlin

Watson M (1987) In Dreams Begin Responsibilities: Moral Imagination and Peace. In: Andrews V, Bosnak R, Goodwin KW (eds.) Facing Apocalypse. Spring, Dallas, pp. 70–95

WBGU (Wissenschaftlicher Beirat der Bundesregierung) (1999) Welt im Wandel: Der gesellschaftliche Umgang mit globalen Umweltrisiken. Jahresgutachten. Springer, Berlin

Webler Th, Levine D, Rakel H, Renn O (1991) The Group Delphi: A Novel Attempt at Reducing Uncertainty. Technological Forecasting and Social Change. 39, 253–263

Wiedemann PM (1993) Tabu, Sünde, Risiko: Veränderungen der gesellschaftlichen Wahrnehmung von Gefährdungen. in: Bayerische Rückversicherung (ed.) Risiko ist ein Konstrukt. Wahrnehmungen zur Risikowahrnehmung. Knesebeck, München, pp. 43–67

Wildavsky A, Dake C (1990) Theories of Risk Perception: Who Fears What and Why? Daedalus, 119, 4, 41–60

Wilkins L, Patterson P (1987, Summer) Risk Analysis and the Construction of News. Communication, 37, 3, 80–92

Zilleßen H (1993) Die Modernisierung der Demokratie im Zeichen der Umweltpolitik. In: Zilleßen H, Dienel PC, Strubelt W (eds.) Die Modernisierung der Demokratie. Westdeutscher Verlag, Opladen, pp. 17–39

Zwick M, Renn O (1998) Wahrnehmung und Bewertung von Technik in Baden-Württemberg. Presentation. Akademie für Technikfolgenabschätzung in Baden-Württemberg, Stuttgart

4 The Importance of Economic Factors in Setting Combined Environmental Standards

4.1
Scope

Scientific enquiry seeks to reveal the nature and quality of combined exposures from pollutants observing how dose-response functions change for individual pollutants as the dose of other agents is varied. Having obtained values for a range of doses, one can then plot isoboles showing combinations of agents with equal impacts (on human health or on other categories of impact target). With deterministic impact phenomena, experimental and epidemiological studies aim to derive health safety thresholds. With stochastic impact phenomena there are no such thresholds, and efforts are restricted to deriving combined dose-response functions.

Scientific findings of this kind are a necessary input to environmental policy, but are not enough when it comes to setting specific environmental targets. Policy choices are also guided to a large degree by legal, economic and social-science factors. Economic enquiry in this context is chiefly concerned with how to design environmental policies incorporating cost-benefit aspects, or with the economic implications of omitting such aspects from a policy programme. In the following, we aim to show how policies differ between single-pollutant and multiple-pollutant phenomena. The conventional economic approach neglects combined effects, generally assuming a given type of adverse effect to be the result of a single factor. This approach is too simplistic for many environmental phenomena.

From an economics viewpoint, the purpose of environmental policy is to ensure that certain exposure levels for human beings, fauna, flora and the climate are not exceeded, and that this is done in due consideration of cost. Environmental economics accentuates the goal of economic efficiency (cost effectiveness). For single pollutants, it shows that in some circumstances fiscal and market instruments (environmental levies and emission certificates) can deliver environmental protection at lower cost than command-and-control regulation and hence are the appropriate choice in such circumstances (Cansier 1996). But for multiple pollutants, the economic efficiency goal takes on even greater importance. In this scenario, at least two substances are responsible for a given adverse effect. The question then is, given a specific exposure limit, which of the substances should be regulated, and to what extent? Should policies target one substance or all of them, and if the latter, in what combination? Economics provides us with a decision rule in the form of the cost minimisation goal. This favours whichever combination of environmental standards has the least total macroeconomic cost while still ensuring that the applicable exposure limit is not exceeded.

Adverse environmental effects can give rise to health damage, material loss and damage (including lost production), and ecological damage. How impacts are rated depends on what impact levels society finds tolerable – that is, on guideline targets. These are the standard against which limit values set by environmental policy are measured. The prime goal of environmental policy is to protect public health, where health is defined in a specific way and can be measured in terms of the relative frequency or severity of a given illness. The (quantified) guideline targets of environmental policy are taken as given when assessing cost-effectiveness.

Since concentrations of pollutants act on human beings through the environmental media (air, water, soil, etc.), it is necessary to apply empirical findings on environmental impacts to real exposures as observed in the field. We assume this can be done satisfactorily and that the dose-response relationships are of the same type as those obtained in the laboratory. In our analysis, we take the experimental doses as being equal to the maximum permissible exposures.[1]

Environmental economics is also concerned with ascertaining socially desirable levels for guideline targets, and recommends that these levels be decided by comparing costs and benefits. A target should be raised when the benefits (adverse effects avoided) exceed the costs. In this way it is possible to obtain the economic optimum for each guideline target (Cansier 1996). It would be interesting to explore how moving from single to multiple substances changes the way in which these optimum targets are determined. We will not pursue this further in the present study, primarily because no reliable way has yet been found of ascribing a monetary value to adverse effects on human beings, fauna, flora and the climate.

Complying with exposure limits entails limiting emissions. The limits for total permissible emissions constitute the operational targets of environmental policy instruments (emission limits, product standards, environmental levies and emission certificates, voluntary commitments by industry). Limiting emissions means taking action to avoid them, and such action has a cost. This cost is indirectly related to the exposure values.

Our analysis initially assumes that all significant factors are known to decision-makers (government or parliament). In reality, environmental policy decisions are uncertain. This is particularly so for combined effects of pollutants, of which relatively little is yet known. It would thus appear appropriate to conduct an additional analysis for the case of imperfect information.

---

[1]   The set of all combinations of doses of different substances which have the same effect on an environmental quality target is termed an "isoquant".

4.2
Cost-effective Environmental Protection with Adequate Information

4.2.1
Adverse Effects

Isoquant curves: Let us take a specific guideline target of environmental policy as given and assume that adverse effects (expressed as the frequency or severity of a disease) can be induced by two substances, A and B. The guideline target must be one that can be stated in quantitative form.[2] We express the impact at alternative doses of A and B with an impact function S(A,B). For our analysis, we need to know what combinations of the two agents remain within the target function. These combinations are plotted by a target isoquant curve. All combinations on the curve meet the target criterion and by this definition are ecologically equivalent. The shape of the curve can vary considerably (see fig. 41):

- With a linear function, the pollutants are mutually substitutable at a constant ratio:

$$(1) \quad -\frac{dA}{dB} = \frac{\partial S / \partial B}{\partial S / \partial A} = \text{constant}$$

  where $\partial S / \partial SA$ and $\partial S / \partial SB$ are the marginal impacts of the substances.
- With a convex curve, the marginal rate of substitution $|dA / dB|$ rises as the quantity of A increases and the quantity of B decreases. Diminishing quantities of A are needed to cancel the adverse effect of each additional unit of B, and diminishing quantities of B are needed to cancel the adverse effect of each additional unit of A.
- With a concave curve, the marginal rate of substitution $|dA / dB|$ diminishes as the quantity of A increases and the quantity of B decreases. Cancelling the impact of each added A requires increasing reductions in B as more As are added, and vice versa.
- An S-shaped or "sigmoid" curve is made up of two adjacent ranges, one concave and one convex. The marginal rate of substitution rises over the first half of the curve and falls again over the second, or vice versa.

Additivity and independence models: The three scientific reference models for evaluating combined effects on organisms are response additivity, dose additivity and independence (see Section 2.2).[3] Effects that go beyond or fall short of these reference standards are referred to as super-additive or sub-additive effects. There are three possibilities:

---

[2] For ease of presentation, we use the same symbol for the type of substance and the quantity.
[3] The models do not take into account reactions between agents in the environmental medium (such as the formation of ground-level ozone).

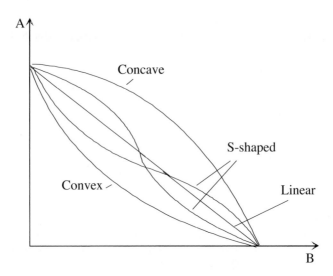

Fig. 4-1 Alternative shapes of the isoquant curve. The isoquant curve for agents A and B is the locus of all pairs of quantities of A and B that are assumed to have the same adverse effect. At a constant level of damage, a given reduction in exposure to A requires a given increase in exposure to B, and vice versa. This rate of substitution is determined by the marginal effects of the agents.

- *Response additivity:* The total effect E of a combination of substances equals the sum of the individual effects E_A and E_B (with the constraint that the sum cannot exceed the maximum adverse effect): $E = E_A (A) + E_B (B)$.
- *Independence:* $E = E_A (A) + E_B (B) - E_A (A) \cdot E_B (B)$
  The total effect is less than that obtained with response additivity.
- *Dose additivity:* Effects are expressed as dose equivalents of a reference substance, for example A. If the dose-response function is linear, the total effect E is defined as $E = E_A(A + u \cdot B)$, where u is the constant coefficient of equivalence. The effect of a given quantity of B is measured with reference to the base level of the given quantity of $\overline{A}$ using the dose-response function for $\overline{A}$. If the function is non-linear, the quantity of $\overline{B}$ at a given dose of $\overline{A}$ is obtained by using this formula (in which X is the upper limit to be found for the quantity of A):

$$(2) \quad \int_{}^{\overline{B}} E_B' (B) dB = \int_{}^{X} E_A' (A) dA \ .$$

The total effect of $\overline{A}$ and $\overline{B}$ is then $E = E_A(X)$.

All three approaches allow us to plot an isoquant curve. If the dose-response functions of the component substances are linear, the dose additivity and response additivity models produce identical linear isoquants. In the low-dose range, we obtain approximately the same results as with independence models (see Section 2.2).

- If *both* agents have a linear, super-linear or sub-linear impact function over the relevant dose range (see the sigmoid function in Section 2.2), the isoquant curve will be linear, concave or convex.
- If one agent has a linear and the other a super-linear (sub-linear) impact function over the relevant dose range, the isoquant curve will be concave (convex). If one function is sub-linear and the other super-linear, the isoquant curve can be concave, convex or linear depending on which component effect predominates.

It is a well established hypothesis that dose-response relationships are linear or slightly super-linear for individual substances in the health-relevant low-exposure range (for chemicals and ionizing radiation). This simplifies our analysis, as it is the low-exposure range that interests us. The isoquant curves for additivity will thus be linear or slightly concave.

Combined exposures: Interactions between component substances either weaken (sub-additivity) or amplify (super-additivity) the overall effect compared with the reference models (additivity and independence). The resulting variation in the shape of the isoquant curves concerns us only over the low-dose range. We obtain two cases:

- Linear reference function: If there is unidirectional or reciprocal synergy between the effects of the component substances (super-additivity), the isoquant curve will be convex. If effects are unidirectionally or reciprocally antagonistic (sub-additive), the curve will be concave.
- Concave reference function: Antagonistic effects increase the concavity of the curve. Synergy between the effects reduces the concavity, eventually producing a straight line or a convex curve.

4.2.2
Abatement Costs

Emission reduction has a cost. This cost generally rises in linear or greater-than-linear proportion with the size of the reduction. A greater-than-linear cost increase can generally be expected when the reduction is large. That is, the cost function is likely to be super-linear with high guideline targets and linear with low guideline targets. We intend to establish the relationship between costs and exposures, under the simplifying assumption that exposures are directly proportionate to emissions.

At constant macroeconomic cost K^* it is possible to achieve different combinations of reductions in A and B. The set of possible pairs of reductions V_A and V_B is stated by an isocost function. The cost of a further unit reduction in A must be matched by cost savings from decreasing the reduction in B, and vice versa. The condition is $K^* = K_A(V_A) + K_B(V)_B$. From this we derive the marginal rate of substitution between the respective abatement costs of A and B:

$$(3a) \quad -\frac{dV_A}{dV_B} = \frac{\partial K_B}{\partial V_B} \Big/ \frac{\partial K_A}{\partial V_A}.$$

Because the respective quantitative reductions are identical with the difference between the initial emissions $\bar{A}$ and $\bar{B}$ and the residual emissions A and B, the rate of substitution can alternatively be expressed as follows:

$$(3b) \qquad -\frac{dA}{dB} = \frac{\partial K_B}{\partial B} \bigg/ \frac{\partial K_A}{\partial A}.$$

If the cost increase with increasing reductions is linear (or super-linear), the iso-cost function is linear (or convex looking from the origin; see K_1 in fig. 4-2).

4.2.3
Cost-efficient Limit Values

Assume a given environmental quality standard is to be complied with (for example prevention of adverse health effects in a case of non-stochastic impact phenomena). This objective corresponds to a specific target isoquant curve. This curve plots the set of possible target-compliant combinations of A and B. The isocost function, on the other hand, plots the combinations that are attainable in prevention terms. The most cost-effective combination of environmental quality standards for A and B is located at the point where the target isoquant curve is touched by an isocost curve with the least possible cost (scenario M in fig. 4-2).

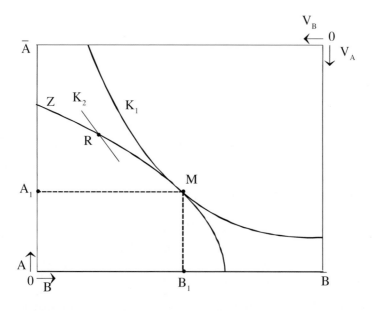

Fig. 4-2 Efficient limit values for agents A and B. In the beginning, exposures to the two agents total $\bar{A}$ and $\bar{B}$. If Z is the target isoquant curve and the cost conditions for preventing A and B are represented by isocost function K_1, the most efficient option is to reduce $\bar{A}$ to exposure level A1 and $\bar{B}$ to exposure level B_1. Selecting a combination such as that at point R would entail a higher cost of environmental protection. B would be reduced too far and A too little relative to the efficient values.

The cost of environmental protection depends on two fundamental conditions: the nature of the relationship between the combined effects and the nature of the relationship between the agents' respective abatement cost and impact per unit.[4]

Significance of relative impacts and abatement cost: Let us begin by taking a given type of combined effect (additive or non-additive). The properties of the efficient solutions vary according to the shape of the effect and the cost curves. The most cost-effective solution may be to regulate both agents (see M in fig. 4-2). The defining condition for an efficient combination is that the agents' relative marginal impacts and relative marginal abatement costs are equal:

$$(4) \quad \frac{\partial S / \partial B}{\partial S / \partial A} = \frac{\partial K_B / \partial B}{\partial K_A / \partial A} .$$

Because the cost rises at a greater-than-linear rate as reductions become larger, achieving cost-effectiveness will generally entail restrictions on both substances (or all substances involved). Policy should not be limited to a single substance even if this is the main cause of the adverse effect in question. With any given pair of pollutants, the reduction in emissions should be greater for the one that has the greater relative adverse effect and the lesser relative abatement cost. All target-compliant combinations other than M in fig. 4-2 would incur a higher macroeconomic cost. If the limit values were to differ in reality – as with scenario R – cost-saving opportunities would go unexploited. The same degree of environmental protection could be reached at lesser cost. In scenario R, the relative impact of B is less (say, 2) than the relative cost of B (say, 4). If we increase B by one unit, we would have to reduce A by two units (assuming A and B are ecologically equivalent). But if we reduce B by one less unit, there is a cost saving that can be reinvested in preventing an additional four units of A, whereas in fact we only need to prevent an additional two units of A. That is, at the specified total cost it is possible to attain a higher target, or the specified target is attainable at lesser macroeconomic cost.

In certain instances the most cost-efficient option may be to regulate a single pollutant. This arises when, at all quantities, the relative marginal abatement cost of the pollutant is less than its relative marginal impact (see also von Ungern-Sternberg 1987). The other factor should only be included in the abatement strategy if the environmental quality target cannot be attained by preventing the main factor alone. The likelihood of this being the case increases as targets become more stringent; that is, with larger reductions in exposures.

In the following we will use the example of carcinogenic effects of radioactivity to illustrate the use of cost-effectiveness principle in determining efficient combined environmental standards. The carcinogenic effects of ionizing radiation can be inhibited or potentiated by other substances (see fig. 4-3 for the possible shapes of isoquant curves). The other substances are not carcinogenic in their own right. Generally, either the effects of ionizing radiation combined with chemotoxic substances act additively or the combined effect is smaller than the sum of the individ-

---

[4]   Where a single agent is more or less the sole cause of an adverse effect and other agents have only a minor contribution to the effect as a whole, regulatory considerations are restricted to the first agent from the outset.

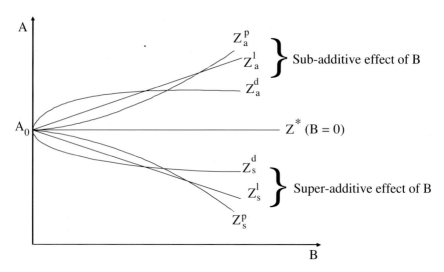

Fig. 4-3 Example of isoquant curves where the adverse effect caused by A is inhibited (potentiated) by agent B. If exposure to B is zero, the adverse effect is assumed to be the limit value A0. With sub-additivity, A must be reduced as the quantity of B increases for the adverse effect to remain constant. The inhibiting effect of B can be linear, sub-linear or super-linear as B increases (see Z_a^l, Z_a^d, Z_a^p). Super-additivity acts analogously.

ual effects. Super-additive effects have been found less frequently (for example with hormones, viruses, and smoking), in most cases at medium to high doses (see section 2.3).

We will look at the case where a second agent inhibits the effect of radiation. What is important is the relationship between the target function and the starting conditions. The target function may only be attainable by reducing A or B. This is assumed to be the case in fig. 4-4. With target function Z_1, the objective is to reduce A (radiation) to A_1; and with target function Z_2, the objective is to reduce B (a chemical) to B_1. Only boundary solutions are efficient. If function Z_1 is typical of cancer risk from ionizing radiation, only the latter ought to be reduced. Taking into consideration the sub-additive effect of the chemical, however, it is enough to reduce radiation to A_1 rather than C (the reduction that would be necessary in the absence of an inhibitor). The chemical reduces the impact of the radiation at zero cost, and this ought to be exploited to the full. If policy were to ignore the interaction and reduce radiation, say, to C, the target would be exceeded, incurring unnecessary cost. We can improve on this policy by relaxing the limits on radiation exposure. (A differentiated policy is needed if the two factors occur together in some places and separately in others.)

A policy may take the interaction into account and choose a combination of environmental standards on the target isoquant curve, for example D, but ignore the cost aspect. At D, the cost is excessive. The inhibiting factor is reduced unnecessarily, which means A is overregulated. It is advantageous to allow the inhibiting factor to take full effect and to reduce the causal factor A to a lesser extent.

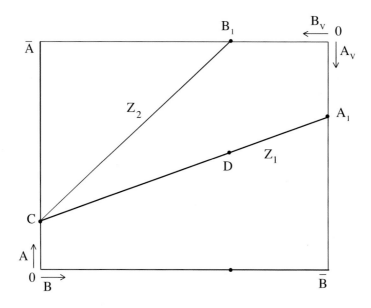

Fig. 4-4 Efficient limit values where the cancer risk triggered by A is inhibited by another chemical B. The sub-additive effect of B reduces the damage caused by A. It would be inefficient not to exploit this mechanism. If effects are sub-additive throughout, exposures to B should be permitted to the extent that they are compatible with the ecological objective. The efficient exposures are determined for target isoquant Z_1 in accordance with A_1 and for target isoquant Z_2 in accordance with B_1. The only cost incurred is from the prevention of A in the former case and from the prevention of B in the latter.

Cost implications of the different types of combined effect: We must distinguish between sub-additive and super-additive effects. The following can be stated:

- The cost of environmental protection is greater with super-additive effects than with additive or independent effects. The potentiation of individual effects is concomitant with an increase in cost. In order to attain the same environmental policy guideline target as would hypothetically be attained with additive or independent effects, policies must stipulate higher limit values for the agents involved. This incurs a higher cost and so tends to make it harder to enforce a given target in the policymaking process. In policy terms it may be appropriate to accept a lower level of environmental protection.
- Where two substances inhibit each other, they effectively deliver abatement at zero cost. If the effects are additive, exposure levels need to be set lower to reap the same benefit as would be obtained if they were sub-additive. Compliance with an environmental target entails less environmental protection and hence lower environmental cost. At the same time, this can give occasion to pursue a more exacting target.

Because super-additive effects tend to produce a convex target isoquant curve, sub-additive effects tend to produce a concave curve and the cost of abatement typically increases at a super-linear rate, there is also another difference: With super-additive effects, the target isoquant curve and the isocost functions can have a similar gradient over a fairly broad range of combinations of exposure values. This facilitates policymaking decisions.

Conclusion: The unnecessarily high cost of environmental protection is a significant barrier to the enforcement of stringent environmental standards. The costs can be reduced with multicausal impact phenomena by combining limit values on the basis of cost-efficiency. Costs are not only a function of the means of abatement; it also depends on the adverse effects of the agents concerned. The two facets should not be considered separately.

The cost-efficiency approach requires policy to have a quantitative focus. Cost comparisons are only possible with quantitative environmental targets (observing acceptable levels of health risk, setting upper bounds for global warming, preventing summer smog, etc.) and exposure values. This supports business and industry demands for policies to make greater use of quantitative as opposed to qualitative targets (such as ones to avoid adverse environmental effects or requiring state-of-the-art preventive environmental management). Operational health risk levels are used, for example, in the Netherlands and the USA (Tegner and Grewing 1996). According to Rehbinder, the accepted view in the debate surrounding hazardous substances in the US is that quantitative risk assessment and ensuing risk/cost analysis is an appropriate means of determining residual risk. The EPA regards risks as tolerable in the range 1:100,000 to 1:1,000,000 (Rehbinder 1991). Under the Clean Air Act, limits for carcinogens must be based on a 1:1,000,000 lifetime probability of contracting cancer (Böhm 1996). In German environmental policy, quantitative targets are still quite rare. They exist in the field of hazard control and for specific phenomena under preventive environmental policies, for example CO_2, SO_2 and NO_X emissions. There are proposals in other areas. The German Council of Environmental Advisors (SRU), for example, has recommended preventing summer smog and reducing the associated cancer risk from currently 1:5,000 in rural districts and 1:1,000 in urban districts to 1:10,000 and 1:1,000,000 respectively (Der Rat von Sachverständigen für Umweltfragen 1994).

4.3
An Empirical Example

The smog problem in urban New York has been investigated using the cost minimization approach (Repetto 1987). Ground-level ozone, which indicates the presence of photochemical oxidants, forms by the interaction of nitrogen oxides (NO_X) and volatile organic compounds (VOCs). The main emitters of VOCs (in Germany) are transport (about 50%) and users of solvents (about 41%). Some 73% of NO_X emissions are produced by transport, 13% by power stations and district heating facilities, and 10% by industry (Umweltbundesamt 1998).

According to Repetto, the isoquants plotting pairs of quantities of NO_X and VOCs that yield equal maximum hourly ozone concentrations have a typical shape that does not significantly change with meteorological conditions. The marginal effect of both precursor pollutants on ozone formation falls with rising concentration. The absolute effect of NO_X on ozone formation diminishes above a certain range. These findings result in convex iso-ozone curves. Repetto observes (for years preceding 1987) that in those US air quality regions where the national ozone standard had been exceeded, there had been efforts to reduce VOCs but not to reduce NO_X. He criticises this policy for failing to consider the cost aspects.

Based on estimates for urban New York, the cost functions for preventing NO_X and VOC emissions have two distinguishing features: an initial range where abatement is possible at negative cost and a sharp rise in cost once straightforward abatement options have been exhausted and more sophisticated and capital-intensive approaches become necessary. Consequently, the isocost functions have a pronounced convex shape (as seen from the origin). The results of the study are as follows:

- If one were to invest, say, $75 million a year in preventing smog, the most cost-effective option would be to reduce NO_X emissions by 33% and VOC emissions by 40%.
- Because the isoquant and isocost curves have a similar shape, the added cost of deviating from the efficient dose combination is not particularly large over a fairly wide range of combinations.
- Because the individual pollutants have significant side-effects – in particular NO_X in the formation of acid rain – these side-effects ought also to be considered when assessing specific strategies.

A more recent study stresses that iso-ozone curves vary with the composition of the VOC component and hence differ between regions (Hall 1998). This would imply different regional standards for NO_X and VOCs under cost minimisation. A further study develops a decision model by which the optimum mix of NO_X and VOC reductions is arrived at in accordance with the goal of minimizing the overall macroeconomic cost of environmental protection (abatement cost and damage cost of smog in urban areas) and taking into account effect-related uncertainties and learning effects (Chao et al. 1994). Using an example with given starting conditions for ozone concentration and the respective emissions and given estimates for the abatement cost functions, Chao et al. calculate the optimum quantitative reductions in NO_X and VOCs for various alternative hypothetical impact functions.

4.4
Choice of Policy Instruments

Policy instruments are used selectively to attain specific environmental policy goals. Because smoking and alcohol consumption are major potentiators of health damage, policies for these substances might focus on increasing education, restricting advertising and raising taxes on alcohol and tobacco. However, these impact phenomena are not usually placed within the remit of environmental policy, and we will not give them further consideration here. In practice, policies for regulating air, water and soil-polluting emissions primarily focus on prescriptions and prohibitions. Environmental levies rarely come into play (an example is the waste water levy in Germany). Certificate schemes are even rarer and to date have only been used in the US (for SO_2 and NO_X).

Multiple pollutants pose a special challenge in selecting policy instruments to implement an efficient abatement strategy, and different instruments are not all equally suited to the purpose. Certificates have advantages when combined effects show linear additivity. Regulators using emission levies or emission limits cannot avoid setting their targets equal to efficient limit values from the outset. The emission levies on two pollutants A and B, for example, must necessarily equal the pollutants' marginal abatement costs at the identified efficient limit values. Cost-minimizing polluters will then adjust their emissions in such a way that the targets are attained.

As regards regulating different emitters of a given pollutant, cost-minimization by individual polluters tends to result in an efficient abatement strategy across the group as a whole. Polluters with relatively low abatement costs tend to reduce their emissions more than those with relatively high abatement costs. This is macroeconomically efficient. In ignorance of the abatement cost faced by each polluter, regulators thus tend to set uniform emission limits for all polluting facilities. Prescriptions and prohibitions are consequently less cost-efficient than emission levies. This observation is unrelated to the multiple pollutant problem, however. On the other hand, prescriptions and prohibitions have the advantage that they give more precise control over emissions. This is particularly important when it comes to preventing excessive pollution within the catchment area of large or locally concentrated emission sources (hot spots).

The policymaking task is easier with emission certificates and linearly additive combined effects. Regulators do not need to know the abatement cost functions. They need not give any thought to efficient limit values. They need to know two things: a combination of emissions that accords with the protection or prevention objective at hand, and the (constant) impact per unit of each pollutant. The regulators then issue certificates for the permitted total quantity of emissions and stipulate that they can be traded or used as permits to emit the other pollutant, in either case in the ratio of the pollutants' impacts per unit (Bonus 1975; Endres 1985).

The most straightforward approach is to express the pollutants in unit equivalents of a reference substance and only to issue certificates for emissions of that substance (see fig. 4-2). The environment agency issues certificates (in 1:1 units) in quantity A^*, which accords with the environmental target (Z_A). The impact functions for agents A and B acting individually are $S_A = a \cdot A$ and $S_B = b \cdot B$. The relationship between permissible combinations of the agents on the target isoquant curve can thus be expressed as $dA = b/a \cdot dB$. Usage of the certificates for emissions of B (Z_B) is stipulated in accordance with this relationship. To emit one unit of B, a polluter must purchase b/a certificates. The certificates are issued by auction. The environment agency sells the permits at the price at which demand equals the supply quantity A^*. Demand for the certificates derives from a cost-minimization rule. Each emitter of A, A_i where $i = 1,...,N$ and of B, B_j where $j = 1,...,M$ wishes to minimize its individual cost. This comprises its abatement cost $V_{A,B}$ and its expenditure for the purchase of certificates. The cost of purchasing certificates incurred by each emitter of A equals the certificate price p multiplied by the emitter's residual emissions of A. Emitters of B calculate the cost of their residual emissions in A-equivalents. This cost equals $p \cdot b / a \cdot B$. We can derive cost functions for each emitter as follows:

$$(5) \ K (V_A^i, Z_A^i) = K(\overline{A}_i - A_i) + p \ A_i$$

$$(6) \ K(V_B^j, Z_B^j) = K(\overline{B}_j - B_j) + p \ b/a \cdot B_j .$$

(where $\overline{A}_i$ and $\overline{B}_j$ are the quantities of emissions produced by each emitter at the outset.)

Taking the certificate prices as given, cost minimization produces:

$$(7) \ p = \frac{\partial K_i}{\partial A_i} \ \text{and} \ p = a / b \cdot \frac{\partial K_j}{\partial B_j} .$$

The emitters attune their emissions (or abatement efforts) to the market price of the certificates until the marginal abatement cost equals the price. The same applies for alternative prices and hence also for the equilibrium price at which there is demand for all certificates issued by the environment agency and hence the condition $A^* = \sum A_i + b/a \sum B$ is met. The resulting structure of demand for the certificates matches the efficient combination of emissions of A and B, which, as we have seen, is governed by:

$$(8) \ \frac{\partial K_j}{\partial B_j} / \frac{\partial K_i}{\partial A_i} = b / a .$$

The market thus ensures that emission certificates issued for A are transformed into the efficient quantity structure for A and B, thus exploiting the advantages of the market mechanism. Policymakers need less information than they do for levies or command-and-control. This is an important advantage of certificates. However, it requires an efficient market. This market efficiency may be assumed to obtain in the domain of global climate protection, for example. Here, the key advantage of (internationally tradable) certificates over a levies scheme would be not needing to know what degree of reduction in CO_2, CH_4 and other greenhouse gases is efficient. This decision can be left to the market. The greenhouse effect per unit of each gas is known. A requirement is that climate policy targets not just CO_2 but all major greenhouse gases, this being necessary for economic reasons (Cansier 1991, Michaelis 1997).[5]

Certificate schemes lose this advantage in the case of non-linearly additive or non-additive combined effects, because in such cases the impact of a unit of each pollutant varies with the quantity of emissions at a given time. It would be asking too much of policymakers to have them adjust the ecologically dictated rate of exchange at which the various certificates are traded to account for changes in unit impacts. Accordingly, we may assume that regulators would have to set efficient

---

[5] On the current state of international climate policy and the chances for and difficulties facing the introduction of tradable permits for greenhouse gases, see Bayer and Cansier, 1999.

limit values for certificates just as with levies and direct regulation. The phenomenon of combined exposures does not suggest a specific set of policy instruments in such cases. How specific protection and abatement objectives are attained then depends, as with single-pollutant phenomena, on the problem at issue. Levies and certificates are particularly well suited for regulating large-area pollution. When it comes to preventing health damage from ionizing radiation, chemicals and the like, on the other hand, command-and-control regulation is the only solution, as protection must be ensured at every exposure site.

4.5
Cost-efficient Limit Values under Imperfect Information

4.5.1
Uncertainty in Decision-making

Relatively little is known about the combined effects of pollutants. Combined effects increase decision-making uncertainty beyond the level that already obtains when determining dose-response curves for individual pollutants and applying laboratory results and epidemiological study findings to real exposures (Der Rat von Sachverständigen für Umweltfragen 1987, Dieter 1995, Hagenah 1996). Yet despite imperfect information, policy decisions on maximum permissible exposures still have to be made, and there is a desire for methodological support from the scientific community. The various models of economic decision theory under uncertain expectations describe the factors that need to be taken into account in complicated circumstances of this nature.

Particular difficulties and uncertainties arise when assessing the nature and intensity of combined effects for two main reasons:

- Because adverse effects can be caused by two or more agents, the effects of many more pairs of quantities must be investigated than with single pollutant phenomena, and this is impracticable for cost reasons. Investigating the effects of only ten different quantities for each of two agents would mean examining a hundred different pairs of quantities. Three agents would mean a thousand experiments. Each of these experiments would require a representative number of fresh laboratory animals. In view of these difficulties, figures for complex mixtures are extrapolated and inferred from results obtained with binary mixtures (see Section 2.4). Hence rather than requiring unequivocal scientific findings, policy must be based on modelling.
- Where the biochemical response mechanisms triggered by individual pollutants are not known, it is not possible to judge the pollutants' overall effects in combination (see Section 2.3). When the various agents operate through different mechanisms, it makes sense to add their *effects*. When the substances operate through identical or similar mechanisms, one should add their *doses*. In such cases, the agents work like dilutions of one and the same substance. Uncertainty about the mechanisms involved would not be a problem here if the dose-

response relationships were linear in the low dose range; however, this is not bound to be the case since linearity is a simplifying assumption made, and reflects the safety margins applied by scientists when extrapolating high doses to lower (realistic) concentrations. Dropping the applied safety margins leaves us with only one "envelope of additivity" (see Section 2.3), and we are forced either to choose one of the alternatives or to take an intermediate solution.

Because response additivity and independence are used as the reference for evaluating interactive combined effects and the latter are not precisely known, the classification of multiple pollutant phenomena as super-linear or sub-linear is likewise uncertain.

These difficulties and uncertainties mean we do not know the precise shape of the target isoquant curves, and are heavily reliant on assumptions. The uncertainty problem is mitigated somewhat in that knowledge of all possible cost and impact-equivalent combinations is not needed in order to make specific recommendations on policy. To satisfy the cost-efficiency principle, it is enough for government decision-makers to obtain (from experts) information on a number of alternative combinations and to make the choice between them on cost criteria.

Scientists endeavour to account for uncertainty by assuming linear and additive dose-response relationships and building a margin for error into the laboratory results:

- For dose-response relationships involving individual substances, figures for low doses are linearly extrapolated from those for high doses. Because the true relationship is probably super-linear, this tends to overestimate rather than underestimate the frequency of adverse effects (see Section 2.2).
- Health safety thresholds for individual pollutants with non-stochastic impact phenomena are arrived at by applying a risk factor to safety levels obtained for animals in the laboratory. The risk factor can exceed five orders of magnitude (see Section 2.4). Only part of the risk factor accounts for uncertainties; the larger remainder reflects the scientifically established greater sensitivity of humans compared with animals and interindividual variations in sensitivity within human populations. The pure safety component is not wholly derived from scientific analysis, but is incorporated by scientists based on their assessment of the risks. Risk factors of this kind are scientifically founded "in principle but not in size". The extreme size of some risk factors is scientifically and politically questionable.
- For combined exposures (where the effects are expected to be the same at the target location in the organism), it is assumed that effects of the substances are additive, corresponding with the effect-additivity, dose-additivity or independence model according to the surmised biochemical process. Additivity ignores possible interactions. If in reality the effects of the agents are mutually antagonistic, the risk is overestimated and relatively high costs of environmental protection are countenanced. If in reality the effects are synergistic, the risk is underestimated.
- With combined pollutants, additional protection is incorporated by using several safety margins simultaneously. For the most part, safety margins do not explic-

itly account for interaction risks, but they may offer some protection in this regard by being set at a generous level. If the effects are sub-additive in reality, safety margins result in (further) overestimation of the actual risk. If the effects are super-additive, it is appropriate to use safety margins in principle, but their size is not made explicit; instead, they are built into the sum total correction factor. A lack of reliable evidence of health impacts through combined exposure given that limits for the various individual substances are observed (medicinal drugs excepted; see Section 2.4) merely implies that the safety margins are not too small. It does not rule out their being larger than needed, in which case a lesser degree of precaution would be acceptable and the cost of environmental protection could be lowered.

These added safety factors for uncertainty are not solely based on scientific analysis. They are assumptions that essentially reflect risk assessments and evaluations. It remains an open question why linearity (in the case of effects of individual pollutants), additivity/independence (in the case of combined effects) and high generic risk factors applied to laboratory findings are considered meaningful conditions for safety – or, alternatively, why other assumptions are not considered sufficient. Besides, it is politically not very realistic to require limits that prevent human exposure beyond all doubt (with 100 per cent probability). People take numerous risks (including health risks) in their everyday lives, and accept a certain probability of suffering harm. Consequently, environmental protection cannot be made absolute. This applies not only for risk prevention in the legal sense of the word, but above all for preventive environmental policies.

The risk factors used in deciding environmental standards under uncertainty should be divulged. Natural scientists can make a fundamental contribution towards determining objective risk factors, but subjective evaluation is outside their domain. The situation is different when it comes to economics, where decisions under uncertainty are a subject of research. Economics shows that in certain circumstances rational decisions are possible with imperfect information. Criteria are derived which ought to be observed when making decisions about risk and which are consequently a useful decision aid when setting limit values. Three situations are distinguished according to the degree of uncertainty:

- The decision-maker has reliable knowledge about possible impact severities and can (subjectively) assign probabilities to them (stochastic model).
- The decision-maker can only state probabilities for the frequency or severity of impacts (fuzzy models).
- The decision-maker only has information about the possible severity of impacts (models subject to uncertainty).

4.5.2
Stochastic Decision Model

In line with the scientific and legal approaches, we assume a policy aim of attaining observance of a specific environmental guideline target with a given probability. The policy stipulates probability $p^* \geq 0$ that impact $S^* \geq 0$ is not exceeded. S^* is the guideline target to be observed. In our example (fig. 4-5), the target isoquant curve

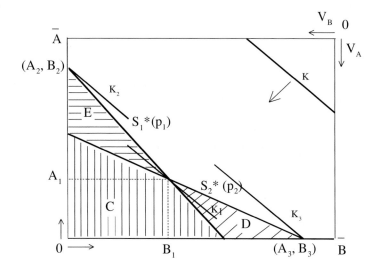

Fig. 4-5 Observance of a protection goal with a specific probability. In the decision-maker's subjective assessment, all combinations of A and B within area C carry a hundred per cent probability that target S* will be observed. If the target is to be observed with certainty ($p_1 + p_2 = 1$), A_1 and B_1 represent the efficient combination of limits for A and B given isocost function K_1. All combinations of A and B in area E (D) carry probability p_1 (p_2) of target S* being observed. For example, if we stipulate a probability of p1, the efficient limits are A_2 and B_2 (if $p_1 > p_2$) or A_3 and B_3 (if $p_1 < p_2$).

is expected to be S^*_1 with probability p_1 and S^*_2 with probability p_2. Area C represents all combinations of A and B for which it is certain that S* will be observed ($p_1 + p_2 = 1$). For combinations in area D, the probability that S* will be observed is p_2, for combinations in area E it is p_1.

The environmental policy objective might be to have one hundred per cent (subjective) probability that all impacts will be prevented (S* = 0 and p* = 1). This requirement matches the scientific concept of limit values with deterministic impact phenomena. It is weakened if we merely stipulate a certain probability p* < 1 of all impacts being prevented. This requirement matches the definition of environmental policy targets in German statute and case law. Exposure limits are specified such that there is a "reasonable probability" that health damage will occur if they are exceeded (Cansier 1994). The policy objective is further weakened if we relax the zero impact constraint and permit a certain level of adverse effects S* > 0. This is unavoidable with stochastic impact phenomena unless emissions are to be prohibited entirely – an extreme solution that would be justified in macroeconomic terms only in exceptional cases – and matches the situation in environmental law with preventive environmental policies. As such policies require reductions in emissions through environmental protection to be held in a reasonable relationship to cost (under the principle of proportionality), society must accept a certain amount of residual emissions and hence of adverse affects.

If we require certainty ($p^* = 1$) that $S^* \geq 0$ will be observed, we look for the least-cost combination in area C. This is found at A_1/B_1. Not only is it unnecessary to lower exposures below A_1 and B_1; doing so would incur excessive macro-economic costs. Note how the safety factors are precisely determined and not explained in terms of exogenously assumed risk margins.

If we vary the environmental policy target and permit a certain impact probability, the solution depends on the ratio of probabilities of occurrence: Let $p_1 < p_2$: To obtain a minimum probability $p^* < 1$ that S^* will be observed, the optimum combination is (A_3, B_3). In area D, the probability of occurrence is greater than p_1. The same result is obtained if the required minimum probability is $p_2^* < 1$. Let $p_1 > p_2$: To obtain a minimum probability $p_1^* < 1$, we look for the least-cost combination in areas C and E. At the assumed slope of the cost function, the optimum combination is (A_2, B_2). If a lower minimum probability $p_2^* < 1$ is required, we look for the least-cost combination in areas C, D and E, and the optimum combination is (A_3, B_3).

4.5.3
Fuzzy Decision Model

Let us assume that our decision-maker knows the set of possible impact severities for a given policy alternative but only has vague assumptions as to their respective probabilities of occurrence. Rather than numeric probabilities, he can only assign them

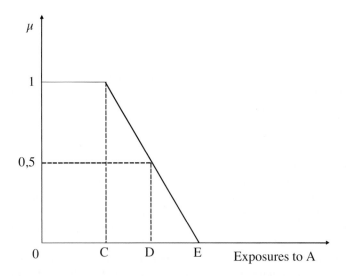

Fig. 4-6 Fuzzy impact interval with a single pollutant. In the exposure range 0 to C, the decision-maker thinks it impossible that a specific severity S^* will be exceeded (rating 1). He does think the target could be exceeded at higher levels of exposure, however, at such levels, the relationship between exposure levels and the probability of compliance with targets is (arbitrarily) assumed to be linear. At exposure level 0D, the decision-maker ascribes a value of 0.5 to compliance with the target. At exposure levels of E or greater he considers compliance with the target to be impossible.

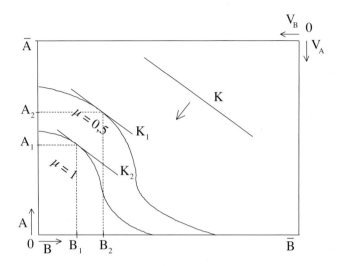

Fig. 4-7 Cost minimisation in a fuzzy model with combined effects. The isoquant curve with $\mu = 1$ plots combinations of agents A and B for which it is subjectively conceivable that impacts will remain at or below the level S*. If the decision-maker requires this degree of certainty, the efficient limit values given isocost function K_2 are A_1 and B_1. If the decision-maker accepts a likelihood of 0.5, on the other hand, the efficient limits are A_2 and B_2.

"degrees of likelihood". We map the problem with an evaluation function $\mu_{S(A,B)}$, which reflects the vague assumptions. In general, one can assign ratings to each impact severity (for more on fuzzy models, see Rommelfanger 1994). The higher the rating assigned to each impact severity, the higher its subjectively adjudged probability of occurrence. Dividing each rating number by the highest rating produces a set of values in the range 0 to 1. We will now proceed to use the values in this range to evaluate the impacts. Severities that are deemed inconceivable are rated 0 and ones the decision-maker considers most likely are rated 1. Imprecise impact ratings of this kind can be represented mathematically in the form of fuzzy impact sets.

By way of example, fig. 4-6 plots a rating function for an individual agent and fig. 4-7 for multiple agents. Ecologically equivalent combinations lie within specific areas. The area containing the set of equivalent combinations is inversely proportional to the stipulated probability that impacts will remain below a specific severity level. If the subjectively greatest possible safety level is required, combinations that come into question are to be found in area $\mu = 1$. On cost criteria, the choice falls to combination (A_1, B_1). This accords with the scientific concept of limit values, which aim to rule out adverse affects given the available knowledge. A policymaker who is satisfied with a probability of occurrence $\mu = 0.5$, on the other hand, would select combination (A_2, B_2).

Fuzzy models can be construed as a weak version of stochastic decision models, requiring less perfect information. It is sufficient for the decision-maker to state "qualitative" severity levels and probabilities of occurrence (such as "low", "medium" and "high"). Qualitative scales of this kind are common, for example, in legal risk theory on preventive environmental policy (Kloepfer 1993). Fuzzy models are thus able to work as a decision aid where stochastic models fail. There is a general demand for such methods since qualitative information is often the only kind available (Munda et al. 1994).

4.5.4
Decision Models Subject to Uncertainty

Let us now assume that the decision-maker can state impact severities but not their probabilities of occurrence. If direct comparison of the policy alternatives fails to yield an optimum combination, we are forced to rely on decision rules like the minimax, minimin and Hurwicz principles. Under the minimax principle, the decision is made in favour of the option with the smallest maximum impact. The decision-maker only considers the worst possible impacts. This brings out a pessimistic or particularly safety-conscious attitude. With the minimin principle the reverse is true. This prefers the alternative with the smallest minimum impact. fig. 4-8 illustrates the two positions. S_1^* and S_2^* are the conceivable isoquant curves with level S^*. Area C contains combinations that are certain to observe S^*. Combinations in areas D and E may observe S^*. The optimist chooses (under the assumed cost con-

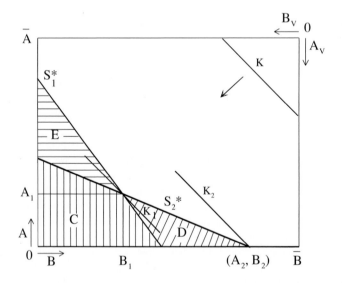

Fig. 4-8 Cost-minimization decision under the minimax and minimin principle. In the example, the decision-maker expects that his ecological objective S^* will be observed by combinations of agents A and B either on isoquant S_1^* or on isoquant S_2^*. He can assign probabilities of occurrence to each of the two estimates. If he is pessimistic, he takes the least favourable result among the alternative combinations of A and B (the combinations in area C) and, assuming the isocost function K_1 and aiming to minimise cost, decides in favour of the combination of limit values A_1, B_1. If he is optimistic, areas C, D and E come into play and the efficient limit values are A_2, B_2.

straints) the least-cost combination in areas C, D and E, i.e. (A_2, B_2); the pessimist chooses the least-cost combination in area C, i.e. (A_1, B_1). This solution accords with the natural science concept of safety standards providing the target $S^* = 0$.

Neither of these decision rules truly lends itself to balanced treatment of environmental risks, since they each take an extreme situation in risk analysis as their starting point. Another approach is the Hurwicz principle, which uses a weighted combination of the greatest and smallest expected impact: $S_H = a \bullet S_{max} + (1 - a) \bullet S_{min}$, where $0 \leq a \leq 1$. The resulting impact function is then confronted with the given protection objective to determine the permissible least-cost environmental standard. This will be closer to solution A_1/B_1 or A_2/B_2 depending on the value assigned to the pessimism parameter (a) or to the optimism parameter $(1 - a)$.

To sum up: in view of the wide variety of uncertainties, decisions about limit values are risk-based decisions. Environmental planning should accordingly be based on systematic risk analysis and make use of economic decision theory. In this point, economic and scientific enquiry should go hand in hand.

The economic decision models clearly show that rational policy decisions are possible with imperfect knowledge. They reveal what factors (relative impacts and relative abatement costs, the extent and measurement of uncertainty, and risk assessment) are important to decisions and how they take effect.

4.6
Conclusions

The cost-efficiency model with multiple pollutants clearly demonstrates that economic methods should be taken into account when setting environmental standards for pollutants in the case of combined exposures and that their relevance is not limited to the selection of policy instruments. The general message is that alternatives should be applied in this decision-making area as well, with the employment of impact-impact and cost-cost comparisons. Since little use is made of such considerations when setting limit values in practice, and since combined effects arise frequently, there is scope to improve current policy in the direction of more cost-effective environmental protection. In the course of reform it may be beneficial (1) to incorporate additional pollutants into policies that have so far focused on individual pollutants (climate protection policy being one example), (2) to exclude all but one pollutant from policies that have so far focused on multiple pollutants (this is likely to be an exception), and (3) to correct the balance of limit values for all agents implicated in a given adverse effect. The economic approach is of benefit even when very little is known regarding ecologically equivalent combinations of agents. For example, if only three ecologically equivalent combinations of a number of implicated substances are known, it is possible to choose between these on the basis of macroeconomic costs. Decision criteria are also available that allow rational decisions to be made about limit values even when there is uncertainty in the data.

Where two or more pollutants act both in combination and separately, separate policies should be followed that apply special limits for combined exposures and otherwise retain the existing limits for individual pollutants (see chapter 5).

The most cost-efficient option in most cases will be to incorporate all significant causal agents into policy. There may be cases, however, in which the most cost-efficient option is to concentrate on a specific pollutant. This may give rise to a conflict with the objective of equal apportionment of burdens, by which all polluters should be called to account. The general applicability rule is integral to the normative legitimization for the "polluter pays" principle. Selective apportionment of burdens causes problems when directly competing enterprises are affected differently, thus distorting free competition. The principle of polluter accountability also applies in liability law. If impacts can be systematically apportioned to polluters, each is held liable for its respective share of the total. If not, all polluters are held jointly and severally liable and any one or any number of them can be called to account for the total (though those who are called to account can claim redress from those who are not). One way of resolving the conflict between efficiency and fairness is to introduce a monetary compensation mechanism between those who come under the policy and those who are unaffected by it. Polluters exempted for efficiency reasons can be induced to share in the cost paid by regulated polluters (the compensation principle; see chapter 5).

4.7
Literature

Bayer S, Cansier D (1999) Kyoto-Mechanismen und globaler Klimaschutz: Die Rolle handelbarer Emissionsrechte, Wirtschaftswissenschaftliche Fakultät der Eberhard-Karls-Universität Tübingen, Tübinger Diskussionsbeitrag 163, March 1999, to be published in: Hamburger Jahrbuch für Wirtschafts- und Gesellschaftspolitik 1999, J.B.C. Mohr, Tübingen 1999
Böhm M (1996) Der Normmensch. Tübingen
Bonus H (1975) Möglichkeiten der Internalisierung externer Effekte als Instrument der Koordination von Unternehmenszielen und gesellschaftlichen Zielen. In: Albach H, Sadowski D Die Bedeutung gesellschaftlicher Veränderungen für die Willensbildung im Unternehmen, Berlin, pp. 207–226
Cansier D (1991) Die Bekämpfung des Treibhauseffektes aus ökonomischer Sicht. Berlin
Cansier D (1994) Gefahrenabwehr und Risikovorsorge im Umweltschutz und der Spielraum für ökonomische Instrumente. In: NVwZ – Neue Zeitschrift für Verwaltungsrecht 7/1994, pp. 642–647
Cansier D (1996) Umweltökonomie 2. edn. Stuttgart
Chao H, Peck SC, Wan YS (1994) Managing Uncertainty: The Tropospheric Ozone Challenge. In: Risk Analysis. Vol.14, pp. 465–475
Der Rat von Sachverständigen für Umweltfragen (1987) Umweltgutachten. Stuttgart
Der Rat von Sachverständigen für Umweltfragen (1994) Umweltgutachten. Stuttgart
Dieter H-H (1985) Risikoquantifizierung: Abschätzungen, Unsicherheiten, Gefahrenbezug In: Bundesgesundheitsblatt. Vol. 38, pp. 250–257
Endres A (1985) Environmental Policy with Pollutant Interactions. In: Pethig R (ed.) (1985) Public Goods and Public Allocation. Bern, pp. 165–199
Hagenah E (1996) Prozeduraler Umweltschutz. Baden-Baden
Hall D-C (1998) Albedo and Vegetation Deman-side Management Options for Warm Climates. In: Ecological Economics. Vol. 24, pp. 31–45
Kloepfer M (1993) Handeln unter Risiko im Umweltstaat. In: Gethmann CF, Kloepfer M (eds.) (1993) Handeln unter Risiko im Umweltstaat. Heidelberg, pp. 55–98
Michaelis P (1997) Effiziente Klimaschutzpolitik. Tübingen
Munda G, Nijkamp P, Rietveld P (1994) Qualitative Multicriteria Evaluation for Environmental Management. In Ecological Economics. Vol.10, pp. 97–112
Rehbinder E (1991) Das Vorsorgeprinzip im internationalen Vergleich. Düsseldorf

Repetto R (1987) The Policy Implications of Non-convex Environmental Damages: A Smog Control Case Study. In: Journal of Environmental Economics and Management. Vol. 14, pp. 13–29

Rommelfanger H (1994) Fuzzy Decision Support-Systeme. 2. edn. Berlin

Tegner H, Grewing D (1996) Haftung und Risikostandards. Strategien im Umgang mit Umweltchemikalien. In: Zeitschrift für Umweltpolitik & Umweltrecht. Vol. 19, 1996, pp. 441–463

Umweltbundesamt (1998) Daten zur Umwelt. Der Zustand der Umwelt in Deutschland. Edition 1998. Erich Schmidt Verlag GmbH & Co., Berlin

von Ungern-Sternberg Th (1987) Environmental Protection with Several Pollutants: On the Division of Labor Between Natural Scientists and Economists.In: Journal of Institutional and Theoretical Economics (JITE) Vol. 143. pp. 555–567

5 Legal Problems of Assessing and Regulating Exposure to Combined Effects of Substances and Radiation

5.1
The Current Legal Situation

5.1.1
Constitutional Duties of Protection

Art. 20a of the German Basic Law (GG) assigns responsibility to the state for protection and conservation of natural resources, in the interests both of current generations and of those to follow. Furthermore, judgements handed down by Germany's Federal Constitutional Court find that the basic right to protection of physical integrity and health (Art. 2(2) sentence 1 GG) makes the state responsible for the protection of life and human health.[1] The protection against dangers and risks to health and the environment provided for under these stipulations has general application. Thus, it covers not only dangers and risks that arise from exposures to individual substances, but also those caused by exposure to the combined effects of multiple substances or of substances and radiation. The Federal Administrative Court made this clear in a later judgement relating to protection from noise pollution.[2] The decision involved the question of whether the segmented assessment method prescribed by the Traffic Noise Abatement Regulation (16th BImSchV), which takes as the starting point for noise abatement only the noise assessed for a new road without taking the prior burden of existing noise into account, is compatible with the authorization to issue regulations under the Pollution Control Act (Art. 43 BImSchG). The Federal Administrative Court judged in favour, but also stated that isolated assessment of differing noise segments reaches its limitations in the constitutional duties of protection under Art. 2(2) sentence 1 GG, which requires that protection against impairments and of health dangers to be provided to a reasonable level of certainty. From a constitutional standpoint, protection against dangers to health requires an all-inclusive assessment that considers in particular the combined effects of multiple substances or of substances and radiation. Then again, apart from the basic requirement to combat exposure to combined effects, the constitution provides no specific guidelines for shaping environmental legislation. With regard to the duty of protection under Art. 2(2) sentence 1 of the German Basic Law, the Federal Con-

---

[1] BVerfGE 49, 89, 141; 56, 54, 73; 77, 170, 214 f.; 79, 174, 201 f.; 88, 203, 251; BVerfG, NuR 1996, 507; NJW 1997, 2509; Steinberg, NJW 1996, 1985 ff.; Steiger, 1998, no. 159 ff.; Murswiek, NVwZ 1996, 222 ff.; Kloepfer, DVBl. 1996, 73.

[2] BVerwGE 101, 1, 4 ff.

stitutional Court constantly finds in favour of an extremely broad discretion on the part of legislators, which – except as regards the protection of human life – can only be tested for constitutional compliance at the limit.[3] The duty of protection is only deemed violated when the state fails to take any protective measures at all, or when the measures taken are wholly unsuitable, are inadequate or fall far short of achieving their protective objective. To put it simplistically, while constitutional duties of protection do not allow exposure to combined effects to be ignored, they contain no specific requirements as to the conditions under which such effects are to be included in legislation and how that legislation should be shaped in each case.

5.1.2
Environmental Protection Law

5.1.2.1
General

Media and substance-related environment laws often contain generally worded objectives and authorizations for action by the state and incorporate risk terminology from police and public-order law in provisions specific to the area being regulated, while often expanding the legal provisions for protection to include risk prevention in anticipation of danger. Wordings like "harmful environmental effects", "harmful soil changes", "harmful effects of substances" or "unacceptable impacts on the balance of nature" provide evidence of media or substance-specific protection, but also imply that there is no need to differentiate between combined effects and the effects of single substances or radiation alone. The relevant literature lacks extensive discussion on this issue, though frequent mention is made of combined effects – if only in passing – in conjunction with incrementally cumulative (summative) effects.[4] There is thus no doubt that, under the pertinent laws, the duty of protection inherently covers exposure to combined effects. However, there is a lack of in-depth debate, and the actual regulatory issue is masked by treating combined effects of different substances in the same way as cumulative emissions of a single substance from different sources. In contrast, there is far more detailed debate on the distantly related issue in noise abatement of whether isolated assessment of individual noise segments is compatible with the Pollution Control Act.[5]

---

[3]   BVerfGE 56, 54, 80 f.; 77, 170, 214 f.; 79, 174, 201 f.; 85, 191, 212 f.; 92, 26, 46; BVerfG, NuR 1996, 507; NJW 1997, 2509.

[4]   See in particular Murswiek, WiVerw 1986, 179, 195 f.; Lübbe-Wolff, 1986, p. 178 f.; Roßnagel in *Gemeinschaftskommentar zum BImSchG*, 1998, § 5, no. 114; Kutscheidt, in Landmann/Rohmer, 1998, § 3, no. 9 b; Jarass, 1999, § 3, no. 8, § 48, no. 23; Reich, 1989, p. 121 f.

[5]   See 5.1.2.2.b.

5.1.2.2
Laws Relating to Environmental Media

The objectives of and legislative powers granted under most important media-specific laws call for the assessment of combined effects.

a) This is generally recognized in terms of air pollution under the Pollution Control Act (BImSchG).[6] Besides Art. 1, 3(1) and Art. 5(1) no. 1. BImSchG dealing with pollution independent of source, Art. 44(2)3 sentence 3 makes clear that the duty of protection under the Act is all-encompassing. Thus, when determining "investigation sites" under the Act (formerly "exposure sites"), consideration must be given to environmental dangers posed by interactions between pollutants. Interactions between pollutants and ionizing radiation should also be taken into account.[7] Then again, there are often restrictions that only allow assessment of combined effects when there is sufficient scientific evidence of their harmful nature.[8] If no such evidence is available, it is assumed that any potential occurrence of combined effects will be adequately dealt with by general emission reductions under the precautionary principle (Art. 5(1) no. 2 BImSchG).[9] This solution matches those recommended in the literature on cumulative damage and applied in existing policies.

The provisions of the Ozone Act – which since 1995 has formed part of the Pollution Control Act (Art. 40a–40e BImSchG)* – deal with a specially problematic situation. The health hazards and harm to vegetation caused by surface ozone do not involve combined effects as such, but rather the effects of a reaction product that occurs in the atmosphere as a result of interaction between secondary substances under certain conditions. The law makes no provision for reducing the secondary substances in view of their part in the ozone creation process, and restricts itself instead to general reduction measures like traffic bans and warnings in anticipation of unacceptable ozone levels.

b) Opinions differ greatly when it comes to noise abatement. While it is often assumed that only assessment of all noise source categories meets the protection requirements under the Pollution Control Act[10], the varying levels of nuisance caused by different types of noise call for more differentiated assessment (say by distinguishing between summative noise pollution from similar sources and pollution from different sources). The Pollution Control Act itself provides justification for source separation where aircraft noise is excluded from its scope of application (Art. 2(3) BImSchG) or where the Noise Abatement Regulation grants powers to set

---

[6] See footnote 5; also OVG Lüneburg, DVBl. 1977, 347, 350; *Ausschußbericht*, Bundestagsdrucksache 7/1513 p. 3, 4.

[7] OVG Lüneburg, DVBl. 1977, 347, 350.

[8] Kutscheidt, in Landmann/Rohmer, 1998, *§ 3,* no. 9b, 9c; OVG Lüneburg, DVBl. 1977, 347, 350.

[9] BVerwGE 69, 37, 42; OVG Münster, NVwZ 1991, 1200, 1202.

* Repealed in 2001 and replaced by EC regulation.

[10] See in particular Koch, 1990, p. 54 ff; idem 1999, p. 218 ff.; idem JUTR 1989, 205 ff.; Petersen, 1993, p. 51 ff., 179 ff.; Feldhaus, JUTR 1993, 29 ff.; idem 1998, p. 181 ff.; Michler, 1993, p. 164 f.; Jarass, 1999, *§ 3,* no. 34f., *§ 5,* no. 16; idem DVBl. 1983, 725, 727; with qualifications: Kutscheidt, NVwZ 1989, 193, 199; idem NWVBl. 1994, 284 ff.; BayVGH, BayVBl. 1990, 84.

noise limits for individual roads and railways (Art. 43, 16th BImSchV). The Federal Administrative Court[11] is of the opinion that noise levels set under the Noise Abatement Regulation are compatible with the Act when the noise caused by a new road is assessed alone without taking the prior burden of existing roads into consideration. The only corrective identified by the Court is via the duty of protection under Art. 2(2) sentence 1 of the German Basic Law.

c) Atomic energy law also assumes that the protection required under Art. 7(2) no. 3 of the Nuclear Energy Act (AtG) includes prevention of harm that can be caused or exacerbated by the combined effects of ionizing radiation and other pollutants. In particular, the exposure minimization requirement might be violated if the site of a nuclear power plant is chosen in such a way that any radioactive pollution it emits as a result of the combined effects of hazardous substances can considerably increase the health risks to local residents. This principle was developed by the Lüneburg Higher Administrative Court, in a well-founded judgement passed in 1978.[12] The Court based its findings on the fact that under the protection objectives of the Nuclear Energy Act it is sufficient for radioactive substances emitted from a nuclear power plant, together with other pollutants emitted at the same time, to be the cause of harm to local residents. This also applies when the effects of combined exposure are largely attributable to chemical pollution, when the presence of even small doses of radiation can significantly increase the harmful effects. Whether the combined effects of chemical and radioactive substances are additive or synergetic is irrelevant. The deciding factor is whether interaction between different substances can cause harm that could not be caused in the same way or to the same extent by any one of the substances on its own. This can also apply when one of the substances merely sensitises human cells to the effects of another pollutant, or when it impairs repair mechanisms. However, violation of the minimum exposure requirement at a given site is only assumed when the expected effects of combined exposure significantly exceed those which could be expected at another site.

In this case, however, the Court based its judgement on the findings of an expert witness and assumed that, according to available scientific knowledge, it was not possible to weigh the extent of the exposure to combined effects in relation to specific pollutant concentrations and radiation doses.

Given the low doses found, the Court believed it highly unlikely that normal operation of the nuclear power plant – even considering the prior burden from emissions of other nuclear power plants and soil contamination from chemical substances – would cause a statistically significant increase in the risk to the population.

The Federal Administrative Court[13] confirmed the lower instance judgement, arguing that at an exposure of 30 mrem (0.3 mSv), the radiological portion of the total cancer risk in relation to chemical substances is negligible. Without expressly

---

[11] BVerwGE 101, 1, 4 ff. rebutting Schulze-Fielitz, UPR 1994, 1, 4 and Michler, 1993, p. 164 f.; on the prior law and on facilities not listed under 16. BImSchV: BVerwGE 87, 332, 357 f.; 59, 253, 268; 52, 226, 236; BGHZ 97, 361, 364 f.: in cases of pre-existing pollution load, unreasonableness is only established if an increase in noise following the opening of a new road or other development is unreasonable.
[12] OVG Lüneburg, DVBl. 1979, 686, 689 f.
[13] BVerwG, NVwZ 1982, 624, 627.

stating it, the court also ruled in favour of the relevance of combined effects to risk prevention under Art. 7 (2) no. 3 of the Nuclear Energy Act (AtG).

Nevertheless, there are cases in nuclear energy law in which radiation segments are considered in isolation. This was the case, for example, after the Chernobyl accident when the question arose as to whether the prior burden from that accident should be considered in licensing nuclear power plants. The courts threw this out on the grounds that the legal system requires that all exposures from normal operations be considered but not those resulting from accidents.[14]

d) The question of the extent to which combined effects should be considered when licensing facilities has also been raised, though not subjected to detailed discussion, in genetic engineering law.[15] It can be assumed that synergetic effects are to be considered in terms of the risk potential to donor and receiver organisms. Winter[16] considers not only additive but also synergetic effects particularly relevant in genetic engineering.

e) In contrast, the conversion of loads or concentrations of different hazardous substances into equivalent risk units under the German Waste Water Charges Act (AbwAG – Annex to Art. 3) cannot be regarded as a method for taking combined effects into account. It is merely a way of assigning monetary values in order to assess charges, for which combined effects as such are irrelevant.

5.1.2.3
Chemicals Law

With some exceptions, chemicals law tends to focus on the effects of individual substances when identifying the properties of chemicals. Combined effects are included in substance risk regulation, however.

a) A single-substance approach is taken in connection with the identification and assessment of harmful properties under the German Chemicals Act (ChemG) and the EU Regulation on the Evaluation and Control of the Risks of Existing Substances (793/93/EEC). Like the Dangerous Substances Directive (67/547/EEC as amended by 92/32/EEC), the Chemicals Act (Art. 7 onwards) requires manufacturers to perform substance-specific risk assessment to identify harmful properties of new chemicals. The EU Directive on Risk Assessment for New Notified Substances (93/67/EEC) likewise makes no provision for assessment of combined effects. The same goes for the Regulation on the Evaluation and Control of the Risks of Existing Substances (793/93/EC) and the subsequent Regulation on Risk Assessment for Existing Chemicals (1488/94/EEC). These do include assessment of substances as parts of preparations (Annex 1 B 3.4 and Annex III 3.3 of Directive 93/67/EEC and Regulation 1488/94/EEC), but this is not the same as assessing preparations as such in terms of combined effects. The only question that arises here is whether a manufacturer can be ordered to submit proof of testing for combined effects as part of

---

[14] BVerwG, NVwZ 1991, 1185; OVG Berlin, NVwZ-RR 1991, 180, 183 (with further references).
[15] See Winter/Mahro/Gintzky, 1992; Hirsch/Schmidt-Didczuhn, 1991, § 7, no. 13; Eberbach, 1997, *GTVO Einleitung*, no. 93 ff., § 4, no. 191 f.
[16] Loc. cit. p. 39 f.

testing under Art. 11(1) no. 3 ChemG. This cannot be ruled out on the face of it since, looking only at the requirements that must be met before such tests can be ordered, there must merely be evidence – a scientifically founded suspicion is enough – that the substance is harmful and that testing is needed for prevention of dangers or precaution. Such risks can also arise in specific applications and so from interactions with other substances. However, testing data that can be ordered under Art. 11(1) no. 2 ChemG are limited to those which can be required of manufacturers in the Level I and II procedure when the quantity thresholds defined in the Act are reached. This appears to rule out an extension of testing to include combined effects. Such effects can only be considered, and in turn invoke restrictions under Art. 11(2) or Art. 17 and 19 ChemG, when supported by data that the competent authority has obtained from other sources when assessing risks arising in the use of the substance concerned.

The provisions of the Chemicals Act on classification, packaging and labelling and on prohibitions and restrictions are wider in scope. Art. 13(2) and Art. 14 ChemG contain stipulations on classifying, packaging and labelling preparations. These allow classification based on mathematical formulae, but prefer classification based either on testing results or established knowledge (Art. 13(2) sentence 2 ChemG). Given the broad scope they derive by reference to the Act's general protection objective, the legislative powers to issue safety regulations under Art. 17 and 19 ChemG in any case allow combined effects to be taken into account. Thus, Art. 18(1) sentence 2 of the German Hazardous Substances Regulation (GefStoffV) requires employers to assess the overall effect of different hazardous substances in the workplace.

b) The licensing procedure under the German Plant Protection Products Act (PflSchG) involves only limited assessment of combined effects. Under the EU Uniform Evaluation Principles Directive (94/43/EC), assessment must include not only products explicitly recommended for use with others in a tank mix, but also products that have the same active substances or residues. Thus, depending on the circumstances, interactions between two or more plant protection products with different active substances but the same metabolites can increase the risk to human health and/or the environment to unacceptable levels. Outside this narrow context, combined effects play no part in licensing procedures for plant protection products.

The Commission's proposal to replace Directive 94/43 makes no attempt to change this.[17]

The legal situation is the same as regards the licensing of biocides, which must be separately regulated in future under the EU Biocidal Products Directive (98/8/EU).

The prohibitions and powers enshrined in the German Foodstuffs and Commodities Act (LMBG) allow the consideration of combined effects.

c) One exception to the principle of single-substance assessment is the German Drugs Act (AMG), whose provisions require manufacturers to identify and label side-effects of interactions with other drugs (Art. 22(1) no. 9 in conjunction with

---

[17] Proposed directive of 6.8.1997, OJ C 240/1; now adopted as Directive 97/57, OJ L 265/87.

Art. 28(2) nos., 2a, 11(1) no. 9 and 11a(1) no. 7 AMG).[18] Combined effects with other drugs can lead to rejection of a license application under Art. 25(2) no. 5 AMG. More usually, however, the licensing agency – the Federal Institute for Health, Consumer Protection and Veterinary Medicine (BgVV) – requires appropriate labelling, user information and prescriber information. The Drugs Act makes no provision for assessment of the combined effects of drugs and other substances.

5.1.3
Secondary Legislation

5.1.3.1
General

Given the paucity of legal debate on combined effects, it is surprising to find that they are included in considerable scope in secondary legislation – in regulations and administrative orders implementing primary statutes. This involves the setting of specific limit or guideline values ("limits" in the following) for the kind of effects that are the focus of this study. Further, there are trigger values for specific substances, above which legal obligations are triggered and which apply to mixtures in some cases. Because classification into harzard categories has legal consequences – as regards labelling and certain statutory duties – these action thresholds may be regarded as lesser-degree limits.

In practical terms, secondary legislation largely assumes multiple substances to have dose-additive effects with linear increases in dose-effect relationships or even simple concentration additivity. Coverage of sub-additive and super-additive effects is deficient, and there is no provision for interactions between substances and radiation, the latter being regulated in other laws.

5.1.3.2
Limits

a) Environmental quality standards attempt to deal with combined effects in part by incorporating safety margins. Thus, section 2.5, sentence 3 of the Technical Instructions on Air Quality *(TA Luft)* states that the emission limits laid down in the Technical Instructions also apply when multiple air pollutants occur simultaneously. The reasoning is that the limits are based on epidemiological research where simultaneous presence of different substances is assumed[19] (this, of course, applies solely in health protection). The same goes for the findings of the German Nuclear Safety Commission (SSK) and the relevant literature on dose limits under the Nuclear

---

[18] Sander, 1994, *§ 2,* no. 10, *§ 1,* no. 9.
[19] OVG Münster, NVwZ 1982, 451 f.; Hansmann, in Landmann/Rohmer, 1998, *Nr. 2.5 TA-Luft,* no. 14 f. and no. 4; Jarass, 1999, *§ 48,* no. 23; contra regarding TA-Luft 1974: OVG Lüneburg, DVBl. 1977, 347, 350.

Safety Regulation (StrSV, 2001 of 20 July).[20] This is in line with the mandate in Art. 45(3) StrlSV* to consider radioactive exposures from all sources. Apart from the stated references to epidemiological findings, the scientific basis for the limit philosophy remains unclear. In practice, safety margins can also be lowered in various ways depending on whether combined effects are anticipated and on their type (additive, super-additive or sub-additive).

The simultaneous occurrence of multiple pollutants is often taken into account by modifying the system of limits. While most are based on emission values, a few systems of environmental quality limit values also target combined effects.

b) In pollution control, this applies particularly when declaring "investigation areas" under Art. 44(2) senctence 3 of the Pollution Control Act (BImSchG), which requires an area to be declared an investigation area when multiple pollutants are present and emissions reach 90% of a set limit in the case of two pollutants or 70% in the case of three pollutants.[21] This rule applies generally, whether or not combined effects are anticipated in a given instance.

An interesting example is the synthesis approach adopted by the *Länder* Committee on Pollution Control (LAI) to limit cancer risk from air pollutants.[22] This uses a unit risk model to determine the unit risk of important carcinogenic air pollutants based on their carcinogenicity (unit risk – the added cancer risk from lifetime exposure to 1 mg/m^3 of a substance) and prevalence in the environment. On this basis, and adopting a policy target for total cancer risk from air pollutants, an individual guideline value is arrived at for each substance. These values are in no way binding, but serve as orientation values for casewise assessment under *TA Luft* in cases where no binding emission values exist. The LAI synthesis approach assumes that, independent of a substance's target organ, the total cancer risk can be determined by adding weighted individual risks (dose-additivity/presence in the environment). This is a scientifically accepted method, including for cases where substances attack different target organs. Further, the basic assumption of linear dose-response relationships in carcinogens, at least at low doses, is in line with available scientific knowledge. The synthesis approach is one of the more science-based approaches to addressing problems of combined effects. Its failure to cover non-homogeneous mixtures can be adequately compensated by combining it with casewise assessment.

In contrast, the 1999 Soil Protection and Contaminated Sites Regulation (BBod-SchV), issued under the Soil Protection Act (BBodSchG), takes a single-substance approach in setting limits.[23] Interactions between carcinogens on a contaminated

---

[20] *Empfehlungen der Strahlenschutzkommission "Synergismen und Strahlenschutz"* of 23.9.1977, Bundesanzeiger 212 of 11.11.1977; Schattke, 1979, p. 101, 115 f.; Hädrich, 1986, § 7, no. 65; contra OVG Lüneburg, DVBl. 1979, p. 686, 689 f. Now Art. 46(3) StrSV of 20 July 2001.

* Now Art. 46(3) StrSV of 20 July 2001.

[21] LAI criteria of 9.10.1991, cited by Hansmann in Landmann/Rohmer, 1998, § 4, no. 17.

[22] Länderausschuß für Immissionsschutz, 1992; commenting thereupon: Franßen, 1992, p. 22 f.; Salzwedel, 1992, p. 24 ff.; Breuer, 1991, p. 158 ff.; Kühling, ZUR 1994, 112 ff.

[23] The limits concerned are concentrations at which further investigation or action is required under Art. 8(1) nos.1 and 2 BBodSchG.

site are dealt with by setting lower tolerance levels and hence lower limits for individual substances.[24]

The German Drinking Water Regulation contains sum total limits for polycyclic aromatic hydrocarbons (PAHs), adsorbable organic halogens (AOXs) and plant protection products, all based on dose-additive effects.[25] In the first two groups, the same values are stipulated for multiple pollutants as for a single substance; the reliance on the (total) carbon content of PAHs probably does not give enough consideration to the risk potential of that substance group. For plant protection products, however, the sum total limit is set at a factor of five. In the case of single substances, the single-substance value must be complied with in each case. The reason for applying a different limit model to plant protection products probably lies in the fact that the single-substance value is already set at the analytical limit and it is neither practicable nor reasonable to set that value as a sum total limit.[26]

Blanket, overall treatment of combined effects is also found in the German Cosmetics Regulation Art. 2(2), Art. 2(3a) and Art. 3 in conjunction with Annexes 2 and 6 stipulate maximum percentages of fluoride, strontium and mercury compounds as components of or as preservatives in cosmetics. These must also be complied with when any of the compounds is mixed with others in the same group. The result is a combined maximum percentage for fluoride, strontium and mercury.

A dose-additive approach to limiting the risk of combined effects is also used in occupational safety and health legislation. Under the Technical Instructions on Hazardous Substances (TRGS 403)[27], in cases where there are multiple harmful substances for which MAK values exist (*Maximale Arbeitsplatzkonzentration*: the maximum permissible concentration of a chemical compound present in the air within a working area, which according to current knowledge does not impair the health of employees or cause undue annoyance), each limit may only be exploited in the proportion the actual substance concentration bears to the MAK value. This model assumes linear dose-response relationships, although it does consider the differing slope of the dose-response curves. Priority is always given to independent scientific assessment of the mixture. The general assessment method is not applicable when completely independent effects occur or could occur. Conversely, the addition method should not be used for substances with a concentration below 10% of the MAK value because it is assumed that they then no longer contribute to the increased risk. The MAK Values Commission has long claimed that the TRGS assessment method is scientifically unacceptable because of the extent to which it fails to take into account the different mechanisms and action sites of different substances.[28] The Commission has

---

[24] See *Ressortabgestimmte fachliche Inhalte einer Bodenschutz- und Altlastenverordnung* of 28.5.1997 (key issues paper), p. 66f.

[25] The hazard levels above the TWV limits, proposed for assessment of short-term peak concentrations, likewise assume dose-additivity; the maximum level proposed for any given substance is set in accordance with the percentage of the mixture represented by that substance; see Dieter/Grohmann/Winter, 1996, p. 199, 201.

[26] On this problem, see Rengeling, 1984, p. 25 ff., 44; Kolkmann, 1991, p. 159 f., 165 ff.; Wegener, IUR 1991, 14, 15 f.

[27] Bundesarbeitsblatt 1989, p. 71; see Arbeitsgemeinschaft für Umweltfragen, 1997, p. 25; Enquetekommission, 1994, p. 472.

[28] Deutsche Forschungsgemeinschaft, 1998, p. 13; likewise with regard to BAT values p. 169; Henschler, 1981, p. 92 f.

yet to offer an alternative solution, although it is currently conducting a study on vapour mixtures and BAT values *(Biologischer Arbeitsstoff-Toleranz-Wert:* biological tolerance value for occupational exposures) for preparations. In practice, however, and particularly in legal literature, the pragmatic approach used in TRGS 403 is generally accepted because when the aim is comprehensive health protection one cannot wait for scientific evidence on the effects of mixtures in the workplace. The (narrow) safety margins for individual substances are inadequate because they do not incorporate uncertainty as regards combined effects.[29] It must be pointed out that, independent of the question of response mechanisms, the assumption of dose-additive effects at low doses usually results in overestimation of combined effects.

The blanket assessment method under TRGS 403 does not apply to carcinogens as no MAK values have been set for them, although the EU Directive on the Protection of Workers from the Risks Related to Exposure to Carcinogens at Work (97/42/EC) makes reference to the classification of preparations according to general requirements (see 5.1.3.3). In compliance with this provision, the handling requirements contained in the Directive also apply to mixtures and preparations that are classified as carcinogenic under the Dangerous Preparations Directive (87/379/EEC).

c) When it comes to emissions, combined effects are often taken into account by setting specific emission limits.

In air pollution control, the German Regulation on Incinerators for Waste and Similar Combustible Material (17the BImSchV) uses a sum total limit for dioxins and furans relative to their respective toxicities and expressed in equivalence factors (Art. 5(1) no. 4). The combined effects of organic substances in changing mixtures are considered in part by setting a total carbon value for the emissions (Art. 5(1) no. 2 (a), 17th BImSchV). However, this approach to regulation is scientifically problematic in that carbon content bears no relation to the harmfulness of a mixture's component substances.

The Technical Instructions on Air Quality *(TA Luft)* lay down sum total limits for a range of air pollutants, particularly carcinogens, anorganic substances in dust or particulate form and certain organic substances. These values aim to provide for interactions between multiple substances, assuming dose-additive effects. For carcinogens, *TA Luft* lays down three risk classes, generalising for different degrees of carcinogenicity with different individual concentration limits. If two or more carcinogens in the same risk category occur together, the individual concentration limit becomes a sum total limit. Two carcinogens in different risk categories are subject to the individual concentration limit for the higher-risk substance plus the limit for the lower-risk substance (section 2.3 *TA Luft*). This ensures that any combination of substances is subject to the stricter limit if the higher-risk substance dominates, but the limit for the lower risk category can only be exploited to the extent that it has not already been used up by the higher-risk substance. This limit system has a loophole when it comes to emissions of carcinogens from all three risk categories. In practice, it is assumed that the concentration limit for substances from the lowest risk category may not be exceeded by the sum total concentration of all the sub-

---

[29] Falke, 1986, p. 176; Beyersmann, 1986, p. 69.

stances.[30] However, it is clear that an increased risk must be accepted with mixtures containing substances from different risk categories, as the limit for the highest risk category may be exhausted while additional emissions of substances from the next-lowest risk category remain admissible. Similar apportionment-based approaches are used for anorganic substances in dust or particulate form and for certain organic substances (priority being given to the provisions on carcinogens under Sections 2.3; 3.1.4 and 3.1.7 TA-Luft).

In water pollution control, numerous wastewater management regulations contain emission values for polycyclic aromatic hydrocarbons (PAHs), and especially for adsorbable organic compound halogens (AOX), that take into account the presence of multiple substances in the same group. Under assumptions of equal toxicity and linear dose-additivity, a sum total limit is set to the value that applies for any individual substance in the group.

d) Digression: Noise. In practice, noise abatement primarily focuses on individual noise sources. Thus, the German Regulation on Traffic Noise Abatement (16th BImSchV) regulates only the relevant assessed noise portion, e.g. the noise from a new road without considering existing noise from other sources. Courts have declared this procedure to be legally correct,[31] but with the caveat that such segmentation is unacceptable when the total noise burden results in health risks because the state's duty of protection then applies under Art. 2(1) sentence 2 of the German Basic Law. The relevant literature sometimes interprets this otherwise.[32] The Regulation on Noise from Sports Facilities (18th BImSchV) goes somewhat further in that it applies a sum total limit (Art. 2(1)) to noise emissions from all sports facilities although external noise is excluded.

A single source assessment method was also used in the 1968 Technical Instructions on Noise *(TA Lärm)*. In determining whether the guideline emission values stated in *TA Lärm* were exceeded, only emissions from the facility under consideration had to be taken into account (Section 2.211), even though some courts based their reasoning on the sum total level from all facilities run by the same operator[33]. Under the protection objectives of the Pollution Control Act (BImSchG), segmented assessment of this kind is now outdated.[34] The 1995 Model Administrative Procedures *(Musterverwaltungsvorschriften)* adopted by the German *Länder*[35] tried to take this into account by basing assessment of facility noise on the receptor rather than the source and setting sum total parameters for all (existing and planned) sources of noise. Compliance with the noise limits also had to be ensured in the event of combined noise (present and future) from different facilities.

In cases of noise from facilities combined with outside sources (particularly traffic), provision was made for the possibility of different qualities of noise by allow-

---

[30] Hansmann, in Landmann/Rohmer, 1998, *TA Luft Nr. 2.3*, no. 17; Junker u.a., 1994, *TA Luft Nr. 2.3*, no. 10; Kalmbach/Schmölling, *TA Luft*, 1994, *Nr. 2.3*, no. 31.

[31] BVerwGE 101, 1, 4 ff.; on existing case law cF. BVerwGE 51, 32; 59, 253.

[32] See references in Footnote 11; Koch, 1999, p. 225 ff.

[33] VG Köln, in Feldhaus, *ES Bundes-Immissionsschutzrecht, GewO § 16-17*; *Runderlaß NW zur TA Lärm, MBl. 1975*, 234.

[34] See references in Footnote 10.

[35] Reproduced in Landmann/Rohmer, 1998, Vol. II, Section 4.1.

ing casewise assessment. This was only to be used, however, where external noise sources did not drown out the noise attributable to the facility and noise exposure was significant, which was deemed to be the case in particular when both noise types reached the applicable exposure limits. The Model Administrative Procedures provided assessment criteria for such cases that relied on the merits of abatement, the level of nuisance and the sources' relative contribution to total noise. For new facilities, the main points for consideration were what the site was used for, dominant noise sources, use history, relative contributions to total noise, noise duration and nuisance level, the willingness of the local residents to put up with the noise and the options for passive noise abatement. Where existing facilities are concerned, additional restrictions were set in place to uphold the status quo.

The Model Administrative Procedures were intended to be used where *TA Lärm* had become obsolete due to developments that had come about since its issue. In the meantime, a new *TA Lärm*[36] has entered into force and is taking the place of the Model Administrative Procedures. The new *TA Lärm* takes a much narrower approach. It also lays down sum total parameters for noise pollution from all facilities that come within its scope (although differing noise characteristics are to be taken into consideration through individual assessment, where appropriate). On the other hand, provisions on external noise are somewhat lacking. Traffic noise – subject to restrictive provisions – is included in the assessment where it involves traffic arriving at and leaving a facility. Other traffic noise and noise from sports facilities, leisure facilities and open-air restaurants is not taken into account at all. An earlier draft went further, however, saying that external noise in excess of 50% of the sum total parameter was unacceptable. *TA Lärm* also allows a gradual increase of noise levels through cumulation of additional exposures over time.

A problem with summative noise pollution is that different types of noise are not always comparable in terms of their nuisance level. This is similar to the problems with combined effects of multiple substances or of substances and radiation. Nonetheless, recognition of differences between noise types hardly justifies segmented assessment. Instead, it calls for casewise assessment.

5.1.3.3
Action Thresholds Giving Rise to Legal Obligations

Legal obligations created in environmental laws or the regulations issued under them often rely on specific values being exceeded as a way of delimiting systems, substances and activities that are considered especially harmful and therefore in need of strict regulation from others that can be regulated less strictly. Such cases again raise the question of how much consideration to give to the simultaneous occurrence of multiple substances and their possible combined effects.

The German Hazardous Accidents Regulation (StörfallV) aims to prevent serious hazards from operational accidents involving major emissions, fires and explosions. Art. 1(2) imposes particularly onerous obligations – particularly the obligation to conduct a safety assessment – for certain facilities in which particularly hazardous substances are present or could occur as the result of an accident, where they

---

[36] GMBl. 1998, 503; commenting thereupon: Feldhaus, 1998, p. 181 ff.

exceed certain quantity thresholds. Two quantity threshold categories are provided for certain types of industrial facilities. The higher quantity threshold applies to all facilities run by the same operator that are less than 500 metres apart or where the distance between them, though greater than 500 metres, is not sufficient for other reasons to rule out, under foreseeable circumstances, the occurrence or escalation of a severe danger. In such cases – though not for facilities run by different operators – the quantities for the various individual facilities must be totalled when assessing significance levels for individual substances, preparations or substance groups (Art. 1(4)). However, the wording of the regulation implies that the addition model is not meant to be used for interactions between different substances or preparations. Germany's Hazardous Accident Administrative Regulation no. 1 (StörfallVwV no. 1, Section 3.6) makes no provision for this type of summative assessment of different substances, though its substance categories, and especially its Annex III applying to storage facilities, take additive combined effects into account in a very general way by setting a joint quantity threshold. The same goes, for example, to plant protection and pest control products and to substances classified in certain hazard categories under the Dangerous Substances Regulation (GefStoffV).

Even if the thresholds contained in GefStoffV are not reached by combined effects or do not take such effects into account, Art.1 (3) StörfallV authorizes the competent authority to impose special safety obligations on operators on a casewise basis to prevent accidents or to limit the impact of an accident. This extends to combined effects. StörfallVwV no. 1 expressly stipulates that if there are multiple substances with similar properties, each below the quantity threshold, the quantities are to be added. If the sum of the individual quantities equals or exceeds the quantity threshold for the risk category concerned, the possibility of an accident cannot be dismissed out of hand (StörfallVwV no. 1, Section 3.3.5). This shows in general that combined effects can be a reason for imposing safety obligations, even in cases where they are not explicitly considered in the generalizing regulatory approach used in a given Regulation.

Under the Seveso II Directive (96/82/EC), StörfallV and StörfallVwV no. 1 must be amended in the near future to make improved provision for combined effects[*]. The Directive lays down threshold quantities that trigger notification requirements and an obligation to draw up a safety plan (Art. 6 and 7), and requires periodic reporting on implementation of such safety plans (Art. 9). Where there is an interaction between certain substances or between substances in certain categories (e.g. highly toxic, toxic and/or environmentally hazardous substances, explosive, highly flammable and/or flammable substances), the thresholds can be reached by addition of the quantities present. The addition formula prescribed in the Directive sets the quantity of each substance present in relation to the threshold quantity, effectively imposing a lower, pro-rata threshold for each substance (Annex 1, Part 2, Note 4). Preparations are classified in accordance with the Dangerous Substances Directive and the Dangerous Preparations Directive (Annex 1, Part 2, Note 1), and as such are likewise subject to the threshold quantities and the addition formula. On the other

---

[*] See now StörfallV of 26 April 2000.

hand, substances that form part of a preparation but are not classified as dangerous because they do not exceed a specific, low concentration are treated as pure substances. That is, the quantity thresholds do not apply to preparations but do apply to their component substances, and so the latter must be incorporated into the addition formula where appropriate. Note that the values mentioned are thresholds that trigger specific obligations. Unlike nuclear energy law, hazardous accident law does not set a limit on the maximum allowable emissions or exposures in the event of an accident.

Accident prevention is also covered by occupational health and safety provisions in the 1992 Technical Instructions on Hazardous Substances (TRGS 514 and 515) for combined storage of hazardous substances. These prohibit or restrict combined storage of highly toxic and toxic substances, and of explosive, highly flammable and flammable substances with other substances. The provisions aim to prevent poisoning as a result of fire, toxic leaks or toxin dispersal, and to limit the hazards of explosions and fires. This largely involves risks from interactions between multiple substances.

5.1.3.4
Classification of Mixtures and Preparations

Classification rules are a generalized way of stating the harmfulness of a substance, a mixture or a preparation (a manufactured mixture). Classification triggers legal obligations like labelling requirements, requirements to take special precautions when using substances, and even requirements to substitute certain substances for others.

a) The classification of water pollutants under Germany's 1999 Administrative Regulation on the Classification of Water Pollutants (VwVwS) serves to protect bodies of water from accidents and incidents. It applies not only to individual substances, but also to manufactured or stored preparations and to mixtures that occur in commercial operations. Imputation of quantities among polluters is thus not an issue.

The Administrative Regulation divides water pollutants into three risk categories. Classification is performed by an independent commission of experts. The amendment of the 1999 regulation to expand the existing process introduced self-classification based on risk categories under the Hazardous Substances Regulation (GefStoffV: R phrases) and on a points scale (target values) for incompletely tested substances.[37] The aim is to avoid automatic assignment to the highest risk category as called for under prior *Länder* regulations.

For mixtures, the Administrative Regulation generally provides for self-classification into the three risk categories using mathematical formulae (no. 2.2 of the Regulation in conjunction with Annex 4).

Combined effects are accounted for here by assuming concentration-additivity and, for each substance in a mixture, taking the typical risk potential and pro-rating for the substance's percentage share of the mixture (dilution effect). While dosage (quantity per unit time) is not considered per se, the assumption of concentration additivity can be regarded as a generalized form of dose-additivity. A distinction is made between

---

[37] See *Amtliche Begründung*, Drucksache des Bundesrats 782/98, p. 56 f., 60.

substances with thresholds and carcinogens. With the former, mixtures containing more than 3% by mass of substances in Water Pollution Risk Category 3 (the highest) are classified in Category 3. This is based on the concept that such substances dominate the harmfulness of the mixture as a whole. The Administrative Regulation also requires mixtures of unknown content to be classified in Category 3. Category 2 applies to mixtures that contain more than 5% of Category 2 substances (based on the sum total) or upwards of 0.2% but not more than 3% of individual Category 3 substances. Category 1 covers mixtures that contain more than 3% of substances in Category 1 (based on the sum total) or upwards of 0.2% but not more than 5% of individual Category 2 substances or individual substances that do not meet all requirements for classification as non-water polluting. Non-water-polluting mixtures are those that contain less than 3% of Category 1 substances and no more than 0.2% of Category 2 or 3 substances, to which no Category 3 substances, no substances of unknown origin and no dispersants (implicated with increased bioavailability) have been added. This classification system for substances that have dose-response thresholds is deficient in that it fails to make separate provision for concentration-additive effects of substances from different risk categories. The classification rules aim to provide for such effects by applying relatively low concentrations – effectively a kind of safety margin.

The Administrative Regulation contains separate provisions on carcinogens. These provisions are based on the same system, but also consider the greater harmfulness of carcinogenic substances by applying lower concentration limits. Mixtures containing Category 3 carcinogens are automatically classified in Category 3 if their concentration lies above 0.1% or, if the Hazardous Substances Regulation sets a lower concentration limit, above that limit (see Art. 35(3), sentence 2 GefStoffV). If Category 2 carcinogens are present, the mixture is classified in Category 2 in the same way. Classification in Category 1 applies when the carcinogens lie below the de minimis limit. Mixtures are only classfied as non-water-polluting if no carcinogens have been added to them, even if they are below the 0.1% mass de minimis limit (relative to the individual substance).

In certain cases, blanket classification of mixtures under the regulations cited above can be corrected by experimental analysis of the mixture as such. The Administrative Regulation contains specific provisions for this purpose. Assessment is conducted using a points scale (target values). Special mixtures, e.g. dye preparations, are classified explicitly under the Administrative Regulation.

b) Because the classification of preparations relates to mixtures that are wholly and intentionally produced by one manufacturer, imputation is not an issue. As already mentioned, classification is a precursor of limits in that classification triggers various obligations for labelling and handling preparations (occupational health and safety; incidents). The classification of preparations is mainly governed by the Dangerous Preparations Directive (88/379/EC) and the Hazardous Substances Regulation (GefStoffV) (Art. 35(3) in conjunction with Annex II), which implements the Directive in German law. In 1996, the European Commission submitted a proposal for amendment[38] of the Dangerous Preparations Directive. The

---

[38] OJ 1996 C 283/1; Common Position, OJ 1998 C 360/1.

amendment is expected to be enacted shortly. Now enacted as Directive 1999/45, OJ 1999 L 200/1. It makes sense, therefore, to include the amendment in this study.

As with classification of water-polluting mixtures, classification of preparations under the Hazardous Substances Regulation is the responsibility of manufacturers. What's more, the wide range of preparations on the market allows only sporadic controls at best. The competent authorities are thus often less than knowledgeable about how the classification rules are applied in practice.

In the classification of preparations, the Hazardous Substances Regulation assumes concentration-additive effects for all substances with the same and some-times with different harmful properties, without considering concentration-(dose)-response relationships. The Regulation requires manufacturers to correct the classi-fication arrived at by this method (excluding carcinogenic preparations), using advanced knowledge from studying the human health impacts of mixtures as such. Correction is also possible when the preparation's impact has been underestimated (as with super-additive effects) or overestimated (as with antagonistic or entirely different effects) (no. 1.3.1(5) GefStoffV). As no separate assessment obligations exist for preparations – and abandonment of the conventional method requires proof – it remains to be seen whether these provisions have any practical signifi-cance. Practitioners report that overestimation of the irritant effects of soaps and cosmetics when using the conventional method is accepted because separately assessing a preparation is too costly. In contrast, manufacturers' product liability provides a certain incentive to investigate any suspicion of super-additive effects. In the first instance, the classification rules assume that a substance classified as haz-ardous is present in an otherwise non-hazardous preparation.

Toxicological assessment of preparations containing toxic substances other than carcinogens, mutagens and reproductive toxins can equally be conducted by testing in the usual manner or by applying the "conventional" formula based on the sub-stances' relative shares of the total and assuming a dilution effect. There are two ways of performing the latter option:

Firstly, in some cases, the rules for classification of an individual substance in Annex I of the Dangerous Substances Directive (67/648/EEC) specify a concentra-tion limit which, if exceeded, requires preparations containing the substance to be classified in the same risk category because the substance can then be said to dom-inate the preparation. Annex I lists concentration limits of between 0.1% and 25% by mass (and above in exceptional circumstances) to take account of the differing toxicity of the various substances.

In the absence of a specific classification, "blanket" concentration limits apply under nos. 1.5.1-1.5.6 GefStoffV, Tables I to VI A. Concentrations are classified by toxic impact severity expressed as risk categories and in some cases subcategories (danger symbols, risk phrases). Provision is made in particular for determining what proportion of a substance in a preparation requires that substance to be classi-fied in the respective risk category.

Contrary to Art. 13(2) sentence 2 of the German Chemicals Act (ChemG), the conventional method is the only admissible approach if carcinogens are involved. If a carcinogen classified in Categories 1 (T+, or highly toxic) and 2 (T, or toxic) is present in a concentration of more than 0.1%, the preparation must at the very least be classified as toxic (T risk symbol R45). Preparations containing Category 3 carcinogens must be classified as harmful to health when the concentration of the substance is above 1%.

The Hazardous Substances Regulation also contains provisions on preparations containing multiple substances that are classified as hazardous. The conventional method is preferred in such cases for administrative simplicity.[39] A preparation is classified as highly toxic, for example, when it contains two or more substances that are so classified and one of them significantly exceeds the above-cited concentration values under Annex I or no. 1.5.1 (Tables I and IA). Otherwise, the percentage weight of each substance is divided by its concentration limit and the quotients added; if the sum is 1 or greater, the substance must be classified as highly toxic. This only applies, however, if its effects are lethal. In the case of non-lethal effects, use is made of either the substance-specific concentration limits for the substance's category (Annex I) or the concentration limits under no. 1.5.1 (Table I). This means that if the concentration of a substance lies below the concentration limits for classification in the highly toxic risk category, the preparation is classified in the next-lowest category without any possible combined effects being taken into account.

The procedure is similar for the "toxic" hazard category, although mixtures of highly toxic and toxic substances are dealt with by using a sum total formula that considers the relative proportions and the amounts by which limits are exceeded. A preparation is classified as toxic if the percentage weight of one, highly toxic substance divided by the concentration limit for highly toxic substances added to the percentage weight of another, toxic substance divided by the concentration limit for toxic substances equals or exceeds 1 – unless the preparation is already classified as highly toxic because it is dominated by highly toxic components. In effect, the assessment method constitutes a blanket dose-additivity model that takes substance dilution in the preparation into account.

In the case of carcinogens, mutagens and reproductive toxins, classification of the preparation and the applicable mandatory R phrase are determined by the concentration limits for individual substances in the respective category or in no. 1.5.6 (Table VI) GefStoffV. Thus, only the individual substance counts and, even if the sum total of two substances in the same category clearly exceeds the limit concentration it is still ignored. In essence, then, there is no provision for mixtures. Instead, the rule makers appear to have assumed that because of the very low concentration limits that have been set (0.1% for Category 1 and 2 carcinogens) the safety margin is sufficient to guarantee correct classification.

The amendments to the EU Dangerous Preparations Directive make no attempt to change this situation, although some of the concentration limits will be reduced. However, there is a move to abandon the pure concentration-additivity model and to incorporate a dose factor into the formula.[*]

The existing preparations classification system completely ignores any harmful impacts on the environment, whereas the proposed new Dangerous Preparations Directive[40] expressly takes environmental hazards into account (mainly hazards to the aquatic environment). Under the proposal (Art. 7), a preparation's potential

---

[39] See Falke/Winter, 1996, p. 568.
[*] This has not been adopted in the new Directive 1999/45.
[40] See footnote 38 above.

harmful impact on the environment is to be determined either by testing of the preparation as such or by the conventional method of applying concentration limits. In much the same way as that used for classifying preparations that are harmful to human health, the system provides for conventional classification based on the presence of one or more hazardous substances in a preparation in a specific, dominant, single concentration. This can be derived either from the concentration limit that has already been stipulated in individual substance classification or – for each different harmful impact on the environment – from the tables in the new directive (Annex III B Tables I to V). Preparations that contain environmentally hazardous substances from the same risk category in single concentrations that do not exceed the stated limits are classified based on a sum total formula that assumes both concentration additivity and dilution effects. A mixture of substances from different risk categories but below the stated concentration limits is classified in the next-lowest category if its concentration limits are exceeded by addition of the proportional substance quantities (by percentage weight). Thus, the respective differing environmental hazardous properties of the substances in the preparation are no longer taken into account.

5.1.4
Imputation of Combined Effects[41]

5.1.4.1
Civil Liability Law

The legal area in which combined effects have been most comprehensively debated to date is that of civil liability law. This involves in particular cease-and-desist orders under Art. 1004 of the German Civil Code (BGB) and damages or compensation under Art. 823 and 906(2) sentence 2 BGB, Art. 2 of the German Water Management Act (WHG) and the German Environmental Liability Act (UmweltHG).

a) The debate in liability law has certain special features. Firstly, interactions between multiple substances are only one of several relevant factors. The problem of assigning liability only arises when emissions of a substance are caused by different individuals or entities. Combined effects are only one aspect of an issue that in liability law also covers summative effects of identical substances from different sources. To this extent, however, the liability law debate also sheds light on general problems of regulating combined effects. Another interesting aspect of liability law is the development of solution models not only for cases of certainty as regards combined effects, but also for cases of uncertainty. Finally, as regards compensation, the establishment of cost-recovery relationships between multiple sources can institute a secondary compensation system that, to a certain extent, can relativate primary liability relationships, i.e. liability of one or another specific source. In so doing, the liability model effectively ascribes prima facie responsibility to the owner of a certain source, but owners of other contributing sources must help foot the bill.

---

[41] On criminal liability for combined effects see Daxenberger, 1997.

b) In general, the first question raised by issues of liability and compensation relates to the rationale for joint and several liability. This is sought in the idea that considering the foreseeable conduct of neighbouring polluters, any (significant) polluter must anticipate that damage will result from its and their conduct combined (a *Verkehrspflicht* – roughly speaking a duty of care – arising from shared risk or from co-responsibility by virtue of shared circumstances).[42] Against this background, proportional (pro rata) liability is seen as the fairest solution in issues of liability and compensation, while joint and several liability, in which the victim may opt to claim against any one of the perpetrators, who in turn have internal recourse to reciprocal recovery, is only a second-best solution that side-steps the actual issue.[43] Even so, legislators have opted for strict, joint and several liability in water management law (Art. 22(1) sentence 2 WHG). Outside water management law, joint and several liability is assumed when the share of cause cannot readily be estimated, which can be the case with either sub-additive or super-additive effects. The same liability rule applies when an individual (significant or unlawful) contribution is not in itself able to cause harm, but takes on a causal role in combination with another.[44] On the other hand, it is not clear whether, under such circumstances, liability arises if an individual contribution in itself does not exceed the threshold of significance or unlawfulness.[45]

c) The real problem of liability law is seen in the uncertainty of proportionate cause, i.e. in cases where it is not clear that each of two sources caused damage in concert. Various liability models have been developed in statute and case law that emphasize victim protection on the one hand and, on the other, aim to limit the responsibility of multiple sources to situations in which joint and several liability obtains by virtue of their spatial proximity. The most comprehensive liability model is offered by Art. 22(1) sentence 2 of the German Water Management Act (WHG), which prescribes joint and several liability for which it suffices that a release from a single source is capable of causing damage in conjunction with other sources.[46] A similar approach is used in Art. 7 of the German Environmental Liability Act (UmweltHG) for cases where damage can have been caused by multiple industrial facilities listed in the Act ("listed facilities"). It suffices for one facility to be capable, on the facts, of causing the damage in conjunction with the others; that some other facility is capable of having been the sole cause is not a defence.[47]

Finally, under Art. 830(1) sentence 1 BGB, the operators of two sources are held jointly and severally liable when multiple pollutants discharged in close proximity and

---

[42] Rehbinder, 1998, Part 8, no. 74 f. (with further references).

[43] BGHZ 66, 70, 76; 71, 102, 107 f.; 72, 289, 298 f.; 85, 375, 387; Rehbinder, loc. cit., no. 75; Landsberg/Lülling, 1991, *§ 1*, nos. 202 f. (each with further references); in favour of joint and several liability: Loser, 1994, p. 123 ff.

[44] BGHZ 66, 70, 76; 72, 289, 298; BGH NJW 1994, 932, 934 (product liability), see furthermore the references in Footnote 43.

[45] In favour of (pro rata) liability: OLG Düsseldorf, NJW 1998, 3720; Kleindienst, 1964, p. 64 ff.; Gottwald, 1986, p. 22 f.; against liability (apparently) BGHZ 70, 102, 108, 111; Hager, in Landmann/Rohmer, 1998, *§ 7 UmweltHG*, no. 207.

[46] BGHZ 57, 257, 259 ff.

[47] Hager, in Landmann/Rohmer, 1998, *§ 6 UmweltHG*, no. 35 ff.; Salje, 1993, *§ 6*, no. 11; also in this direction: Quentin, 1994, p. 296 f., 301 ff.; contra Schmidt-Salzer, 1992, *§ 7*, no. 37 ff.

in the same period cause damage and each source is capable of being the sole cause. In all other cases, particularly under general law (Art. 906(2) sentence 2 and Art. 823 BGB), opinion sways towards proportional liability with estimation of the respective shares and only resorts to joint and several liability when estimation proves impossible.

To sum up, the rules on proven or suspected combined effects from multiple sources are basically the same. Problems do arise, however, in justifying any form of liability where there is no more than a suspicion of combined effects. The solution to such problems lies in the proof provisions under liability law (simplified evidence rules or reversal of the burden of proof). In each case, certain indications are needed as regards contributory cause (capability or even high probability) to ease the burden on the victim to provide proof of (contributory) cause or even to lay the burden of proof on a potential perpetrator.[48]

d) A highly controversial issue in cease-and-desist claims is whether such a claim is valid if and to the extent that individual contributions in themselves are not unlawful or significant. Prevailing opinion rejects liability and, moreover, comes to the conclusion that where there is a combination of contributing factors, claims demanding a reduction in any one factor are limited to the degree to which that factor is unlawful or significant.[49] The counter-argument takes the stance that each source must reduce its contribution to the point at which the combined effect falls below the threshold of unlawfulness or significance.[50] This theory, which in effect means joint and several liability, needs to be expanded to incorporate rights of recourse. Depending on the victim's choice, the theory lays full responsibility on each of the multiple contributors to reduce their own contribution, while the other contributors pay towards the cost incurred.[51]

e) Combined effects also play a role in civil liability law as part of investigation obligations. In cases of producer liability on the basis of unlawful conduct and of liability by virtue of contractual obligations to provide advice, the product monitoring obligation also covers combined effects when, by its own actions, a manufacturer makes it possible for such effects to occur or where its obligations to provide advice explicitly cover the use of two substances. Where significant evidence exists for combined effects in such cases, the manufacturer must warn users accordingly.[52] In cases of environment liability on the basis of unlawful conduct,

---

[48] See Hager, in Landmann/Rohmer, 1998, *§ 7 UmweltHG*, no. 21 ff.; Salje, 1993, *§§ 1, 3*, no. 125 ff.; Balensiefen, 1994, p. 166 f., 190 ff.; 237 ff.; Loser, 1994, p. 242 ff.; Rehbinder, 1998, Part 8, no. 76; Gerlach, 1989, p. 255 ff.; in favour of liability solely for that part of the damage which the polluter's own emissions are capable of causing: BGH, NJW 1994, 932, 934.

[49] See BGHZ 66, 70, 73, 74 ff.; also BGH NJW 1976, 799 on liability in a consecutive polluter chain; Herrmann, 1987, p. 537 ff. (with further references); Gerlach, 1989, p. 201 ff.

[50] LG Köln, NJW-RR 1990, 865; Staudinger/Roth, 1996, *§ 906,* no. 249; Balensiefen, 1994, p. 112 f.; Soergel/Baur, 1989, *§ 906,* no. 123 f.

[51] By way of parallel, the decision BGH NJW 1997, 2234 upheld a cease-and-desist order in a case of contributory wilful conduct on the part of the victim subject to the latter contributing towards the cost. This corroborates the solution described above, since recourse among jointly and severally liable parties (Art. 426 BGB) is qualified by contributory wilful conduct or contributory negligence (Art. 254 BGB).

[52] BGHZ 96, 167, 173 ff. re. Honda: duty to monitor products for combined effects in industry; BGH VersR 1977, 918, 920: transferred duty to give advice on the use of two plant protection products.

potential combined effects must also be accounted for by means of further assessment. Under the *Kupolofen* ("cupola furnace") doctrine in German case law, an operator can no longer rely on the pollution limits set by the state if, in given circumstances, the operator has cause to believe that compliance with those limits is not sufficient to prevent damage being caused by emissions. In this event, the operator must conduct additional assessment.[53] The Federal Court of Justice, in a new judgement, expanded this principle to include summative effects.[54] The judgement provides support for general justification of assessment requirements as regards combined effects – at least in terms of the more onerous research requirements for manufacturers partially developed in the *Kindertee* ("infant tea") judgement.[55]

f) It is reasonable to assume that health hazards can be caused by combined effects of substances that stem partly from external sources and partly from the victim's private sphere (e.g. smoking or alcohol consumption). In such cases, the contributory negligence rule under Art. 254 BGB (or similar provisions in other liability laws) applies and allows pro-rating of the damage caused. Up to now, this case scenario has played no part in civil liability law because cases of health damage are generally covered by social insurance.

5.1.4.2
Law On Occupational Health and Safety

German law relating to occupational health and safety reaches an entirely different conclusion on liability. The major liability standards – Art. 8 of Book VII of the Code of Social Welfare Law (SGB VII) for occupational accidents and Art. 9 SGB VII for occupational illnesses – hold welfare insurance bodies liable for occupational accidents involving injuries that have occurred as a result of work-related activities. The generally accepted interpretation – which despite amendments to the cited legislation has remained unchanged since the Reich Insurance Code (RVO) – applies the significant factor theory in occupational accident law. Under this theory, it is enough for an occupational accident or exposure to hazardous substances in the workplace to be a significant proportional cause of the total damage.[56] A contributory cause from pollution in the private sphere does not rule out the responsibility of social accident insurance if the occupation-specific cause is significant. A qualitative assessment must be carried out as to what extent the personal factors are significant. The mere fact that personal factors are largely to blame does not rule out a significant contribution from occupation-specific risk. If the existing risk from personal factors, e.g. from smoking, is already considerable, the occupational risk can still make a significant contribution to the cause as the activity can conceivably

---

[53] BGHZ 92, 143.
[54] BGH, NJW 1997, 2748, 2749.
[55] BGHZ 116, 60 (with reference to advertisements giving an impression of harmlessness).
[56] See Schöpf, 1995; Köhler, MedSach 1996, 101 ff.; Krasney, 1988, p. 67 ff.; Bley, 1988, p. 214 f.; Wannagat, 1997, *§ 8 SGB VII*, no. 41 f.; Erlenkämper/Fichte, 1996, p. 81 ff.; Schulin, 1991, p. 136 f.; precedents include BSGE 1, 72, 76; 1, 150, 157; 12, 242, 245; 13, 40 ff.; 13, 175, 176; 45, 176, 178; 48, 224, 226; 62, 220, 221 f.; BSG SozR 2200 *§ 548 RVO Nr. 77* and *Nr. 81*; 2200 *§ 551 RVO Nr. 1*; Sgb. 1987, 425; BSG DStR 1998, 50 (also published in EzS 40/561).

result in a considerable increase in the risk to smokers. Significance is only negated in cases where the contribution from occupational risk is negligible.[57] The reason for applying the significant factor theory lies in the fact that welfare law serves to protect the victim in the first instance and thus leans towards a wider interpretation of occupational risk.

5.1.4.3
Public Law

Issues of state responsibility for hazards have received a lot of attention lately, particularly as regards the problems of contaminated sites. However, combined effects do not play a special part either in case law or in the public law literature.

The issue of summative emissions is discussed outside the law on contaminated sites. The combined effects of substances can be tackled at three levels through

- measures for the prevention of dangers,
- post-incident cleanup measures and
- hazard research measures.

Under general police law principles, the point at which a third party is held liable as a polluter is when a danger has already occurred (accident) or when a specific danger is expected to occur in the immediate future. The third party must at least be a joint contributor to the cause of the dangerous circumstances[58] and its behaviour must be imputable.

An individual substance that in itself poses an immediate danger independently of potential hazardous reactions with other substances suffices for intervention by law enforcement agencies. Combined effects are only taken into account when it comes to naming the polluter to be held liable for carrying out the cleanup measures and thus for meeting the primary cost of those measures.

Substances that are not inherently hazardous at a given concentration and only develop hazardous properties when they interact with other substances can trigger preventive measures when, in specific circumstances, they can be expected to cause a danger. There must be imminent danger that a specific substance or emissions thereof can cause a dangerous situation.[59] An example of imminent danger is when two substances that can have hazardous combined effects if mixed with oxygen are stored side by side and there is damage to the containers they are stored in.

The theory of direct causation[60] holds liable only the individual or entity who, after assessment of all the circumstances of the specific case, has exceeded the risk

---

[57] See BSGE 13, 175 re. silicosis and privately induced cancer; BSG, SozR 2200 *§ 548 RVO Nr. 77:* alcohol-linked incapacity to drive annuls insurance cover if it predominates over other circumstances to such an extent that in the eyes of the law it is deemed the sole significant cause of an accident; Erlenkämper, SGb 1997, 355, 357 f.

[58] Lisken/Denninger, 1995, E 60.

[59] Götz, 1995, no. 140.

[60] The theory of direct causation predominates in case law; see VGH Kassel, DÖV 1986, 441; OVG Münster, NVwZ 1985, 355, 356.

limit and is thus directly responsible for causing the danger.[61] The prevailing interpretation[62] requires that the cause presents a risk in excess of the usual, even if it remains within permissible limits. Obviously, this formula provides no answers to the difficult issue of imputation of contributory cause. But what can be concluded is that where there are multiple polluters, each polluter who has made a significant contribution to the occurrence of a danger can be held liable.[63] Under pollution control law in particular, the general interpretation is that minimal pollution contributions cannot be imputed.[64]

Apart from prevention of dangers and cleanup, measures can also be taken for hazard research on combined effects. These are generally seen as part of prevention of dangers.[65] Such measures are considered when evidence exists for the occurrence of a hazard[66] or a danger actually exists but the polluter cannot be clearly identified.[67] Thus, hazard research measures regularly serve to establish whether a situation is indeed dangerous or to identify the polluter who actually caused the danger.

The great uncertainty surrounding causation and source where combined effects are involved, and the cost risks to public agencies who falsely assume a danger or the relevance of a contributory cause,[68] appear to have excluded from the debate hazard research measures for contaminated sites involving combined effects. This situation can only be expected to change if liability for the cost burden for hazard research measures is detached from the issue of identifying the polluter and the cost charged to the suspected polluter. While this finds acceptance[69] in some cases, at least where a danger is suspected, it is largely rejected.

Where multiple polluters are identified (the rule where combined effects are concerned), the question arises as to which of the polluters is liable. General police law makes provision for all, several or just one polluter to be held liable.[70] The competent authority must thus exercise its discretion in compliance with its obligations. The courts are, in part, of the opinion that the last in the pollution chain[71] should be held liable, but this is an indicative opinion at best.[72] Case law and the relevant literature put forward various theories that essentially express the proportionality principle: questions of who is best able to clean up the site quickly and effectively, who is actually and legally able to do so, for whom does the duty of prevention of a danger pose the mildest impact, who carries the greatest burden of cause and who is

---

[61] OVG Münster, NVwZ 1985, 355, 356.
[62] See Ronellenfitsch, VerwArch 1986, 435, 438; Schoch, JuS 1994, 932, 937.
[63] VGH Mannheim, NVwZ-RR 1994, 565; VGH Kassel, NVwZ 1992, 1101; VG Darmstadt, NVwZ-RR 1994, 497; also OVG Hamburg, BB 1990, 662.
[64] Feldhaus/Schmitt, WiVerw 1984, 1; Hansmann, in Landmann/Rohmer, 1998, *Nr. 2.2.1.1 TA Luft*, no. 17; Winter, 1986, p. 133; Koch, 1999, p. 222 f.
[65] Thus VGH Mannheim, NuR 1995, 547; VGH München, BayVBl 1986, 590, 592.
[66] Breuer, NVwZ 1987, 751, 754; Schink, DVBl 1986, 161, 166.
[67] Nierhaus, JUTR 1994, 369, 375; OVG Koblenz, NVwZ-RR 1992, 238.
[68] See BGH, NJW 1992, 2639; Götz, 1995, no. 155.
[69] Classen, JA 1995, 608, 612.
[70] Contra: OVG Hamburg, BB 1990, 662: liability always limited to contributory share.
[71] VGH Mannheim, DVBl 1950, 475, 477; now contradicted by prevailing opinion (see OVG Münster, UPR 1984, 279, 280).
[72] Lisken/Denninger, 1995, E 105 (with further references).

closest to the damage.[73] This list allows easy identification of specific parties. If, for example, only one polluter is legally able to clean up a site, the cleanup order is issued against that polluter. But general selection models cannot be prescribed for dangers that are caused by combined effects. Instead, each case must be taken in on its merits.

An obligation to meet the costs need not necessarily imply liability as polluter. It would be extremely unfair if the cost burden were assigned to only one of two co-polluters held liable solely on the grounds of its ability to cause a danger.[74] Some *Länder* have provided in their own legislation for proportional cost sharing (Art. 21(1) sentence 2 of the Waste Management and Contamination Act of the State of Hesse (HAbfAG)). The issue of multiple polluters being obliged to compensate one another is controversial and obligations of this type have largely been rejected in the past.[75] Those who oppose this standpoint argue that it contravenes the fair allocation of burdens that ensues from the principles of equality (Art. 3(1) GG) and proportionality.[76] While the issue of effectiveness in preventing dangers is a particular priority in polluter selection, other criteria should be considered when it comes to selecting those who are to bear the costs. It would seem fairer to assess the respective contribution to the dangers and use that as a basis for assigning the cost burden.[77] An obligation to pay compensation is contained in soil pollution law under Art. 24(2) of the German Soil Protection Act (BBodSchG), patterned on various *Länder* contamination laws.

In pollution control law, facility licensing is usually conducted on the principle of priority, while the procedures for existing facilities and non-licensed facilities involve casewise decisions based on the proportionality principle (see Art. 17(2) and 24 BImSchG). No special criteria for combined effects have been developed to date.

Imputation issues are far more complex when it comes to the related problems of cumulative noise pollution under the Pollution Control Act.[78] The 1995 Model Administrative Procedures *(Musterverwaltungsvorschriften)* adopted by the German *Länder* (see 5.3.1.2 d above) contained generally interesting provisions for the apportionment of noise abatement burdens to sources and took as their starting point the facilities listed under the Technical Instructions on Noise *(TA Lärm)*. Compliance with emission limits – which also applied for cumulative noise – was to be achieved by entering into agreements with operators, by imposing conditions for the event that a new facility was added, or by a quota system that assigned each facility a noise quota based on the area of the site as a percentage of the area occupied by all relevant facilities. Except where the noise added by a new facility was negligible or was masked by noise from other sources, the quota system could also

---

[73] Drews/Wacke/Vogel/Martens, 1986, § 19, 6; Lisken/Denninger, 1995, E 105 ff.
[74] In this sense already Kormann, UPR 1983, 281, 287.
[75] BGHZ 98, 235; 110, 313, 318; commenting thereupon: Lisken/Denninger, 1995, E 110 (with further references).
[76] Kormann, UPR 1983, 281, 285; Kloepfer/Thull, DVBl 1989, 1121, 1128; Lisken/Denninger, 1995, E 111.
[77] Kormann, UPR 1983, 281, 287; Lisken/Denninger, 1995, E 112 f.
[78] On this subject in general, see Petersen, 1993, p. 180 ff. and the remaining references in footnote 10.

be used where noise from existing facilities was already at or above the limit and the existing facilities reduced their noise emissions accordingly. This was to be decided according to the provisions for existing facilities, including the proportionality rule under Art. 17 BImSchG. The Model Administrative Procedures generally required facilities to be adjusted in line with technical progress. If necessary, additional noise abatement measures were to be required for facilities which had been licensed subject to the condition that such measures could be imposed or which exceeded their noise quota.

The special assessment to be undertaken in the event of noise from facilities being combined with noise from other sources under the Model Administrative Procedures was subject to criteria based on the interest in abatement, the level of nuisance and the relative noise levels. Existing facilities were given priority. Reduction measures had to be based on the criteria that applied generally for facilities of the same type and were limited to the noise quota applicable to facilities covered by *TA Lärm*. The burden of reduction measures on operators was further limited by setting the maximum reduction that could be imposed at 3 dB(A) below the exposure limit cited in *TA Lärm*.

The new *TA Lärm* uses a simple approach to apportion reduction burdens among multiple facilities. The priority principle applies except in negligible cases (such as additional noise or noise ly over the limit), in cases where noise is masked by other facilities, and where compliance will be achieved by the operator adjusting other facilities. A new facility may not be constructed if limit values are exceeded due to existing or additional loads, although the way is left open for special assessment if emissions and exposures from other facilities are expected to improve. Where multiple existing facilities are involved, the competent authority must use its discretion in deciding which operators must reduce noise, basing its decision on the proportionality principle and taking into consideration existing noise abatement plans, the effectiveness of the measures, the costs involved, relative noise levels, and the presence and degree of any fault based on wilful conduct.

Outside of *TA Lärm*, the idea of polluter groups being jointly responsible for total pollution is rejected on all counts, and so existing pollution is not seen as grounds for remediation. Instead, based on the principle of proportionality, existing pollution is added to the pollution emanating from a new source and assessment is based on the extent to which the new source can reasonably be accepted.[79]

5.1.5
Lifestyle-based Substance Risk

Substance risk arising from personal lifestyles, particularly from the consumption of alcohol and from smoking, is regarded as general lifestyle risk that cannot be regulated by the state unless it endangers a third party. Third-party risk in road traffic caused by alcohol consumption is dealt with by the criminal code and road traffic law together with applicable sanctions (Art. 315a, 315b and 316 in conjunction with Art. 24 of the German Criminal Code and Art. 24a and 25 of the German Road

---

[79] BVerwGE 87, 332, 357 f.; 52, 226, 236.

Traffic Act); while third-party risk from passive smoking is covered by as yet rudi-
mentary rules on the protection of non-smokers in public buildings, in public transport facilities and in the workplace.[80] According to the Federal Constitutional
Court[81], the German constitution assigns no duty to provide greater protection. In
its last legislative period, the *Bundestag* rejected a bill on passive smoking.

Self-risk from personal choice, on the other hand, is accepted in principle. Up to
now, the only attempts to reduce self-risk from alcohol consumption or smoking
have included taxation on alcohol and tobacco by means of a special consumption
tax, a declaration requirement (alcohol) and a prescribed warning (tobacco)
together with restrictions or even bans on advertising. It comes as no surprise,
therefore, that combined effects resulting from the consumption of alcohol or
smoking and exposure to chemical substances have found no place in health or
environmental policy to date.

5.2
The Situation in Other Countries

5.2.1
USA

America's main media-specific environmental laws, particularly the Clean Air Act
(CAA), the Clean Water Act (CWA), the Safe Drinking Water Act (SDWA), the Comprehensive Environmental Response, Compensation and Liability Act (CERCLA)
and the Nuclear Energy Act (NEA) are all largely based on a combination of environmental quality (exposure) and emission-related strategies, although greater emphasis
is placed on emissions. The powers under these laws to set environmental quality and
emissions standards and generally to prevent risk are usually targeted at comprehensive protection of human health and the environment, and thus allow adequate coverage of combined effects. In practice, however, the focus is on individual substances.
This is legally admissible since the agencies responsible for enforcing the laws, especially the Environmental Protection Agency (EPA), have broad discretionary powers
that are only subject to judicial review as regards the limits of their extent. Additionally, some regulatory powers relating to hazardous substances emphasize the health
and environmental risks arising from those substances and can be interpreted such
that consideration of combined effects does not accord with the wishes of the legislators. Combined effects play no role in the legal literature. There are, however, exceptions where combined effects have been incorporated into regulation.[82]

American nuclear protection law takes no account of interactions between radiation and hazardous substances.[83] Regulation of hazardous substance emissions usu-

---

[80] For a compilation of the relevant regulations, see BVerfG, NJW 1998, 2961; BayVerfGH,
BayVBl 1988, 108; on labour law in particular, see BAG NJW 1999, 162; BVerwG, NJW
1985, 876.
[81] BVerfG, loc. cit.
[82] Overy/Richardson, ELR 1995, 10657; cf. Travis et al., Health Physics 56 (1989), 527, 530 f.
[83] Overy/Richardson, ELR 1995, 10657, 10661 ff.

ally takes a single-substances approach. This applies, for example, to emission standards for hazardous substances under the Clean Air Act and Clean Water Act, and for quality standards for drinking water under the Safe Drinking Water Act. Under the Comprehensive Environmental Response, Compensation and Liability Act (CERCLA), however, combined effects are taken into account in risk assessment for cleanup measures for contaminated sites.[84] The EPA requires cleanup activities for carcinogens with a cumulative individual risk of more than 10^{-6}. As regards interactions between other hazardous substances the cleanup requirement kicks in where the total risk is found to cause significant health hazards. The total risk is assessed based on the risk indices and using an additive model.

Cleanup standards are generally derived from existing environmental standards issued under other laws, where they are applicable and appropriate to the circumstances. As regards combined effects of carcinogens, which are not considered in such standards, the aim is to reduce the individual risk from 10^{-4} to 10^{-6}, and otherwise to reduce the burden in such a way as to avoid any noteworthy risk of significant health impairment.

Under the 1986 Emergency Planning and Community Right to Know Act (EPCRA) (Section 311 et seq.), the rules on notifying the public about the risks arising from hazardous accidents contain specific provisions on mixtures containing hazardous substances. A mixture is also deemed harmful if it contains 1% by weight or volume of a substance classified as hazardous or 0.1% of a carcinogenic substance. However, the quantitative action thresholds are calculated separately for each substance. Mixtures are subject to annual reporting requirements and to other requirements when the quantities of substances in a mixture exceed the thresholds. The safety data sheet required by the Act can be used to report a mixture as such or list its individual harmful components.

In contrast, the main chemicals laws – the Toxic Substance Control Act (TSCA), the Federal Insecticide, Fungicide and Rodenticide Act (FIFRA), the Federal Food, Drugs and Cosmetics Act (FFDCA) and the Occupational Health and Safety Act (OSHA) – do allow consideration of combined effects. The reporting process for new substances under Section 5 TSCA is, however, restricted to single substances. Reporting requirements for substances already in circulation under Section 4 TSCA can also be applied to mixtures, including preparations, when the effects of the mixture can not be determined or predicted more efficiently or more cost-effectively by assessing its components. In general, the law emphasizes the importance of cumulative and synergetic effects. The substantive regulatory powers under Sections 6 and 7 TSCA also cover mixtures. The plant protection products and foodstuffs provisions contained in the 1996 Food Quality Protection Act require the EPA to assess the cumulative health risk that arises from exposure to pesticides and other substances with the same response mechanism. The EPA recently issued a policy document on the procedure for identifying such substances.[85] The Occupational Health and Safety Act also covers mixtures with regard to occupational illnesses. Thus, occupational health and safety law

---

[84] 40 C.F.R. § 300. 430 (e)(2)(i) (1993); Overy/Richardson, ELR 1995, 10657, 10666; Rummel, 1996, p. 120 f.
[85] 64 Fed. Reg. 5796 (1999).

contains mixture provisions as part of the General Industry Standards of the OSHA and the Threshold Limit Values (TLVs) recommended as guideline values by the American Conference of Government Industrial Hygienists (ACGIH).[86] This legislation follows the same simple model of linear dose(more exactly: concentration)-additivity found in Germany's Technical Instructions on Hazardous Substances (TRGS 403) (where the model originated), under which each hazardous substance in a mixture can only exploit the portion of the limit that applies to it. The mixture rule does not apply, however, if there is evidence to assume that the substances concerned have entirely different response mechanisms. With this exception, the provisions assume additivity, though subject to proof of different forms of combined effect. For mixtures with a large number of (changing) components of which only one can be measured, the ACGIH recommends a reduction of the guideline value for the measurable component, with the reduction factor based on the number and toxicity of other components and the relative quantities in which they typically occur.

In the face of inconsistent procedures for assessing and regulating mixtures, the EPA has attempted to guide at least risk assessment by issuing guidelines since the mid-1980s: under the EPA's 1986 guidelines, toxicological risk assessment must also include mixtures and preparations (mixture rules).[87]

The EPA gives priority to mixture assessment derived from toxicological data obtained for a specific mixture or for mixtures with similar properties – furnace gases and diesel soot are two examples. If this type of assessment cannot be conducted, the dose-additivity model should be used where the mixture's components are known or can be assumed to have the same response mechanism. In all other cases, the effect-additivity model should be applied. If there is evidence of interaction (super-additive effects or antagonism), a special assessment is conducted in accordance with available knowledge or based on assumptions on the nature of the interaction. Finally, the EPA assumes for carcinogens that the different additivity models do not yield significantly different results at low doses.

The EPA recently issued assessment guidelines for ecotoxicological risk assessment. These also take mixtures into account[88] and are based on a dose-additivity model which divides the concentration by LC 50 for each component and adds the quotients. According to the EPA, while a hypothesis of dose-additivity or of similar effects is generally justified in the case of identical response mechanisms, empirical evidence gathered in fish toxicity tests shows that it is often also justified in the case of different response mechanisms. On the other hand, the EPA stresses that a quota method is not a suitable means of quantifying risks and does not adequately consider exposure intensity, secondary and reciprocal ecosystem effects, or uncertainties.

---

[86] 40 C.F.R. § 1910, 1000 (d)(2)(1983); American Conference of Governmental Industrial Hygienists, 1998, p. 81 ff.
[87] 51 Fed. Reg. 34014 (1986).
[88] 63 Fed. Reg. 26845 (1998).

5.2.2
Switzerland

From a legal standpoint, Switzerland's environment laws are quite favourable to adequate assessment of combined effects in hazardous substance regulation. Under Art. 8 of the Swiss Environmental Protection Act (USG) (revised in 1994), environmental effects must be assessed in their own right, in combination and in terms of their interactions.[89] Nevertheless, these provisions have yet to play a significant role in practice. Apart from chemicals law, there is a lack of impetus for incorporating combined effects into existing regulations.[90] The problems are likewise neglected in the legal literature.

The 1985 Air Quality Regulation (LRV), whose system is closely modelled on the classification system in Germany's Technical Instructions on Air Quality *(TA Luft)*, contains sum total limits for interactions of harmful anorganic substances (including heavy metals) in dust or particulate form, organic substances, and carcinogens from the same risk category (Sections 51, 71 and 82). The sum total limits correspond to the limits for substances occurring in isolation. Only organic substances are subject to a mixture rule for substances from different risk categories: in addition to the limit for the higher risk category, there is an overall limit corresponding to the limit for the lowest risk category.

In contrast to German water pollution law, Annex 2 of the 1998 Swiss Water Pollution Regulation (GSchV) prescribes only single-substance values for polycyclic aromatic hydrocarbons (PAHs) and adsorbable organic halogens (AOX). The 1998 Regulation on Water-Polluting Liquids (VWF) includes mixtures without stipulating specific rules for dealing with them. Liquids are classified into one of two risk categories according to the percentage volume of a given substance in the liquid, without regard to differing degrees of harmfulness between substances. This also applies to substance mixtures. On the other hand, the 1998 Soil Protection Regulation (BSV) stipulates not only limits for dioxins and furans expressed in the form of toxic equivalents, but also sum total limits for sixteen PAHs to allow for their additive effects. In the case of assessment and remediation values these are ten times the values prescribed for benzo(a)pyrene, and the preventive limit is five times the single substance limit.

As a guideline for imposing risk assessment obligations (Art. 4), the 1991 Hazardous Accidents Regulation (StFV) lays down thresholds that, in part, take account of interactions of stored and other substances. While the thresholds prescribed for highly toxic substances (Annex I 1 Section 3) contain no mixture provisions, they do apply to substances in preparations and products. The general thresholds (Annex I 1 Section 4) cover all substances that belong to specific risk categories or subcategories. The risk from combined effects is thus considered in blanket form in a similar way to that under the Seveso II Directive.

---

[89] Vallender/Morell, 1997, p. 129 f.; Rausch, in Kölz/Müller, 1996, *Art. 8*, nos. 1 ff.; Bundesrat, *Botschaft zu einem Bundesgesetz über den Umweltschutz vom 31.10.1979* (79.072), p. 37; Knebel/Sundermann, UPR 1983, 8, 11.
[90] In addition to the references in footnote 89, see the general commentary with no more than passing reference to combined effects in Saladin, KritV 1989, 27, 34 f.; Koechlin, 1989, p. 73.

The Regulation on Environmentally Hazardous Substances (StoffV), which is functionally similar to Germany's Chemicals Act, covers both manufactured substance mixtures (described as products) and incidental mixtures (Art. 3(3) StoffV). In general, manufacturers are also required to perform risk assessment for mixtures (Art. 12 StoffV) and in doing so must consider the classification of components of preparations together with the results of available studies; in certain circumstances they may also be required to carry out their own studies (Art. 16(1), (2) StoffV). Manufacturers must conduct additional studies if there is evidence that an interaction between different substances can have significant adverse effects on the environment or on humans through the environment (Art. 16(4) StoffV). The Regulation also includes combined effects in the assessment of existing substances for which there is evidence of an environmental risk. Under Art. 15(2) (b) StoffV, the competent authority may request that a manufacturer performs a comprehensive environmental risk assessment or additional studies where substances are involved that pose an increased risk to the environment, in combination with other substances. This power has not been used in practice to date.

In occupational health and safety, Switzerland uses the same blanket dose-additivity-based sum total formula as Germany for mixtures that affect the same target organs or whose components have synergetic effects. MAK values can only be exploited in the ratio the actual substance concentration bears to the MAK value. This sum total assessment does not apply when the components affect different target organs or do not mutually increase their combined effects (Section 1.1.6 of the Swiss Accident Insurance Institute's 1997 "Limits in the Workplace").

5.2.3
Netherlands

The Netherlands has been debating how to incorporate the combined effects of hazardous substances into environmental standards and individual decisions since about 1985. In an official report[91] released in 1985, the Dutch Health Council took the stance that in the case of substances that have dose-response thresholds, there was no need to assess combined effects below ADI or TDI levels. The report did note that combined effects were possible at higher concentrations, especially around the NOAEL (no-observed-adverse-effect level), but rejected preventive action in the form of an added safety factor for want of adequate scientific supporting evidence. This position was also generally followed in the 1986–1990 Indicative Environmental Programme.[92] This programme developed the concept of quantifying an unacceptable risk and an acceptable residual risk that has since been a feature of Dutch risk policy. For substances lacking dose-response thresholds, the unacceptable risk and the acceptable residual risk are set at 10^{-6} and 10^{-8} per year respectively (applied to lifelong exposure these figures would increase by a factor of 100, i.e. 10^{-4} and 10^{-6}). A single-substance approach is taken. For substances that have dose-response thresholds, a limit is to be set giving consideration to a safety margin; a substance lies in the area of unacceptable risk if it exceeds this limit, while

---

[91] Health Council of the Netherlands, 1985.
[92] Environmental Programme 1986-1990, p. 175 ff.

the residual risk is set at 1% of the limit. In the "grey area" between unacceptable risk and acceptable residual risk, further risk reduction measures are required subject to observance of the proportionality principle. This can also include assessment of exposures to multiple substances.

Where the combined effects of substances are concerned, environmental policy in the Netherlands today is largely shaped by the Dutch Environment Ministry's report *Omgaan met Risico's* (Dealing with Risks) published as part of the 1989 National Environment Plan.[93] The report sets out a strategic concept for risk quantification that includes combined effects. For substances that lack response values, the annual values for unacceptable risk and acceptable residual risk remain at 10^{-6} and 10^{-8} respectively. This last target has also found a place in environment protection law; Art. 5.1.5. of the Environmental Protection Act mandates that environmental quality standards that are still inadequate (because they do not yet lie in the region of acceptable residual risk) must be revised every eight years. The report also presents a special sum total limit of 10^{-5} for unacceptable risk and of 10^{-7} for residual risk per year for the combined effects of substances that lack dose-response thresholds. In practice, the sum total limit only comes into effect for "cocktails" (mixtures containing a large number of substances) because otherwise the sum total limit of 10^{-5} would never be exceeded given a single-substance limit of 10^{-6}. For substances that have a dose-response threshold, the report generally follows the 1986–1990 Environmental Programme, with a 1% residual risk target (though new substances are subject to more stringent requirements than existing substances). Regarding combined effects, the report maintains that it is impossible to set a special risk limit because of the diverse nature of possible effects. Thus, combined effects are only to be assessed in risk reduction measures in the "grey area" between unacceptable and residual risk. In effect, this means a binding safety factor of 100 that is based solely on the principle of proportionality and yet has general application, that is, also applies to single substances. Recently, the introduction of an additional safety factor has been discussed that would reduce the limit by a further power of ten.[94] The report does not mention combined environmental impacts, for example of ecotoxic substances.

Omgaan met Risico's is merely a declaration of policy intent on behalf of the government. It does not rank in the league of a National Environment Plan, which is subject to a duty of consideration.[95] The question of whether policy plans of this type can ever justify action against polluters has been dealt with in various court decisions.[96] In the meantime, doubt has been expressed in particular as to whether a fixed risk value for acceptable residual risk in the amount of 1% of the risk or limit is valid from a scientific or economic standpoint. Parliament especially is of the opinion that use of the minimization requirement (ALARA principle, Art. 8.11 of the Dutch Environmental Protection Act) without a fixed lower limit for risk

---

[93] See Henschler et al., Food & Chem. Toxicol. 34 (1996), 1183, 1185.

[94] Omgaan met Risico's, 1989, p. 18 ff.; cf. Stallen/Smit, 1993, p. 7 ff.; van Zorge, Food & Chem. Toxicol. 34 (1996), 1033, 1034 f.; Tauw milieu et al., 1990, p. 28 f.; Beroggi et al., 1997, p. 14, 24 f.

[95] See van Zorge, loc. cit. (with references).

[96] Tweede Kamer 1992–93, 22660 *Nr. 2*; see also Technische Commissie Bodembescherming, 1997, p. 23.

reduction is sufficient[97] – including for combined effects. The government has reiterated its position,[98] but in practice leans towards simple application of the minimization rule.

Finally, Dutch policy on quantifying the risk from combined effects is based on a simple model of linear dose-additivity that ignores the possibility of different forms of combined effects or of such effects being absent.

5.3
Appraisal

5.3.1
Constitutional Foundations

Combined effects play a largely subordinate part in German environmental law today. Different substances and different types of radiation are mainly treated in isolation. At the same time, the state must generally consider total exposures and hence include combined effects in environmental law under its constitutional obligation to protect the individual (Art. 2(2) sentence 1 GG) and its constitutional objective of protecting the environment (Art. 20a). Unlike the environmental protection objective under Art. 20a, however, the duty of protection under the first sentence of Art. 2(2) generally means protection from extant dangers and not precaution.[99] Within these bounds, the focus on the individual accords with the constitutional obligation to protect.[100] Basic rights being individual rights, the state duty to protect accrues to the individual and not to a notional average human being. To make a person of average susceptibility the sole standard of protection would be unconstitutional since protection must also extend to the infirm. However, a certain degree of generalization (aggregation) is allowed for practical reasons – mostly by identifying "vulnerable groups" such as children, senior citizens and allergy sufferers, with the result that people with unusual illnesses and dispositions fall through the state safety net and must rely on self-help.[101] Additionally, while vulnerable groups must be selected by objective criteria, the state has a certain amount of policymaking discretion in doing so.

In contrast, environmental protection must follow a collective (aggregated) approach, particularly when it is based on Art. 20a of the Basic Law. Environmental protection does not target individual specimens of flora and fauna, but the environment as a whole as a "natural resource"; that is, the natural ecosystem's ability to

---

[97] Staatsrat (Verwaltungsabteilung), 29.4.1994, Milieu en Recht 1995, 19.

[98] Staatsrat (Rechtsprechungsabteilung), 11.11.1991, no. S 03.91.3217; 23.9.1992 no. S 03.92.2167 (partially reproduced in Stallen/Smit, 1993, p. 9 and 10); see also Staatsrat, 6.3.1991, Milieu en Recht 1992 no. 6; 23.5.1991, Milieu en Recht 1991 no. 49 (on administrative orders where there only is a duty to take combined effects into account).

[99] See BVerfGE 49, 89, 141; 53, 30; 56, 54, 73, 80; 77, 170, 214 f.; BVerfG, NuR 1996, 507; NJW 1997, 2509; Steiger, 1998, no. 184 f.; contra Böhm, 1996, p. 114.

[100] BVerfGE 88, 203, 252 on the protection of unborn life; Steiger, 1998, no. 166; Wulfhorst, 1994, p. 99 ff.; Böhm, 1996, p. 129 ff.

[101] Böhm, loc. cit.; Wulfhorst, loc. cit.

function including its biotic and abiotic elements and life processes.[102] This leaves considerable scope for deciding whether a given change to the environment is "significant". The state also has lawmaking power regarding the protection of natural resources in private ownership, under Art. 14(1) sentence 2 GG. Despite the underlying civil (ownership) element in such cases, strict individual protection of the kind accorded to human health does not apply here, and in many such cases protection is deemed collective simply because the privately-owned protected resource in question is a piece of landed property as such.

The state's constitutional duty of protection and its constitutional objectives are fleshed out by specific policies, and the legislature is left considerable scope as to how it does so.[103] The conventional environmental law focus on individual substances and individual types and qualities of radiation is not in itself unconstitutional where there is a lack of scientific evidence regarding combined effects or where safety margins built into the limits set for individual substances can be expected to cover risks of combined effects. Constitutional constraints do not come into play except where combined effects pose an identifiable danger to human health or the environment.

Another question is whether the constitution places the state under a duty to undertake research for combined effects. Risks to human health and the environment resulting from gaps in scientific knowledge normally belong to the residual risk we all have to accept as part of the burden of civilization.[104] But the state's constitutional duties to protect human health and the environment include a duty to undertake systematic research into questions of scientific uncertainty. Issues of this kind have so far been debated from another perspective under the heading of "proceduralization" in matters involving the precautionary principle: Where the state, acting in accordance with the precautionary principle, imposes freedom-constraining duties on enterprises and consumers because there is (merely) a scientifically founded suspicion of risk, it must initiate or fund research to reduce the scientific uncertainty and, if the findings so dictate, must relax the legislation in favour of enterprises and consumers.[105] In principle, there is also a corollary duty to ascertain whether existing levels of protection suffice. With regard to the state's constitutional duty of protection inferred from the basic constitutional rights, this corollary follows from a (controversial) recent case-law development known as the *Untermaßverbot* – a proscription of laws and other acts of state that are inadequate to their purpose, such that the state is duty-bound to provide appropriate and effective protection of the basic rights at issue in a given situation.[106] The state's constitutional duty of protection thus includes a duty to advance knowledge about combined effects by systematically initiating or funding research.

The duties to advance knowledge about risk that derive from the state's constitutional duty of protection are restricted to instances where there is reason to suspect

---

[102] Steiger, 1998, no. 56 ff.; Murswiek, 1999, Art 20 a, no. 29 f.; Kloepfer, DVBl. 1996, 73, 76.
[103] See 5.1.1 above.
[104] BVerfGE 49, 89, 143; BVerwGE 55, 250, 254; BVerwG, NJW 1970, 1890, 1892.
[105] Wahl/Appel, 1995, p. 126 ff.; Scherzberg, VerwA 1993, 484, 497 ff.; Ladeur, 1995, p. 142 ff., 151 ff., 156 ff., 206 ff.
[106] BVerfGE 88, 203, 254; BVerfG, NVwZ 1996, 507; Steiger, 1998, no. 187 f.

that combined effects pose significant danger. The broader constitutional objective of environmental protection, on the other hand, carries with it an intrinsic, general obligation for the state to reduce environmental risks by initiating or funding risk research, including research into combined effects. Note that in both instances we are dealing merely with general, for the most part programmatic obligations lacking any stipulation as to the choice of policies used to meet them. That is, the state is called upon to include combined effects in state-sponsored research, but how it discharges this obligation is its own affair.

5.3.2
Summary Appraisal of the Current Legal Situation and Reform Needs

The objectives and discretionary powers embodied in current German laws relating to environmental media do not generally allow combined effects to be disregarded in that they adopt the general danger concept used in police and public-order law (or policy-area-specific derivations of that concept) and widen the duty of protection to include precaution against risk. At the same time, for practical reasons, most legislation relating to environmental media focuses on individual substances. Where it does include combined effects in environmental standards, action thresholds or classification rules, it usually follows a fairly simplistic approach that bypasses the complexity of dealing with different types of combined effects – either attempting to regulate such effects with environmental standards by incorporating them in safety margins, or assuming that all such effects exhibit dose-additivity regardless of the actual type of dose-response relationship involved (an assumption of linearity, with or without allowance for variations in the slope of the dose-response function). Classification rules and action thresholds in environmental media protection either assume a single preponderant substance or use simple additive formula; in some cases, rules for mixtures are dispensed with entirely, which is equivalent to using safety margins.

German substance-related environmental laws take a more varied approach. Some strictly target individual substances, as with risk assessment under the Chemicals Act (ChemG). Others include preparations that contain the same active agents or metabolites in risk assessment, risk evaluation and the resulting regulation. Finally, a few substance-related laws or parts of them expressly include combined effects in risk assessment, risk evaluation and regulation – although for practical reasons they are selective regarding the substances they cover. This applies to medicinal drugs law, to the classification of preparations in chemicals law, and to prohibited and restricted substances in drinking water and cosmetics under food and consumer goods law. The classification rules assume simple concentration additivity and make exclusive use of danger categories in considering differing harmful effects, though a mixture can be reclassified after it has been subjected as such to assessment or if there is evidence of greater-than-additive combined effects. Product standards for drinking water and cosmetics use simple sum total values. There is no systematic coverage at all for combined effects of substances that come under different laws or involve lifestyle-related exposures.

As regards imputation of financial burdens resulting from combined effects, the conceptual approach followed in civil liability law appears up to the task, though its rationale is based on the presence of multiple polluters rather than that of combined effects as such: the latter are only a special case of a more general solution. Except in social accident insurance – where the insurance funds pay all claims resulting from combined effects except only for marginal contribution of occupational hazards – liability law has developed various ways of imputing combined effects to multiple actors and, in particular, looks to combined effects in cases of uncertainty. Retrospective liability law offers two methods of fairly apportioning the burden once co-responsibility has been established: first, proportionate damages awards matching each polluter's estimated share of the total blame and, second, joint and several liability towards the victim combined with reciprocal redress among contributing polluters. However, neither of these methods can be transferred to public law without qualification. The availability of redress among polluters – a constant feature of liability law – is not inconceivable in public law, as can be seen from the legislation on contaminated sites, but it is unlikely to be acceptable as a general means of assigning responsibility in prospective regulation. On the other hand, the priority principle favoured by public law for this purpose is unsatisfactory in legal policy terms.

If we assume that combined effects can be expected to occur in many cases rather than being the exception, the German legislation currently in force is in general need of reform. Working from this premise, it is necessary to develop rules of law that foster the identification of combined effects and make greater and more appropriate allowance for them when assessing and regulating risk. However, a general regulatory model is hard to conceive of since systematic assessment of all possible combined effects is impossible.

The reform proposals – based on draft codes of environmental law submitted by the Professorial Commission on the Environmental Code and the Expert Commission on the Environmental Code – endeavour in very general form to consider the problems of combined effects when setting environmental standards. Both draft codes allow for the consideration of multiple pollutants in setting environmental standards (Art. 145(4) of the professorial draft and Art. 12(4) of the expert commission draft). The reform proposals do not, however, specify in detail how this is to be achieved.

The call to make suitable allowance for combined effects must, however, observe health and environmental priorities. It would not appear appropriate to advance the issue of combined effects until risks from individual substances are suitably regulated. This applies, for example, for the effects of air pollutants on vegetation (see Section 2.5) and accidents (regarding which there are no action thresholds for individual substances, let alone for mixtures). It also applies for substance risks resulting from personal lifestyle choices (tobacco and alcohol) – even though this is an area where super-additive effects are found in individual cases.

5.4
Reform Proposals I:
Assessment of Combined Effects

5.4.1
Testing of New Substances

Existing chemicals law provides only limited scope for requiring potential polluters to assess combined effects. As described earlier, Section 11(1) no. 2 of the German Chemicals Act (ChemG) provides that if a substance is suspected to be hazardous and to necessitate precautionary testing, the Notification Authority can require the manufacturer or importer to submit testing data regardless of whether the quantity thresholds laid down in the Act have been reached or exceeded. The test reports are restricted, however, to evidence regarding the hazardous nature of the substance and must remain within the scope of the – exclusively substance-specific – Level I and II tests under the Act. There are no express powers to require later assessment of combined effects. It is doubtful whether powers of this kind could be incorporated into the Act without infringing the EU Dangerous Substances Directive (67/547 EEC as last amended by 92/32 EEC). Under Article 16(1) of the Directive the competent authorities can, if necessary for evaluation of the risk caused by a substance, ask for further information and/or tests concerning that substance or its transformation products. A testing requirement relating to effects that a notified substance gives rise to solely in interaction with another substance would appear on the face of it to be compatible with this wording. By "dangers", however, the Directive appears to mean only the hazards inherent to a given substance. The Risk Assessment Directive (93/67/EEC) issued under the Dangerous Substances Directive likewise targets individual substances. The primary directive thus probably rules out having combined effects assessed at the expense of manufacturers for preparing restrictions under Art. 17 ChemG.

The existing gaps in the testing of new substances could be closed by amending the Dangerous Substances Directive so that it requires data on combined effects to be ascertained in the base set of tests and if necessary expanded on in subsequent confirmatory testing. A possible approach would be to require that manufacturers investigate evidence of combined effects in the exposure scenarios they must already test for (the workplace, intended applications and intended areas of use; see Art. 3(4) (b), (c), (e) and (i) of ChemPrüfV, the German regulation on testing data). This addition to initial testing would accord with the established system providing it were limited to merely ascertaining if there is evidence of combined effects so as not to place a disproportionate burden on manufacturers regarding the extent of the required testing. Yet if initial testing is going to be extended, one might question the logic of putting less likely phenomena such as combined effects before other major impact categories that are currently likewise omitted from testing, for example the subchronic and chronic toxicity of individual substances. In any event, Art. 7(2) (with Annex VIII) and the second paragraph of Art. 16(1) of the Dangerous Substances Directive should be amended such that the competent authorities can order tests for combined effects either as part of Level I and II testing or at a later date. The Risk Assessment Directive would have to be amended accordingly. Finally, there would be knock-on changes in substance classification (risk phrases).

5.4.2
Testing of Existing Substances

Art. 11(1) no. 2 ChemG does not cover existing substances, assessment of which is governed by the EU Regulation on the Evaluation and Control of the Risks of Existing Substances (793/93/EEC). This targets individual substances in isolation. Art. 3 of the Regulation requiring submission of data on, among other things, environmental pathways and the environmental fate of substances is arguably open to broader interpretation, and Art. 10 on risk evaluation speaks of risks to man and the environment in general. The EU Regulation on Risk Assessment for Existing Substances (Regulation 1488/94/EC) issued for this purpose, however, does not appear to allow assessment to include combined effects.

There is a need here for clauses that leave open the possibility of imposing a duty on manufacturers and importers to test for combined effects and to incorporate these tests into risk assessment.

5.4.3
Classification

Conventional classification rules allow, per se, a preparation or other mixture to be put in a different harzard category if the effects of substances contained in it are super-additive or sub-additive or differ entirely from those of the substance alone. The classification rules in the German Hazardous Substances Regulation (Gef-StoffV) do not, however, oblige manufacturers to test for indications that reclassification may be appropriate. Manufacturers have a commercial incentive to do such tests if they anticipate sub-additive or entirely different effects that would place a preparation in a lesser hazard category, though they may prefer to remain with the existing, higher hazard category if the cost of testing outweighs the potential savings. In the case of super-additive combined effects – at least with manufactured preparations – manufacturers may incorporate such effects into testing out of a desire to avoid civil liability claims or loss of consumer confidence in a product because of inappropriate labelling.

5.4.4
Research Obligations

Civil-law research obligations along the lines of the *Kindertee* (infant tea) judgement[107] are restricted to specific circumstances, in particular where manufacturers' advertising gives the impression that a product is harmless. These obligations are not generally applicable. However, civil liability can have implications for the gen-

---

[107] BGHZ 116, 60.

eration of knowledge about risk by producers, though the implications differ according to the type of liability.[108] Where there is a core realm of certainty bounded by uncertainty, manufacturers may have an incentive under both strict and fault-based liability to push back those bounds by doing or funding research. Where combined effects are wholly uncertain, on the other hand, fault-based liability can achieve little since manufacturers cannot be blamed for failing to research them. This is why the courts do not recognize product or environmental liability in civil law cases involving combined effects unless there is evidence that the precautions taken (such as observing exposure limits for individual substances in isolation) are insufficient.[109] Strict liability (except where it excludes unforeseeable risks as in product liability) does theoretically have the potential to push forward research into areas of uncertainty. The precondition is that manufacturers must have a specific direction in which to conduct research. For example, a manufacturer may focus on characteristic exposure scenarios for a number of particularly hazardous substances. It would be unrealistic, however, to expect strict liability to result in systematic research into combined effects.

5.4.5
Conclusions

Overall, a strengthening of industry testing and research obligations regarding combined effects is generally an appropriate and suitable means of improving knowledge of such effects. But industry obligations of this kind have their limits.

A health and environmental policy research programme is needed to direct decision makers' attention to the problems of combined effects, to fund research on priority areas (such as widespread highly toxic substances) and to make it possible to coordinate the activities of public agencies (particularly in areas that are covered by a number of different laws and consequently are subject to overlapping competencies).

Where public agencies rather than manufacturers are responsible for classification (as with water pollutants) administrative orders could be issued to make agencies aware of the need for investigation so that they classify mixtures more appropriately at least where there is evidence of combined effects (or similar effects) that are not exclusively dose-additive.

---

[108] See Schwarze, 1996.
[109] BGHZ 80, 186, 195; 96, 167, 172.

5.5
Reform Proposals II: Regulation

5.5.1
General

5.5.1.1
Limits to the Law in Regulating Combined Effects

By its very nature, the law only offers limited help in choosing fair and appropriate regulatory strategies. It places limits on policymaking but does not itself aim to improve policy. Except when it comes to supreme rights such as the inviolability of human life, the *Untermaßverbot* developed in recent German constitutional case law relating to the state duty of protection – the rule that the state must furnish appropriate and effective protection[110] – enjoins only manifest failures to protect, say, human health and the environment. It does not force the state to find a solution that is "optimal" and fairly distributes the onus of compliance. Makers of statute law and, within the bounds set by statutes, issuers of secondary legislation are normally allowed considerable freedom to choose the regulatory models they think fit. Conversely, environmental policy is prevented from overstepping the mark by the proportionality principle enshrined in the rule of law.

When considering the regulation of combined effects, we need to distinguish between what is to be protected (human life, the environment), the desired level of protection (prevention of dangers, precaution, sustainability), the conceptual approach to solving the problem (the regulatory strategy) and the policy instruments used to implement the strategy. We will not go into further detail regarding the choice of policy instruments here. Against the background of current environmental law and the prevailing selection of policy instruments used to implement it, the present study takes direct regulation as its starting premise. The possibility of replacing command-and-control with market-based or flexible mechanisms raises a number of prima facie questions that have less to do with combined effects and are beyond the scope of this study.

Ideally one would be able to distinguish between proven combined effects and mere evidence of such effects and could logically choose different strategies and in particular different policy instruments for the two. But in reality there is a grey area where a "corridor" of established knowledge is flanked by mere evidential data. For example a combined effect can be known to exist though there is uncertainty about its response mechanism. Nonetheless, for our present purposes we will assume that the ideal distinction holds.

---

[110] Footnotes 3 and 99 above.

5.5.1.2
Prevention of Dangers and Precaution Against Hazards

Given proven knowledge of combined effects, regulation can serve the purposes of both prevention of dangers and precaution against hazards.* Where there is insufficient knowledge as to the existence and nature of such effects but their existence cannot be ruled out, state regulation is governed entirely by the precautionary principle. Since all major environmental laws serve prevention as well as precautionary purposes, in general there are adequate statutory powers for precautionary regulation of combined effects.

The distinction between prevention of dangers and precaution implies different degrees of rigour in applying the state's constitutional duty of protection, and also different degrees of acceptability for trade-offs between the duty of protection and other (for example, economic) concerns as to how planned changes affect industry's investment in existing technology, as to how the rights of other parties are affected, and in many cases as to the regulatory strategies themselves. The difference hinges upon whether the probability of loss or damage is deemed to lie above the threshold of "sufficient probability" ("reasonableness"), or below it though above that of acceptable residual risk.[111] There is no natural dividing line between prevention of dangers and precaution, and drawing one involves a value judgement. Some environmental laws, and in particular the German Nuclear Energy Act (AtG), apply a uniform risk concept embracing both prevention of dangers and precaution on the basis of which risk must be virtually eliminated.[112]

A particularly important question relates to the limits of prevention of dangers and precaution. From a legal point of view, excessive protection must be avoided because it comes at the expense of those who are required to comply and incurs a loss to the economy as a whole. The conventional understanding is that the proportionality principle does not apply in prevention of dangers until it comes to the choice of policy instruments, after the decision whether to regulate has already been made. Where they serve preventive purposes, as they often do, it is only in rare cases that statutory powers to set environmental standards expressly sanction taking economic interests into account. In practice, however, economic interests already play a significant part at this level; witness the divergence of environmental standards proposed by scientists and those consented to by politicians. The situation is different at the level of instruments where the legitimacy of taking economic concerns into account when establishing the proportionality of specific measures is undisputed.

There are arguments for and against "economicizing" the concept of danger. Doing so can bring trade-offs into the open – some of which already exist in practice anyway – and can help focus efforts on priority environmental problems or on

---

*   "Danger" in the German terminology is a risk of injury which is sufficiently probable and therefore unacceptable, "risk" or "hazard" is a risk below the threshold of danger which shall be reduced for reasons of precaution.

[111] See Rehbinder, 1998, Teil 4, no. 21 ff.; Wahl/Appel, 1995, p. 84 ff.; Petersen, 1993, p. 213 ff.

[112] BVerwGE 72, 300, 315 f.; BVerwG, NVwZ 1991, 1185, 1186; Breuer, NVwZ 1990, 211, 214; on atomic energy law see Wahl/Appel, 1995, p. 100 ff.

the problems that can be addressed at least cost. But it runs contrary to the constitutional hierarchy of values, at least when it comes to protecting human health.[113] A look at the peculiarities of the policymaking process (in which environmental concerns still tend to be considered last of all) likewise comes out in favour of retaining the status quo. Explicit powers to make allowance for economic interests can also be construed as an open invitation to water down environmental standards. In all this we should remember that while the border between prevention of dangers and pecaution has a grey area subject to value judgement, in certain respects danger itself is an open concept: danger is relative to the extent that it depends not only on probabilities of occurrence, but also on the nature and scope of the possible damage. Moreover, the requirement that a danger must be "significant" in order for it to be prevented leaves room for interpretation, in particular as regards safeguarding well-being (protection from nuisance), safeguarding property and protecting the environment in the narrow sense of the word. Even the protection of human health does not have rigid limits if one considers psychosomatic complaints and immune deficiencies, which generally come under harmful effects but are given significance thresholds.[114] Value judgements of this kind prevent overprotection without strict comparison of costs and benefits.

Where other societal concerns are brought into the equation – in particular the cost of regulation – the proportionality principle does not require formal cost-benefit analysis, cost-risk analysis or other quantitative methods, but neither does it forbid them.[115] As the American experience shows[116], the advantages of greater transparency and formalization in decision making are matched by disadvantages in policymaking – in particular slower decision processes, lower output in terms of the number of limits set, and greater risk of decisions failing in the courts (in some cases this has prompted a flight to contractual arrangements).[117] Then there are open questions of evaluation. One is tempted to view the reluctance to adopt formalized substantive decision criteria primarily as an immunization strategy against judicial review. Yet it is driven by the legitimate motive of not wishing to stretch the

---

[113] Steiger, 1998, no. 171; but see BVerfGE 56, 54, 80, in which the court does not entirely rule out that "in finding a balance between conflicting interests, human exposure to adverse effects and dangers may be deemed reasonable" under the proportionality principle; however, the decision relates to nuisance from noise.

[114] See Kunig, 1992, *Art. 2*, no. 63; Starck, 1985, *Art. 2*, no. 130; Möllers, 1996, p. 35 ff.; going too far: BVerwG, NJW 1995, 2648, 2649.

[115] See BVerwGE 81, 12; Rose-Ackermann, 1995, p. 147 ff.; also Ossenbühl, 1988, *§ 63*, no. 7 rebutting a constitutional duty to pass optimum legislation.

[116] See AFL-CIO v. Hodgson, 449 F.2d 467 (D.C. Cir. 1974); AFL-CIO v. American Petroleum Institute, 448 U.S. 607 (1980); AFL-CIO v. OSHA, 965 F.2d 962 (11th Cir. 1992); Corrosion Proof Fittings v. EPA, 947 F.2d 1201 (5th Cir. 1991); also American Paper Institute v. EPA, 660 F.2d 954 (4th Circ. 1981); Chemical Manufacturers' Ass'n v. EPA, 570 F.2d 177 (5th Cir. 1989); Mendeloff, The Dilemma of Toxic Substance Regulation, 1988; the courts are far more cautious where the law does not require a cost-benefit comparison; see Reserve Mining Co. v. EPA, 514 F.2d 492 (8th Cir. 1975); Ethyl Comp. v. EPA, 541 F.2d 1 (D.C. Cir. 1976); Lead Industries Ass's v. EPA, 647 F.2d 1130 (D.C. Cir. 1980).

[117] As with the Environmental Protection Agency's Industrial Toxics Program (33/50 Program); EPA, Office of Pollution Prevention and Toxics, A Progress Report: Reducing Toxic Risks Through Voluntary Action, July 1991; idem, Draft: 33/50 Beyond the Numbers, 1996; see Lewis, Elni Newsletter 1998, 11.

decision process on a Procrustean bed such that policymakers lose substantive control over individual decisions.

The state has a lesser obligation to act for precautionary purposes than for prevention of dangers. What's more, questions of proportionality and hence economic aspects already come into play in the initial decision whether to act at all.[118]

The precautionary principle is also tempered by the rule of law with regard to grounds for taking action. Prevailing opinion seems to affirm grounds for action – that is, circumstances in which precautionary policies are admissible – only if there is evidence of a specific hazard.[119] In its decision of 19 December 1985 regarding the Wyhl nuclear power station, however, the German Federal Administrative Court[120] ruled for atomic energy law that "the potential hazards that can be (or rather must be) considered include hazards that cannot be ruled out merely because current knowledge can neither confirm nor deny certain causal relationships, such that there is not an extant risk, but only a suspected hazard or 'cause for concern'." Generally, then, a well-founded suspicion – that is, merely scientifically plausible, more-or-less established evidence – can provide sufficient grounds for precautionary action. This extends to uncertainties surrounding combined effects. The degree of plausibility required here is the result of a value judgement that takes into account the importance of what is to be protected and the type of hazard in question. Prevention takes on higher priority when it comes to severe, possibly irreversible effects. No empirical findings are then required. Cause for concern can derive from theoretical models that assume identical mechanisms in the case of combined effects. Precautionary action cannot be justified, however, by pure speculation as to effects unfounded in plausible scientific reasoning.[121] The case law here evidently follows a model of knowledge based on established science.[122]

The proportionality principle is of particular importance when it comes to choosing precautionary policies. The options are restricted to policies that are proportionate to the assumed risk.[123] In other words, plausible cause for concern can only be transformed into cause for action by the state as long as the action taken is not manifestly out of proportion to the assumed environmental benefit. In this instance, the proportionality test is applied only at an abstract level. Special weight is given here to the interests of industry, hence there are additional limits to the nature and scope of the policies used; an appropriate policy might thus be a rational reduction plan that follows assigned priorities with allowance for substitutability and time to adjust.[124]

---

[118] Winter, 1995, p. 12 ff., 25.

[119] BVerwGE 69, 37, 43 f.; Ossenbühl, NVwZ 1986, 161, 164, 166 ff.; Kloepfer/Kröger, NuR 1990, 8, 12 ff.; Feldhaus, UPR 1987, 1, 8.

[120] BVerwGE 72, 300, 315; likewise BVerwG, NVwZ 1992, 984, 985; Petersen, 1993, p. 285 ff.; Wahl/Appel, 1995, p. 126 ff.; Rehbinder, 1991, p. 269, 279 f.

[121] BVerwGE 92, 185, 196; BVerwG, NVwZ 1992, 984, 985; Wahl/Appel, 1995, p. 92 f.; 112; Roßnagel, in *Gemeinschaftskommentar zum BImSchG*, 1998, § 5, no. 250.

[122] This is particularly clear in BVerfG, NJW 1997, 2509.

[123] BVerwGE 69, 37, 45; Ossenbühl, 1986, p. 167 f.; Wahl/Appel, 1995, p. 135 ff.

[124] See Winter, 1995, p. 1, 20 ff.; Ladeur, NuR 1994, 8, 13 f.; Di Fabio, 1994, p. 145.

5.5.1.3
Environmental Quality, Emission Limits and Minimization

Prevention of dangers and precaution can be combined with various regulatory strategies. Under the systematic approach used in the German Pollution Control Act (Art. 5 (1) nos. 1, 2 BImSchG) and Water Management Act (Art. 6 and 7(1) WHG), it would seem natural to classify strategies that relate directly to the quality of environmental media (air and water quality standards and targets) as prevention of dangers and strategies to reduce emissions (emission limits) as prevention of dangers. But this would be only approximately correct. Quality-based strategies can also be used as a means of precaution. Quality-based precautionary strategies are expressly incorporated into Germany's atomic energy and radiation laws (dose limits), its Soil Protection Act (threshold values giving cause for concern), and its chemicals laws.[125] Thus, for example, dose limits (and likewise the unit risk model for carcinogenic air pollutants used by LAI, the *Länder* Committee on Pollution Control) are based on the assumption that individual risk must be reduced to a quantified tolerable level. Nor is precaution restricted to the emission reductions that are attainable with current technology. Although Art. 5(1) no. 2 BImSchG focuses on the "best available technology", this is not the only precautionary strategy permitted under the Act. Particularly with regard to carcinogens and certain other hazardous substances (Technical Instructions on Air Quality *(TA Luft)* nos. 2.2.1.5, 2.3, 3.1.6 and 3.1.7), the emission limits for hazardous substances, which at least partly serve precautionary purposes, are broken down into hazard classes.[126] *TA Luft* also contains precautionary exposure limits in some instances (no. 2.2.1.4(3)). On the other hand, the technical guideline concentrations in occupational health and safety law are precautionary (indoor) air quality limits based on what is technically possible.[127]

Precaution in proportion with risk in the form of emission or exposure limits is not always the exclusive aim. Particularly hazardous substances are often additionally subject to a minimization rule (nos. 2.3 sentence 1 and 3.1.7(7) *TA Luft* and Art. 36(5) of the German Hazardous Substances Regulation (GefStoffV)) and in radiation control law to the ALARA principle ("as low as reasonably achievable" – Art. 28(1) sentence 1, Art. 46 sentence 1 and Art. 48(1) no. 2 of the German Nuclear Safety Regulation (StrlSV)). There is not an objective difference between the two rules, since the minimization rule is likewise tempered by the proportionality principle.[128] Emission limits and minimization are combined to reduce exposures to a specific level in general, and to achieve further reductions in individual cases where this is technically possible and not entirely out of proportion with the ends attained.

An alternative to the minimization rule might be a general reduction of individual risk to a specific, quantified level[129] (say to 1/10th of the risk threshold, for carcinogens to 1/100th of the concentration level below which there is no observable carcinogenic

---

[125] See Rehbinder, 1998, part 4, no. 155; Art. 7 and 8(2) BBodSchG; Art. 17(1) and 19(1) ChemG, Art. 16(2) and 36(2), (3) and (5) GefStoffV.
[126] Rehbinder, 1998, part 4, no. 35, 150; see Böhm, 1996, p. 36 ff.
[127] See Falke, 1986, p. 180 ff.
[128] No. 2.3 TA Luft; BVerwG, NVwZ 1991, 1187.
[129] See Cansier, NVwZ 1994, 642; Akademie der Wissenschaften zu Berlin, 1992, p. 223 f., 370 ff.; but also Rehbinder, NuR 1997, 313, 321 f.

effect, or to a risk measure of between 10^{-5} and 10^{-6}).[130] Unlike the minimization rule, this approach has the advantage that it lays open the acceptability threshold for residual risk and enables a systematic balancing of the social utility and cost of additional safety. However, it does not accord with the tradition of German environmental law.

5.5.1.4
Conclusions

Current environmental law is flexible enough to incorporate combined effects through appropriate regulation strategies, in both prevention of dangers and precaution. To the extent that regulatory strategies and the means used to meet the protection level required of state environmental policy (and in particular the means used to implement the proportionality rule) are open to fundamental criticism[131], such criticism ought to be elucidated first with reference to risks from individual substances before turning to risks of combined effects.

5.5.2
Regulatory Strategies for Combined Effects

5.5.2.1
Limits to the Regulatory Coverage of (Possible) Combined Effects

The state's constitutional obligations and duties to protect and the objectives and powers it is ascribed in specific laws require it to pay due consideration to combined effects when legislating on substances and radiation. The scope of this requirement is limited by the current state of scientific knowledge and by the rule of law, under which purely speculative risks cannot be addressed with policies that impinge upon freedom. This all means in the first instance that regulatory policies cannot be justified on no more than a blanket assumption that because all substances can theoretically interact there is always cause for concern at least about response-additive combined effects. Normally, exposure to two or more hazardous substances does not give rise to combined effects because they attack different organs, they do not influence each other's toxicokinetic properties, and they each have entirely different effects. A worst-case regulatory strategy for combined effects is thus inadmissible. Even considering the difficulty of telling apart scientifically plausible, hypothetical and speculative risk, this caution seems justified in principle since the most problematic individual substances are often provided for by limits with built-in safety margins and in some cases by minimization requirements even though there is little interaction at low doses. Systems of limits (such as Germany's Technical Instructions on Hazardous Substances, TRGS 403) and classifications that assume combined effects to be possible wherever there is an exposure to two or more substances are admissible for reasons of administrative economy to the extent that they incorporate adequate correctives and the procedures and proof needed to invoke these correctives are not excessive. A simple assumption that combined effects exist may thus conflict with the rule of law.

---

[130] See Fischer, 1996, p. 65 ff.
[131] See Cansier, NVwZ 1994, 642; Schuldt, 1997.

The multitude of theoretically possible combined effects makes it necessary to select candidates that urgently call for regulation. An all-inclusive approach is impracticable. Possible selection criteria include the type of effect, the level of plausibility that interactions will occur, and ubiquity/exposure levels. First priority thus goes to toxic and highly toxic agents, carcinogens, mutagens, reproduction toxic and ecotoxic substances (Art. 3a (6), (7) and (12)-(15) of the German Chemicals Act (ChemG)). Knowledge regarding substances in these categories can be augmented by studying structure-response relationships with chemically and physically related substances. Production quantities and commonly used mixtures such as preparations and crop protection products can be included as indicators of typical exposures. The resulting selection is narrowed down by the rule that there should be no regulation without scientifically plausible evidence for combined effects.

Finally, a special problem with regulating combined effects is that people can undergo simultaneous exposure through different environmental pathways. It is thus necessary to incorporate cumulative risk. This must be done on a casewise basis; it cannot be done by simple apportionment because of the differing regulatory traditions for the different environmental pathways, for example regarding the conceptual approach to limits and safety margins.

5.5.2.2
Regulatory Strategies

The available regulatory strategies are many and varied. Among the ones most worthy of further consideration are the following:

- Case-wise risk assessment
- Limits (exposure or risk-weighted emission limits) for individual substances with safety margins for combined effects
- Limits for individual components that dominate a mixture
- Special (exposure) limits for combined effects
- Quantitative reductions of emissions to the point of minimization[132]

There is considerable executive discretion as to the choice of strategies and how they are implemented. This is especially the case when it comes to implementing strategies based on mandatory limits in law: when limits are stipulated in a regulation *(Verordnung)*, judicial review is restricted in Germany to ascertaining that they are constitutional and accord with the statutory powers used when setting them;[133] when limits are set forth in an administrative rule *(Verwaltungsvorschrift)* under powers to concretise broad statutory terms ("adverse environmental effects", "precautions", etc.), precedent again binds the courts to uphold them providing they accord with the value principles of the primary statute and their scientific basis is not manifestly out of date (except when a court judgement is required in an atypical case).[134] Though its foundations are disputed and the line followed by case law is

---

[132] See Murswiek, VVDStRL 48 (1990), 207, 215 f.
[133] BVerwGE 61, 256, 262 ff.; Ossenbühl, 1988, § 65, no. 78; Gusy, NVwZ 1995, 105.
[134] BVerwGE 55, 250, 256 ff.; 72, 300, 314 ff.; 78, 177, 180 ff.; 81, 185, 190 ff.; BVerwG, NVwZ 1988, 824; NVwZ-RR 1994, 14; Breuer, DVBl. 1978, 28, 34 ff.; idem, NVwZ 1988, 104, 108 ff.; Wahl, NVwZ 1991, 409; Gusy, NVwZ 1995, 105; Rehbinder, 1998, part 4, no. 271 (with further references).

unclear, it is safe to assume that this doctrine will continue to prevail in practice (except where EU directives prescribe that limits be implemented by a regulation). In either case the executive has considerable discretion for generalization according to the protection needs of receptors, risk assessment and the acceptability of residual risk. This makes it possible to pursue different strategies for combined effects.

5.5.2.3
Case-wise Risk Assessment

In scientific terms, case-wise assessment of combined effects is preferable to any generalized approach. It can allow for different types of combined effects and largely avoids overstating or understating combined risk. It is mainly used in substance approval (e.g. drug licensing) and in decisions to ban or restrict hazardous substances. Case-wise assessment can also be used as part of a classification-based approach to determine criteria for assignment to a specific class of combined effect – in particular whether the effects are dose-additive or independent. Accordingly, safety margins, limits for specific substances, action thresholds and classification rules for combined effects should be based where possible on case-wise assessment of the combined effects. There are practical limits to the use of this regulatory model, however. The diversity of substances in circulation and of their possible combined effects means that research into such effects cannot keep up with the practical need for manageable regulation. Hence generalized classification must still be used to some extent, however scant its scientific basis. What's more, the complexity of combined effect phenomena makes case-wise assessment problematic in the licensing of industrial facilities, where public agencies work on a tight schedule and are often not well enough informed about the subject matter to make objective judgements.

Case-wise assessment can act as a necessary corrective wherever administrative economy or an urgent health safety need calls for the use of generalization when setting safety margins, limits, action thresholds and classification rules. Of course, a mere power or even an obligation to incorporate a corrective does not mean it will actually be used, though when it comes to substance classification, product liability gives manufacturers an incentive to look for evidence of effects that are more severe than the (usually dose-additive) effects assumed when the classification was first devised.

Where existing limits or other regulatory requirements do not include combined effects, case-wise assessment to allow for known or suspected effects of this kind is admissible; indeed, compliance with the objectives of the pertinent legislation may require it. No. 2.2.1.3 of the Technical Instructions on Air Quality *(TA Luft)*, for example, provides for a special investigation in cases such as carcinogens where no exposure limits are set and there is sufficient evidence of adverse environmental effects despite compliance with emission limits and the minimization rule.[135] This must include combined effects. Case-wise assessment is also possible with existing licensed industrial facilities (e.g. under Art. 17 of the German Pollution Control Act (BImSchG)). Constitutionally speaking there is thus broad legislative scope in this

---

[135] See BVerwG, NVwZ 1998, 1181.

regard, but this scope is considerably narrowed in statutory law by the stricture of preventing dangers and providing adequate precaution. Much of the primary legislation on environmental media and specific substances demands adequate consideration of combined effects, but a lack of statutory instruments to this end does not in itself contravene that legislation to the extent that provision is made for risk assessment in specific instances.

The interventionary powers under environmental legislation mostly leave an option open for case-wise correction of limits and other regulatory requirements that extend to combined effects. The exception (where admissible) is where limits or other requirements are stated exhaustively; any judicial review can then only be based on noncompliance with the constitutional duty to provide protection (conceivable for example in the case of highly super-additive combined effects).[136]

5.5.2.4
Safety Margins

Another possible approach – and one that to some extent accords with current practice – is to allow for combined effects by building a safety margin into limits that apply for individual substances. The *prima facie* problem with this kind of strategy is that current limit-setting practice lacks openness: the reasoning behind the safety margins is not disclosed. One can generally assume that safety margins are meant to account for variations in individual susceptibilities and for possible sources of error when extrapolating from animal experiments to humans and from high to low concentrations.[137] In other words, safety margins have been used to date mostly for reasons other than allowing for combined effects.[138]

Any appraisal of a safety margin strategy must start by looking at the bounds to the precautionary principle that are set by the rule of law. These preclude the merely prophylactic use of special safety margins for combined effects where there is no evidence that they occur. If such evidence exists but is not sufficiently certain, a safety margin is an admissible strategy for combined effects if the substances concerned at least have a dose-response threshold and the safety margin gives adequate protection from them when they occur individually. The situation is different with carcinogens and moreover in occupational health and safety where regulation operates close to the NOAEL. A safety margin strategy is especially likely to be considered where a set of substances very frequently occur in a specific combination; even if the resulting margin is too generous where one of the substances occurs individually, this is still an appropriate and practicable generalizing strategy because a system of limits with built-in safety margins cannot be applied differently when substances occur on their own and in combination. Depending on the form it takes, a safety margin for combined effects can drastically lower the maximum permissible level compared with a genuine limit for combined effects based on linear dose-additivity. This raises the question whether such a large increase in safety mar-

---

[136] See BVerwGE 101, 1, 4 ff.
[137] Dieter, BGesBl. 1995, 250; Dieter/Konietzka, Regul. Toxicol. Pharmacol. 22 (1995), 262; Fischer, 1996, p. 55 ff., 63 f.; Wagner, Columbia Law Review 95 (1995), 1613, 1622 ff.; Latin, Yale Journal on Regulation 5 (1988), 89, 98 ff., 103 ff.
[138] But see Winter, 1986, p. 134; Englert, 1996, p. 84.

gins is really necessary. There are good arguments in favour of (at most) modest safety margins for combined effects. There is little interaction at the low dose ranges to which most limits (unlike classification rules) restrict exposures. An admissible response to the hypothetical lower level of interactions at these dose ranges would thus be reduced safety margins. The safety margins would have to be increased again, of course, if there were any signs of super-additivity.

A further question is whether special safety margins for combined effects are needed in the first place. The safety margins for individual substances already keep exposures well below the NOAEL and additionally reduce the risk of combined effects below the already low level found at low doses. Moreover, epidemiological studies have so far produced no evidence that safety margins set for individual substances are inadequate when combined effects are assumed to exist. This is the basis of *TA Luft* in matters of health safety. However, the data from past studies is far from sufficient to permit generalization. All the same, special safety factors could conceivably be applied solely where there is evidence of super-additive effects.

5.5.2.5
Dominant Components

Separate regulatory requirements for combined effects become superfluous when one component of a mixture is so dominant that it can be said to be characteristic of the mixture as a whole. This is particularly the case where a substance has a dose-response threshold and the limit based on the threshold is exceeded. Exceeding the limit for the individual substance invokes a ban or restriction or affects the classification of the mixture. Because adding a substance to a mixture has the same effect as diluting the substances already in the mixture, classification rules often stipulate a special concentration threshold above which a substance is considered dominant. The alternative is a set of general limits on the presence of a substance in a mixture, with different limits for substances of differing toxicity. Classification proceeds accordingly if it is not possible to specify a dose-response threshold for a given substance but the substance is already classified in some other way (e.g. as carcinogenic). This approach seems appropriate.

5.5.2.6
Limits for Combined Effects

An alternative to combined-effect-specific safety margins supplemented by special limits for dominant components comprises limits or guideline values for combined effects ("limits" in the following) or policy instruments that approximate to limits such as action thresholds or classification rules. An extreme (and rarely needed) case is a ban on a specific mixture or safety advice for users. Limits for combined effects are appropriate for carcinogens or where there is evidence of super-additive effects, and in occupational health and safety where the safety margin approach cannot be or is not applied. When it comes to substances for which it is not possible to identify a dose-response threshold, the decision whether to opt for (exposure) limits, with or without an added minimization rule, cannot be made separately for combined effects: a tolerable individual risk level would have to be set first. This apart, limits for combined effects should be preferred from a proportionality standpoint in any event if the substances concerned occur both on their own and in combination and if a specific frequency distribution cannot be identified.

The current state of scientific knowledge does allow known combined effects to be quantified to a certain extent, and the phenomenon of low interaction at low doses can be taken into account when doing so. This leaves broad scope for generalization. When regulators and standard-setters extend limits, action thresholds and classification rules to include known combined effects, however, the resulting regulatory requirements and standards must meet the protection aims of the law under which they are issued. Thus, if super-additive effects are known to exist, a regulatory requirement cannot be based on dose-additivity alone without at least including a corrective (say, a special limit) for the known super-additive effects. In principle, the reverse must apply for antagonistic effects under the proportionality rule; the need to allow for antagonistic effects – for example by setting a lower limit – is often negated, however, by the possibility of exposure to one of the substances on its own. That is, a corrective for antagonistic effects is only appropriate if individual exposure can be ruled out.

In addition to substance-specific toxicity, regulators should also consider substance prevalence and exposures. This is done in current legislation – relating to bioavailability – for water pollutants to correct classifications that are based on the assumption that effects are exclusively dose-additive. The synthesis approach used to set exposure targets for combined effects of carcinogenic air pollutants likewise uses (typical) exposure situations to weight individual components of a mixture. However, the main motive for looking at exposure situations with combined effects is more often to apportion reductions among polluters than to protect human health and the environment, since exposure scenarios are already built into the limits set for individual substances.

A central question raised by limits for combined effects is how far regulatory requirements and standards can follow a prescriptive model of linear dose-additivity – in some cases even ignoring the slope of the dose-response curve – when there is evidence of combined effects among specific substances, substances in specific classes or danger categories but the nature and mechanism of the effects is unknown. Given the broad scope for generalization open to the executive, the solution must merely be a reasonable one. Several aspects developed earlier (in Sections 2.2 to 2.4) come into play here. With substances that have similar chemical response mechanisms, dose-additive (or similar) effects are evidently common and hence unexceptional. The dose-additivity hypothesis is relatively conservative – errs on the side of caution – because for a given pair of substances it includes combinations of doses that are separately harmless even though there is little proof of interaction in the low dose range. Nor does the response linearity assumption understate risk at low doses relative to the S-shaped curves that are more often found in reality. At medium and high doses, a corrective can be applied – and is applied in practice – by assuming, where there is a dominant substance with a dose-response threshold and a target based on the threshold, that the dominant substance is characteristic of the mixture and hence that additivity is no longer an issue. With carcinogens, response linearity at low doses is a scientifically plausible hypothesis in any case, and with differing target organs the linearity model describes the calculated total cancer risk. For substances with a dose-response threshold, however, the model of independent combined effects (relative response additivity) should also be given consideration as an alternative prescriptive model. This is a more realistic explanatory model where substances have different mechanisms – as is more often the case – but does not yield a significantly different evaluation of the risk of combined effects at low doses. At medium and high doses it is generally less conservative than the dose-additivity model. The (relative) response-additivity model is less useful when setting regulatory requirements because it is impracticable to determine the effects of components in advance.

The appropriate response in these circumstances would appear at first sight to be a dual assessment rule that stipulates dose-additivity or independent combined effects as the prescriptive model according to the context (substances with identical/similar or different response mechanisms, carcinogens, or substances that have a dose-response threshold). This approach is useful as a framework for case-wise risk assessment, as shown in particular by the US Environmental Protection Agency's assessment guidelines. The rules could be used as a basis for substance approval and decisions on bans and restrictions. However, it does not solve the common problem of how to answer a real need to set limits when there is insufficient knowledge of the context. If a decision were made not to use prescriptive models, state environmental policy would have to revert to case-wise assessment wherever there is a suspicion that combined effects exist but inadequate knowledge as to their nature, and policy would be less effective as a result. This justifies the choice of a single prescriptive model even if it only matches part of "reality". There is considerable policymaking discretion in the choice of model.

Although the fundamental assumption of linear dose-additive effects is scientifically untenable, a prescriptive model based on it is admissible for the reasons described above. The independent combined effects model could likewise be chosen – save as regards carcinogens – but does not have any inherent advantage over the dose-additivity model.

Rigid regulatory provisions that stipulate a specific model such as linear dose-additivity are to be avoided, however. Adequate and reasonable correctives must be provided to allow for different assessments of risk. The linear dose-additivity assumption underlying the current legislative provisions can understate impacts on health and the environment if all combinations of substances are permitted without regard to the slope of the dose-response curve. In certain cases, use of this model can disregard a real possibility of individual combinations of substances having super-additive effects even at low doses. Environmental standards that do not incorporate such correctives are not legally inadmissible per se, but run the risk of not being applied in specific cases.[139]

5.5.2.7
Emission Restrictions and Minimization

The next question is whether general restrictions on the quantities of hazardous substances brought into circulation are generally admissible as a precautionary measure addressing possible combined effects. Active reductions of this kind accord with the "substance flow management" programme initially developed in

---

[139] See BVerwG, NVwZ 1998, 1181 (on Art. 20, 17th BImSchV).

chemicals policy and later extended to mass substances and materials as natural resources in the Rio process.[140] The primary focus here is on environmental risks and resource consumption relating to individual substances. Others, however, recommend an emission reduction strategy as a precautionary measure to reduce or even eliminate as yet unknown risks from combined effects.[141] In the Netherlands, emission reductions are effectively used as a means of preventing unknown combined effects by requiring observance wherever possible of a specific level of residual risk for such effects (10^{-7} per year).

It is not possible to agree with such a blanket approach. Without any indication that combined effects are there to be regulated and how they are to be dealt with, the risk is merely speculative and fails to justify encroachments on the freedom or property of those who cause it (though this does not preclude such encroachments being justified by risks from individual substances).[142] Wherever emissions are reduced or a minimization rule is adopted as an expression of the precautionary principle, the eventuality of combined effects is covered. Moreover, it is possible to base substantial restrictions on the mere persistence or bioaccumulation of an individual substance by regarding these as a relevant risk in themselves regardless of toxicity and by assuming dose-additive effects. Though favoured by some,[143] this approach to risk assessment has yet to become established in environmental policy.

Where it is plausible (though not yet reasonably probable) that combined effects exist, it may be admissible and appropriate to reduce combined risks in parallel with risks from individual substances by limiting the quantities brought into circulation. This is especially the case where the primary environmental legislation concerned is not solely committed to technology-based precautionary measures but also requires or at least permits risk-oriented policy instruments. These include risk-based emission limits for combined effects. The minimization rule can be used in addition for particularly hazardous substances, though it does not generally differentiate between combined risks and risks from individual substances. An approach of this kind is followed for the combined effects of carcinogenic air pollutants in *TA Luft* (the Technical Instructions on Air Quality). To combat cumulative effects, the approach should be supplemented by a system of exposure-based limits like that proposed by the *Länder* Committee on Pollution Control (LAI).

Another approach would be to use the minimization rule to limit combined effects in cases where it does not apply to risks from individual substances. However, since there is a lack of systematic cost-benefit analyses for the minimization rule and in view of the financial burden on the affected enterprises, this approach would only be appropriate for very severe adverse effects that have so far defied quantitative assessment and so make it impossible to apply emission limits.

---

[140] See Held, 1991; Enquete Kommission, 1994; idem, 1998; Umweltbundesamt, 1997; Misereor/BUND, 1996.

[141] For example Müller-Herold/Scheringer, 1998.

[142] For example when precaution is enforced in the form of keeping well below critical exposure limits (which makes sense with human and environmental toxins that accumulate); on this point see Lübbe-Wolf, 1998, p. 53 ff.

[143] See ENDS Report 269 (1997), p. 21 ff.; Klöpffer, USWF - Z. Umweltchem. Ökotox. 1989, 43; idem, USWF - Z. Umweltchem. Ökotox. 1994, 61; von Gleich, USWF - Z. Umweltchem. Ökotox. 1998, 367.

5.5.3
Policy Addressees (Imputation Models)

5.5.3.1
Basic Imputation Models

The policy instruments used to reduce combined effects give rise to serious problems of allocation and apportionment. The question is which emissions or uses of a substance to reduce and which party should shoulder the burden. Allocation and apportionment problems are worse where multiple parties are involved. The problem of inter-party imputation is characteristic of substance combinations that are not intentional (emissions, discharges and accidents). Conceivably, however, there are also cases (in occupational health and safety, and with accidents) where one and the same party is responsible for multiple substances. With these cases, and with manufactured preparations, the issue is reduced to intra-entity allocation and apportionment problems whose resolution raises fewer difficulties.

In principle there are six conceivable regulatory models for combined effects caused by one or more parties:

- Voluntary reductions (voluntary model)
- Apportionment of reduction obligations among multiple substances or polluters according to (stipulated or flexible) quotas (quota model)
- A ban on or obligations to reduce new emissions or usage of a substance (priority principle)
- Obligations to reduce existing emissions or usage of a substance (remediation model)
- Obligations to reduce emissions regardless of when they occur or to reduce usage based on one substance dominating a mix of pollutants (dominant component model)
- Obligations for one polluter to reduce emissions or usage combined with compensation from other parties (compensatory model)

5.5.3.2
Voluntary Model

If there is only one polluter – as is the case with manufactured preparations and with emissions in the domain of occupational health and safety – the polluter might be left freedom of choice as to whether and to what extent it reduces emissions of a given substance (voluntary model). The same approach can be used if there are multiple polluters who each emit multiple pollutants provided the limits system exclusively targets emissions and does not take total exposures into account. A voluntary arrangement is only possible within the bounds of the effect categories determined by the type of combined effect and the limits set for individual substances. In practice, however, combined effects are frequently assumed to be dose-additive and to have a linear dose-response relationship, with the implication that different combinations can have identical effects. This widens the scope for voluntary arrangements on the face of it but is only legally admissible if it incorporates adequate correctives to account for the combined effects that are anticipated or must be dealt with in reality.

5.5.3.3
Quota Model

Quota arrangements aim to apportion reduction obligations among multiple substances or polluters (for example in proportion to their shares of total emissions). As with voluntary arrangements, the use of quotas is constrained by the type of combined effect and by limits set for individual substances. Quotas are used in particular where environmental standards or classification rules stipulate sum total values for multiple substances based on assumed dose-additivity. In most cases, however, what is used is a set of "pseudo-quotas" in that the regulatory requirements concerned do not allow each substance to exploit its allotted range of pollution levels to the full. The result is apportionment by other criteria, for example in accordance with a voluntary arrangement or the priority principle. This is not a problem in allocation or distribution terms where a single party is responsible for the production or use of multiple substances that have combined effects, as the producer or user has the possibility of optimizing the composition of the preparation or mixture along technical and business lines. Where there are multiple parties, especially as regards exposures, inter-party allocation and apportionment problems arise. So far this has not been regarded as much of a problem since genuine quota arrangements with multiple polluters are rare in current law. Only the synthesis approach used to maintain air quality constitutes a genuine, broad-based quota arrangement – applying to all geographically adjacent emitters of a given substance – as emitters of each substance can exploit a guaranteed range of pollution levels to the full; that is, their emissions must not be reduced relative to emissions of other substances.

Individual quota arrangements are also possible, in particular for cases of existing pollution. Subsequent orders to existing permit-holders under Art. 17(1) of the German Pollution Control Act (BImSchG) must take account of the nature, quantity and hazard levels of emissions and exposures from different sources (cf. *TA Luft* no. 2.6.1.2). This allows apportionment according to polluter shares of the total, including for combined effects. On the other hand, the cost to the parties must be also considered under the proportionality principle, though not with cost-effectiveness in mind (identifying the best cost-avoider), but to avoid imposing an unreasonable burden on individual polluters.

The new *TA Lärm* (Technical Instructions on Noise) goes into greater detail on the aspects that are to be taken into account here (see 5.1.4.3 above), requiring consideration of existing noise reduction plans, the efficacy of the steps taken, to what extent a polluter contributes to the noise, plus whether and to what extent the polluter is at fault. Conversely, the former Model Administrative Procedures of the German *Länder* accorded greater weight to the quota concept (polluter share of total noise).

The debate surrounding advance quotas in the prevention of noise pollution[144] from new facilities shows that plausible legal apportionment criteria are hard to

---

[144] See in particular Petersen, 1993, p. 180 ff. and the references in footnote 10.

draw up. This apart, it is doubtful whether quotas based on exposure forecasts are still admissible as a danger prevention method under Art. 5(1) no. 1 of the Pollution Control Act (BImSchG).[145] Accordingly, the new *TA Lärm* favours the priority principle, whereas the Model Administrative Procedures had proposed general quotas based on the land area occupied by facilities as a percentage of the total area used by all potential emitters.

Building combined effects into the safety margin set for an individual substance does not constitute a separate regulatory model in allocation and distribution terms. An extra-large safety margin to allow for the possibility of other substances occurring at the same time is effectively a prospective quota arrangement. The regulated substance may not be able to fully exploit its allotted range of pollution levels because of the latitude allowed for other substances. Conversely, it is possible that a safety margin set for an individual substance is not large enough to adequately cover combined effects. This would be nothing other than a regulatory shortfall.

5.5.3.4
Priority Principle

Combined effects are often dealt with according to the priority principle used in most approval procedures, with the result that existing polluters have an advantage over newcomers. This is the case for example when a sum total limit is made equal to the individual substance limit. In decisions on individual cases, pollution law ascribes existing pollution loads to newcomers, who must bear the entire reduction burden even if their share of the total is insignificant when viewed in isolation. There is not a general duty to reappraise existing polluters who do not pose a risk in their own right.[146] The situation is the same with combined effects of chemical and radiation emissions, which are governed by different laws.[147] Here, the separate agencies regulating the two types of pollution each tend to take the other type as given and to order unilateral reductions in the type they regulate. This concentrates the reduction burden on new emissions and new uses and lowers the burden on existing emissions and uses. However, the resulting allocation and apportionment problems are mitigated at the expense of existing polluters by blanket reductions in emissions that follow from application of the precautionary principle; this makes way for newcomers (whether we consider this an objective of the precautionary principle or merely a result of applying it is immaterial). The trend towards integrated environmental agencies promotes fair implementation. Compensatory arrangements should be created for the remaining area of conflict. This is particularly important when setting limits.

---

[145] Petersen, 1993, p. 182 ff.
[146] BVerwGE 55, 250, 267 ff.; Jarass, 1999, § 5, no. 16, 19.; Petersen, 1993, p. 50 f., 56 f.; Murswiek, 1985, p. 299 f.
[147] OVG Lüneburg, DVBl. 1977, 347, 350; 1979, 686, 689 f.

5.5.3.5
Remediation Model

As an alternative to the priority principle, another way of legislating for combined effects is to impose remediation obligations on existing facility operators, dischargers and soil polluters, or equivalent reduction obligations on existing users of hazardous substances. In German environmental law, remediation obligations are often imposed to ensure compliance with environmental quality standards, to respond to changes in existing environmental quality standards, emission standards or technology, or to take new production or usage restrictions into account (Art. 17(1) of the Pollution Control Act (BImSchG); Art. 7a(3) of the Water Management Act (WHG); Art. 4(3) of the Soil Protection Act (BBodSchG); Art. 4(3) sentence 3 and Art. 5 of the Recycling Act (KrW-/AbfG)). In principle this applies for combined effects as well to the extent they are covered by the applicable secondary legislation. Allowance must be made for the business concerns of affected enterprises in accordance with the proportionality principle, in particular by allowing adjustment periods and where necessary facility-specific standards (Art. 17(2) BImSchG; Art. 7a(2) WHG; BBodSchG loc. cit.).

Remediation obligations are rarely imposed to make way for a new facility, new discharges or a new use for a hazardous substance. They can be used for existing facilities and discharges when technology has advanced (Art. 17(1) BImSchG; Art. 7a(2) WHG; Art. 35 KrW-/AbfG). Where general review rules do not exist, public agencies can use the discretion frequently accorded them to make way for newcomers. This takes account of newcomers' interests, though only indirectly. The proportionality principle sets limits to the obligations that can be imposed on existing operators (e.g. in Art. 17(2) BImSchG). Within these limits, the Act stipulates that only the existing operator's cost of compliance is to be compared with the benefits to the environment; it does not allow comparison between the newcomer's abatement cost and the existing operator's remediation cost, as would be more logical in efficiency terms. This severely narrows the scope for using remediation obligations as a regulatory model to solve imputation issues under current law. And in all this, of course, current remediation practices and other corrective methods used to enforce technology-based precaution likewise make way for newcomers and may also mitigate problems of combined effects on their behalf. In any case, if reform is accepted as necessary, it must start with the remediation obligations relating to individual substances.

5.5.3.6
Dominant Component Model

The dominant component approach described earlier (5.5.2.4) is also an imputation model. Regardless of the relative sizes and chronological order of polluters' contributions to the cocktail and hence to the resulting environmental and health risks, the risks are ascribed to whoever makes, uses or emits the dominant substance. What matters is that risk assessment for the mixture is based on the dominant effects of the one substance. The imputation process can also be seen as an application of the "polluter pays" principle, however.

5.5.3.7
Compensatory Model

A common way of solving imputation problems in liability law when there are multiple polluters is to impose environmental obligations on one or more of the group and have the remainder pay them (partial) compensation. This compensatory model also works for combined effects involving multiple polluters. In their external relations with the public, all polluters who carry a share of the blame are jointly and severally liable towards victims; in their internal relations with each other, there is an arrangement where the polluter who is required to pay damages is compensated by the others in proportion to their respective shares of the total (see 5.1.4.2 above). Article 24(2) BBodSchG introduced the compensatory model into German public law with regard to remediation: the competent authority has discretionary power to decide who should carry out and pay for remediation; the remaining polluters must pay towards the cost (see 5.1.4.3 above).

The only other compensatory arrangements are to be found in chemicals law, in a special case relating to the re-use of licensing or notification documents and of records of tests ordered subsequent to initial licensing.[148] Licensing or notification documents which a producer has specially drawn up for the purpose of licensing or notification and which relate to animal experiments can be re-used for later applicants (subject, for a time, to the earlier applicant's consent). The later applicant must then refund half the expense saved (Art. 20a of the Chemicals Act (ChemG) and Art. 13 of the Plant Protection Products Act (PflSchG)). The same laws also provide for a compensation arrangement where two or more applicants or reporting parties must submit identical test reports. If the parties cannot agree among themselves, the competent authority can order one of the parties to draw up the documents concerned and the remainder must pay towards the cost (Art. 20a ChemG and Art. 14 PflSchG). This latter arrangement is also provided for in drugs law (Art. 24b of the Drugs Act (AMG)). The aim is to prevent unnecessary experiments on animals while respecting the interests of the first applicant in maintaining its market lead.

The compensatory arrangements currently provided for in German law are isolated instances and as such do not constitute a policy trend towards changing the imputation schemes normally used in public environmental law. All the same, they provide arguments in favour of desirable changes and indications as to how these changes can be legally implemented.

5.5.3.8
General Imputation Principles

As described earlier, German law lays down general principles for the choice of regulatory model only to a limited extent since in most cases legislators are left considerable discretion in this regard. The same applies for the choice of groups targeted by rules on combined effects. In particular, there is no strict constitutional requirement that environmental policy must be fairly balanced when addressing multiple polluter groups who

---

[148] On these problems see Denninger, GRUR 1984, 627; Scholz, 1983; Albach, BB 1984, Supplement to Issue 29.

contribute towards combined effects.[149] Neither the equal treatment principle (Article 3(1) of the German Basic Law) nor the proportionality principle guide state regulation of combined effects in a specific direction since there will often be sufficiently weighty objective reasons to differentiate[150] and the proportionality principle does not demand an optimum solution. The equal treatment principle and proportionality principle merely set certain limits on the exercise of policy choice. The equal treatment principle comes into play where different groups of polluters who contribute towards combined effects are in competition with each other and unequal apportionment of the reduction burden would distort that competition. The proportionality principle may be infringed if comparatively severe reduction requirements are imposed on a specific polluter group merely out of an unreasoned governmental desire to exempt other groups.

Within this framework it is possible, firstly, to lend more weight to the "polluter pays" principle, which favours a quota arrangement based on shares of total emissions or total risk; this sort of arrangement is also generally preferable in efficiency terms since ordering only one polluter to reduce emissions or usage leads to excessive reductions at higher cost. Secondly, stricter requirements can be imposed on whichever polluter group is the best cost-avoider. Compensatory arrangements that impose the entire reduction burden on one group and share the cost among the remainder are then admissible. They meet efficiency and apportionment requirements equally well, though as a rule they are not legally mandated and are also hard to implement in law, as the provisions in chemicals law show (Art. 20a of the Chemicals Act (ChemG); Art. 13 and 14 of the Plant Protection Products Act (PflSchG); and Art. 24b of the Drugs Act (AMG)). In the special case where one party is responsible for several chemicals that give rise to combined effects, voluntary arrangements may be appropriate providing that substance-specific or sum total limits are complied with and it is acceptable in environmental or health terms for concentrations of the chemicals concerned to occur in different combinations. The same applies for multiple sources where only emission limits are laid down.

Decisions can take account of economic factors in various ways – for example with regard to market losses suffered by manufacturers, adjustment and retrofitting costs, effects on the European single market, possible job losses, the effect on consumer prices, the availability of chemical substitutes, and so forth. In principle, where the purpose of restrictions is to prevent dangers, the economic cost is only considered at the instrument selection stage. However, this includes selecting who the restrictions are to be imposed on.

From a legislative policy point of view, voluntary arrangements are preferable if only because they leave the decision to free enterprise and relieve government of the need to apportion reductions. They also come closest to meeting efficiency requirements. If a particular combined effect or if market failure (usually because there are multiple polluters) makes it difficult or impossible to implement voluntary arrangements, preference should be given to enforcing the "polluter pays" principle with proper quotas as the fairest approach. As an alternative that also takes economic efficiency aspects into account, compensatory arrangements should be tried out that impose the reduction bur-

---

[149] Sachverständigenrat für Umweltfragen, 1994, no. 78; Wahl/Appel, 1995, p. 205 f.; Rehbinder, 1998, part 4, no. 29; see BVerfGE 47, 109, 118 f.; 48, 346; 71, 206, 217 f.
[150] On these problems in general see Jarass, NJW 1997, 2545 ff.

den on the best cost-avoider but share the cost among all polluters. Like the voluntary approach, this is only appropriate when it is acceptable in environmental and health terms for concentrations to occur in multiple combinations. The priority principle – the most commonly used approach in existing practice – can only be justified in terms of administrative practicability and is clearly inferior in all other respects.

5.5.3.9
Relationship between Regulatory and Imputation Models

We have already touched upon relationships between regulatory and imputation models when discussing voluntary arrangements and safety margins, and will now go on to consider these relationships in general terms. Not all the regulatory strategies introduced in 5.5.2 allow the full choice of imputation models when it comes to apportioning the reduction burdens they entail. To this extent, the regulatory logic (the environmental policy choice) narrows the imputational logic (the decision on who to regulate). There is free or nearly free choice in the latter issue only in the case of individual limits and individually assessed combined effects, and even then restrictions apply for voluntary arrangements. The safety margin regulatory model is incompatible with imputation based on a dominant substance and with compensatory arrangements, while the minimization approach is effectively a quota arrangement but precludes all other imputation models. The dominant component approach is similarly unyielding towards most imputation models; a (unilateral) compensation scheme is theoretically conceivable but impractical and does not accord with the "polluter pays" rule. The relationships are illustrated in Table 5-1.

Table 5-1 Relationships between regulatory strategies and imputation models

Imputation model / Regulatory strategy	Voluntary	Quota	Priority	Remediation	Dominant component	Compensatory
Case-wise assessment	(+)	++	++	+	(+)	+
Safety margins	(+)	+	++	+	–	–
Dominant component	–	–	–	–	++	(+)
Limits	(+)	++	++	+	–	+
Minimization	(+)	+	–	–	–	–

++   Highly possible   +   Possible
(+)   Possible with reservations   -   Not possible

5.6
Recommendations

The following are recommendations for the future architecture of environmental law regarding combined effects:

- Statutory powers to set limits, action thresholds and classification rules should make clear that combined effects must be taken into account.
- Efforts should be directed at supplementing EU law on chemicals with a requirement for manufacturers and importers of new substances to investigate evidence of combined effects. Combined effects should also be included in the assessment of existing substances.
- Policies to prevent combined effects must be based on a scientifically plausible assumption that such effects are possible. All-purpose safety margins or mixture requirements for limits, action thresholds (trigger values) and classification rules should always come with a corrective for the common event that different impact and response mechanisms rule out the existence of combined effects. Where there is a plausible suspicion that combined effects may exist, it is acceptable to assume for practical purposes that they exhibit linear dose-additivity or independence provided once again that there are adequate correctives for instances where the assumption does not hold, and in particular where combined effects prove to be super-additive or wholly absent.
- Of the various strategies for regulating combined effects – case-wise risk assessment, safety margins, dominant components, combined-effect-specific limits, and emission restrictions – the first is superior from a scientific perspective. Its applications include licensing procedures, bans and restrictions on the marketing of substances, and the development of specific limits; however, it cannot be used everywhere and so must not be allowed to obstruct the search for generally applicable solutions.
- Incorporating special (additional) safety margins for combined effects into substance-specific limits is an acceptable strategy for substances which have a dose-response threshold (outside of occupational health and safety) and which frequently occur in specific combinations with other substances; in such cases, the special safety margins constitute a generalization that accords with the proportionality principle. Because, and to the extent that, substance-specific limits restrict exposures to low doses in any case, the margins already built into these limits may well suffice without an additional margin for combined effects. Also, the low degree of interaction that is assumed to exist at low doses may restrict the size of the additional safety margin that is needed for combined effects.
- In general, however, preference should be given to special limits for combined effects (and other policy instruments that approximate to limits, like action thresholds and classification rules). This is particularly the case with carcinogens and in occupational health and safety. Current knowledge allows known combined effects to be quantified at least to a certain extent. Otherwise, rules and regulations can be based on the prescriptive model of dose-additive or independent effects wherever it is impracticable to use case-wise assessment or to identify the conditions for applying one model or the other; appropriate correctives must be provided, however, particularly for the case of super-additive effects.

- If one component dominates a mixture to the extent that it can be said to be characteristic of the mixture as a whole, only the limit for that component is important.
- It is not admissible to use blanket emission limits in response to mere speculation that combined effects might exist. This strategy is acceptable to the extent it is used for risks from individual substances as well.
- Regarding imputation of the regulatory burden entailed by combined risks, a voluntary arrangement that leaves apportionment of the burden to the polluter should be used wherever possible, and in particular where only one polluter is involved. This presupposes, however, that multiple combinations of the substances concerned are acceptable in environmental and health terms. Elsewhere, the aim of fair apportionment would suggest giving greater weight to the "polluter pays" principle, for example by using a quota arrangement where each substance is allowed to exploit its allotted range of pollution levels to the full. Reduction obligations combined with compensation by other parties could be tested as an alternative that simultaneously takes cost-efficiency aspects into account. Imputation models relating to the time at which hazardous activities are first engaged in (such as the priority principle or the remediation model) should only be considered as an alternative.

5.7
Literature

Akademie der Wissenschaften zu Berlin (1992) Umweltstandards. de Gruyter, Berlin, New York

Albach H (1984) Ökonomische Wirkungen von Lösungen der Zweitanmelderfrage, Betriebs-Berater (BB) 39. Suppl. 18 to issue no. 29

American Conference of Governmental Industrial Hygienists (1998) 1998 TLVs and BEIs. Threshold Limit Values for Chemical Substances and Physical Agents. Cincinatti

Arbeitsgemeinschaft für Umweltfragen (1987) Grenzwertfindung im Arbeitsschutz- und Umweltrecht. Bonn

Balensiefen G (1994) Umwelthaftung. Nomos, Baden-Baden

Beroggi G, Abbas J, Stoop M, Aebi (1997) Risk Assessment in the Netherlands. Akademie für Technikfolgenabschätzung in Baden-Württemberg, Stuttgart

Beyersmann D (1986) Gibt es naturwissenschaftliche Grundlagen bei Stoffkombinationen? In: Winter G (ed.) Grenzwerte. E. Schmidt, Berlin, pp. 65–71

Bley H (1988) Sozialrecht. 6. edn. Luchterhand, Neuwied

Böhm M (1996) Der Normmensch. Mohr Siebeck, Tübingen

Breuer R (1988) Gerichtliche Kontrolle der Technik. Neue Zeitschrift für Verwaltungsrecht (NVwZ) 7, 104–115

Breuer R (1991) Rechtliche Bewertung krebserzeugender Immissionen. In: Neuere Entwicklungen im Immissionsschutzrecht. Umweltrechtstage Nordrhein-Westfalen, MURL, Düsseldorf, pp. 158–181

Breuer R (1978) Gefahrenabwehr und Risikovorsorge im Atomrecht. Deutsches Verwaltungsblatt (DVBl.) 93, 28–37

Breuer R (1987) Rechtsprobleme der Altlasten. Neue Zeitschrift für Verwaltungsrecht (NVwZ) 6, 751–761

Breuer R (1990) Anlagensicherheit und Störfälle – Vergleichende Risikobewertung im Atom- und Immissionsschutzrecht. Neue Zeitschrift für Verwaltungsrecht (NVwZ) 9, 211–222

Bundesministerium für Umwelt, Naturschutz und Reaktorsicherheit (1997) Ressortabgestimmte fachliche Inhalte einer Bodenschutz- und Altlastenverordnung vom 28.5.1997 (Eckwertepapier). Bonn

Cansier D (1994) Gefahrenabwehr und Risikovorsorge im Umweltschutz und der Spielraum für ökonomische Instrumente. Neue Zeitschrift für Verwaltungsrecht (NVwZ) 13, 642–647

Classen C-D (1995) Gefahrerforschung und Polizeirecht. Juristische Arbeitsblätter (JA) 27, 608–614

Daxenberger M (1997) Kumulationseffekte. Grenzen der Erfolgszurechnung im Strafrecht. Nomos, Baden-Baden

Denninger E (1984) Die Zweitanmelderproblematik im Arzneimittelrecht. Gewerblicher Rechtsschutz und Urheberrecht (GRUR) 87, 627–637

Deutsche Forschungsgemeinschaft (1998) Senatskommission zur Bewertung gefährlicher Arbeitsstoffe. MAK- und BAT-Werte-Liste, VCH, Weinheim

Di Fabio U (1994) Risikoentscheidungen im Rechtsstaat. Mohr Siebeck, Tübingen

Dieter HR Konietzka (1995) Which Multiple of a Safe Body Dose Derived on the Basis of Safety Factors Would Probably be Unsafe? Regul. Toxicol. Pharmacol. 22, 262–267

Dieter H (1995) Risikoquantifizierung Abschätzungen, Unsicherheiten, Gefahrenbezug. Bundesgesundheitsblatt (BGesBl.) 250–257

Dieter H, Grohmann A, Winter W (1996) Trinkwasserversorgung bei Überschreitung von Grenzwerten der Trinkwasserverordnung. in: Transparenz und Akzeptanz von Grenzwerten am Beispiel des Trinkwassers. E. Schmidt, Berlin, pp. 189–210

Drews B, Wacke G, Vogel K, Martens W (1986) Gefahrenabwehr. 9. edn. Heymanns, Köln, Berlin, Bonn, München

Eberbach W, Lange P, Ronellenfitsch M (1997) Recht der Gentechnik und Biomedizin. C.F. Müller, Heidelberg

Empfehlungen der Strahlenschutzkommission. Synergismen und Strahlenschutz vom 23.9.1977, Bundesanzeiger 212, 11.11.1977

ENDS Report 269 (1997) Sweden sets the Agenda for tomorrow's chemicals policy. ENDS Report 269, Juny, 21–25

Englert N (1996) Ableitung von Grenzwerten für Stoffe in der Luft. In: Transparenz und Akzeptanz von Grenzwerten am Beispiel des Trinkwassers. E. Schmidt, Berlin, pp. 78–87

Enquete-Kommission „Schutz des Menschen und der Umwelt" des Deutschen Bundestages (1994) Die Industriegesellschaft gestalten. Economica, Bonn

Enquete-Kommission "Schutz des Menschen und der Umwelt" des Deutschen Bundestages (ed.) (1998) Konzept Nachhaltigkeit, Economica, Bonn

Erlenkämper A (1997) Arbeitsunfall, Schadensanlage und Gelegenheitsursache, Sozialgerichtsbarkeit (SGb) 44, 355–364

Erlenkämper A, Fichte W (1996) Sozialrecht. 3. edn., Heymanns, Köln, Berlin, Bonn, München.

Falke J, Winter G (1996) Management and regulatory committees in executive rule-making. In: Winter G (ed.) Sources and Categories of EU Law. Nomos, Baden-Baden, pp. 541–582

Falke J (1986) Rechtliche Kriterien für und Folgerungen aus Grenzwerten im Arbeitsschutz. in: Winter G (ed.) Grenzwerte. Werner, Düsseldorf, pp. 164–198

Feldhaus G (1993) Die Schwierigkeiten mit der Immissionssummenbewertung nach TA Lärm. Jahrbuch für Umwelt und Technikrecht (JUTR) 1993, UTR 21, 29–47

Feldhaus G (1987) Rechtliche Instrumente zur Bekämpfung von Waldschäden, Umwelt- und Planungsrecht (UPR) 7, 1–11

Feldhaus G (1998) 30 Jahre TA Lärm – Auf dem Weg zum gesetzeskonformen Lärmschutz. In: Koch HJ (ed.) Aktuelle Probleme des Immissionsschutzrechts. Nomos, Baden-Baden, pp. 181–190

Feldhaus G, Schmitt O (1984) Kausalitätsprobleme im öffentlich-rechtlichen Umweltschutz – Luftreinhaltung, Wirtschaft und Verwaltung (WiVerw.) 9, 1–22

Fischer M (1996) Vergleich und Bewertung des Krebsrisikos durch Luftverunreinigungen. In: Transparenz und Akzeptanz von Grenzwerten am Beispiel des Trinkwassers. E. Schmidt, Berlin, pp. 65–77

Franßen E (1992) Krebsrisiko und Luftverunreinigung. In: Gesellschaft für Umweltrecht. Dokumentation zur 16. wissenschaftlichen Fachtagung. E. Schmidt, Berlin

Koch HJ, Scheuing D (1998) Gemeinschaftskommentar zum Bundes-Immissionsschutzgesetz. Werner, Düsseldorf

Gerlach J (1989) Privatrecht und Umweltschutz im System des Umweltrechts. Duncker & Humblot, Berlin

Gleich A von (1998) Ökologische Kriterien der Technik- und Stoffbewertung: Integration des Vorsorgeprinzips. UWSF - Z. Umweltchem. Ökotox. 10, 367–373

Gottwald P (1986) Kausalität und Zurechnung - Probleme und Entwicklungstendenzen des Haftungsrechts. In: Kausalität und Zurechnung, Karlsruher Forum (Suppl. to VersR 1988), Verlag Versicherungswirtschaft, Karlsruhe

Götz V (1995) Allgemeines Polizei- und Ordnungsrecht. 12. edn., Vandenhoeck & Rupprecht, Göttingen

Gusy C (1995) Probleme der Verrechtlichung technischer Standards. Neue Zeitschrift für Verwaltungsrecht (NVwZ) 14, 105–112

Hädrich H (1986) Atomgesetz. Nomos, Baden-Baden

Health Council of the Netherlands (1985) Establishment of health-based recommendations for setting standards for non-carcinogenic substances, Den Haag

Held M (ed.) (1991) Leitbilder der Chemiepolitik. Campus, Frankfurt, New York

Henschler D, Bolt HM Jonker D, Pieters MN, Groten JP (1996) Experimental Designs and Risk Assessment in Combination Toxicology: Panel Discussion, Food & Chem. Toxicol. 34, 1183–1185

Henschler D (1981) Offene Probleme und Lösungsmöglichkeiten – Risikoabschätzung bei gesundheitsgefährlichen Arbeitsstoffen. In: Deutsche Forschungsgemeinschaft. Wissenschaftliche Grundlagen zum Schutz vor Gesundheitheitsschäden durch Chemikalien am Arbeitsplatz, Bonn, pp. 92–96

Herrmann E (1987) Der Störer nach § 1004 BGB. Duncker & Humblot, Berlin

Hirsch G, Schmidt-Didczuhn A (1991) Gentechnikgesetz. C.H. Beck, München

Jarass HD (1983) Schädliche Umwelteinwirkungen – Inhalt und Grenzen eines Kernbegriffs des Immissionsschutzrechts, Deutsches Verwaltungsblatt (DVBl.) 98, 725–732

Jarass HD (1997) Folgerungen aus der neuen Rechtsprechung des BVerfG für die Prüfung von Verstößen gegen Art. 3 I GG, Neue Juristische Wochenschrift (NJW) 50, 2545–2550

Jarass HD (1999) Bundes-Immissionsschutzgesetz. 4. edn., C.H. Beck, München

Junker A, de la Riva C, Schwarze R (1997) Genehmigungsverfahren und Umweltschutz. DWD, Köln

Kalmbach S, Schmölling J (1994) TA Luft. 4. edn., E. Schmidt, Berlin

Kleindienst B (1964) Der privatrechtliche Immissionsschutz nach § 906 BGB. Mohr Siebeck, Tübingen

Kloepfer M (1996) Umweltschutz als Verfassungsrecht: Zum neuen Art. 20a GG. Deutsches Verwaltungsblatt (DVBl.) 111, 73–80

Kloepfer M, Thull R (1989) Der Lastenausgleich unter mehreren polizei- und ordnungsrechtlich Verantwortlichen. Deutsches Verwaltungsblatt (DVBl.) 104, 1121–1128

Kloepfer M, Kröger H (1990) Zur Konkretisierung der immissionsschutzrechtlichen Vorsorgepflicht. Natur und Recht (NuR) 12, 8–16

Klöpffer W (1989) Persistenz und Abbaubarkeit in der Beurteilung des Umweltverhaltens anthropogener Chemikalien. USWF - Z. Umweltchem. Ökotox. 1, 43–51

Klöpffer W (1994) Kriterien zur Umweltbewertung von Einzelstoffen und Stoffgruppen. UWSF - Z. Umweltchem. Ökotox. 6, 61–63

Knebel J, Sundermann A (1983) Der Entwurf eines schweizerischen Umweltschutzgesetzes. Umwelt- und Planungsrecht (UPR) 3, 8–12

Koch HJ (1989) „Schädliche Umwelteinwirkungen" – ein mehrdeutiger Begriff? Jahrbuch für Umwelt- und Technikrecht (JUTR), UTR 9, 205–215

Koch HJ (1990) Der Erheblichkeitsbegriff in § 3 Abs. 1 BImSchG und seine Konkretisierung durch die TA Lärm. In: ibid. (ed.) Schutz vor Lärm. Nomos, Baden-Baden, pp. 41–60

Koch HJ (1999) Die rechtliche Beurteilung der Lärmsummation nach BImSchG und TA Lärm 1998. in: Festschrift G. Feldhaus. C.F. Müller, Heidelberg, pp. 215–233

Koechlin D (1989) Das Vorsorgeprinzip im Umweltschutzgesetz. Helbing & Lichtenhahn, Basel, Frankfurt/M.

Köhler T (1996) Das Berufskrankheitenrecht in der aktuellen Diskussion. Der medizinische Sachverständige (MedSach), 101–104

Kolkmann J (1991) Die EG-Trinkwasserrichtlinie. E. Schmidt, Berlin

Kölz A, Müller HU (1996) Kommentar zum Umweltschutzgesetz. Schulthess, Zürich

Kormann J (1983) Lastenverteilung bei Mehrheit von Umweltstörern. Umwelt- und Planungsrecht (UPR) 3, 281–288

Krasney O (1988) Die Kausalgrundsätze des geltenden Berufskrankheitenrechts. In: Kolloquium Krebserkrankungen und berufliche Tätigkeit. Süddeutsche Eisen- und Stahl-Berufsgenossenschaft, Mainz, pp. 67–73

Kühling W (1994) Zum Risikoschutz für kanzerogene Luftverunreinigungen. Zeitschrift für Umweltrecht (ZUR) 5, 112–114

Kutscheidt E (1994) Eine neue TA-Lärm – Zur Bindung der Länder an Verwaltungsvorschriften des Bundes. NWVBl. 8, 281-287.

Kutscheidt E (1989) Rechtsprobleme bei der Bewertung von Geräuschimmissionen. Neue Zeitschrift für Verwaltungsrecht (NVwZ) 8, 193–199

Ladeur K-H (1994) Berufsfreiheit und Eigentum als verfassungsrechtliche Grenze der staatlichen Kontrolle von Pflanzenschutzmitteln und Chemikalien. Natur und Recht (NuR) 16, 8–14

Ladeur K-H (1995) Umweltrecht in der Wissensgesellschaft. Duncker & Humblot, Berlin

Länderausschuß für Immissionsschutz (1992) Krebsrisiko durch Luftverunreinigungen. Düsseldorf

Landmann R, Rohmer G (1998) Umweltrecht, Kommentar. C.H. Beck, München

Landsberg G, Lülling W (1991) Umwelthaftungsgesetz. Verlag Bundesanzeiger, Köln

Latin H (1988) Good Science, Bad Regulation, and Toxic Risk Assessment, Yale Journal on Regulation 5, 89-148

Lewis S (1998) US Environmental Protection Agency's 33/50 Programm. Elni Newsletter, 11–18.

Lisken HF, Denninger E (1995) Handbuch des Polizeirechts. 2. edn., C.H. Beck, München

Loser P (1994) Kausalitätsprobleme bei der Haftung für Umweltschäden. Haupt, Bern, Stuttgart, Wien

Lübbe-Wolf G (1998) Präventiver Umweltschutz – Auftrag und Grenzen des Vorsorgeprinzips im deutschen und europäischen Recht. In: Bizer J, Koch HJ (eds.) Sicherheit, Vielfalt und Solidarität. Nomos, Baden-Baden, pp. 47–74

Lübbe-Wolff G (1986) Die rechtliche Kontrolle incremental summierter Gefahren am Beispiel des Immissionsschutzrechts. In: Dreier H, Hofmann J (eds.) Parlamentarische Souveränität und technische Entwicklung. Duncker & Humblot, Berlin, pp. 167–188

Michler HP (1993) Rechtsprobleme des Verkehrsimmissionsschutzes. Werner, Düsseldorf

Misereor/BUND (ed.) (1996) Zukunftsfähiges Deutschland. – Studie des Wuppertal Institut für Klima, Umwelt, Energie. Birkhäuser, Basel, Boston, Berlin

Möllers T (1996) Rechtsgüterschutz im Umwelt- und Haftungsrecht. Mohr Siebeck, Tübingen

Müller-Herold U, Scheringer M (1999) Zur Umweltgefährdungsbewertung von Schadstoffen und Schadstoffkombinationen durch Reichweiten- und Persistenzanalyse. Report for Europäische Akademie zur Erforschung von Folgen wissenschaftlich-technischer Entwicklungen, Graue Reihe No. 18

Münch I von, Kunig P (1992) Grundgesetz. 4. edn., C.H. Beck, München

Mangoldt H von, Klein F, Starck C (1985) Grundgesetz. 3. edn., Vahlen, München

Murswiek D (1986) Zur Bedeutung der grundrechtlichen Schutzpflicht für den Umweltschutz, Wirtschaft und Verwaltung (WiVerw) 11, 179–204

Murswiek D (1990) Die Bewältigung der wissenschaftlichen und technischen Entwicklungen durch das Verwaltungsrecht. Veröffentlichungen der Vereinigung Deutscher Staatsrechtslehrer (VVDStRL) 48, 207–229

Murswiek D (1996) Staatsziel Umweltschutz (Art. 20 a GG). Neue Zeitschrift für Verwaltungsrecht (NVwZ) 15, 222–231

Murswiek D (1985) Die staatliche Verantwortung für die Risiken der Technik. Duncker & Humblot, Berlin

Nierhaus M (1994) Störererforschung und Kostentragungspflicht bei Verdachtslagen – dargestellt am Beispiel von Boden und Grundwasserverunreinigungen. Jahrbuch für Umwelt- und Technikrecht (JUTR), UTR 27, 369–404

Omgaan met Risico's (1989) Tweede Kamer 1988-1989. 21137, 5, Omgaan met Risico's, De risicobenadering in het milieubeleid, Den Haag

Ossenbühl F (1986) Vorsorge als Rechtsprinzip im Gesundheits-, Arbeits- und Umweltschutz. Neue Zeitschrift für Verwaltungsrecht (NVwZ) 5, 161–171

Ossenbühl F (1988) Verfahren der Gesetzgebung. In: Isensee J, Kirchhof P Handbuch des Staatsrechts, Vol. III. C.F. Müller, Heidelberg, pp. 351–385

Ossenbühl F (1988) Autonome Rechtsetzung der Verwaltung. In: Isensee J, Kirchhof P Handbuch des Staatsrechts, Vol. III. C.F. Müller, Heidelberg, pp. 425–462

Overy D, Richardson A (1995) Regulation of Radiological and Chemical Carcinogens: Current Steps Toward Risk Harmonization, Environmental Law Reporter 25, 10657–10670

Petersen F (1993) Schutz und Vorsorge. Duncker & Humblot, Berlin

Quentin A (1994) Kausalität und deliktische Haftungsbegründung. Duncker & Humblot, Berlin

Rehbinder E (1998) Ziele, Grundsätze, Strategien, Instrumente. In: Salzwedel J et al. (eds.) Grundzüge des Umweltrechts. E. Schmidt, Berlin, 04

Rehbinder, E. (1998) Privates Immissionsschutzrecht. In: Salzwedel J et al. (eds.) Grundzüge des Umweltrechts. E. Schmidt, Berlin, 08

Rehbinder E (1997) Festlegung von Umweltzielen – Begründung, Begrenzung, instrumentelle Umsetzung, Natur und Recht (NuR) 19, 313–328

Rehbinder E (1991) Prinzipien des Umweltrechts in der Rechtsprechung des Bundesverwaltungs-gerichts – Das Vorsorgeprinzip als Beispiel. FS. Sendler H. C.H. Beck, München, pp. 269–284.

Reich A (1989) Gefahr – Risiko – Restrisiko. Werner, Düsseldorf

Rengeling H-W (1984) Umweltvorsorge und ihre Grenzen im EWG-Recht. Heymanns, Köln

Ronellenfitsch M (1986) Die Eigensicherung von Flughäfen. Verwaltungs-Arch (VerwA) 77, 435–456

Rose-Ackermann S (1995) Umweltrecht und -politik in den Vereinigten Staaten und der Bundesre-publik Deutschland. Nomos, Baden-Baden

Rummel UD (1996) Inhaltliche Anforderungen an Maßnahmen zur Altlastensanierung. Nomos, Baden-Baden

Sachs M (ed.) (1999) Grundgesetz. 2. edn. C.H. Beck, München

Sachverständigenrat für Umweltfragen (1994) Umweltgutachten 1994. Metzler-Poeschel, Stuttgart

Saladin P (1989) Probleme des langfristigen Umweltschutzes. Kritische Vierteljahresschrift (KritV) 4, 27–55

Salje P (1993) Umwelthaftungsgesetz. C.H. Beck, München

Salzwedel J (1992) Rechtsgutachten. In: Länderausschuß für Immissionsschutz. Krebsrisiko durch Luftverunreinigung. Düsseldorf

Sander A (1994) Arzneimittelrecht. Kohlhammer, Stuttgart, Berlin, Köln, Mainz

Schattke H (1979) Rechtsfragen im Zusammenhang mit der Konkretisierung der Strahlenschutz-grundsätze. 6. Atomrechtssymposion. Heymanns, Köln, pp. 101–124

Scherzberg A (1993) Risiko als Rechtsproblem. Verwaltungs-Archiv (VerwA) 84, 484–513

Schink A (1986) Wasserrechtliche Probleme der Sanierung von Altlasten. Deutsches Verwaltungs-blatt (DVBl.) 101, 161–170

Schmidt-Salzer J (1992) Umwelthaftungsgesetz. Recht u. Wirtschaft, Heidelberg

Schoch F (1994) Grundfälle zum Polizei- und Ordnungsrecht. Juristische Schulung (JuS) 34, 932–937

Scholz R (1983) Konkurrenzprobleme bei behördlichen Produktkontrollen. Heymanns, Köln

Schöpf U (1995) Multikausale Schäden in der gesetzlichen Unfallversicherung. E. Schmidt, Berlin

Schuldt N (1997) Rationale Umweltvorsorge. Economica, Bonn

Schulin B (1991) Sozialrecht. 4. edn. Werner, Düsseldorf

Schulze-Fielitz H (1994) Rechtsfragen der Verkehrslärmschutzverordnung (16. BImSchV). Um-welt- und Planungsrecht (UPR) 14, 1–8

Schwarze R (1996) Präventionsdefizite der Umwelthaftung und Lösungsmöglichkeiten aus ökonomischer Sicht. Economica, Bonn

Soergel HT, Baur JF (1989) Bürgerliches Gesetzbuch. 12. edn. Vol. 6, Kohlhammer, Stuttgart, Berlin, Köln

Stallen PJM, Smit PWM (1993) Het omgaan met risico's, Studie in opdracht van het Ministerie van Economische Zaken, Directie Regio's. Bedrijfsomgeving en Milieu. s'Gravenhage, Arn-hem

Staudinger J von (1996) Kommentar zum Bürgerlichen Gesetzbuch. 13. edn. [§§ 903–924], Vol. 3, J. Schweitzer, Berlin

Steiger H (1998) Verfassungsrechtliche Grundlagen. In: Salzwedel J et al. (eds.), Grundzüge des Umweltrecht. E. Schmidt, Berlin, 02

Steinberg R (1996) Verfassungsrechtlicher Umweltschutz durch Grundrechte und Staatsziel. Neue Juristische Wochenschrift (NJW) 49, 1985–1994

Tauw milieu et al. (1990) NOBIS Project: Risk Reduction – Environmental Merits Costs (REC-Method), o.O.

Technische Commissie Bodembescherming (1997) Ecologische risico's van Bodemverontreinig-ing. Den Haag

Travis C, Richter Pack S, Hattemer-Frey H (1989) Is Ionizing Radiation Regulated More Strin-gently Than Chemical Carcinogens? Health Physics 56, 527–531

Umweltbundesamt (1997) Nachhaltiges Deutschland. Wege zu einer dauerhaft-umweltgerechten Entwicklung. E. Schmidt, Berlin

Umweltprogramm 1986-1990, Ministerium für Wohnungswesen, Raumordnung und Umwelt (VROM). Umweltprogramm der Niederlande 1986–1990, Den Haag

Vallender K, Morell R (1997) Umweltrecht. Stämpfli, Bern

van Zorge JA (1996) Exposure to Mixtures of Chemical Substances: is there a Need for Regulations?. Food & Chem. Toxicol. 34, 1033–1036

Wagner W (1995) The Science Charade in Toxic Risk Regulation. Columbia Law Review 95, 1613–1723

Wahl R (1991) Risikobewertung der Exekutive und richterliche Kontrolldichte – Auswirkungen auf das Verwaltungs- und gerichtliche Verfahren. Neue Zeitschrift für Verwaltungsrecht (NVwZ) 10, 409–418

Wahl R, Appel I (1995) Prävention und Vorsorge. Von der Staatsaufgabe zur rechtlichen Ausgestaltung. In: Wahl R (ed.) Prävention und Vorsorge. Economica, Bonn

Wannagat G (1997) Kommentar zum Recht des Sozialgesetzbuchs – gesetzliche Unfallversicherung. Heymanns, Köln, Berlin, Bonn, München.

Wegener B (1991) Pflanzenschutzmittel im "Europäischen" Trinkwasser – Schwierigkeiten bei der Umsetzung der EG-Trinkwasserrichtlinie. Informationsdienst Umweltrecht (IUR) 2, 14–17

Winter G (1986) Gesetzliche Anforderungen an Grenzwerte für Luftimmissionen. In: ibid. (ed.) Grenzwerte. E. Schmidt, Berlin, pp. 127–141

Winter G, Mahro G, Gintzky E (1992) Grundprobleme des Gentechnikrechts. Werner, Düsseldorf.

Winter G (1995) Maßstäbe der Chemikalienkontrolle im deutschen Recht und im Gemeinschaftsrecht. In: ibid. Risikoanalyse und Risikoabwehr im Chemikalienrecht. E. Schmidt, Berlin, pp. 1–63

Wulfhorst R (1994) Der Schutz „Ungewöhnlich empfindlicher" Rechtsgüter im Polizei- und Umweltrecht. Duncker & Humblot, Berlin

6 Appraisal and Recommendations

6.1
Summary Appraisal

The purpose of environmental standards is to prevent exposure to harmful effects from pollutants. Up to now, the vast majority of environmental standards have dealt with exposures to individual substances, and exposures to two or more substances are only rarely taken into account. Combined exposures exacerbate the already complex issue of assessing exposure to harmful substances and thus deserve special attention.

Reference to effects is a basic principle in setting environmental standards. This applies to both individual and combined exposures. Science based medical studies provide a basis for appropriate risk assessment.

Also generally applicable is the *balancing principle:* the comparative assessment of options as a fundamental element of environmental policy decision-making. The balancing principle embraces the principles of transsubjectivity and consistency: the former promotes decision-making criteria independent of individuals, legal entities and context; the latter, the consistency of relevant habits of action. In its "meta" role, the balancing principle also aims to guide selection of the most appropriate basic principle of environmental policy (protection, precaution, cost-effectiveness, etc). Application of the balancing principle requires that risk be defined as it is in systematic risk assessment, i.e. in generally verifiable identification of pollutants and probability of occurrence, as opposed to the concept of perceived risk.

In terms of the sheer number of pollutants, particularly chemical substances, sequences of interaction, dose ranges and other supporting circumstances, the variety of possible combinations is so great as to be beyond estimation. Furthermore, the criteria used in setting limit values for exposures to individual substances vary enormously. Thus environmental standards focus on dose-response thresholds wherever possible; the application of safety margins keeps emissions significantly below those thresholds in most cases. Minimization requirements apply where such thresholds cannot be identified. Along with other factors, the broad scope of safety margins, embracing five orders of magnitude, prohibits the use of simple mathematical models to regulate combined exposures.

In the absence of systematic and all-encompassing research on the effects of combined exposures, the only feasible solution appears to be to collate and assess knowledge by way of mechanistic models. Knowledge of response mechanisms in the development of adverse effects is crucial to these models. Given this knowledge, conclusions can be reached for new combined exposures to pollutants by analogy and extrapolation. The shapes of dose-response curves following both exposure to individual substances and combined exposures are an important assessment tool in this regard. Dose-response relationships can be described using sigmoid curves or polynomial functions, although sigmoid and linear relationships coincide in the low dose range.

If pollutants attack different target organs and their harmful effects develop separately as a result, their effects are said to be mutually independent. This is also the case when the pollutants attack the same target organ but no interaction takes place in the development of the harmful effect. Unproblematic cases of this kind account for a significant number of combined effects.

Assessment becomes considerably more difficult when interactions occur. Such cases involve specific mechanisms and are thus relatively rare. Interactions can occur in absorption and distribution, in the metabolism and biokinetics of organs and cells, and at various stages of response development. This is a particular aspect in the causes of cancer which involve a large number of damage phases over a period of years or decades.

In processes that involve interactions, both sub-additive and super-additive effects can occur. The latter deserves particular attention.

Where interactions occur, it is often the case in practice that super-additive effects are weaker at low doses and no longer detectable at limit levels. In such cases, the dose-additivity model leads to a more conservative approach than the independence model. This results in more stringent assessment than under the independence model and can thus lead to overestimation of risk.

In cases where it is not necessary to take toxicokinetics into account when assessing combined exposures, the following processes and substances are of great importance to interactions in the damage development phase of any two pollutants and the potential for super-additive effects:

- Substances that hinder repair of damaged DNA following exposure to genotoxic substances or ionizing radiation.
- Substances that influence the regulation of cell proliferation following the impact of genotoxic pollutants or ionizing radiation. Substances that stimulate cell proliferation in particular (e.g. substances that affect hormones) can increase the likelihood of cancer.

In interactions between chemical pollutants, these processes can be considerably influenced by additional toxicokinetic changes.

Specific combined exposures have stronger super-additive effects at higher doses. This is proven for smoking and asbestos, for smoking and radiation from radon and its radioactive decay products with regard to certain types of cancer, and also for combined exposures in connection with alcohol consumption.

Experience with exposures from natural sources can be extremely helpful in assessment. This is the case, for example, with ionizing radiation: the impact of each toxic substance released into the environment develops in combination with radiation exposures from natural sources.

The vegetation of Central Europe and North America is currently at greatest risk from ozone, oxygen-bearing sulphur compounds and nitrogen exposures. Combinations of these pollutants cause not only sub-additive, but additive and super-additive responses in plants. Super-additive effects occur mainly in cases of weak damage, with the additive effects often being only marginally exceeded. In the main, the combined effects observed usually match independent effects. One exception is the combination of sulphur dioxide and nitrogen dioxide, which often shows a super-additive effect.

The varying effect characteristics of the selected components lead to qualitative changes in plant responses to minor variations in concentrations. The use of guideline values to protect vegetation from combined emissions is based on results from three methodological approaches:

- Comparative experimental studies on the effects of two or three components, either individually or in combination.
- Comparative epidemiological studies in exposure sites on the effects of filtered and non-filtered air.
- Models for use of yet tolerable concentration-time patterns.

The results of these studies give no evidence of a need to set guideline values for combined exposures from selected components that are subject to existing guideline values for individual components, for the following reasons:

- In using guideline values for individual components, the responsible expert committees more or less explicitly take account of the results of studies on combined exposures.
- In combination tests, the lowest tested concentrations of pollutant gases that still lead to sub-additive, additive or super-additive effects exceed the guideline levels for individual components in standards laid down to protect vegetation.
- Plants exposed under chamber conditions are usually subject to a higher deposition rate – a higher effective dose from the same ambient pollutant concentration than would be observed in the field. As the use of guideline values for individual components is largely based on experimental and epidemiological studies under chamber conditions, a safety margin should be built into guideline values arrived at by these means.

When setting environmental standards, consideration must also be given to the acceptance of the methods employed. Analysis of risk perception focuses mainly on constructed reality, with the emphasis on notions and associations that help people comprehend their environment. The majority of people perceive risk as a complex, multi-dimensional phenomenon in which loss expectations often play only a subordinate role. Risk assessment is dependent on supporting circumstances which, rather than being random, are subject to a degree of regularity. A range of factors have been identified in the contextual conditions of risk perception. Especially controversial factors include risk sources that are deemed particularly low risk in technical risk assessment but which are a major source of concern among the general public.

There have so far been few empirical studies that describe perceptual response patterns to combined risks. A survey conducted as part of this study shows a wide gap in the assessment of combined environmental risks between expert opinion and the laypeople questioned in the survey. Combined exposures are seen as a *lurking danger* that involves great uncertainty. As the majority of people place considerable belief in reports on the health risks from combined environmental pollutants, scientists and regulatory authorities will have great difficulty in convincing a sceptical public otherwise.

This is not just a matter of how people perceive the impact of the physical effects of pollutants on human health. Psychological and psychosomatic processes are also involved. Another provocative topic is *multiple chemical sensitivity syndrome*. Perception of combined exposures sparks feelings of fear and unrest among individuals. This in turn leads to psychosomatic reactions in certain cases. These illnesses are then observed and seen as evidence of the assumed relationship between combined emissions and damage to human health.

Combined effects also lend themselves as subjects for political mobilization. Awareness of real illnesses, diminishing trust in experts and disagreement between risk experts and laypeople lead to demands for policy-makers to implement more stringent regulations even though science based medical studies provide absolutely no evidence of damage to human health. Such demands can come at significant cost to the economy.

In economic terms, combined exposures pose the special question as to which pollutants to regulate and to what extent. A decision must be made on whether to focus on an individual pollutant or on all pollutants; and in the latter case on which combinations. Effective environmental protection once again calls for alternative thinking in this regard. The costs of emission control must be weighed against the harmful effects of the agents involved (the balancing principle). Combinations must be identified and selected for which the total cost of achieving the objectives of environmental protection are lowest (the cost-effectiveness principle). In general, the most efficient way is to incorporate all significant contributing pollutants into regulation because more stringent control of a specific pollutant means progressively increasing costs and diminishing effectiveness in pollution reduction.

Environmental policy can already be made more effective with only limited knowledge of combined exposures. Selection by economic criteria is still possible even if only two equivalent combinations of multiple contributing pollutants are known.

Where the most cost-effective method proves to be focusing on reductions in an individual pollutant, there may be a conflict between cost-effectiveness and fair distribution of the reduction burden. Under the polluter pays principle, all sources of environmental pollution must be held liable. Problems arise with one-sided imputation when businesses that are in direct competition with each other are affected to different degrees, resulting in a distortion of competition.

It is important to find out whether combined exposures have additive, super-additive or sub-additive effects. Mutual attenuation or amplification of individual effects is equivalent to a cost reduction or rise in cost respectively. Policy mistakes as a result of limit values that are set too low or too high can be avoided.

Because combined exposures are often inadequately researched, decisions on exposure limits in this area are frequently made under conditions of uncertainty. Rational decisions are possible given certain framework conditions, even with limited knowledge. Various criteria can be applied as decision aids when setting limits. A number of typical situations can be distinguished according to the degree of uncertainty: reliable knowledge of the range of possible impact severities combined with (subjectively derived) probabilities of occurrence (stochastic models); knowledge of probabilities for risk frequency and severity of impact only (fuzzy models); knowledge of the range of possible impact severities only (uncertainty model).

Judicious choice of policy instruments can help ensure that environmental policies are successful in reaching the corresponding limit values. Where multiple pollutants are involved, the choice of policy instruments is additionally constrained by the desire to bring about the most efficient possible combination of reductions. Emission certificates tradable permit have certain advantages over command and control regulation and environmental levies in this regard. Permits simplify the regulatory task of environmental policy where substances have linearly additive effects. Only two things need to be known: a combination of emissions that is compatible with the environmental policy objective, and the relative harmfulness of each pollutant involved. Tradable permits are issued for emission quantities that are considered acceptable, and can be traded at an exchange rate reflecting the relative harmfulness of the respective agents. Market mechanisms ensure that trading in tradable emission permits produces an efficient quantity structure for those agents. This presupposes a well functioning market in tradable permits, of course. International climate protection is an area in which the benefits of emissions certificates can usefully be exploited.

The tradable permits solution is of no benefit in the case of non-linear combined effects because it is not practicable to have a separate permit for combined effects. As in the case of individual exposures, the best way to achieve specific protection and precaution objectives depends on the problem at hand.

Legal aspects of combined effects of substances and radiation have not played a major role in German environmental law. However, the state is under a fundamental obligation to consider overall exposures and hence also combined effects in environmental legislation. This is derived from the state's duty of protection enshrined in the German constitution. The usual environmental law focus on individual substances or on individual categories of radiation or radiation characteristics does not, therefore, contravene the constitutional duty per se provided that scientific evidence of combined effects is absent or it can be assumed that the safety margins for individual substance-related limit values cover the risks from combined effects.

The majority of environmental laws contain no special provisions on combined effects, but do not entirely rule out their consideration. Conversely, the German Medical Drugs Act expressly requires consideration of interactions between multiple substances. Secondary legislation stipulating limit or guideline values, action thresholds and classification rules often takes account of combined effects, either by including such effects in safety margins, regarding one substance as dominant, or assuming combined effects to be concentration-additive or dose-additive.

Chemicals testing, however, exclusively relates to individual substances. The state's duty to protect the individual and its responsibility for environmental protection also include an obligation to systematically reduce scientific uncertainty as regards combined effects. If on the grounds of a scientifically founded suspicion of risk the state subjects businesses and consumers to restrictive obligations under the precautionary principle, then it is also duty-bound to narrow the scope of scientific uncertainty by doing further research.

As regards imputation of the financial burden of reducing combined effects, civil liability law appears to offer an adequate solution, even though the focus is then on a group of polluters rather than on combined effects as such. As the law relating to contaminated sites shows, while the possibility of recourse among polluters is always an option under liability law, and is not completely ruled out in public law, it does not lend itself as a generally applicable concept for imputation of responsibility in prospective regulation. On the other hand, the priority principle that dominates public law is unsatisfactory in legal policy terms. Overall, the applicable law appears in general need of reform if one assumes that in many cases combined effects are not the exception but the rule.

From a legal standpoint, when regulating for combined effects, it is not the job of the legal framework to impose an "optimum" solution on the state. Application of the protection principle (prevention of dangers) and the precautionary principle (precaution against hazards) leads to an appropriate solution if their legal limits are observed. In general a distinction must be upheld between proven and assumed combined effects. The precautionary principle can be applied in cases where combined effects are merely suspected, but this presupposes at a minimum that the suspicion is plausible.

Blanket risk reduction strategies in the absence of evidence for combined effects are thus problematical. Where indications of combined effects exist, the dose-additive or independent effects models are generally acceptable. However, sufficient room for correction must be allowed for the event that other effects, particularly super-additive effects, are identified.

Special safety margins for combined exposures built into individual-substance limits are an acceptable strategy in the case of substances that have a dose-response threshold. Special limits for combined effects are generally to be preferred. This is especially the case in occupational health and safety, and for substances that lack dose-response thresholds.

Wherever possible, a voluntary solution for combined exposures should be selected that leaves the decision on apportionment to the polluter. This, of course, presupposes that multiple combinations of the substances concerned are acceptable from both a health and an environmental standpoint. Apart from that the polluter pays principle is preferable from the point of view of fair distribution. An alternative that also takes in economic efficiency aspects might be to impose reductions on one polluter while requiring others to share in the cost.

As regards the relationship between regulation and imputation, the regulatory logic (environmental policy choice) narrows the imputational logic (the choice of the addressee of regulation). Not all of the available regulatory models are compatible with the imputation models considered in this study. The most favourable regulatory models in this regard are case-wise assessment and special limits; these can be used in conjunction with various imputation models.

6.2
Possible Solutions and Recommendations for Action

(1) Precautionary measures for combined effects should be based on science based medical studies, because policies should only be implemented when plausible assumptions on the possibility of combined effects exist. All-purpose safety margins or mixture rules for limit values, action thresholds and classification rules should always come with a corrective for the common event that effects can be assumed to be completely independent rather than combined. For practical reasons, where there is a plausible suspicion of a combined effect it is possible to assume the dose-additive combined effects model, which often coincides with the dose-effect curves of the independent model in the low dose range. In turn, there must be corrective options for the event of super-additive combined effects.

(2) From an economics standpoint, methods based on decision theory should be incorporated when setting safety margins for limits. In general, it is both vital and desirable to advance scientific knowledge on possible combined effects thus to provide an empirical basis on which to set special limits for those effects.

(3) Where science based medical studies exist on combined effects, priority should be given to special limits. This applies particularly to carcinogens. A purely speculative suspicion of combined effects should not be used to apply blanket emission limits.

(4) To fulfil the environmental policy desiderata described above, science based medical research is needed to enhance knowledge and understanding of the effects of combined exposures. Particular importance should be attached in this context to the response mechanisms in damage development following the impact of toxic agents and interactions in the investigated development chains. This applies in particular to the development of stochastic effects (carcinogenesis, mutagenesis and teratogenesis in certain developmental phases of the mammalian organism) whose dose-response relationships lack a threshold dose. Based on this, other combinations of toxins can be assessed by analogy provided the toxins can be assigned to specific response mechanisms.

(5) Studies on response mechanisms must include the identification of dose-effect relationships because interactions between effect chains of combined exposures are largely dose-dependent. Because the effects of toxic agents are often unmeasurable at low doses, it is necessary to examine the potential for extrapolation from the medium and high dose ranges into the low dose range for these combined effects. Medium and high dose ranges can however be significant in cases of incidents or accidents in and around industrial facilities, and so interactions in these dose ranges should not be neglected.

(6) As DNA repair and regulation of cell proliferation are of particular importance in the impact of genotoxic agents on human health, especially in the incidence of cancer and mutations, particular attention should be given to these processes during examination of the mechanisms activated by combined exposures. In such cases there may be super-additive effects that deserve special consideration.

(7) Individual sensitivity to toxic agents varies enormously. It is thus necessary to consider this variance when setting limits in general and especially where combined exposures are involved. This applies particularly to high exposures to multiple agents in connection with incidents and accidents at industrial facilities.

(8) Consideration of the time span is also of great importance in this context, because depending on the response mechanisms, the interval between two individual exposures can vary dramatically and allow super-additive effects to occur. Substances that affect hormones can increase the incidence of cancer some considerable time after the initial impact of a genotoxic substance. Additional data are needed if we are to understand these mechanisms, especially at low dose ranges for long-term exposures.

(9) The need to consider diverse combinations of varying composition makes it difficult to formulate representative policies based on experimental and epidemiological data. It makes sense, therefore, to look for general solution models that cover the greatest possible number of relevant combinations.

(10) Specific, frequently occurring mixtures of similar composition (e.g. halogenated dioxins and furans, organic-based pyrolysis products) should be regulated by an appropriate limit for the respective mixtures themselves; that is, they should be examined for their harmful effects in their entirety. If a mixture occurs in constant proportions – for which proof must be supplied in each individual case – a dominant substance can be taken as a reference point for the limit value. Example: benzo(a)pyrene in tar, soot and smoke.

(11) In the case of conventional toxins, dose-response thresholds must be proven and founded. Limits for individual substances are largely set with safety margins (uncertainty factors), e.g. in standards for food contaminants (additives, residues), ambient air standards, and the majority of drinking water standards. Safety margins normally protect against combined adverse effects given that the limits set for the individual substances are complied with. In such cases – the vast majority of relevant combinations – there is no need for special limit values for mixtures.

(12) If limit values are close to the dose-response threshold, as for example with MAK values for hazardous substances in the workplace, the individual substances should be aggregated as additive elements relative to their proportions in the mixture. Exceptions should be allowed where there is enough knowledge of the specific response mechanisms involved (including metabolism and toxicokinetics) to rule out adverse (synergetic) interactions.

(13) In certain rare combinations (almost always of a binary nature), super-additive amplification of effects can still occur even if the limit values for the individual substances are complied with. Such occurrences are generally predictable in cases where the toxicokinetics and mechanisms of the individual substances are known. Thus, as in medicinal drugs legislation, it is necessary to require additional analysis when investigating possible exposure scenarios in the development and introduction of new substances or the reprocessing of existing substances. The additional analysis would comprise appropriate testing and inference to identify possible combinations and their part in the total severity of harmful effects on health and the environment. The results must

then be taken into account when setting limits. Efforts should be directed at supplementing EU law on chemicals with a requirement for manufacturers and importers of new substances to investigate evidence of combined effects. Combined effects should also be included in the assessment of existing substances.

(14) Reliable dose-response thresholds can neither be identified nor theoretically founded for typical carcinogens and mutagens. However, limit values are necessary for these substances in combination because of their significance in health safety. But these limits must be set using principles other than those used for conventional substances. The first step is to negotiate a social consensus on the acceptable level of the total *residual risk*. The second step is to distribute the residual risks from the components in a mixture so that the proportions add up to the acceptable total risk. *Unit risk* could be used as a standard measure (the probability of contracting cancer in a standardized subset of the total population assuming 1 μg pollution per m^3 of air breathed in a lifetime). The *Länder* Committee for Pollution Control used a similar model to assess carcinogenic air pollutants. The model can overestimate risk because it assumes a linear dose-response relationship down to the lowest dose range and because the components of a mixture usually attack different target organs making their effects in such cases mutually independent, which means they have no additive effect on the organism or biological system. This has to be weighed against the benefits of ease of use and the desirability of erring on the side of caution. Exceptions should be allowed for certain simple and easily understood mixtures for which sound knowledge exists regarding organ-specific tumour formation and response mechanisms, e.g. for vehicle exhaust fumes containing benzene (leukemia from damage to bone marrow stem cells) and soot particles (lung cancer from damage to mucous membranes in the lower respiratory tract).

(15) To identify combined effects on plants, chamber-free systems must be used to expose plants and plant communities in exposure sites to filtered and unfiltered air, and to develop and systematically apply prescribed emission combinations. Experiments with pollutant pathways from outdoor air to action sites in plants via stomata must be stepped up, as must studies on the question of the extent to which global changes in the form of increased CO_2 concentrations and changes in UV radiation influence the quantitative relationship between ecotoxicologically significant emission components and their effects on plants. In particular, the development of active biomonitoring methods should be promoted to assist assessment of what proportion of the total effect is due to individual components in pollutant mixtures.

(16) As for environmental policy decisions in general, the assessment of combined exposure to risks that have been quantified by science based medical studies should be used as a guideline for measures to be taken. However, perceptions of risk should also be included in the decision-making process, using risk balancing as a measurement tool.

(17) From a rational standpoint, it would thus be highly worthwhile to systematically identify the various dimensions of intuitive risk perception and to measure empirically observed risk against them. This is based on the traditional

understanding of democratic decision theory, that the dimensions or special aspects of intuitive risk perception in society should be seen as legitimate elements of regulatory policy, but that the assessment of the various risk sources in each dimension must be conducted using scientific methods. Thus the results of risk perception should not be made a dominant principle of decision-making, but should instead serve to supplement systematic risk perception.

(18) Exposure to combined pollutants triggers a particularly high degree of fear that scientific expertise can only allay to a certain extent. In recognition of this, it is the duty of policy-makers to combine scientific knowledge and residual uncertainty on possible combined effects with the opinions and desires for change of those affected by the risk, and to incorporate them into holistic policies that focus both on available knowledge and on basic norms.

(19) If multiple combinations of substances enable compliance with environmental objectives, then reduction of pollutants with relatively low avoidance costs and/or relatively high degrees of harmfulness makes more sense than reducing substances with relatively high avoidance costs and/or a relatively low contribution to pollution. If concentrating avoidance on one substance presents the most cost-effective strategy and that strategy results in a competitive advantage for non-regulated emitters, then provision could be made for compensation in the form of an arrangement by which those not covered by regulation compensate the affected businesses. If, in contrast, a component of a mixture is so dominant that it can be said to be characteristic of the mixture as a whole, then the applicable limit should be applied for that substance only. The desire for fair imputation would appear to favour use of the polluter pays principle, for example by way of quota solutions with a guaranteed range of permitted exposure levels.

(20) To reduce the uncertainties in assessing and regulating combined exposures, research activities must be promoted and intensified in all areas covered by this study. Priority should not be given to arbitrarily selected combinations under arbitrary conditions, but rather to systematic study with practical relevance to the development of mechanisms and their assessment.

Glossary

Acceptability	Normative concept used to lay down binding criteria for the tolerability of specific states or actions; an option is acceptable if it meets the applicable acceptance criteria
Acceptance	Toleration of a state or action in practice (empirical concept); an option is accepted if people act in accordance with it
Acute injury	→ Damage
Additive	Effects acting in the same direction leading to a greater overall effect
Additivity	Additive model of combined effects Dose additivity: Where the combined effect is the sum of equieffective doses of the individual substances Effect additivity, response additivity: Where the combined effect is the sum of effects of the individual substances Antagonism Subtractive effect; reduction of the effect of one agent by another; the combined effect is less than the effect of the agent on its own, and the effect of the agent on its own is greater
Autecology	Branch of ecology concerned with the relationships between an individual or a population of a species and their environment
Background (value)	Exposure level permanently present e.g. due to natural sources
Benefit	Measure of the degree to which needs are fulfilled
Biocenosis	Community of plants and animals in a defined living space-biotope
Bioindication	Detecting of air pollution effects by means of organisms present in the area concerned (passive biomonitoring) or by organisms exposed following a standard procedure (active biomonitoring)
Biokinetics	Uptake and dispersion of (radioactive) substances. Commonly used term in radiobiology. Equivalent to pharmacokinetics and toxicokinetics; more general: chemical kinetics
Cardinal scale	A scale used to measure proportions between individual, numerical quantities; uniquely defined except for similarity transformations
Causality	Cause-effect relationship

Chronic damage	→ Damage
Combined exposure	→ Exposure
Concentration	Volume or mass of one substance per unit volume or mass
Consumers	In ecology, heterotrophic organisms; that is, organisms such as humans and animals that are bound to obtain their food and energy from organic substances
Cost efficiency	Attaining an environmental target at the lowest possible cost to the national economy
Critical levels (ecotexicology)	The concentration of an air pollutant, above which, according to current knowledge, direct damage to receptors like plants, ecosystems or organic matters can occur
Critical loads (ecotexicology)	The quantitative level of deposition of an air pollutant, below which, according to current knowledge, no significant negative effects on specific, sensitive elements of ecosystems is expected
Damage (ecotoxicology)	All effects of air pollutants that impair the utility of plants or their gene pool as measured by economic, ecological or subjective criteria
Danger	Law: condition, circumstance, act or process that with reasonable probability will cause harm to mankind, the environment, or another protected asset
Destruents, decomposers	Organisms decomposing the detritus i.e. dead organic matter like shed leaves, cadavers and excrements down to the mineral level, thereby closing the bio-geochemical substance cycles. Destruents include lower animal species e.g. woodlice and earthworms as well as bacteria and fungi
Dose	Radiobiology: energy absorbed per unit mass Pharmacology and toxicology: dispensed quantity of a substance, for example in mg/kg body weight Ecotoxicology: concentration × time (Quantity of radiation or harmful substances taken up = dose quantity = time integral of the dose rate)
Dose additivity	→ Additivity
Dose quantity	→ Dose
Dose rate	Radiation dose per unit time
Dose-response curve	Plots the empirical or computed relationship between the dose or concentration and the effect of an agent
Ecosystem	Geographically distinct part of the biosphere in which plants (producers), animals (consumers) and (destruents) combined with the abiotic factors soil and climate, forming a structural and functional unit capable of self-regulation. This capability,

from which stems the "ecological stability" of ecosystems, is a central objective of environmental protection

Ecotoxicology
: The science concerned with the distribution and effects of chemical compounds on plants and animals, on levels ranging from the cell through organ, organism, community and ecosystem

ED_{50}
: Dose leading to an effect incidence of 50%, often used as a curve parameter

Effect additivity
: $\rightarrow$ Additivity

Effect direction
: Generally: the direction of an effect e.g. stimulation or inhibition
For combined effects: same-direction (synergism = amplification) or opposite-direction (antagonism = attenuation)

Emission
: Discharge of pollutants into the air, water and soil

Emission certificates
: Tradable emission rights

Environmental quality target
: States in concrete terms a desired environmental quality condition, based on a number of quality dimensions and subject to the chosen environment policy paradigm and guidelines based thereon

Environmental standard
: Stipulation of statute law, case law, public agency decree or standards (e.g. DIN) that implements indeterminate legal concepts (such as "harmful effect", "precautionary measures", "due care", and "state of the art") in the form of specific prohibitions, prescriptions and permissions by operationalizing and standardizing measurable quantities
Means of attaining or maintaining a specific environmental quality; statement in concrete terms of environmental quality targets.

Equieffective
: Effective to an equal degree or with equal incidence

Exposure
: Exposure to pollutants or harmful agents including radiation
The condition of being subjected to a hazard
Combined exposures: Exposure to mixtures of pollutants or harmful agents
Zero exposure: Complete freedom of exposures

Exposure areas
: Areas under high exposure

Genotoxic
: Capable of modifying the genome

Guideline value
: Environmental standard that takes the form of a recommended value and is used as a reference standard by public agencies when evaluating concentrations and exposures in relation to environmental media and protected assets

Injury
: All effects of plant exposure regardless of their consequences for the utility of the plants; cf. *damage*

Acute injury	Immediate injury, e.g. necroses of leaf organs and flowers, following a short-term exposure to a high concentration of air pollutants
Chronic injury	Slow injury, e.g. bleaching (chloroses) and discolouring of leaf organs and flowers, following long-term exposures to low concentrations of air pollutants
Latent injury, hidden non visible injury	Harm affecting the growth, the vitality and/or the resistance of plants against other stressors Hazard Inherent condition of being capable of causing harm
Immission	Impact of pollutants on objects
Immission rate	→ Dose rate
Immission value	→ Dose
Independent	Generally: mutually independent, different effects Specifically: mutually independent, similar effects which remain unchanged when combined and which equal the sum of the relative effects (at low doses, the sum of the individual effects).
Initiation	Triggering of an effect; primary lesion in the DNA, leading to carcinogenesis
Interaction	Reciprocal effect of agents involved in a combined exposure, for example at molecular and/or cellular level
Interval scale	A scale defined unequivocally except for positive linear trans-formations
Intraspecific and interspecific competition	Competition between individuals of the same species or of different species respectively
Irreversible	Persistent, irreparable; e.g. damage to the DNA or chemical reactions leading to covalent bonds
Isocost curve	For a given pair of agents, the locus of all pairs of emission quantities that have the same combined prevention cost
Isoquant curve	For a given pair of agents, the locus of all pairs of emission quantities that cause the same amount of harm
K-strategists (K = capacity of the environment)	Organisms which, at a low rate of reproduction and with constant population sizes, are well adapted to the capacity of their environment by using it most effectively and evenly
Latent injury	→ Injury
Latent period	Time interval between the occurrence of a damage and its effect e.g. between the initiation and the development of cancer
Limit value, limit, limiting value	Quantification of an environmental quality target; unlike a threshold, a limit is prescriptive, that is, it defines a boundary to the scope of permissible action by reference to the environmental quality target in question

Marginal cost of prevention	Additional cost of avoiding an incremental increase in emissions
Marginal effect of an agent	Additional effect of an incremental increase in dose
Marginal rate of substitution between the prevention cost of two agents	Ratio of the marginal prevention cost of one agent to the marginal prevention cost of another
Marginal rate of substitution between two agents	Ratio of the marginal effect of one agent to the marginal effect of another
Mechanism	Effect chain or part of an effect chain
Multiplicative combined effect	Effect resulting from the multiplication of the relative risks ascribed to the agents involved in a combined exposure (epidemiology)
Mutation	Hereditary modification to DNA, consisting of either a change in an individual base (point mutation) or in the structure of a gene or an entire chromosome (e.g. a deletion or translocation)
Mycorrhiza	A close partnership between fungi and roots of higher plants, beneficial to both partners (symbiosis)
Nominal scale	A scale in which items are classified according to specific criteria; the classification is unaffected by one-to-one transformation of the numbers assigned to the classes
Non-stochastic effect	Effects that have a threshold dose and vary in severity according to dose; examples: erythema formation, cataracts, acute radiation syndrome, and fertility disorders
Oligotrophic ecosystems	Ecosystems with soils poor in nutrients, e.g. neglected grassland communities or heathlands
Ombrotrophic plants	E.g. mosses, mainly living on nutrients absorbed from the air, hence particularly sensitive to anthropogenic trace substances in the air
Open top chamber	Standard, open chamber in which plants are exposed to air pollutants
Ordinal scale	A scale in which items are classified into sequence relative to a specific reference value; the classification is unaffected by monotonic (order-preserving) transformation of the numbers assigned to the classes
Peak level (of exposure)	Maximum of immission occurring within a larger assessment area or a long time interval
Pollutant	Any harmful agent apart from radiation

"Polluter pays" principle	The principle that the cost of environmental protecion should be borne by the emitters of pollution
Pollution	→ Exposure
Precursors (ecotoxicology)	Substances like NO_x and reactive hydrocarbons from which, under sunlight, photo-oxidants like ozone are formed in the troposphere
Preventive environmental management (PEM)	Actions and regulations to prevent harm to the environment
Producers	In ecology, a term for autotrophic organisms (primarily, green plants), which are able to produce organic substances from inorganic components using chlorophyll and sunlight in photosynthesis
Promotion	Driving the proliferation of initiated cells during carcinogenesis
Protected asset	Human beings, fauna, flora, soil, water, air, cultural assets and other assets that are subject to protection
(Environmental) protection objective	Law: the general nature and degree of protection to be provided for each protected asset
Residual risk	Residual risk after steps have been taken to reduce risk and acceptance has been established
Response additivity	→ Additivity
Reversible	Reparable, temporary effects, e.g. reactions of messenger substances and hormones mediated by weak bonding forces
Risk	Generally: the probability of harm being caused Systematic risk: definition of risk demanding that any concept of risk should be (i) generally applicable and (ii) comparable with regard to the allocation of relevant actions Technical: quantified adverse effect × probability of the effect arising Legal: Possibility of an adverse effect arising below the danger threshold Economic: Uncertainty of an event
Risk analysis	Systematic method of characterizing and where possible quantifying a risk in terms of the probability of (a) the event arising and (b) the magnitude of the consequences EPA: identification, investigation, analysis and evaluation, and management
Risk assessment	Method of assessing the acceptability of risk-assessment findings with reference to personal or collective criteria
Risk management	Set of activities and approaches used to arrive at decisions regarding the acceptability of risks
Risk perception	The process of subjectively perceiving, processing and evaluating risk-related information originating in personal experi-

	ence, direct observation, things read and seen in the media, and communication with other individuals
r-strategists (r = rate of reproduction)	Organisms with high rates of reproduction, quick succession of generations and a strong tendency to migrate from their environment
Safety factor, safety margin	Factor that is introduced in order to lower a threshold derived from primary data, allowing for uncertainties (such as the possibility of transmission to humans), differing susceptibilities between species and between individuals, measurement errors, and combined exposures with other agents
Stochastic effect	Effect whose probability of occurrence – rather than its severity – should be regarded as a function of dose, without any threshold; examples: induction of cancer, leukemia, and genetic mutations
Sub-additive	The combined effect is less than additive (in an additivity model)
Subtractive	Effects acting against each other are called subtractive
Succession	The succession in time of certain communities residing in an ecosystem
Super-additive	The combined effect is greater than additive (in an additivity model)
Symbiosis	A permanent or temporary, closely interdependent partnership of two different types of organism
Synecology	Branch of ecology concerned with the relationships between communities of organisms (biocenoses) and their environment
Synergism	Where the combined effect is stronger than the effect of agent x (is synergetic with regard to x); used with reference to the agent that has the stronger effect when acting alone
Target	Structure or organ attacked by an agent or a mixture of agents (cf. *protected asset*)
Threshold value	Exposure parameter value below which damage is no longer to be expected (cf. *critical deposition* and *critical concentration*)
Transsubjective	Context- and person-invariant
Utility	→ Benefit
Vulnerable group	Group at risk from harmful agents
Zero exposure	→ Exposure

Authors

Bücker, Josef, Dr. rer. nat. Dipl.-Biol., studied biology at the University of Münster; doctorate in the field of immission effects on plants (1991, University of Essen); research associate under Professor Dr. Robert Guderian (s.b.), research mainly in the fields of applied ecology, ecotoxicology and immission ecology (1991-1996); since then: staff scientist with the *Trägerverein zur Nutzung erneuerbarer Energien* Haspe e.V., Hammerstraße 10, D-58135 Hagen.

Cansier, Adrienne, Dipl.-Mathem., studied maths and economy at the University of Tübingen; research associate with the chair for business management at Tübingen, specialized in marketing; mail address: Universität Tübingen, Wirtschaftswissenschaftliche Fakultät, Nauklerstr. 47, D-72074 Tübingen.

Cansier, Dieter, Professor Dr. rer. pol., Dipl.-Volksw., studied at the University of Hamburg; habilitation at the University of Freiburg i. Br.; since 1977 professor of economics (speciality: finance) at the University of Tübingen; main research interests: finance policy and environmental economy; mail address: Universität Tübingen, Wirtschaftswissenschaftliche Fakultät, Melanchthonstr. 30, D-72074 Tübingen.

Gethmann, Carl Friedrich, Professor Dr. phil. habil., lic. phil., studied philosophy at Bonn, Innsbruck and Bochum; since 1990 professor of philosophy with emphasis on applied philosophy (University Duisburg-Essen); director of the Europäische Akademie Bad Neuenahr-Ahrweiler GmbH since 1996; fields of research: linguistic philosophy and philosophy of logic, phenomenology, practical philosophy and technology assessment.

Guderian, Robert, Professor (em.) Dr. agr., studied agricultural sciences at Humboldt-Universität Berlin; staff scientist with the *Landesanstalt für Immissions- und Bodennutzungsschutz des Landes NW* (1958 to 1977); professor of applied biology at the University of Essen (1978 to 1994); fields of research: applied ecology, ecotoxicology, especially immission effects on plants; mail address: Berghausweg 6, D-45149 Essen.

Hanekamp, Gerd, Dr. phil. Dipl.-Chem., studied chemistry at Heidelberg and Marburg and at the École Nationale de Chimie in Lille; doctorate in philosophy at the Philipps-Universität Marburg (1996); staff scientist with the Europäische Akademie Bad Neuenahr-Ahrweiler GmbH since 1996; fields of research: philosophy of science, linguistic philosophy, culturalist theory of social sciences.

Henschler, Dietrich, Professor Dr. med.; professor of toxicology and pharmacology at the University of Würzburg from 1964; chair of his institute at Würzburg from 1965; emeritus since 1994; fields of research: mechanisms of toxic effects of chemicals, especially mechanisms of cancerogenesis, dose-response relationships for toxic effects, systematics of limit values for toxic substances; chair of several consultation committees of the Deutsche Forschungsgemeinschaft (DFG) and with ministries and government agencies; member of the management board of the European Agency for the Evaluation of Medicinal Products (EMEA) in London; mail address: Institut für Toxikologie der Universität Würzburg, Versbacher Straße 9, D-97078 Würzburg.

Pöch, Gerald, Professor Dr. med., studied medicine at Graz; doctorate 1961; assistant at the Institute of Pharmacology of the Faculty of Medicine at the University of Graz; from 1964 at the Institute of Pharmacology and Toxicology of the Science Faculty at the University of Graz, from 1973 as lecturer, from 1979 as senior lecturer, professor of pharmacology and toxicology, emeritus since 1998; mail address: Edelsbachstraße 73, A-8063 Eggersdorf bei Graz.

Rehbinder, Eckard, Professor Dr. jur., studied jurisprudence at FU Berlin and the Johann Wolfgang Goethe-University, Frankfurt am Main; professor of civil law, business law and comparative law at the Bielefeld University (1969); from 1972 professor of economic, environmental and comparative law at the University of Frankfurt am Main; co-director of the *Institut für ausländisches und internationales Wirtschaftsrecht* and the *Forschungsstelle für Umweltrecht;* since 1987 member, since 1996 chair of the *Rat von Sachverständigen für Umweltfragen;* fields of research: environmental and business law; mail address: Fachbereich Rechtswissenschaft der Universität Frankfurt am Main, Senckenberganlage 31, D-60054 Frankfurt am Main.

Renn, Ortwin, Professor Dr. rer. pol., studied economy, sociology, social psychology and journalism at Cologne; professor of *Technik- und Umweltsoziologie* (sociology of technology and the environment) at the University of Stuttgart; spokesman of the board of *Akademie für Technikfolgenabschätzung* (technology assessment) in Baden-Württemberg; member of the science council of the Federal Government "Global Environmental Changes" (WBGU); fields of research: risk analysis, technology assessment, environmental sociology and economy, participation research; mail address: Akademie für Technikfolgenabschätzung, Industriestraße 5, D-70565 Stuttgart.

Slesina, Marco, studied biology, postgrad. at the Johann Wolfgang Goethe-University, Frankfurt in the working group for neurochemistry under Professor Zimmermann; thesis (prov.): "Analysis of the Cellular Dynamics of a large Ribonucleic Protein Particle" (Vault) in Living Cells"; mail address: Johann Wolfgang Goethe Universität Frankfurt am Main, Fachbereich Biologie, Arbeitskreis Neurochemie, Marie-Curie-Straße, D-60439 Niederursel.

Streffer, Christian, Professor Dr. rer. nat. Dr. med. h.c., studied chemistry and bio-chemistry at Bonn, Tübingen, Munich, Hamburg and Freiburg; professor of medical radiation biology at the University of Essen (from 1974), emeritus since August 1999; member of the board of the *Institut für Wissenschaft und Ethik* of the universities of Bonn and Essen; fields of research: radiation risk especially during the prenatal development of mammals, effects of the combination ionizing radiation and chemical substances, experimental radiotherapy of tumours, especially predictive tests for human tumours; mail address: Auf dem Sutan 12, D-45239 Essen.

Wuttke, Kerstin, Dr. rer. nat., studied biology at Ruhr-Universität Bochum; doctorate at the University of Essen (1993), research associate at Universitätsklinikum Essen, Institut für Medizinische Strahlenbiologie (1991–1998); specialities: radiobiology, especially biological dosimetry; cytogenetics, immunology; mail address: Alte Kirchstraße 46, D-45327 Essen.

This study is a translation of Volume 5 „Umweltstandards" of the series „Wissenschaftsethik und Technikfolgenbeurteilung" (ethics of science and technology assessment) published by the Europäische Akademie zur Erforschung von Folgen wissenschaftlich-technischer Entwicklungen Bad Neuenahr-Ahrweiler GmbH. Further volumes are:

Vol. 1: A. Grunwald (Hrsg.) Rationale Technikfolgenbeurteilung. Konzeption und methodische Grundlagen, 1998

Vol. 2: A. Grunwald, S. Saupe (Hrsg.) Ethik in der Technikgestaltung. Praktische Relevanz und Legitimation, 1999

Vol. 3: H. Harig, C. J. Langenbach (Hrsg.) Neue Materialien für innovative Produkte. Entwicklungstrends und gesellschaftliche Relevanz, 1999

Vol. 4: J. Grin, A. Grunwald (eds) Vision Assessment. Shaping Technology for 21st Century Society, 1999

Vol. 5: C. Streffer et al., Umweltstandards. Kombinierte Expositionen und ihre Auswirkungen auf den Menschen und seine natürliche Umwelt, 2000

Vol. 6: K.-M. Nigge, Life Cycle Assessment of Natural Gas Vehicles. Development and Application of Site-Dependent Impact Indicators, 2000

Vol. 7: C. R. Bartram et al., Humangenetische Diagnostik. Wissenschaftliche Grundlagen und gesellschaftliche Konsequenzen, 2000

Vol. 8: J. P. Beckmann et al., Xenotransplantation von Zellen, Geweben oder Organen. Wissenschaftliche Grundlagen und ethisch-rechtliche Implikationen, 2000

Vol. 9: G. Banse, C. J. Langenbach, P. Machleidt (eds) Towards the Information Society. The Case of Central and Eastern European Countries, 2000

Vol. 10: P. Janich, M. Gutmann, K. Prieß (Hrsg.) Biodiversität. Wissenschaftliche Grundlagen und gesellschaftliche Relevanz (2001)

Vol. 11: M. Decker (ed) Interdisciplinarity in Technology Assessment. Implementation and its Chances and Limits (2001)

Vol. 12: C. J. Langenbach, O. Ulrich (Hrsg.) Elektronische Signaturen. Kulturelle Rahmenbedingungen einer technischen Entwicklung (2002)

Vol. 13: F. Breyer, H. Kliemt, F. Thiele (eds) Rationing in Medicine. Ethical, Legal and Practical Aspects (2002)

Vol. 14: T. Christaller et al. (Hrsg.) Robotik. Perspektiven für menschliches Handeln in der zukünftigen Gesellschaft (2001)

Vol. 15: A. Grunwald, M. Gutmann, E. Neumann-Held (eds) On Human Nature. Anthropological, Biological, and Philosophical Foundations (2002)

Vol. 16: M. Schröder et al. (Hrsg.) Klimavorhersage und Klimavorsorge (2002)

Vol. 17: C.F. Gethmann, S. Lingner (Hrsg.): Integrative Modellierung zum Globalen Wandel (2002)

Vol. 18: U. Steger et al.: Nachhaltige Entwicklung und Innovation im Energiebereich. (2002)

Printing: Mercedes-Druck, Berlin
Binding: Stein+Lehmann, Berlin